FARMING SYSTEMS RESEARCH AND DEVELOPMENT

Also of Interest

Small Farm Development: Understanding and Improving Farming Systems in the Humid Tropics, Richard R. Harwood

Successful Seed Programs: A Planning and Management Guide, edited by Johnson E. Douglas

Food Security for Developing Countries, edited by Alberto Valdés

Wheat in the Third World, Haldore Hanson, Norman E. Borlaug, and R. Glenn Anderson

Irrigated Rice Production Systems: Design Procedures, Jaw-Kai Wang and Ross E. Hagan

Managing Pastures and Cattle Under Coconuts, Donald L. Plucknett

Readings in Farming Systems Research and Development, edited by W. W. Shaner, P. F. Philipp, and W. R. Schmehl

Small-Scale Processing and Storage of Tropical Root Crops, edited by Donald L. Plucknett

Tomatoes in the Tropics, Ruben L. Villareal

† *Managing Development in the Third World,* Coralie Bryant and Louis G. White

Managing Renewable Natural Resources in Developing Countries, edited by Charles W. Howe

Cassava, James Cock

Azolla as a Green Manure: Use and Management in Crop Production, Thomas A. Lumpkin and Donald L. Plucknett

The Role of Centrosema, Desmodium, *and* Stylosanthes *in Improving Tropical Pastures,* edited by Robert Burt, Peter Rotar, and James Walker

† Available in hardcover and paperback

Westview Special Studies in Agriculture/Aquaculture Science and Policy

Farming Systems Research and Development: Guidelines for Developing Countries

W. W. Shaner, P. F. Philipp, and W. R. Schmehl

Farming systems research and development (FSR&D) is an approach that is being used increasingly to meet the need for greater food production and a better standard of living for small-scale farmers in developing countries. This book synthesizes the FSR&D procedures used by national governments and international research centers around the world, emphasizing methodologies that have proved successful in practice.

The authors describe the characteristics and objectives of FSR&D, then present information on selecting target and research areas, problem identification and development of a research base, research design, on-farm research, extending research results, and implementation and training procedures. They emphasize that the FSR&D approach requires a clear understanding of farmers and their families, farmers' conditions, and governmental staffing and organizational capabilities, and in one chapter discuss how to determine whether an FSR&D approach is in a particular country's best interests. Appendixes present detailed examples of procedures described in the text, covering a variety of countries with different cropping and livestock systems, environmental conditions, and research and development capabilities.

Dr. Shaner, project director of the Farming Systems Research and Development Methodology Project and associate professor of engineering at Colorado State University, has been chief economic advisor to the Technical Agency, Ministry of Planning, Ethiopia, and associate director of the Consortium for International Development. He has written *Project Planning for Development* (1979) and edited with Professors Philipp and Schmehl *Readings in Farming Systems Research and Development* (Westview 1982). **Dr. Philipp** is professor emeritus in the Department of Agricultural and Resource Economics at the University of Hawaii. **Dr. Schmehl** is professor and associate department head of agronomy at Colorado State University.

A CONSORTIUM FOR INTERNATIONAL DEVELOPMENT STUDY

FARMING SYSTEMS RESEARCH AND DEVELOPMENT

Guidelines for Developing Countries

W. W. Shaner
P. F. Philipp
W. R. Schmehl
Editorial Assistance, Donald E. Zimmerman

Westview Press / Boulder, Colorado

Westview Special Studies in Agriculture/Aquaculture Science and Policy

This book was prepared for the United States Agency for International Development (Contract No. AID/DSAN-C-0054), September 1981.

Published in 1982 in the United States of America by
Westview Press, Inc.
5500 Central Avenue
Boulder, Colorado 80301
Frederick A. Praeger, President and Publisher

Library of Congress Catalog Card Number: 82-050365
ISBN 0-86531-389-X
ISBN 0-86531-425-X(pb)

For further information about this document, please contact J.K. McDermott, USAID S&T/AGR, Department of State, Washington, D.C. 20523; W.W. Shaner, College of Engineering, Colorado State University, Fort Collins, Colorado 80523; or J.L. Fischer, Consortium for International Development, Suite 1500, 5151 E. Broadway, Tucson, Arizona 85711-3766.

Printed and bound in the United States of America.

CONTENTS

PREFACE xv

ACKNOWLEDGMENTS xvii

1. EXECUTIVE SUMMARY 1

1.1. Background 3
1.2. Nature of FSR&D 3
1.3. FSR&D Activities 4
1.3.1. Target and Research Area Selection 5
1.3.2. Problem Identification and Development of a Research Base 5
1.3.3. Planning On-Farm Research 5
1.3.4. On-Farm Research and Analysis 6
1.3.5. Extension of Results 6
1.4. Issues of FSR&D Implementation 6
1.4.1. Timing 6
1.4.2. Organizational Flexibility 6
1.4.3. Staffing Requirements 7
1.4.4. Training 7
1.4.5. FSR&D Costs 8
1.4.6. Governmental Support 8
1.5. What This Book Offers 8

2. INTRODUCTION 11

2.1. Purposes of FSR&D 13
2.2. Definition of FSR&D 13
2.3. Additional Definitions and Comments 16
2.3.1. Farming 16
2.3.2. Households 16
2.3.3. Small-Scale Farming 16
2.3.4. Enterprises 16
2.3.5. Farming Systems 16
2.3.6. Cropping and Livestock Systems 16
2.3.7. Mixed Systems 17
2.3.8. Cropping Systems Research 17
2.3.9. Livestock and Mixed Systems Research 17
2.3.10. Commodity-Oriented Research 17
2.4. Characteristics of FSR&D 18
2.4.1. Farmer Based 19

2.4.2. Problem Solving . . . 19
2.4.3. Comprehensive . . . 19
2.4.4. Interdisciplinary . . . 19
2.4.5. Complementary . . . 19
2.4.6. Iterative and Dynamic . . . 19
2.4.7. Responsible to Society . . . 20
2.5. Objectives of These Guidelines . . . 20
2.6. Users of These Guidelines . . . 21
2.7. Scope . . . 21
2.7.1. Applicability . . . 22
2.7.2. Coverage . . . 22
2.7.3. Country Situations . . . 22
2.8. Approach to the Guidelines . . . 23
2.9. Contents of the Book . . . 23
References . . . 23

3. CONCEPTUAL FRAMEWORK . . . 25
3.1. FSR&D Activities . . . 27
3.1.1. Target and Research Area Selection . . . 27
3.1.2. Problem Identification and Development of a Research Base . . . 28
3.1.3. Planning On-Farm Research . . . 28
3.1.4. On-Farm Research and Analysis . . . 28
3.1.5. Extension of Results . . . 29
3.1.6. Collaboration . . . 30
3.1.7. Feedback . . . 31
3.2. Timing of Activities . . . 31
3.2.1. New FSR&D Programs for Predominantly Cropping Systems . . . 31
3.2.2. Ongoing FSR&D Programs for Predominantly Cropping Systems . . . 32
3.2.3. New FSR&D Programs for Predominantly Livestock Systems . . . 32
3.2.4. Closing Comments on Timing . . . 33
3.3. Staffing . . . 33
3.3.1. Field Teams . . . 33
3.3.2. Regional Headquarters Team . . . 33
3.3.3. National Headquarters Team . . . 34
3.3.4. Others . . . 34
3.3.5. Involvement of the Principal Groups in FSR&D . . . 34
3.4. FSR&D Strategies . . . 35
3.4.1. How Much Change? . . . 35
3.4.2. How Soon to Attempt Change? . . . 36
3.4.3. Where the Ideas for Change Originate? . . . 37
3.4.4. What Type of Research? . . . 37
3.4.5. A Partial Resolution of These Issues . . . 38
3.5. Summary . . . 38
References . . . 38

4. TARGET AND RESEARCH AREA SELECTION . . . 41
4.1. Amount of Data to Collect . . . 43
4.2. Selection of Target Area and Subareas . . . 43
4.2.1. Definitions . . . 43
4.2.2. Criteria and Methods . . . 44
4.2.3. Information Needed . . . 46
4.2.4. Actual Selection . . . 49
4.3. Selection of the Research Area . . . 51
4.3.1. Definition . . . 51

4.3.2. Criteria and Methods 51
4.3.3. Staffing 53
4.3.4. Information Needed 53
4.3.5. Collection of Information 54
4.3.6. Selection by Stages 55
4.3.7. Actual Selection 55
4.3.8. Abandoning a Research Area 56
4.4. Early Identification of Opportunities for Action 56
4.4.1. Research Opportunities 57
4.4.2. Government Action 57
4.5. Summary 57
References 57

5. PROBLEM IDENTIFICATION AND DEVELOPMENT OF A RESEARCH BASE 59

Part 1: Identification of Problems and Opportunities 61
5.1. Definitions of Problems and Opportunities 61
5.2. General Approach 61
5.3. Variations in Problem Focus 62
5.4. Understanding the Farmers' Situation 62
5.4.1. Describing the Farmers' Environment 64
5.4.2. Description of the Farming System 64
5.4.3. Analyses of the Farming System 67
5.5. Analysis of Problems and Opportunities 69
5.6. Setting Priorities for Problems and Opportunities 70
5.6.1. Seriousness of a Problem 70
5.6.2. Potential for Solving the Problem 71
5.6.3. Importance of the Problem in the Research Strategy 71
5.7. Summary of Problem Identification 72

Part 2: Development of a Research Base 72
5.8. Collecting Data: General Comments 72
5.9. Assembling Secondary Data 72
5.10. Collecting Primary Data 73
5.10.1. Informal Methods 73
5.10.2. Formal Methods 77
5.11. Combining Data Collection Methods 81
5.12. Data Management 82
5.13. Summary of Research Base Development 82
References 82

6. PLANNING ON-FARM RESEARCH 85
6.1. Laying the Groundwork 87
6.1.1. Reviewing Priority Problems and Opportunities 87
6.1.2. Appraising the Organization's Capabilities and Resources 88
6.1.3. Appraising Present Technologies 88
6.1.4. Setting Assumptions About Near-Term Conditions 88
6.1.5. Categorizing and Setting Research Priorities 88
6.1.6. Developing Hypotheses for Testing 89
6.1.7. Establishing Research Collaboration 89
6.2. Making Preliminary Analyses of On-Farm Experiments 89
6.2.1. Alternative Solutions 89
6.2.2. Farmers' Conditions 89
6.2.3. Perspectives 90

6.2.4. Technically Viable Designs . . . 91
6.2.5. Estimating Values . . . 91
6.2.6. Eventual Consequences . . . 92
6.3. Considering Alternative Research Activities and Methods . . . 92
6.3.1. Research Activities . . . 92
6.3.2. Research Methodologies . . . 95
6.3.3. Applying Methods to Activities . . . 97
6.4. Finalizing Plans for On-Farm Experiments . . . 97
6.4.1. Deciding on the Design Conditions . . . 97
6.4.2. Searching for Improvements . . . 98
6.4.3. Setting Design Standards . . . 99
6.4.4. Gathering Additional Data . . . 102
6.5. Conducting Regional Planning Workshops . . . 103
6.5.1. Nature and Purpose . . . 104
6.5.2. Attendance . . . 104
6.5.3. Activities . . . 104
6.6. Summary . . . 106
References . . . 106

7. ON-FARM RESEARCH AND ANALYSIS . . . 109

Part 1: On-Farm Research . . . 111
7.1. Researcher-Managed Trials . . . 111
7.1.1. Research on Crops . . . 111
7.1.2. Research on Livestock . . . 113
7.2. Superimposed Trials . . . 114
7.2.1. Research on Crops . . . 114
7.2.2. Research on Livestock . . . 115
7.3. Farmer-Managed Tests . . . 116
7.3.1. Research on Crops . . . 116
7.3.2. Research on Livestock . . . 121
7.4. Team Organization . . . 122
7.4.1. Team Leader's Activities . . . 122
7.4.2. Assignment of Resources . . . 123
7.4.3. Review Sessions . . . 123
7.4.4. Integration with Local Organizations . . . 123
7.5. Summary . . . 123

Part 2: Analysis . . . 123
7.6. Concepts of Analysis . . . 124
7.6.1. An Integrative Approach . . . 124
7.6.2. Prediction Versus Acceptance . . . 124
7.6.3. Partial Budget Analysis Versus Whole Farm Analysis . . . 125
7.7. Illustrative Designs and Analysis Procedures . . . 126
7.7.1. Illustrative Designs . . . 126
7.7.2. Biological Results . . . 126
7.7.3. Net Benefits . . . 128
7.7.4. Economic Criteria . . . 131
7.7.5. Financial Feasibility . . . 139
7.8. Acceptability of New Technologies . . . 139
7.8.1. Analysis of Farmer-Managed Tests . . . 139
7.8.2. Acceptability Index . . . 141
7.9. Sociocultural Feasibility . . . 141
7.10. Other Analysis Procedures . . . 142
7.10.1. Further Data Analysis . . . 142
7.10.2. Long-Term Investments . . . 142
7.10.3. Analyses from Society's Point of View . . . 143
7.11. Analysis Workshops . . . 143
7.12. Summary . . . 144
References . . . 144

8. EXTENSION OF RESULTS 147
8.1. Integration of Extension into FSR&D 149
8.2. Staffing and Organizing for Extension's Involvement 149
8.2.1. Extension Specialist in Farming Systems 149
8.2.2. Proposed Organization of Extension at Three Levels 150
8.3. Details of Extension's Involvement in FSR&D 151
8.3.1. Target Area Selection 151
8.3.2. Subarea and Research Area Selection 151
8.3.3. Problem Identification and Development of a Research Base 151
8.3.4. Planning On-Farm Research 151
8.3.5. On-Farm Research and Analysis 152
8.3.6. Extension of Results 152
8.4. Multi-locational Testing 153
8.4.1. Nature and Participation 153
8.4.2. An Example from Southeast Asia 154
8.5. Pilot Production Program 155
8.6. Problems in Extension 156
8.6.1. Ties Between Extension and Research 156
8.6.2. Training 157
8.6.3. Orientation 157
8.6.4. Organization 158
8.6.5. Budgets 158
8.7. Conclusions 158
References 158

9. DECIDING ON AN FSR&D APPROACH 161
9.1. Basic Issues in Research and Development 163
9.1.1. Are the Activities of the Research and Development Process Consistent with the National Development Goals? 163
9.1.2. Is the Research and Development Process Producing Results that are Relevant to Small Farmers' Needs? 163
9.1.3. Some Answers 163
9.2. Development Policy and the Role of FSR&D 164
9.3. Farmers' Needs and the Role of FSR&D 165
9.3.1. Modifications to Existing Systems 165
9.3.2. Introductions of New Systems 166
9.3.3. Decisions on How Much Change 166
9.4. Organizational Capacity and the Role of FSR&D 166
9.5. FSR&D and Supporting Organizations 166
9.6. Adoption of an FSR&D Approach 167
9.6.1. Emphasis on Applied Research 168
9.6.2. Adaptability to Ongoing Development Programs 168
9.6.3. Requirements for Skilled Personnel 169
9.7. Cost-Effectiveness of FSR&D 169
9.7.1. Comparison of Expenditures 169
9.7.2. Comparison of Rates of Adoption 169
9.7.3. Comparison of Numbers of Farmers Affected 169
9.7.4. Conclusions on Cost-Effectiveness 170
9.8. A Concluding Comment 170
References 170

10. IMPLEMENTATION 171
10.1. Deciding on an Approach 173
10.1.1. Project Approach 173

10.1.2. Program Approach 173
10.1.3. Project Versus Program Approach 173
10.1.4. Government Support 174
10.2. Organizational Structure 174
10.2.1. For the Project Approach 175
10.2.2. For the Program Approach 175
10.2.3. A Generalized Organizational Diagram 176
10.3. Staffing 178
10.3.1. Field Teams 178
10.3.2. Research Specialists 178
10.3.3. Extension's Input 179
10.3.4. Team Leadership 179
10.3.5. Approach When Trained Staff are Severely Limited 180
10.4. Off-Site Management 180
10.4.1. General Considerations 180
10.4.2. Personnel Management 181
10.5. The Roles of the Field Team 182
10.5.1. The Field Team and the Farmers 182
10.5.2. The Field Team and the Research Specialist 182
10.5.3. The Field Team and the Extension Service 182
10.5.4. Functional Assignments for the Field Team 182
10.5.5. Composition and Organization of the Field Team 183
10.6. Interdisciplinary Teamwork 184
10.6.1. A Model for Interdisciplinarity 185
10.6.2. Application of Interdisciplinarity to FSR&D 188
10.7. Getting Started 190
10.8. Evaluation of Projects 191
10.8.1. Types of Evaluation 191
10.8.2. Which Types of Evaluations to Use? 192
10.8.3. Developing Evaluation Procedures 192
10.8.4. A Caveat on Evaluations 192
10.9. Project and Program Management: A Two Country Perspective 193
10.9.1. An Example from Honduras 193
10.9.2. An Example from the Philippines 193
10.10. A Summary Perspective 193
References 194

11. TRAINING 195
11.1. Initial Exposure to FSR&D Concepts 197
11.2. Development of National Training Programs for FSR&D 197
11.2.1. Programs for Researchers 197
11.2.2. Programs for Extension 199
11.2.3. Programs for Technicians 200
11.3. Complementing Activities 201
11.3.1. Non-Degree Training 201
11.3.2. Graduate Degree Training 201
11.3.3. Short-Term Activities 201
11.4. International and Regional Centers 202
11.5. University Programs in the United States 203
11.5.1. Objectives for University Programs 204
11.5.2. Approach for University Programs 204
11.6. Training Materials on FSR&D 204

11.7. Summary . . . 204
References . . . 205

ACRONYMS . . . 207

GLOSSARY . . . 211

APPENDIXES . . . 219

P-A. Project Contributors: Field Contacts, Participants in the Workshops and Pretesting, and Reviewers . . . 223

2-A. Mathematical Modeling . . . 231

3-A. Review of FSR&D Activities . . . 237

4-A. Illustration of the Use of General Farm Data to Help Establish Recommendation Domains . . . 243
4-B. Grouping Farmers into Homogeneous Populations . . . 243
4-C. Selection of FSR&D Areas . . . 246
4-D. Climatic Zones in Southeast Asia . . . 248
4-E. Physiographic Regions in Southeast Asia . . . 249
4-F. Research Area Selection in ICRISAT's Village Level Studies . . . 249

5-A. Physical Resources of the Research Area Affecting Biological Production . . . 255
5-B. Land Types and Land Evaluation . . . 255
5-C. Marketing Factors Affecting Small Farmers . . . 259
5-D. The Sociocultural Environment . . . 261
5-E. An Example of Sociological Research . . . 262
5-F. Decision Making by Small Farm Families . . . 264
5-G. Decision Trees: A Method for Learning About Farmers' Decisions . . . 265
5-H. Describing Existing Cropping Systems . . . 267
5-I. Sample Forms for Describing On-Farm Resources Used in Crop Production . . . 269
5-J. Conceptual Models . . . 270
5-K. Mixed Cropping and Livestock Systems . . . 275
5-L. Data Collection in a Rural Setting . . . 278
5-M. Gathering Data About Women . . . 281
5-N. Assessment of Secondary Data . . . 283
5-O. Analysis of the Content of Informal Interviews . . . 284
5-P. Illustrative Tables for Collecting Data During Reconnaissance Surveys . . . 284
5-Q. Summary of the *Sondeo* Methodology Used by ICTA . . . 289
5-R. Guidelines for Pre-Survey Sequence . . . 293
5-S. Suggestions for Dealing With Farmers' Recollection of Information . . . 296
5-T. Validity From the Social Science Perspective . . . 297
5-U. Questionnaire Design . . . 299
5-V. Sampling . . . 303
5-W. Selecting, Training, and Supervising Interviewers . . . 306
5-X. Farm Record Keeping . . . 309
5-Y. Monitoring and Observational Activities . . . 316
5-Z. Data Management . . . 316

6-A. Selection of Variables for Climatic Monitoring of a Research Area . . . 321
6-B. The Land Equivalent Ratio (LER) . . . 323
6-C. A Guide for Locating On-Farm Experiments . . . 324
6-D. Field Designs and Statistical Procedures for On-Farm Experiments . . . 325
6-E. Example of a Procedure for Designing a Cropping Pattern Experiment . . . 335
6-F. Field Assignments . . . 337

7-A. Forms for Collecting Data for On-Farm Cropping Experiments 341
7-B. Forms for Collecting Data for On-Farm Livestock Experiments 345
7-C. Data Collection Form for Farmer-Managed Tests 348
7-D. Field Design for Farmer-Managed Cropping Test 349
7-E. Field Design of a Farmer-Managed Cropping Pattern Test 350
7-F. Example of a Mixed Crop-Livestock Research Project 350
7-G. Estimating Net Benefits from Alternative Treatments 352
7-H. Analysis of Cropping Pattern Research in Indonesia 354

8-A. Memorandum of Agreement to Establish a Pilot Production Program in the Philippines 359
8-B. Letter of Understanding between ICTA and DIGESA 360

10-A. An Agricultural Research Project for the Senegalese Institute for Agricultural Research (ISRA) 365
10-B. Alternative Organizational Diagrams for Farming Systems Research and Development 369
10-C. Team Building 374
10-D. Summarized Checklist for Successful Interdisciplinarity 375
10-E. USAID's Logical Framework 376

11-A. Honduran Training Program in FSR&D 383
11-B. Outline of an In-Service Training Program in FSR&D Prepared by ICTA for DIGESA 388
11-C. Training in FSR&D at Selected International Centers for Agricultural Research 389
11-D. A Cropping Systems Training Program at IRRI 393
11-E. Six Principles for Technical Change 397
References Cited in the Appendixes 399

INDEX 407

PREFACE

In recent years, policy makers have been paying more attention to the problems of small farmers in developing countries with the idea of increasing their production and standard of living. The policy makers' objectives are twofold: (1) to help those whose welfare is materially below the rest of society, and (2) to help a country increase its agricultural production. With adequate agricultural policies, these two objectives are mutually reinforcing. For example, increased food production gives farm households additional food for consumption and surpluses for sale. Farmers can then use the money from these sales to buy items they do not produce, and the buyers of farm products benefit from the increased supplies.

By focusing on these two objectives, leaders in developing countries work toward other national objectives. For example, increasing production on small farms may (1) improve a nation's self-sufficiency in food production, (2) supply more raw materials for industry, and (3) improve the nation's foreign exchange position. Improving small farmers' production should narrow the range of incomes among groups in the country, reduce political instability, slow rural-to-urban migration, lessen the need for food relief, and so on.

In developing countries, farmers with limited resources often do not adopt new technologies because (1) their conditions are not like those where the technologies were developed, (2) they do not have resources to purchase the required inputs, (3) the technologies do not apply to the crops grown or the livestock raised on their farms or the way they operate, or (4) they do not know about the new technologies. For whatever reason, development of new technologies sometimes leaves small farmers worse off than before. This happens when large farmers adopt new technologies and small farmers do not.

But conditions are changing. Recently, more national and international research organizations are directing their attention to the conditions and problems of small farmers. As a result, small farmers and their environments are better understood; more research is applied to solving small farmers' problems; and the extension service works with better technologies. An approach now being applied more widely to make research relevant for small farmers is called farming systems research and development (FSR&D) or sometimes simply farming systems research (FSR).

With more countries interested in FSR&D, the United States Agency for International Development (USAID) contracted with the Consortium for International Development (CID) to write a set of guidelines on FSR&D methodology. CID made Colorado State University (CSU) the lead university, entered into agreements with other CID schools, and subcontracted part of the work to the University of Hawaii.

The result is this set of guidelines for those in the developing countries who wish to learn about FSR&D and who might apply the procedures. In keeping with this emphasis, the contract required the writers to synthesize

> ". . . a set of integrated, multidisciplinary farming systems R&D methodologies adapted to the personnel and financial constraints of the LDC's, packaged for easy delivery in the form of a comprehensive handbook or handbooks to LDC institutions."[1]

These methodologies refer to ways for gathering data on farming systems and farmers' environments, identifying problems and opportunities for improvement, developing research data for farmer and experimental plot conditions, generating alternative farming systems, diffusing improved technologies, and receiving feedback for further improvements.

To produce these guidelines, we synthesized data on FSR&D and related practices from documents and

---

[1]USAID. 1978. Farming systems R&D methodology contract. USAID, Washington, D.C.

meetings with individuals and representatives of organizations throughout the world. Because the published literature includes few documents on farming systems concepts and procedures, we relied heavily on visits to organizations with agricultural research programs for small farmers. Early in our study, we identified issues in FSR&D and discussed them at a workshop that a group of internationally recognized practitioners in FSR&D attended.

The review process entailed critiques of three preliminary drafts of the guidelines. Our first draft was reviewed during a two-day workshop in June 1980 attended by a small group of FSR&D practitioners and the project team. Our second draft was pretested at a two-week session in August 1980 attended by a group from the developing countries, the project team, and observers. We distributed the third draft worldwide in December 1980, and received comments from more than 30 reviewers. In Appendix P-A we provide further comments on these activities and the names of contributing individuals and organizations.

To conclude, we stress that FSR&D concepts and procedures are evolving rapidly. Moreover, the early concentration on crops is giving way somewhat to allow more consideration of livestock—either as pure livestock systems or in combination with crops. Furthermore, we anticipate other topics will be integrated into the FSR&D approach such as mathematical and ecological modeling, agro-forestry, and soil and water conservation. Consequently, this book of guidelines will require updating as more is learned and better procedures become available. With this in mind, we welcome your suggestions.

W. W. Shaner
P. F. Philipp
W. R. Schmehl

Fort Collins, Colorado

ACKNOWLEDGMENTS

We, the three co-authors of these guidelines, are grateful to the other members of the team; to consultants; to advisers; to the governments of Guatemala and Honduras, and New Mexico State University for hosting the pretesting sessions; to the participants in the pretesting and their parent organizations; to the reviewers of the various drafts and outlines; to those who met with team members during their travels and provided insight and information on agricultural research and extension; to the national governments and international centers that cooperated; to USAID's project monitor; and to the "pioneers" in farming systems research and development for their vision and courage in developing new approaches to solve the problems of poor farmers around the world. While we acknowledge the contributions of the above, we do not hold them responsible for any of this book's shortcomings.

As co-authors, we had the responsibility for writing the book of guidelines, but we could not have completed the manuscript without help from others. This help came in conceptualizing the approach, gathering and analyzing data, drafting chapters and appendixes, and reviewing others' work. Below, we list the project staff and their principal inputs. By being an enthusiastic and effective member of a team effort, each person contributed much more than the contributions listed.

Gary Hansen drafted Chapters 9 and 10 on the merits of the FSR&D approach and on organization. Tom Trail drafted Chapters 8 and 11 on extension and training. Richard Tinsley drafted sections on the cropping systems research that went into Chapters 6 and 7. Helen Henderson drafted sections on sociocultural aspects for the whole book. Howard Stonaker drafted sections on livestock systems in Chapters 6 and 7. James Meiman drafted the section on interdisciplinary teamwork and Tom Sheng drafted the section on evaluation in Chapter 10.

Contributors to the appendixes were George Beal on sociocultural research, Jen-hu Chang on agroclimatology, Michael Read on experimental design and statistical procedures, Tom Sheng on mathematical modeling and other topics, Don Zimmerman on analysis of data and research results, and Ann Perry-Barnes and John Roecklein on a variety of topics.

Don Zimmerman was project editor. Robert Dils was responsible for the pretesting. We obtained help in gathering data on farming systems in Francophone West Africa from Derrick Thom, in Central America from Albert Ludwick, and in Indonesia from Martin Waananen. Tom Sheng was research associate. Ann Perry-Barnes was research assistant. Michael Read and John Roecklein were graduate research assistants.

Other project staff included Dale Rosenbach for the book's format and artwork. Carol Marander drafted the figures. William Shaw created the sketches. Don Zimmerman coordinated the preparation of the photographs. Miriam Palmer was responsible for the bibliography. Jan Schweitzer prepared the index. Marilee Long helped with the editing. Patty Sheng helped with the preparation of the drafts. Veryl Meyers typed the final copy of the text. Regular and temporary secretaries were Debbie Bartow, Cheryl Buster, Lillian McKee, Lori Neubauer, Peggy Neff, Vicky Lynn, Christine Stanley, and Imogene Wood.

The project's advisory committee included Gerald Burke, Frank Conklin, Jack Keller, Shelley Mark, Gerald Matlock, Martin Waananen, and James Meiman as chairman.

Consultants, who contributed greatly to the project, were Peter Hildebrand, David Norman, and Robert Waugh on FSR&D in general; Ramiro Ortiz on the farming systems program in Guatemala; George Beal and Edward Knop on sociology; and Elmer Remmenga on experimental design and statistics.

Kenneth McDermott—USAID's project monitor—briefed the project team at the outset, assisted in coordinating the interactions between the project and USAID, and, most importantly, contributed substantially to our understanding of farming systems concepts.

Finally, we wish to thank all those who provided us with photographs to illustrate farming activities and conditions in developing countries. Those photographs selected for each of the chapters were supplied by the following photographers and organizations: Chapter 1: Food and Agriculture Organization of the United Nations (FAO) photographs by A.E. Deutsch of the International Plant Protection Center (IPPC), Corvallis, Oregon; Douglas Horton of the International Potato Center (CIP),

Peru; International Crops Research Institute for Semi-Arid Tropics (ICRISAT), India; and Tom Sheng, FSR&D Project, Colorado State University (CSU); Chapter 2: Tom Sheng and W.W. Shaner, FSR&D Project; Donald Sungusia, Tanzanian government; Wayne Freeman, International Agricultural Development Service (IADS), Nepal; and ICRISAT; Chapter 3: International Center for Agricultural Research in Dry Areas (ICARDA); FAO; and IPPC; Chapter 4: IADS, Nepal; Tom Sheng and P.F. Philipp, FSR&D Project; and ICRISAT; Chapter 5: P.F. Philipp and Tom Sheng, FSR&D Project; ICARDA; Douglas Horton, CIP; and IADS, Nepal; Chapter 6: ICARDA; IADS; Dan Lattimore, CSU Egyptian Project; and Tom Sheng, FSR&D Project; Chapter 7: P.E. Hildebrand, Agricultural Science and Technology Institute (ICTA), Guatemala; ICARDA; ICRISAT; Tom Sheng and P.F. Philipp, FSR&D Project; FAO; and IPPC; Chapter 8: ICRISAT; ICARDA; and Douglas Horton, CIP; Chapter 9: ICRISAT; FAO; IPPC; and ICARDA; Chapter 10: Douglas Horton and Francis Tardieu, CIP; and Tom Sheng and W.W. Shaner, FSR&D Project; and Chapter 11: Douglas Horton, CIP; ICRISAT; and ICARDA.

Chapter 1
EXECUTIVE SUMMARY

This book provides guidelines for farming systems research and development (FSR&D) as applied to conditions in developing countries. The purpose of the guidelines is to assist national governments interested in helping poor farmers—primarily small-scale farmers with limited resources. Therefore, the guidelines discuss the nature of FSR&D, processes and methodologies appropriate for various conditions, and alternative means for implementation. Because most of the applied work in FSR&D has been with cropping systems, this book of guidelines emphasizes cropping systems research. By synthesizing implemented and successful approaches, these guidelines have a strongly applied orientation.

As a synopsis of the principal features of the guidelines, this executive summary is intended for those who wish a quick review of FSR&D's principal features. This summary contains brief sections on the background of FSR&D, its nature and activities, issues of implementation, and the contents of this book.

1.1. BACKGROUND

Considerable attention is currently being given to improving the lot of small farmers in developing countries. An important way of helping them is through agricultural research, extension, and related programs specific to their needs. A better approach for such efforts became necessary because farmers' conditions were not improving adequately. Research and development programs had often been undertaken without having small farmers in mind or without knowing much about them. In contrast, the FSR&D approach starts and ends with small farmers and thereby focuses specifically on their conditions and aspirations.

While much of FSR&D has been directed toward farmers with limited resources, the approach has relevance for improving agricultural research and development in general. Some argue FSR&D is simply a modified version of farm management that has been widely practiced in the United States during the 20th century. While this claim has merit, the general feeling among those actively engaged in FSR&D is that FSR&D is new—at least as applied to the needs of small farmers in developing countries. The accomplishments of some national and international research organizations support the contention that improved technologies can be designed for and will be adopted by small farmers.

1.2. NATURE OF FSR&D

A common thread among alternative approaches to FSR&D is the selection of relatively uniform sets of conditions for conducting research and implementing change. FSR&D allows researchers to (1) both intensively investigate the individual conditions of small farmers and (2) make an impact on large numbers of farmers. This result is accomplished by selecting reasonably uniform physical, biological, and socioeconomic environments, where farmers' cropping and livestock patterns and management practices are similar. Improved technologies developed for farmers in these research areas are expected to be applicable to farmers operating elsewhere under similar conditions.

The FSR&D approach typically uses interdisciplinary teams, whose composition varies according to the task. Field teams conduct on-farm research and are aided by (1) disciplinary specialists in the physical, biological, and social sciences who may operate out of regional or national headquarters or experiment stations, (2) extension specialists, and (3) others concerned with agricultural production.

Together, they study

- physical conditions such as rainfall, temperatures, and land forms
- biological factors such as production potential and pest problems
- socioeconomic conditions such as the size and nature of landholdings, farmer and community customs, markets, and local services
- the farming system.

The farming system is the complex arrangement of soils, water sources, crops, livestock, labor, and other resources and characteristics within an environmental setting that the farm family manages in accordance with its preferences, capabilities, and available technologies. Farmers manage the household's resources involved in the production of crops, livestock, and nonagricultural commodities (e.g., handicrafts), and may also earn income off the farm.

Farms are classified according to major characteristics—e.g., grazing systems, permanent cultivation on rain-fed land, or irrigated farming—and the environment—e.g., agroclimatic zone, soils, and terrain. Re-

searchers classify farms according to the area, the needs of the study, and the available information.

FSR&D focuses on the interdependencies among the components under the farmers' control, and between these components and the physical, biological, and socioeconomic environments. Also, FSR&D identifies and generates improved technologies and adapts, tests, and promotes them.

The various production activities are subsystems of the whole farming system. For example, crop production is a subsystem of the whole farm and is, in turn, made up of individual cropping activities. The study of a cropping system comprises everything required for the production of one or more crops, including interactions between different crops. More specifically, research on cropping systems concentrates on

- crops and cropping patterns
- alternative management practices in different environments
- interactions between crops
- interactions between crops and other enterprises
- interactions between the household and environmental factors beyond the household's control.

A similar description could be given for livestock systems research.

Thus, FSR&D can be summarized as being farmer-based, problem solving, comprehensive, interdisciplinary, complementary, iterative, dynamic, and responsible to society. The approach is

- farmer-based because FSR&D teams pay attention to farmers' conditions and integrate farmers into the research and development process
- problem solving in that FSR&D teams seek researchable problems and opportunities to guide research and to identify ways for making local services and national policies more attuned to the farmers' needs
- comprehensive in that FSR&D teams consider the whole farming activity (consumption as well as production) to learn how to improve the farmers' output and welfare, to identify the flexibilities for change in the environment, and to evaluate the results in terms of both farmers' and society's interests
- interdisciplinary in that researchers and extension staff with different disciplinary backgrounds work with farmers in identifying problems and opportunities, searching for solutions, and implementing the results
- complementary because it offers a means for using the outputs of other research and development organizations and for giving direction to others' work
- iterative in that FSR&D teams use the results from research to improve their understanding of the system and to design subsequent research and implementation approaches
- dynamic in that oftentimes FSR&D teams introduce relatively modest changes in the farmers' conditions first and the favorable results encourage more significant changes later
- responsible to society in that FSR&D teams keep the long-run interests of the general public—both present and future—in mind, as well as those of the farming groups immediately affected.

While much of the above is true of other forms of agricultural research and development programs, the combination of these factors distinguishes FSR&D from other approaches. Even more, FSR&D is systems oriented in that the researchers study the farmers' conditions at the outset, keep these conditions in mind during research and implementation, and use their knowledge of these conditions in evaluating the results. In this sense, FSR&D departs from reductionism, which is an approach that breaks the whole into parts and studies them more or less independently. Furthermore, FSR&D uses acceptance by the whole family as its key measure of success, rather than some abstract or narrowly defined criteria of effectiveness.

1.3. FSR&D ACTIVITIES

The approach to FSR&D varies according to the organization's mandate, which may be for certain commodities or which may be localized, countrywide, or international. Approaches also vary by the physical, biological, and socioeconomic characteristics of the target areas and groups, as well as by the preferences of FSR&D administrators and researchers. Some approaches are comprehensive, taking many factors as variable, including public policy; but more frequently, FSR&D works within existing conditions or assumes only modest changes in the existing conditions.

The basic FSR&D activities are target and research area selection, problem identification and development of a research base, planning on-farm research, on-farm research and analysis, and extension of results. Each of these is summarized below.

1.3.1. TARGET AND RESEARCH AREA SELECTION

Using national and regional objectives, key decision makers—including those from the FSR&D team—select one or more target areas. Then, the FSR&D team divides the target area into subareas with relatively uniform characteristics and selects a research area representative of the selected subareas. The team continues by choosing the target group—farmers who have common environments and common production patterns and farming practices. This group of farmers might be those with a particular cropping, livestock, or mixed (e.g., crops and livestock) pattern; alternatively, the approach could be based more on environmental conditions. Such classifications are usually adequate for identifying problems and opportunities of sufficient magnitude to justify the research effort. Where practical, the FSR&D team tries to apply the research results to farmers operating under similar conditions beyond the target area.

1.3.2. PROBLEM IDENTIFICATION AND DEVELOPMENT OF A RESEARCH BASE

The FSR&D team identifies and ranks problems and opportunities according to such criteria as the short-run and long-run significance to the farmers and society, availability of suitable or potentially suitable technologies, and ease of implementation. Besides ideas arising out of the previous activity, the team commonly identifies problems and opportunities through quick reconnaissance surveys of the area. The study of livestock systems tends to take longer and may involve aerial photography, satellite imagery of rangelands, and monitoring of development programs to learn how herding societies function over time. A subject with considerable and yet untapped potential is research on mixed farming systems in which the researchers consider the influence of crops and livestock on each other.

In the process of identifying problems and opportunities, the team gains considerable knowledge about the area. This knowledge and the collected data form the initial research base for developing improved technologies for the area's small farmers.

1.3.3. PLANNING ON-FARM RESEARCH

Once the FSR&D team has identified and ranked problems and opportunities, gathered preliminary data,

and set out hypotheses, it plans the on-farm research activities. Early in the process, the team needs to decide the extent to which the farmers' environment can be changed. For the most part, the team takes resource availability, support services, and government policy about as they are. But, an important part of FSR&D is to identify where and how much change of this type is possible. Given an understanding of this, the team then considers opportunities for improving farmers' conditions.

On-farm research emphasizes alternative cropping and livestock patterns, management practices, and other activities of the farm household. The team incorporates the farmers' conditions into the design procedures by working closely with farmers. The team meets with farmers in their fields and learns farmers' terms such as those for farmers' activities and units of measure. Researchers also learn how the farm household divides its activities, which members perform which activities, who has responsibility for the different family decisions, who controls which resources, how members tend the family's crops and livestock, and how they market their surplus production. Farmers, in turn, take part in the research experiments and evaluate the results. This collaborative style calls for integration of experiment station and other research and development personnel who are specialists in (1) disciplines such as entomology, economics, and soil conservation; (2) commodity topics, such as plant breeding and cattle production; and (3) extension.

Furthermore, the team designs record keeping systems, special studies, climatic monitoring, and surveys to provide additional information about the farmers and their environment. Often the team initiates recording of farmers' activities early in the FSR&D process to develop a continuing base of information on farmers' productive activities throughout the cropping and livestock seasons. The team uses special studies of selected topics, such as cultivation practices, to help fill in gaps in its knowledge about the area. The team needs information on the environment, including climatic data, to help design research and interpret the results from crop and animal experiments. Also, the team uses long-run studies of farm households, local conditions, and related topics to provide a sound basis for understanding the situation and implementing change.

Before finalizing the research plan, the team evaluates the proposed technological changes. It does this to learn if the results are biologically feasible and in the interests of the farmers and society. Finally, the team assesses the extent to which local support systems and national policies will accommodate the new technologies.

1.3.4. ON-FARM RESEARCH AND ANALYSIS

Most national FSR&D programs emphasize applied research by conducting much of the research on farmers' fields. Three types of biological production experiments are common: researcher-managed trials to experiment under farmers' conditions where control of the experiment is important; farmer-managed tests to learn how farmers respond to the suggested improvements; and superimposed trials to apply relatively simple researcher-managed experiments across a range of farmer-managed conditions.

The researchers initiate experiments, studies and other activities, and gather data. Then, they analyze the results in terms of the statistical meaning of biological performance, actual resource requirements, economic and financial feasibility, and sociocultural acceptability. They estimate the overall impacts on both farmers and society. Researchers study the acceptability of the experiments to farmers through observations of farmers' actions, talking with farmers, and in other ways. Finally, the researchers examine the opportunities for improving support services and government policies.

1.3.5. EXTENSION OF RESULTS

Throughout the research process, the FSR&D team maintains contact with support organizations in the area. Extension plays an especially important role in the process. Inputs from extension should occur at all levels of FSR&D—from initially identifying areas to the broad implementation of results. FSR&D practitioners generally recommend that the extension staff be trained in FSR&D and become regular members of the field and regional teams.

Extending the results involves multi-locational testing—an activity that spreads the improved technologies more broadly than the previous on-farm trials and tests. Multi-locational testing helps define the specific conditions by applying the results on a broad scale. In this process, extension agents learn the details of the technologies and how to apply them.

Another means of extending research results is through pilot production programs—an activity that applies the improved technologies on a scale large enough to effectively test the area's support systems. This activity provides further insight into the needs for modifying the technology, altering the support system, or both. However, the concept of FSR&D is that the derived technologies should fit the farmers' and environmental conditions sufficiently well so that few adjustments are needed at this stage.

Once these steps have been taken, the country can broadly apply the new technologies among the groups for which they have been designed.

1.4. ISSUES OF FSR&D IMPLEMENTATION

Some of the issues concerning FSR&D implementation relate to the time required to obtain results, organizational flexibility, staffing requirements, training, FSR&D costs, and governmental support.

1.4.1. TIMING

The general approach to FSR&D is rapid initiation of on-farm experiments combined with adjustments in the program's direction as results provide feedback. With adequate planning, researchers often start experiments without missing a cropping season. Sometimes they try exploratory experiments to learn how farmers respond to new opportunities; at other times, researchers conduct trials to screen locally available technologies for their applicability to specific farmers' conditions. Under favorable conditions, some research results may be ready for widespread diffusion to farmers within a few seasons. However, more fundamental changes in farmers' cropping patterns and management practices normally take longer.

The approach being developed for livestock systems is an exception. For larger animals such as cattle, the environment, livestock systems, and growth stages often require more careful study than most crops or small animals.

1.4.2. ORGANIZATIONAL FLEXIBILITY

FSR&D is primarily a modification of existing research and extension methods; therefore, the approach is adaptable to a variety of situations, as illustrated by the following possibilities. A country can implement FSR&D through a semiautonomous government corporation that has more flexibility in operations, budgeting, and personnel management than ministerial research and development organizations. A country can implement FSR&D through a ministry of agriculture if the ministry is responsible for research and extension. A country can apply FSR&D to the activities of an experiment station in which

one or more teams trained in FSR&D methods work closely with experiment station staff. Or, a country can build FSR&D into a project to increase production; in such a case, FSR&D methods can improve the efficiency of the overall project.

Each approach has its advantages and disadvantages, so the approach selected depends on the situation. Here, we emphasize that FSR&D, whether in whole or in part, can be and has been implemented in a variety of ways.

1.4.3. STAFFING REQUIREMENTS

FSR&D strongly emphasizes working with farmers in their fields. To the extent that this emphasis is new, those currently at research stations or at regional or national headquarters will require some reorientation. This reorientation includes research methodology as applied to field conditions and methods for working with the whole farm family—male and female, young and old. Where appropriate, females may need to be added to the research and extension staff.

However, FSR&D does not replace existing research or extension; rather, it builds on the existing base. Consequently, experienced researchers and extension specialists usually remain in their existing organizations and much of the field staff consists of young professionals trained specifically for FSR&D's purposes. Enough senior staff members will be needed—whether nationals or expatriates—to guide the younger members of the staff until they gain adequate experience.

One approach is to begin FSR&D activities in one or two regions and, after several years of experience, to choose leaders from these teams when moving to new areas. Heads of FSR&D programs must also train staff to replace those who periodically leave the program.

1.4.4. TRAINING

An early activity when implementing an FSR&D approach is to train the staff about the objectives, processes, and methodologies of FSR&D. Training materials will need to be collected from ongoing programs elsewhere and augmented by new materials appropriate for the country. During this early stage, the International Agricultural Research Centers (IARCs) and organizations with similar activities can be especially helpful.

The principal objectives of the training are to

- acquaint team members with on-farm techniques
- give them guidance and experience working as an interdisciplinary team
- instill in the team members an enlightened appreciation of small farmers as a useful source of information and as valuable partners in the research and implementation process.

Where members of the FSR&D team are recent graduates, in-service field training under the guidance of experienced staff is needed.

Initially, program leaders may want to take advantage of production and farming systems training at one or more of the IARCs and any regional center specializing in applied agricultural research. With such training as a base, in-country training programs for both research and extension personnel can then be developed and implemented. Some staff members may be selected and sent abroad for further academic training.

In training, as well as in other aspects of FSR&D programs, national governments may want to consider using expatriate staff experienced in FSR&D. As the program matures, the expatriates can be phased out gradually. In one case, this occurred about six years after the program began.

1.4.5. FSR&D COSTS

A discussion on the relative costs of FSR&D centers on expenditures, rates of adoption, and breadth of coverage. This discussion must be general since carefully quantified appraisal of FSR&D's costs, relative to other research and development approaches, has not, to our knowledge, been made. While firm estimates are not available, those closely associated with FSR&D generally feel that the approach is cost effective. The reasoning follows.

The first of the three issues concerns expenditures for facilities and operating costs. To the extent that FSR&D reduces experiment station activity, costs of expanded installations, operations, and the accompanying staff will be lowered. In its place will be more work on farmers' fields by generally less expensive staff. However, the field work requires increased expenditures for vehicle purchase and maintenance, field equipment, per diem, and incentives. Overall, the combined initial and recurring costs of FSR&D appear to be less than the costs of comparable levels of activities on experiment stations, when administrators consider the costs of building, staffing, and equipping the stations. However, such comparisons are of limited value since FSR&D replaces only a portion of experiment station activities.

The second issue concerns the generation of new technologies acceptable to farmers. This too is not a straightforward issue, because the target group for FSR&D is sometimes different from that of general agricultural research. Proponents of FSR&D, however, point to the high levels of adoption of improved technologies by small farmers targeted by the FSR&D process.

The third issue centers on the range of applicability of research results. Opinion differs about how widely FSR&D can be applied. Traditional research, by its nature, often has general and wide applicability. FSR&D is designed to be more specific, but it may also be applied broadly if the team can identify environmental conditions sufficiently wide ranging and target groups in sufficiently large numbers. FSR&D practitioners expect work in categorizing research areas to eventually make it easier to locate situations in which the new technologies generated by FSR&D will have broad applicability. Eventually, the study of environmental gradients will permit a better understanding of the relationship between research results and the conditions leading to these results, but this latter possibility, especially when speaking of national programs, lies in the future.

1.4.6. GOVERNMENTAL SUPPORT

Because FSR&D concentrates on field activities, the government will need to take steps to allow team members to effectively carry out this work. Materials for conducting experiments need to be available at appropriate times, otherwise the experiments may not be completed. Reliable transportation is essential, especially where the terrain and weather conditions make travel difficult. The FSR&D team needs adequate servicing and spare parts for its vehicles. Finally, incentives are often required to attract and hold qualified staff. Incentives such as the recognition of team accomplishments will be needed to overcome the uncertainties of working in a new and different program and the hardships of living and working in remote areas.

FSR&D does not place great demands on the government, but these demands must be met to create and maintain the momentum necessary to sustain an effective FSR&D effort. Where the central or regional organization cannot meet some of the above requirements, the organization should give the field teams adequate local autonomy.

1.5. WHAT THIS BOOK OFFERS

To repeat, this book of guidelines describes an approach to agricultural research and development for governments of developing countries interested in improving the output and welfare of small farmers. We present the FSR&D activities, methods, and illustrations of various approaches in the main body of this book and elaborate on these points in the appendixes. We emphasize cropping systems research because most experience lies here; however, we include materials on livestock systems. Systems concepts are included, but few analytical tools for

systems analysis such as simulation or linear programming are included because we found few examples of their use in national FSR&D programs.

This book of guidelines is for those in the developing countries who must decide whether to accept FSR&D and bear the responsibility for its implementation. This book is also for the expatriate who aids in this process.

In designing FSR&D activities, administrators must decide on the approach, methods, organization, staffing, training, and ways to secure technical assistance and funds. The book should aid such individuals in making reasoned decisions on these topics. Because of the diversity of conditions and the wisdom of allowing those in a country to make their own decisions, the book does not prescribe how a country should implement FSR&D activities. Instead, the book presents general concepts, offers alternatives that have worked in different countries, and provides the reader with sources of additional information.

In conclusion, undertaking an FSR&D approach that modifies a more traditional approach to agricultural research presents a considerable challenge to any country. Existing institutions and individuals may feel threatened by the change. False starts are possible. Still, if the enthusiasm of those who have been most active in the FSR&D movement is any indication of its validity, the effort is justified.

Chapter 2

INTRODUCTION

During the past decade, considerable attention has been focused on the plight of the rural poor in the developing countries. One aspect of this emphasis has been to direct agricultural research specifically to the needs and aspirations of farmers with limited resources. Historically, these have been small farmers who have not adequately benefited from agricultural research because the research was not specific enough for their needs. Instead, research in the less developed countries has typically been undertaken for farmers who have more resources and who often produced for export.

Generally, technologies offered to the small farmers have come from a top-down approach. By that, we mean the research would be largely initiated and conducted on experiment stations and then offered to small farmers to accept or reject. As a result, farmers rejected many of the proposed changes because the suggested improvements were unprofitable or too risky, or the farmers lacked adequate inputs or suitable markets. In short, the technologies were not suitable because the researchers did not know or consider the conditions of small farmers.

Therefore, research, extension, and other programs are needed to correct these deficiencies, if small farmers in developing countries are to be helped. One approach that considers farmers' conditions specifically is called farming systems research and development (FSR&D), or simply farming systems research (FSR). In this book we use the term FSR&D to emphasize the integration of research and the development of technology for dissemination through extension and by other means. The FSR&D approach provides a means for dealing with the close interaction of the many on-farm activities that characterize subsistence farming.

This book of guidelines was written for those with national programs in developing countries who wish to orient part of their research efforts toward benefiting small farmers. These guidelines concern FSR&D processes and procedures and include examples from specific situations. This chapter covers the purposes and definition of FSR&D, additional definitions and comments, characteristics of FSR&D, objectives and users of these guidelines, the guidelines' scope, and the approach and contents of the guidelines.

2.1. PURPOSES OF FSR&D

As with other national approaches to agricultural research and extension, FSR&D's purpose is to generate more appropriate technologies for farmers and, where possible, to improve policies and support services for farm production, to raise farm families' welfare, and to enhance society's goals. But more specifically, FSR&D aims at increasing the productivity of farming systems by generating technologies for particular groups of farmers and by developing greater insight into which technologies fit where and why. This latter purpose concerns using scientific methods for generating hypotheses and then, by deduction, determining which technologies to use in a particular farm setting. Such an approach contrasts with an empirical approach that through trial and error arrives at suitable technologies for the conditions of specific farmers (Harwood, personal communication).

We include the farm family in the above description because the collective interests of the family are important, not just the interests of the head of the household. Furthermore, we include agricultural production because FSR&D concentrates on increasing crop and livestock yields and overall farm output. And we include family welfare because improved welfare is the ultimate goal of individual families just as societal interests are the ultimate concern of an enlightened government.

2.2. DEFINITION OF FSR&D

FSR&D is an approach to agricultural research and development that

- views the whole farm as a system
- focuses on (1) the interdependencies between the components under the control of members of the farm household and (2) how these components interact with the physical, biological, and socioeconomic factors not under the household's control.

Farming systems are defined by their physical, biological, and socioeconomic setting and by the farm families' goals and other attributes, access to resources, choices of productive activities (enterprises), and management practices.

The systems approach applied to on-farm research considers farmers' systems as a whole, which means

1) studying the many facets of the farm household and its setting through close and frequent contact with household members on their farms

2) considering problems and opportunities as they influence the whole farm
3) setting priorities accordingly
4) recognizing the linkages of subsystems within the farming system and considering them when dealing with any part of the system
5) evaluating research and development results in terms of the whole farming system and the interests of society.

The FSR&D team implements the FSR&D process by

1) selecting areas and groups of farmers with reasonably similar characteristics as targets for research and development
2) identifying and ranking problems and opportunities and setting forth hypotheses for alternative solutions
3) planning experiments, studies, and procedures for data collection
4) undertaking experiments on farmers' fields, in conjunction with other research, to identify or generate improved technologies suitable for farmers' conditions
5) coordinating the on-farm experiments and studies with commodity and disciplinary-oriented research
6) evaluating the acceptability of the results of these experiments to the targeted farmers and society
7) extending the results widely to farmers within and outside the target area
8) focusing attention on ways to improve public policy and support services to assist both the targeted farmers and those operating under similar conditions.

The distinction between FSR&D and "conventional"[1] research can be summarized in the following way. FSR&D looks at the interactions taking place within the whole farm setting and measures the results in terms of farmers' and society's goals. Traditionally, con-

[1]The authors encountered differences of opinion as to the meaning of conventional research, consequently, the use of the quotation marks.

ventional research separates tasks into progressively narrower subject areas to be studied more or less independently and then evaluates results by standards within the discipline, not by their contribution to the whole (Dillon, 1976). Furthermore, FSR&D places relatively more importance, than in the past, on integrating the social sciences into the research and development process. This is accomplished by considering such factors as farmers' preferences, community norms, markets, public policies, and support services (Norman, personal communication).

FSR&D's comprehensiveness can be illustrated by discussing the differences between FSR&D's results and those of a single disciplinary approach. For example, a breeder may seek to obtain the highest physical yield for a single crop through variety and fertilizer trials. In contrast, an FSR&D approach integrates the breeder's work by considering more objectives and means of improvement. For example:

- An earlier maturing variety might be sought that allows time for planting a second crop, even though the yield from such a variety is less than from other varieties.
- Net profits from fertilizer application could be increased by reducing the application rate to a lower level than is needed to produce the maximum biological yield.
- Recognizing farmers' aversion to risk could suggest a less profitable crop whose yields are more stable during unfavorable growing conditions.
- Social and cultural study could explain why some farmers accept improvements and others do not, so that the resulting technologies could be applied to more farmers.
- Integrating the extension service into the FSR&D process could result in the extension staff suggesting modifications to the technologies; these changes, in turn, could help the extension service serve farmers more effectively.

Not all aspects of a farming system must be addressed for the process to be considered FSR&D. Crop-

ping and livestock systems and even commodity research may qualify. What is needed is for the research on subsystems—e.g., cropping systems—to be taken within the context of the whole farm. Such an approach for cropping systems requires a study of the farming system to verify

- that research into cropping systems is justified
- that the research on the subsystems and the resulting recommendations fit within the overall system
- that the final evaluation is within the whole farm context.

Finally, some improvements in farmers' conditions may not result from breakthroughs in agricultural technologies; instead, the improvements may result from identifying and implementing more suitable agricultural policies and support services.

2.3. ADDITIONAL DEFINITIONS AND COMMENTS

To help clarify the above concepts, we will now define additional terms used in this book. They are farming, households, small-scale farming, enterprises, farming systems, cropping and livestock systems, mixed systems, cropping systems research, livestock and mixed systems research, and commodity-oriented research.

2.3.1. FARMING

Farming is an activity carried out by households on holdings that represent managerial units organized for the economic production of crops and livestock (Ruthenberg, 1971).

2.3.2. HOUSEHOLDS

The household is a social organization in which members normally live and sleep in the same place and share their meals. They may or may not be a joint family. A joint family is one consisting of two or more lineally related kinfolk, their spouses, and offspring.

Women may be heads of households in various ways, as (1) recognized heads of households such as when they are widowed or divorced, (2) acting heads such as when their husbands are away for extended periods, or (3) informal heads such as when they have command over resources and make decisions on their initiative. Even when they are not heads of households, women usually have a recognized and important role through their contribution of labor, management, marketing, and ownership of resources. At times, individuality among males and females leads to competition within the households as when husbands sell firewood to their wives (Venema, 1978) or when wives and husbands lend each other money *with interest* (Robertson, 1975-76).

Most farm households in developing countries strive to produce a dependable and continuous food supply and many of their other needs such as clothing and shelter, and surpluses for sale. To do this, the members of the household engage in several on-farm enterprises using primarily their labor. Furthermore they are cautious about adopting changes that threaten their ability to maintain a reliable food supply. Members customarily have duties within the household according to sex, age, and relationship that are dictated by custom and practical considerations.

2.3.3. SMALL-SCALE FARMING

In this book of guidelines, we emphasize small-scale operations in which the farmers frequently have difficulty obtaining sufficient inputs to allow them to adequately use the available technology as would medium-scale and large-scale commercial farmers. Small does not necessarily refer to the area of land held because some farmers that meet our definition of small-scale farming have access to considerable amounts of land—as do pastoralists and shifting cultivators. Such small-scale farmers are unable to easily raise their levels of production because of limited resources and technologies suitable for their needs.

2.3.4. ENTERPRISES

Enterprises mean activities undertaken to produce an output that contributes to total production or income of the farm family. Enterprises in FSR&D typically concern crops, livestock, processing or otherwise upgrading agricultural commodities produced on the farm, productive nonagricultural activities carried out on the farm such as handicrafts, and productive off-farm activities of the household members.

2.3.5. FARMING SYSTEMS

For this book of guidelines, we consider a farming system as a unique and reasonably stable arrangement of farming enterprises that the household manages according to well-defined practices in response to the physical, biological, and socioeconomic environments and in accordance with the household's goals, preferences, and resources. These factors combine to influence output and production methods. More commonality is found within the system than between systems. The farming system is part of larger systems—e.g., the local community—and can be divided into subsystems—e.g., cropping systems.

Decisions as to classification depend on the needs for analysis and decision-making. A system with a greater cash income—e.g., when off-farm employment or sale of handicrafts is possible—is different from one with lesser cash income. Even though the same crops, patterns, and management practices may be followed, farmers' reactions to change will vary because of different capabilities, attitudes, and other factors (Harwood, 1979).

2.3.6. CROPPING AND LIVESTOCK SYSTEMS

These are subsystems within the farming system. A cropping system, a set of one or more crops, comprises all

components required for production, including the interactions between other household enterprises, and the physical, biological, and socioeconomic environments. Livestock systems can be defined similarly.

2.3.7. MIXED SYSTEMS

Cropping, livestock, and possibly other enterprises are present within the farming system.

2.3.8. CROPPING SYSTEMS RESEARCH

Research on cropping systems concentrates on crops and cropping patterns, alternative management practices in different environments and interactions between crops, between crops and other enterprises, and between the household and environmental factors beyond the household's control. The procedures are similar to farming systems research, but the breadth of cropping systems research is generally less. However, when the initial analysis considers the whole farm situation and then focuses on cropping systems as the best area for research, the differences between the approaches of cropping systems research and of FSR&D are few.

2.3.9. LIVESTOCK AND MIXED SYSTEMS RESEARCH

The approach to livestock systems research and mixed systems research follows a process similar to cropping systems research except for the procedures that reflect the inherent differences between cropping and livestock systems—e.g., fewer numbers of animals than plants. In mixed systems research, the team focuses directly on the interactions between crops, livestock, and possibly other enterprises.

2.3.10. COMMODITY-ORIENTED RESEARCH

Commodity-oriented research focuses on one or more crops or animals by studying them in detail. Commodities selected for emphasis should be the result of prior investigation demonstrating their importance to the farming system. While often conducted along disciplinary lines such as breeding, physiology, and pathology, commodity-oriented research supports FSR&D best when it keeps the needs of the whole system in mind and takes advantage of FSR&D results as a source of information for making its programs more relevant to the farmers' circumstances. Commodity-oriented research organizations such as the International Maize and Wheat Improvement Center (CIMMYT), International Rice Research Institute (IRRI), and International Center for Tropical Agriculture (CIAT) work mainly with farmers and research organizations in those areas of their specialties that offer the best potential for improving the farming system.

2.4. CHARACTERISTICS OF FSR&D

Further comments on the nature of FSR&D should help in understanding the emphasis placed on this approach. FSR&D considers the *farmers* and their *problems* in a *comprehensive* manner using an *interdisciplinary* approach that *complements* existing research and development activities, and is *iterative*, *dynamic*, and *responsive to society*. Many of these characteristics have their origin in farm management that has been practiced in the United States since the early 1900's. However, the emphasis on a systems approach that considers the whole in terms of the parts and evaluates results in terms of farmers' and society's goals is relatively new in developing countries. Moreover, the new emphasis in developing countries is on research on farmers' fields using interdisciplinary teams.

In time, formal reference to FSR&D may fade away as the process and procedures are absorbed into agricultural research and development programs. But, in the meantime, considerable opportunity remains for improving FSR&D concepts, developing FSR&D procedures, and expanding the FSR&D approach into areas such as livestock and mixed-farming where experience and literature are limited.

2.4.1. FARMER BASED

FSR&D starts with farmers and learns about their environments, resources, methods of production, problems and opportunities, aspirations, and how they react to change. The FSR&D team designs experiments with these factors in mind, carries out studies in farmers' fields, and judges the results by farmers' standards. While other factors enter into the process, FSR&D strongly emphasizes obtaining a clear picture of farmers and their environments. Much can be learned by literally "walking in the farmers' footsteps." Farmers may not have accurate technical explanations of their problems nor know the range of opportunities for improving their conditions, but learning more about farmers helps the researchers produce better technologies and extension workers promote FSR&D's results more effectively.

Because the farm household both consumes and produces, the values of the marketplace and the farm household are mixed. Farmers tend to be cautious about change—especially involving their subsistence crops—but they will change when suitable opportunities arise. They have multiple goals and the community's norms influence farmers in varying degrees. FSR&D practitioners consider farmers rational according to the farmers' values and perceptions of alternatives. But, individual farmers have different values, perceptions, skills, and resources. Thus, some farmers produce more and accept change more readily than other farmers.

2.4.2. PROBLEM SOLVING

While some farming systems research applies broadly and is long-run, a national FSR&D program tends to be applied to specific, short-run objectives, as when adapting available technologies. FSR&D identifies problems on farms and introduces improvements that frequently require little governmental support. The approach identifies farmers' constraints and distinguishes between those constraints that are within and those that are beyond their control. For example, farmers can often implement changes in varieties, planting distances, methods of applying fertilizers, and time of weeding, but may have difficulty if a solution calls for more rapid plowing requiring oxen that are either not available or too costly. In the first case, we can look internally at ways to introduce change to the farmers. In the second case, we must look externally to changes that will make oxen available or else redesign the technology.

2.4.3. COMPREHENSIVE

FSR&D studies the whole farm setting to identify problems and opportunities, notes their interrelationships, sets research priorities responsive to farmer and societal goals, carries out experiments, proposes changes in light of this comprehensive perspective, measures results in terms of impacts on the farmers and society, observes farmer acceptance of change, and transfers acceptable research results to implementing organizations. The FSR&D team uses a whole-farm perspective to identify the most relevant problems and to evaluate the acceptability of results.

Even though FSR&D views the farming system and its environment comprehensively, some aspects may not be researched or considered for change—for example, farmers' values and social customs or the level of support services. As part of FSR&D's strategy, the FSR&D team decides which areas offer the greatest potential for change and are the most suitable for research and development (Sec. 3.4. in Chapter 3).

2.4.4. INTERDISCIPLINARY

Because of the comprehensive approach and interactions of many technical and human factors, FSR&D teams should be interdisciplinary. By interdisciplinary, we mean frequent interactions among those from different disciplines who work on common tasks and come up with better results than had they worked independently. As a minimum, both technical and social sciences should be represented on a team with leadership strong enough to integrate the disciplines and direct their efforts toward team objectives. Moreover, where cultures discourage communication between unrelated men and women, field teams may have to have members from both sexes if they are to adequately communicate with male and female farmers (Staudt, personal communication).

2.4.5. COMPLEMENTARY

FSR&D replaces neither commodity nor disciplinary research nor extension. On the contrary, FSR&D requires a continuing inflow of improvements from such research and close contact with farmers through extension. For example, if a shorter season variety is needed to overcome a problem associated with planting time, researchers will have more reason to believe that success in identifying a suitable variety will lead to its acceptance. Or, if extension is having difficulty introducing change to its clientele, FSR&D provides a means for bringing farmers' problems to researchers. In view of these advantages, the direction that FSR&D gives to commodity and disciplinary research, coupled with FSR&D's influence on extension, is as important as improvements introduced to farmers participating directly in FSR&D experiments.

2.4.6. ITERATIVE AND DYNAMIC

The FSR&D approach calls for a conceptual understanding of the farming system and its environment from the very beginning. This framework provides the basis for gathering data and directing the course of the research and development effort. Initially, the system may not be well understood, but the conceptualization improves as the FSR&D team gathers data and gains experience.

FSR&D's iterative nature is shown by the process by which the team begins by acting on partial information, gains insight through studies and experimentation, and

modifies its actions. This process continues until research and extension staff are satisfied that changes can be broadly implemented. Such an approach encourages the FSR&D team to begin working within a whole farm framework from the outset, rather than working haphazardly or waiting for excessive precision before initiating on-farm research. In this way, FSR&D seeks to provide *better* solutions to farmers' conditions, but not necessarily the *best* solutions.

Solutions to one set of problems usually generate opportunities for further research. FSR&D is dynamic in that objectives and approaches for future work can be adjusted in light of the accomplishments. For example, FSR&D might initially work with only slight modifications in farmers' existing cropping and livestock patterns. After the farmers grow accustomed to change, greater modifications to their farming systems could be tried.

2.4.7. RESPONSIBLE TO SOCIETY

As with other national programs that rely on private initiative, FSR&D needs to produce results acceptable to small farmers and society. Consequently, FSR&D operates from the farmers' and society's viewpoints. The two can be brought into accord by identifying issues of possible conflict and agreement, by measuring possible impacts of alternative courses of action, and by devising appropriate incentives and restrictions. For example, farmer groups in Kenya receive government, technical, and financial assistance in organizing and constructing terracing, interceptor ditches, and other forms of soil and water conservation. Without some government assistance, the farmers would often not be inclined to do this work, which is in their and society's long-run interests.

2.5. OBJECTIVES OF THESE GUIDELINES

A few very capable persons have been able to apply FSR&D concepts quickly and accurately. They succeed because of their considerable experience and gift for proposing practical solutions to complex problems. These persons are extremely useful, but too few of them are available to the developing countries. A number of organizations and individuals are currently setting down their thoughts and findings on this subject in the attempt to institutionalize the FSR&D approach. In this way, FSR&D can be described, taught, learned, and applied on a much broader scale. This book of guidelines is part of that effort. These guidelines:

- show the general process for FSR&D programs at the national level
- provide procedures that illustrate how to carry out the FSR&D process under different conditions
- supply an overview of the current state of FSR&D so that readers will understand its nature, its relationships to other agricultural research, and its potential for contributing to increased production and the welfare of small farmers.

2.6. USERS OF THESE GUIDELINES

This book of guidelines was written specifically for those in the developing countries who are responsible for deciding on and implementing an FSR&D approach and for those with a general interest in FSR&D. This audience includes five groups.

The first group consists of top decision makers, possibly at the ministerial or cabinet level, who decide whether or not an FSR&D approach will be undertaken. This group's members need to understand the advantages and disadvantages of FSR&D, as well as its implications, such as

- effectively reaching the intended farmers
- staffing and training requirements
- length of time necessary to start
- institutional and policy relationships with ongoing programs in agriculture and other areas
- relative costs
- special governmental support.

We prepared Chapter 1, a summary of the principal features of these guidelines, for this group.

The next group of potential readers includes those who are responsible for administering FSR&D activities. They might be technical directors and their seconds in command. Although such individuals need to know about FSR&D's broad implications, they will be more concerned with understanding the entire process, how the parts fit together, how others have organized programs and projects, and the strengths and weaknesses of alternative approaches for particular situations. They will need to know enough about the procedures to select appropriate staff and integrate their activities. We provide this information in the remaining chapters.

The third group consists of regional directors, field team leaders, and technical researchers who carry out the various FSR&D tasks under the direction of the technical directors and their assistants. We expect them to be interested in individual chapters and the appendixes.

The fourth group includes expatriates who may be technical advisers to each group. They could be involved in all phases such as advising on whether to accept an FSR&D approach, aiding in program and project design, training staff, implementing FSR&D activities, and assisting in evaluation of results. For them, all parts of the book—the summary, main chapters, and appendixes—should be useful.

The last group comprises all the rest who have an interest in FSR&D, including others within the developing countries, members of international agricultural research and development organizations, agricultural researchers, and others with similar interests outside the developing countries.

2.7. SCOPE

This book of guidelines was written to convey the approach to FSR&D generally being followed by or recommended for national governments. By being general, this book provides information on concepts applicable to a variety of situations; however, this book alone will not provide enough information for designing an FSR&D approach for a particular country. Because of the breadth of FSR&D, many topics are included in this book. Some are developed in considerable detail, while others are not. In this section on the scope of FSR&D, we discuss the topics

included in this book and why we emphasized them. At times, we have said little about a topic because information and experience are lacking.

2.7.1. APPLICABILITY

These guidelines reflect the state of the art of FSR&D at the national level. They encompass a broad range of conditions, which can help the reader understand FSR&D's general process. The illustrations provide the reader with the opportunity to see how some of the principles have been applied to specific situations. With these concepts and illustrations, those in developing countries should be in a position to design an FSR&D approach specifically for their conditions. For example, the FSR&D team needs to select an overall strategy responsive to the country's goals, resources, stage of development, and types of farming systems. Moreover, once the approach has been designed, training is needed. These guidelines include the basis for selecting training materials, but we do not provide detailed instructions on training in FSR&D.

2.7.2. COVERAGE

The farming system consists of subsystems—e.g., cropping systems—and in turn is part of larger systems—e.g., the local community. Fig. 2-1 depicts the general categories of factors that influence small-farmer production and welfare. The figure shows that the farming system is made up of crops, livestock, and other on-farm subsystems. The farming system is greatly influenced by

Figure 2-1. Farmers' setting and scope of the guidelines.

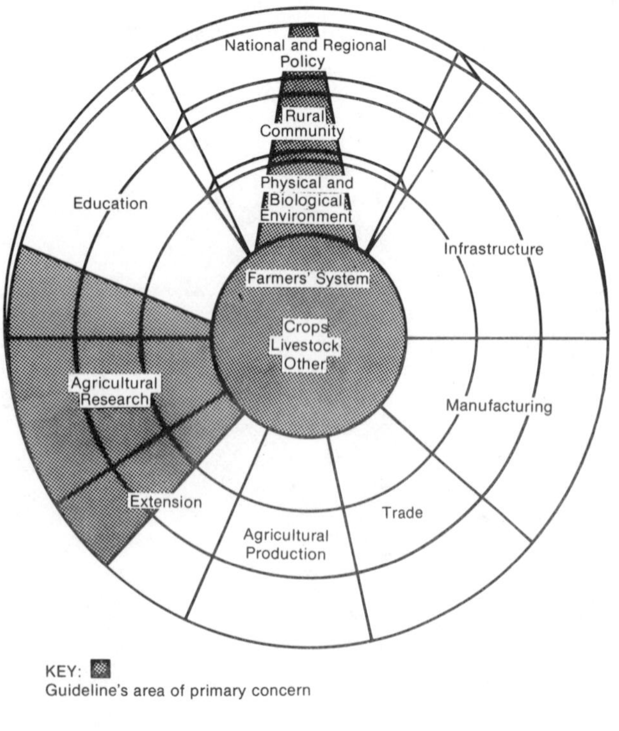

the physical setting and the rural community where the farm household members trade, socialize, and otherwise take part in local affairs. The rural community, in turn, is part of a regional and national setting. Government's national and regional policies and institutions reach the household level to help the members through research, extension, and agricultural production programs, and by provision of infrastructure. Private and government enterprises—large-scale agriculture, manufacturing, trade, etc.—also interact with the farm household by providing services and supplies, facilitating the sale of farm output, and offering off-farm employment.

Because FSR&D is comprehensive, these guidelines are concerned directly and indirectly with all activities that touch upon the farmers' lives. But, for practical reasons, we have included FSR&D procedures primarily on (1) agricultural research, (2) extension's part in transferring the technologies, (3) organization, and (4) training. Also, many of the procedures relate to cropping systems because most experience is there.

In this book, we use only general concepts of systems analysis and provide little discussion of the techniques of systems analysis such as linear programming, simulation, or other forms of mathematical modeling. This omission should not be construed to mean that we think that such modeling is not useful. We have excluded mathematical modeling because our review of FSR&D practices did not reveal at this time its significance among national programs. In contrast, interesting work is occurring at some of the international centers, such as the conceptual and detailed livestock models of the International Livestock Centre for Africa (ILCA), the simulation and optimization models applied by IRRI, and the ecology-based models of the Tropical Agricultural Research and Training Center (CATIE). We describe this work briefly in Appendix 2-A. Included in some of the modeling work at IRRI and other institutions are studies of relatively narrow topics, such as soil-water-plant relationships.

Except for a few cases, we have not included, in this book, procedures that are either commonly available in the literature or are narrowly confined to a discipline and are not central to the FSR&D process. For example, we considered experiment station procedures as being generally well-known and soil-water-plant relationships as narrowly confined to a discipline and not central to FSR&D methodology. For this same reason we have not provided detailed procedures on topics such as marketing, transportation, credit, and extension at the agent's level.

2.7.3. COUNTRY SITUATIONS

As noted above, the guidelines in this book apply to a broad range of situations encountered in different countries throughout the world. They apply to countries that contemplate instituting FSR&D activities, as well as to those with full-fledged FSR&D programs. They apply to semiautonomous research corporations that conduct FSR&D or ministeries that have FSR&D as part of a broader program. And they apply to production programs in which upgrading agricultural research is important.

2.8. APPROACH TO THE GUIDELINES

In this book, we have

- described alternative approaches to various situations
- suggested reasons for the differences in approaches
- identified references where research procedures are reasonably standard
- included detailed descriptions of procedures that are not standard or where we felt emphasis was needed.

We have included numerous references to the FSR&D work of national and international groups throughout the book. These references should prove useful for details about specific concepts and procedures. Especially helpful in preparing this book of guidelines were the published and draft guidelines prepared by CIMMYT (Perrin et al., 1976; Byerlee et al., 1980), and the International Rice Research Institute (Zandstra et al., 1981), two state-of-the-art papers on farming systems (Gilbert et al., 1980 and TAC, 1978), and a paper on cropping systems research in Indonesia (McIntosh, 1980).

2.9. CONTENTS OF THE BOOK

We have organized the remainder of this book of guidelines so that readers begin with a review of the overall approach to FSR&D, and then learn the details of each of the principal activities. With that background, readers should be able to decide whether or not FSR&D is suitable for their situations. Assuming that some readers will be interested, we then provide information on how to implement the approach. The appendixes elaborate on the approach.

More specifically, Chapter 3 contains the conceptual framework of the principal features of FSR&D. The next five chapters elaborate on FSR&D's major activities: Chapter 4 explains target and research area selection, Chapter 5 presents problem identification and development of a research base, Chapter 6 discusses planning on-farm research, Chapter 7 presents on-farm research and analysis, and Chapter 8 discusses extension of results. Then, Chapter 9 considers whether or not an FSR&D program is in a country's best interests. The next two chapters explain implementation: specifically, Chapter 10 presents alternative organizational approaches for FSR&D, interdisciplinary teamwork, staffing, and other management issues, and Chapter 11 reviews training as a way of implementing new or improved FSR&D programs. The appendixes contain detailed information supporting the various chapters, some general procedures, and worldwide examples and illustrations. Most chapters end with references and suggested readings.

CITED REFERENCES

Byerlee, D., M.P. Collinson, R.K. Perrin, D.L. Winkelmann, S. Biggs, E.R. Moscardi, J.C. Martinez, L. Harrington, and A. Benjamin. 1980. Planning technologies appropriate to farmers: concepts and procedures. CIMMYT, El Batan, Mexico.

Dillon, J.L. 1976. The economics of systems research. Agric. Sys. 1:1:5-22.

Gilbert, E.H., D.W. Norman, and F.E. Winch. 1980. Farming systems research: a critical appraisal. MSU Rural Dev. Paper No. 6. Dep. of Agric. Econ., Michigan State Univ., East Lansing, Mich.

Harwood, R.R. 1979. Small farm development: understanding and improving farming systems in the humid tropics. Westview Press, Boulder, Colo.

McIntosh, J.L. 1980. Cropping systems and soil classification for agrotechnology development and transfer. *In* Proc. Agrotech. Transfer Workshop. 7-12 July 1980. Soils Res. Inst., AARD, Bogor, Indonesia and Univ. of Hawaii, Honolulu.

Perrin, R.K., D.L. Winkelmann, E.R. Moscardi, and J.R. Anderson. 1976. From agronomic data to farmer recommendations: an economics training manual. Inf. Bull. 27. CIMMYT, El Batan, Mexico.

Robertson, C. 1975-1976. Women and change in marketing conditions in the Accra area. *In* Rural Africana. Winter 1975-1976. Michigan State Univ., East Lansing, Mich.

Ruthenberg, H. 1971. Farming systems in the tropics. Clarendon Press, Oxford, UK.

Technical Advisory Committee (TAC). Review Team of the Consultative Group on International Agricultural Research. 1978. Farming systems research at the international agricultural research centers. The World Bank, Washington, D.C.

Venema, L.B. 1978. The Wolof of Saloum: social structure and rural development in Senegal. Center for Agricultural Publishing and Documentation, Wageningen, Netherlands.

Zandstra, H.G., E.C. Price, J.A. Litsinger, and R.A. Morris. 1981. A methodology for on-farm cropping systems research. IRRI, Los Banos, Philippines.

OTHER REFERENCES

IRRI. 1977. Symposium on cropping systems research and development for the Asian rice farmer. IRRI, Los Banos, Philippines.

De Tray, D.N. 1977. Household studies workshop. *In* A/D/C Sem. Rep. No. 13. The Agricultural Development Council, Inc., New York.

Staudt, K.A. 1978. Agricultural productivity gaps: a case study of male preference in government policy implementation. Development and Change 9:439-457.

Chapter 3
CONCEPTUAL FRAMEWORK

FSR&D is a process that involves a set of interrelated activities. While individual programs reflect specific conditions within a country and the preferences of its leaders, the various approaches to FSR&D have much in common. Thus, the conceptual framework described in this chapter shows this commonality. And, this framework illustrates the nature and interrelationships of each of these activities and should help the reader understand the material presented in subsequent chapters. The following sections briefly describe FSR&D activities, timing of activities, staffing requirements, and strategies. The chapter ends with a summary of the conceptual framework.

3.1. FSR&D ACTIVITIES

A natural sequence for the FSR&D process, as outlined in Fig. 3-1, includes

- target and research area selection
- problem identification and development of a research base
- planning on-farm research
- on-farm research and analysis
- extension of results.

Experiment station collaboration, which interacts primarily with the first four activities, is set off to the side to emphasize its supporting role in on-farm research. Also, extension is shown as collaborating with each of the five principal FSR&D activities. Results from the last two activities—on-farm research and analysis, and extension of results—feed back to the earlier activities.

While these activities are shown as being sequential and discrete, in practice, an activity often overlaps with other activities, some activities are not taken in the sequence indicated, and some are repeated. In fact, the iterative nature of FSR&D usually calls for several repetitions, which leads to progressively improved results. While flexibility of the approach exists, omitting any one of the activities may jeopardize the value of the final results.

The FSR&D approach typically uses interdisciplinary teams and the team composition varies according to the task and available staff. During FSR&D's early stages, disciplinary representation should be broad and flexible to allow adequate response to problems and opportunities concerning the targeted group of farmers (Hildebrand, personal communication). Field teams, normally residing in or near the research area, conduct on-farm research aided by specialists and others concerned with agricultural production. The specialists come from the physical, biological, and social sciences, and include extension. These specialists may also reside in the research areas, or operate out of regional or national headquarters or experiment stations. Section 3.3. provides additional details on staffing requirements.

3.1.1. TARGET AND RESEARCH AREA SELECTION

The FSR&D team aids national decision makers in selecting target areas and target groups of farmers. The team subsequently divides the target areas into subareas and establishes the boundaries of the research area. In FSR&D, the government chooses, implicitly or explicitly, target areas or target groups of farmers to receive increased attention. These choices are generally based on national policies reflecting governmental objectives such as the better use of resources, raising the income of poor farmers, and greater domestic food production. A response to these objectives is more appropriate agricultural technologies, policies, and services.

The FSR&D team usually divides the target areas or target groups of farmers into subareas according to common physical, biological, and socioeconomic characteristics. Such a stratification separates environmental conditions and farming systems into reasonably homogeneous segments. By working with these homogeneous segments, the FSR&D team is able to develop improved technologies for farmers operating under similar conditions throughout the target area. And, sometimes the technologies are suitable for farmers outside the target area. Stated somewhat differently, appropriate stratification according to environmental conditions and farming systems enables the FSR&D team to identify farmers who are expected to benefit from the same recommendations (Byerlee et al., 1980).

The number and location of research areas throughout a country depend on FSR&D goals and resources and the characteristics of the areas. Because the relationships between farmers' practices and their environments take time to understand, the research areas should be broad enough to allow the team to adjust its approach as new information is obtained.

Figure 3-1. The five basic activities of on-farm research in FSR&D.

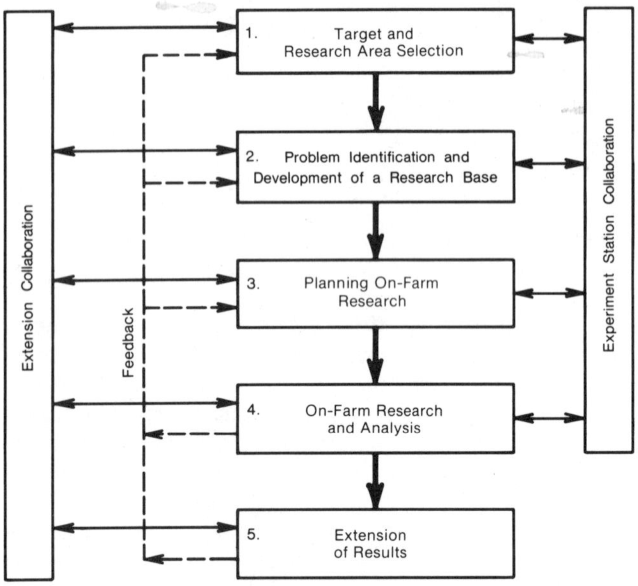

3.1.2. PROBLEM IDENTIFICATION AND DEVELOPMENT OF A RESEARCH BASE

After the research area has been selected, the FSR&D team moves to more careful and detailed studies of farming systems and the area's characteristics. The team studies, analyzes, and ranks farmers' problems and opportunities and either acts upon them immediately or plans further studies and experiments.

The team gathers information about the farmers and their environments through a review of secondary information[1] and through direct observation and discussions with members of farm households. Many experienced FSR&D practitioners strongly endorse rapid surveys at the outset of FSR&D activities in an area. However, FSR&D teams sometimes conduct experiments to help understand the farming system and to identify problems. As part of the next set of activities, the FSR&D teams plan more elaborate surveys, studies on special topics, monitoring activities, and experiments.

While starting research in the right direction is important, the nature of FSR&D allows the team to adjust its approach as information is gained from experiments, studies, and other forms of research. So, rather than delay until a precise plan of action can be prepared, FSR&D teams are advised to begin research early. In fact, FSR&D practitioners urge that a cropping season not go by without conducting some form of on-farm experiment. Livestock systems, on the other hand, are more complex and may require more careful planning.

3.1.3. PLANNING ON-FARM RESEARCH

The research team begins the research design and planning activity by

1) reviewing the priorities given to problems and opportunities and the hypotheses for solution
2) reviewing previous research findings
3) seeking whatever help is needed and available from regional, national, and international sources
4) deciding whether to accept the current environmental conditions as they are, or to assume some degree of change.

With its research agenda set, the team analyzes current farming practices and environmental conditions as they specifically relate to the proposed research program. Team members investigate such topics as alternative cropping and livestock patterns, management practices, the number and types of experiments, and the level to set nonexperimental variables.

The FSR&D team works collaboratively with farmers in their fields. In this way, a solid base is developed for the team to understand farmers' conditions and to design and implement appropriate experiments. From this base, FSR&D activities can also be more effectively coordinated with the work of other research and development organizations.

Moreover, the researchers design record keeping systems, special studies, climatic monitoring, and surveys. The FSR&D team needs these back-up activities to

1) better understand the research area
2) implement the on-farm experiments
3) measure progress
4) evaluate the results.

Before finalizing the research plan, the team makes a preliminary analysis of possible impacts of the proposed technological changes on the farmers and the environment. This analysis provides advanced information on likely biological performance, resource requirements, economic and financial feasibility, and sociocultural acceptability. Included in economic and financial feasibility are requirements placed on support systems and the need for any changes in national and regional policies.

3.1.4. ON-FARM RESEARCH AND ANALYSIS

During this set of activities, the FSR&D team conducts on-farm experiments, initiates studies, gathers data, helps to coordinate supportive research, and analyzes the results.

The FSR&D team commonly conducts three types of biological experiments: researcher-managed trials, farmer-managed tests, and superimposed trials. Researchers use

[1]Secondary information is published or unpublished data collected for purposes other than the current activity.

the first of these experiments to develop new technologies under farmers' conditions. Such experiments are used where control is important. The second of these approaches seeks to learn how farmers respond to suggested improvements by allowing the farmers to introduce the new technologies in their fields. The third approach involves relatively simple researcher-managed experiments applied across a range of farmer-managed conditions.

In conducting these experiments, the FSR&D team cooperates with extension in establishing contacts with farmers and local groups, identifies the locations for on-farm experiments, secures resources, and coordinates activities with the experiment station. Agreements need to be reached among farmers, researchers, and cooperating groups concerning who supplies which inputs, how the outputs will be distributed, and who bears the risk of loss. In addition, the team initiates special studies and surveys, keeps records of farmers' practices and experimental results, monitors local conditions, samples yields, and analyzes experimental results.

The team's analysis of experimental results involves evaluation of biological performance, actual resource requirements, economic and financial feasibility, and socioeconomic acceptability. These analyses are similar to those made during the planning of on-farm research. The difference is that the analyses of on-farm experiments are based on actual biological results, whereas the analyses made during planning (Sec. 3.1.3.) are based on estimates of biological performance. For farmer-managed tests, the economic, financial, and sociocultural analyses incorporate farmer reactions directly.

The researchers also estimate the acceptability of new technologies by noting farmers' reactions to the proposed changes during and following the experiments. Results from these analyses assist the team in evaluating

1) the readiness of the technologies for diffusion
2) the need for improved support services and government policies.

3.1.5. EXTENSION OF RESULTS

The FSR&D team can promote technologies acceptable to farmers and society at large[2] in several ways. For example, the technologies can be tested in multi-locations, incorporated in pilot production programs, turned directly over to the extension service, or promoted through other organizations.

Through multi-locational testing, the FSR&D team extrapolates the improved technologies to other locations within the target area. Extrapolation is usually limited to environments similar to those encountered in the research area. The FSR&D team can facilitate this process if, at the outset, it adequately describes the research area and subareas.

FSR&D team members continue to be actively involved in multi-locational testing by assisting extension staff in identifying the conditions under which the technologies apply. In some cases, researchers may have to modify the recommendations slightly to accommodate moderate differences in farmers or environmental conditions. In assisting with these tests, extension agents not previously associated with the FSR&D team will learn the details of the technologies and how to apply them.

Up to this point, the team's appraisal of the ability of the support systems to accommodate the introduced technologies is based on analytical studies. Before turning the technologies over for widespread diffusion, the team may want to initiate a pilot production program within the target area. The purpose of the program is to test the improved technologies under conditions similar to those likely to be encountered when the technology is broadly diffused. Based on results, the team makes whatever changes in the technology that are necessary, or alternatively suggests changes in the support services or government policies. These changes should be small, however, since the FSR&D approach is intended to produce technologies that fit farmers' and environmental conditions.

Sometimes extension or production programs can diffuse improved technologies without additional testing.

[2]Technologies acceptable to society at large are those that take into account not only the long-run interests of the farmers receiving the technologies, but also the interests of present and future generations indirectly affected. Generally, decision makers with a regional and national perspective are called upon to make this judgment.

This might occur when the new technologies do not call for major changes or when the government prefers not to spend the time and funds for such refinements. However, we caution FSR&D teams that bypassing multi-locational tests and pilot production programs increases the risk of failure.

To summarize, researchers, extension staff, and those from production programs have mutually reinforcing roles during the diffusion process. Researchers can help advise others about ways to identify conditions suitable for the new technologies; they can also suggest alternate recommendations in light of the slight variations that occur throughout the research and extrapolation areas. Extension and production staff who become familiar with the improved technologies can then take over when widespread diffusion begins. The close links established throughout the FSR&D process aid in this transfer of responsibility and implementation.

3.1.6. COLLABORATION

Effective collaboration with other organizations closely associated with the needs of small farmers is critical to successful FSR&D activities. The two most relevant groups will usually be experiment station staff and the extension service. We show the linkages between the five FSR&D activities and these two organizations in Fig. 3-1.

Experiment station staff and facilities may be involved in some or all of the following ways. The target and research areas may be selected because of available knowledge of the agricultural potential generated by the presence of a research station. Experiment station personnel usually aid the field teams in identifying farmers' problems. The FSR&D team frequently brings such personnel directly into the FSR&D process because of their knowledge and experience in the research process. Thus, experiment station personnel will be able to help carry out research on farmers' fields, to conduct other research in support of FSR&D's needs, and to help analyze and interpret the results. Sometimes, the stations provide the housing, office, storage, training, and related needs of the FSR&D field teams.

Similarly, the extension staff can help the FSR&D team understand the characteristics of the research areas and particularly the target groups of farmers. Furthermore, this staff can aid in problem identification by introducing the FSR&D team to individual farmers and local leaders. As members of the FSR&D team, extension specialists can advise the researchers on farmers' conditions when the team designs experiments. Then, when the team undertakes research on farmers' fields, the extension specialists can help select farmers and conduct the experiments. Finally, extension specialists trained in FSR&D can transfer the improved technologies to the extension service and other promotional organizations.

3.1.7. FEEDBACK

FSR&D does not require that all aspects of the farming system be studied or that a plan of research be completed before on-farm research begins. As long as the team has prepared the general framework of analysis, on-farm research can start and the results used to refine the approach. The framework of analysis needs to include a conceptualization of the farming system and a preliminary identification of relevant problems for research.

Moreover, for practical reasons, the approach to FSR&D seeks to uncover improvements that are substantially better than existing practices, but not necessarily the best practices. Initially, FSR&D tends to focus on readily identifiable and researchable problems for study and experimentation. This pragmatic approach starts quickly and uses feedback from experiments and other sources to produce better results than from more complex and lengthy approaches. As the FSR&D program matures, more fundamental constraints to improved farmer production may come to light. Then, the FSR&D team may work on more complex problems.

As Fig. 3-1 illustrates, the feedback goes from activities 4 and 5 to activities 1, 2, and 3. As the team analyzes studies and experiments and evaluates and extends improved technologies, it gains new insights into the target and research areas and about farmers' problems and opportunities. This information, in turn, may be used

- for selecting new target areas
- for redefining subareas or the research area
- for improving research designs and planning
- for altering the approaches to on-farm research and analysis.

Such information eventually becomes part of the general body of knowledge of the area.

By setting experiment station and extension off to the side in Fig. 3-1, we stress the central role of on-farm research. Such emphasis focuses attention on farmers' problems and opportunities. At the same time, we do not underestimate the importance of effective collaboration with experiment station and extension staff. Effective integration of the efforts of experiment station, extension, and field team staff is essential to FSR&D.

Finally, analysis is listed as part of activity 4 because of its importance at this point in the FSR&D process (Fig. 3-1). The available information on biological performance, resource requirements, economic and financial viability, and farmer response gives the FSR&D team the first good indication of how well farmers are accepting the new technologies. Actually, the FSR&D team analyzes each activity to evaluate results and plan the next activity.

3.2. TIMING OF ACTIVITIES

The time required to generate and transfer improved technologies varies because of several factors: types of farm enterprises, research team's knowledge of the area, the backlog of suitable agricultural technologies, strength of research and extension programs, need for training staff, perceived urgency for improvements, governmental support, success in identifying better technologies, and so on. For example:

- Some researchers working on irrigated lands will want to obtain an understanding of the water balance—i.e., accounting for water inflow, outflow, uses, and losses—before initiating irrigation experiments.
- Research results can be obtained more quickly in areas suitable for growing the same crop more than once per year than for single season crops that are part of a three-year rotation.
- Livestock experiments involving pastoralists are generally longer and more drawn out than cropping experiments, and experiments with tree crops generally take even longer.

3.2.1. NEW FSR&D PROGRAMS FOR PREDOMINANTLY CROPPING SYSTEMS

Table 3-1 presents a representative schedule for FSR&D activities for cropping systems in new areas. This schedule allows on-farm experiments to begin 21 weeks after initiating target area selection. Time could be reduced if the target area has been selected or if conditions are sufficiently uniform that subareas do not have to be identified. Alternatively, the time could be increased if the area's characteristics, potential, or problems are complex. The six weeks for gathering and reviewing secondary data, conducting reconnaissance surveys, and ranking of the possibilities for improvement represent a fairly rapid, and yet, a reasonable schedule according to FSR&D practitioners. Note, however, that this schedule presumes that researchers already know the FSR&D process.

The team uses the time for planning on-farm research for a detailed analysis of the alternatives suggested during the previous set of activities; deciding on the types, locations, and numbers of experiments; planning data gathering; establishing contacts with cooperating farmers; acquiring materials; assigning tasks for team members; and other activities.

The schedule in Table 3-1 assumes

- good sequencing of activities
- no serious problems
- an available cadre of capable researchers
- synchronization with the cropping seasons.

Delays in recruiting and training staff, securing equipment, having to build facilities, obtaining permission to proceed, securing operating funds, and the like would extend the schedule.

The following example illustrates the time the IRRI staff needed to help the Government of the Philippines initiate a cropping systems program in a new area:

IRRI helped launch a new program in the Philippines in 4-1/2 months between the time the research area was selected and

Table 3-1. Representative schedule for initially completing the first three activities in the FSR&D process for cropping systems in a new area.

Activities		*Time (weeks)*
1. Target and Research Area Selection		
Selection of general target area		1.5
Division into subareas		2.5
Selection of research area		2.0
	Total	6.0
2. Problem Identification and Development of a Research Base		
Gathering and reviewing secondary data		2.0
Conducting reconnaissance surveys		2.0
Analyzing results and setting priorities		2.0
	Total	6.0
3. Planning On-Farm Research		
Reviewing identified problems and gathering additional information*		1.5
Designing experiments		2.5
Planning for research, including identification of collaborating farmers		5.0
	Total	9.0

* Special studies and surveys, which are identified during this review, may be designed and initiated after the nine weeks indicated for this step.

experiments commenced in farmers' fields. The sequence follows: (1) the decision to go to the field was made on the 15th of November; (2) the research team looked at the farm setting and farmers' practices in the area, including such factors as soil classification and fertility, climate, major crops grown, varieties, yields, types of farming enterprises, farm and family size, markets, and credit; (3) the team began to identify soil types and to plan the season's activities in terms of cropping patterns, treatments, factors to hold constant, and so on by January 8th; and (4) the team started the on-farm experiments by April 1st. In the meantime, the staff was acquired, offices were set up, equipment secured, and other tasks necessary for field operations were completed. The approach was timed so that experiments began with the first available cropping season (Zandstra, personal communication).

3.2.2. ONGOING FSR&D PROGRAMS FOR PREDOMINANTLY CROPPING SYSTEMS

Once the FSR&D program has been initiated and on-farm experiments begun, the team usually repeats a series of trials and tests until improved technologies are identified and passed on to the farmers. We have not included the time for on-farm research and analysis and diffusion of results in Table 3-1 because conditions vary widely. For example, on-farm research and analysis might take 18 weeks for experiments with short season crops. If two or three crops are sequenced, a year might be required. Experiments with sugar cane could take two years and rotations of single-season crops in temperate climates would take longer.

The overall program may take several years to develop technologies with broad applications. The Agricultural Science and Technology Institute (ICTA) in Guatemala often begins with researcher-managed trials, follows with farmer-managed tests, and then evaluates farmer acceptability.

As experiments are completed each season, ICTA teams quickly analyze the results so that they can plan next season's experiments. And, ICTA teams use the time between seasons to conduct reconnaissance surveys before moving into new areas. According to Hildebrand (personal communication), the following activities associated with problem identification and planning on-farm research can be completed in 5 to 10 weeks at the rate of 1 to 2 weeks per activity:

1) gathering and analyzing background information
2) conducting reconnaissance surveys, identifying problems and opportunities, and setting priorities
3) locating collaborating farmers
4) designing experiments and identifying farmers' fields
5) obtaining experimental materials.

We consider such a rapid schedule as attainable once an FSR&D team has gained experience in the region and has mastered FSR&D methodologies.

On rice-based systems, IRRI follows another approach that takes about three years to adequately identify new cropping patterns. Using both researcher-managed trials and farmer-managed tests, IRRI's researchers begin the first year with several cropping patterns with few replications of experiments and by the third year they concentrate on the most promising patterns using more replications. Improved technologies may then pass through multi-locational testing and pilot production programs, before being handed over to promotional organizations (Zandstra et al., 1981).

3.2.3. NEW FSR&D PROGRAMS FOR PREDOMINANTLY LIVESTOCK SYSTEMS

The staff of the International Livestock Centre for Africa (ILCA) is considering a substantially different approach to FSR&D because livestock systems are distinct from cropping systems (deHaan, personal communication).

Researchers studying the productivity of herds and flocks may want to classify substantial numbers of animals (perhaps 2,000) by age and sex and to consider such factors as livestock sales and cow-oxen ratios. Cow histories can be obtained from interviews with farmers to trace such factors as animal growth from birth through calving and progeny history. These histories provide information on fertility and mortality rates, but little on the causes for these rates. To learn about the causes for these rates, record keeping and analysis should continue for at least 18 months, or more when variable annual rainfall influences herd and flock conditions. Other ways to learn about livestock systems are through modeling and monitoring changes accompanying livestock development programs. Modeling helps to identify important features of the system for research. Monitoring seeks to learn, among other things, about the sociocultural responses to the induced change.

These characteristics of livestock systems, which often depend heavily on grazing for animal nutrition, suggest a longer period for data gathering than for cropping systems. However, not all livestock or mixed systems take so long to research before improved technologies can be introduced. For example, research can be conducted faster on mixed-system relationships and factors such as manure and stubble, supplementary feeding from surplus crops, livestock as traction and transportation, competing and complementary land use, and opportunities for labor specialization (Delgado, 1978).

3.2.4. CLOSING COMMENTS ON TIMING

Experienced researchers have opposing views about the time required to identify improved technologies. Some researchers caution against rapid initiation of FSR&D activities. They argue that

- The team requires time to gather base-line data for use in measuring research accomplishments.
- The team needs time to understand the farmers and for farmers to become accustomed to the team.
- The team must be cautious when introducing new technologies to farmers; otherwise, farmers might be harmed by the change, thereby causing farmers to lose confidence in the team.

While recognizing the validity of these arguments, we interpret the emphasis of FSR&D practitioners as pressing for early initiation of on-farm experiments. Rather than start off by collecting base-line data, farm records can be used for measuring FSR&D impacts. Farm records are accounts usually kept by one or more members of a farm household of the inputs and outputs for a single crop or animal type. Researcher-managed trials can acquaint the researchers with farmers' conditions before subjecting the farmers to the risks of farmer-managed tests. Furthermore, FSR&D practitioners generally feel that experimentation provides a good way to learn about the farming system. Some practitioners argue that early improvement in farmers' conditions helps to interest the farmers. In fact, frequent researcher contact with farmers without producing tangible benefits usually dampens the farmers' interests in further cooperation.

To conclude, FSR&D's approach needs sufficient flexibility to adapt to local conditions and to take advantage of new opportunities. Having said this, we recommend initiating experiments early and relying on FSR&D's checks and balances to guard against actions that might harm the farmer.

3.3. STAFFING

This section provides information on those who carry out the FSR&D process. The principal groups are teams at the field, regional, and national levels. Whether all of these teams will be actively involved depends on how FSR&D activities are organized. The team make-up will vary according to the types of FSR&D programs, the availability of staff, and which FSR&D activity is being undertaken. In this section we assume a national program with teams at each of the three levels and sufficient personnel to implement the approach.

3.3.1. FIELD TEAMS

The field teams work with farmers in their fields. Such teams often consist of agronomists (called *ingenieros agrónomos* in Latin America), economists, and technical assistants. Where livestock is important, an animal science specialist should be part of the team; where irrigation is practiced, an irrigation engineer can be a key member of the team; and where women are responsible for growing important crops or performing critical operations, field teams should include women.

These teams are assisted from time to time by specialists in disciplines such as extension, sociology, entomology, and pest management. Often, such field teams range from two to five professionals supported by technical assistants. As a minimum, FSR&D practitioners generally recommend that the team has a representative from the physical or biological sciences and another from the social sciences. Based on his experience in two Central American countries, Waugh (personal comunication) recommends staffing the field team with agronomists who also specialize in another discipline useful to the field team. For example, agronomists might take short courses or classes at the master's level in economics, plant physiology, diseases of one of the commodities, soil fertility, statistical procedures, and so on. Furthermore, for this example, one of the members should specialize in general agricultural production and serve as liaison between the FSR&D team and the extension service.

These teams report to the regional headquarters and may live there or in the area where they work. The technical assistants often live in villages within or near the research area. Experience has shown that one member of the field team can often manage 15 to 20 researcher-managed cropping trials and more farmer-managed tests. In round figures, a team of five might manage about 100 experiments during a season.

3.3.2. REGIONAL HEADQUARTERS TEAM

The regional staff may include a regional director, the field teams, commodity and disciplinary specialists assigned to the area, and support staff. Commodity specialists are generally assigned to those areas where

crops and livestock of their specialty are most important to the country. These specialists may work in other regions as the need and opportunity arise. If a country does not have enough specialists for each region, they might receive short-term assignments in several regions. Normally, experiment station staff is part of the regional FSR&D team.

3.3.3. NATIONAL HEADQUARTERS TEAM

The staff at national headquarters administers the FSR&D program and may include senior management, a technical director, commodity coordinators, heads of the disciplines, the director of experiment stations if the experiment stations are the responsibility of the FSR&D program, commodity and disciplinary specialists, and other technical staff. This team may be located at some central headquarters or dispersed throughout the country.

3.3.4. OTHERS

Other groups and organizations become involved with the FSR&D staff from time to time such as farmers' organizations, regional and local planning and administrative organizations, production organizations, educational and training organizations, key decision makers at the national level, and international organizations in agricultural research.

3.3.5. INVOLVEMENT OF THE PRINCIPAL GROUPS IN FSR&D

The seven principal groups listed in Fig. 3-2 are involved in the FSR&D process to varying degrees depending on the activity and individual country situations. Others are involved as well, but to a lesser degree.

Target area selection requires high-level decisions, which explains why national level staff is involved. Dividing the target area into subareas and selecting research areas, a more technical matter, requires those knowledgeable about the area. Because problem identification brings the FSR&D team into direct contact with farmers and the surrounding environment, the team seeks support and suggestions from regional and local authorities, local businessmen and leaders, and a cross section of local farmers. Planning on-farm research calls

Figure 3-2. The sequence of principal groups involved in the FSR&D process.

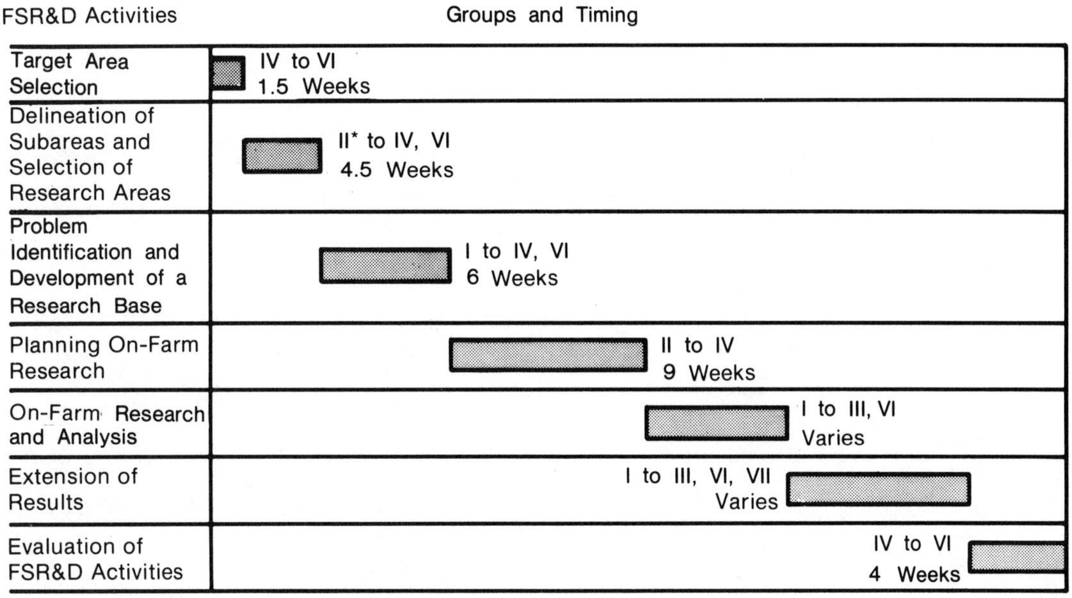

*Provided field teams have been selected

Key: I. Farmers and Farmers' Organizations; II Field Team; III Regional Headquarters Team; IV. National Headquarters Team; V. Key National Decision Makers; VI. Extension Service; VII. Production Oriented Groups.

for sound technical input from the field team, and commodity and disciplinary specialists.

On-farm research concentrates on activities on farmers' fields and can profitably use help from the extension service when supervising farmer-managed tests. Extending the results moves the center of activities to the extension service and other promotional organizations, but requires inputs from research to assist in the transfer process. Finally, evaluation of FSR&D activities generally requires involvement of national and regional level staff from the major organizations concerned with FSR&D.

3.4. FSR&D STRATEGIES

Earlier sections of this chapter on FSR&D activities did not dwell on *how much* change to attempt, *how soon* to attempt change, *where* the ideas for change originate, nor *what type* of research to consider. The possibilities are numerous and FSR&D practitioners have not settled on a common approach. Consequently, we advise those beginning FSR&D activities to be flexible so they can make changes as they gain experience and new information comes to light. In this section we discuss some of the issues to consider in deciding on an FSR&D strategy.

3.4.1. HOW MUCH CHANGE?

Agricultural researchers frequently debate the issue of how much change to introduce into farmers' systems such as changes to the farmers' cropping and livestock patterns, management practices, and the farming environment. At one extreme, nearly everything is, initially, considered subject to change. Changes could range from improved seed and pest management to better farm-to-market roads and price supports for grains. The resulting farming systems could be quite different from those prevailing before introducing such changes. Such an approach might be followed if the FSR&D teams answer "yes" to these questions:

- Are farmers willing to consider major changes to their systems?
- Is the research team capable of dealing with the complexities of the whole farm?
- Is the government prepared to respond to suggested changes across a broad range of possibilities?
- Are few opportunities apparent for materially improving the farmers' conditions without having to initiate major change?

The other extreme takes the farmers' cropping or livestock patterns about as they are and works on improving the efficiency of farmers' practices. The reasons for working with existing patterns are

- The above four conditions are not present.
- Some form of quick results are needed to capture the interests of the farmers, the researchers, and funding organizations.
- Farmers with limited resources are evaluated as be-

ing particularly cautious about accepting new crops or livestock on a significant scale, especially when the change reduces production of one of their traditional food sources.

Deciding on which approach depends on accurately appraising the farmers' responsiveness to change, the capabilities of the FSR&D team, the flexibilities within the farming system and the environment, and the urgency for change.

This discussion can be summarized by what has been described as a comparison of "farming systems in the large" and "farming systems in the small" (Harrington, 1980). The ratio of variables to parameters is high for the former and low for the latter. CIMMYT emphasizes "farming systems in the small" because this approach focuses on those maize or wheat systems where the chances for improvement are high. CIMMYT's researchers consider both the requirements of maize or wheat and how these crops fit into the farmers' system. In contrast, those following a "farming systems in the large" approach would not settle on any particular part of the system without first establishing that the parts selected offer the best chances for improvement.

3.4.2. HOW SOON TO ATTEMPT CHANGE?

How soon to attempt change frequently comes up when discussing FSR&D. Because of the diversity of farming systems and the risks of doing something that will cause irreversible damage, some researchers delay research and implementation of results until a careful study of farmers' conditions is completed. These researchers feel that this approach allows them time to gain a good understanding of farmers' conditions and the research potential, to gather base-line data for evaluating the effectiveness of their programs, and to design an integrated research and development program. Pursuing this approach might take one or more years before on-farm research begins.

Other researchers associated primarily with cropping systems research propose to introduce change quickly using on-farm experiments to gain an understanding of how the farmers and the environment respond to change. In the process, the FSR&D team will uncover opportunities for further research and improvement. Opportunities for change by this process may be less than optimum, but such opportunities often interest farmers. Moreover, introducing change in this way sometimes yields greater returns from scarce research funds than more precise and drawn out procedures. In addressing a meeting of agricultural economists, Collinson (1979) argued that with resources remaining limited, "low cost/rapid coverage approaches seem to be an essential starting point for a bread and butter contribution from the profession."

Most FSR&D practitioners in cropping systems whom we have met prefer the quicker approach. They argue that waiting reduces benefits to present-day farmers and that such costs are greater than the costs of misdirected action. Moreover, the FSR&D approach contains adequate controls for keeping the research on track through frequent experiments, close interaction with farmers, and feedback from research and development results. Finally, record keeping of farmers' activities in the target area can serve as one of the bases for evaluating the effectiveness of the FSR&D approach.

3.4.3. WHERE THE IDEAS FOR CHANGE ORIGINATE?

Much of the research in developing countries has been conducted on experiment stations, which generated technologies that were subsequently brought to farmers for acceptance or rejection. In contrast, FSR&D emphasizes early contact with farm households to learn about their conditions so that improvements can be developed in response to their needs. A critical element of this approach is understanding farm households through on-farm research in which farmers collaborate. Experiment station and extension staff are integrated into the process. The proper mix of on-farm and experiment station research depends on each situation, including the ability of the station to respond to research needs identified through the on-farm research.

3.4.4. WHAT TYPE OF RESEARCH?

Another aspect of the foregoing issue is the debate over the desired mix of site specific and generally applicable research. Several writers have described this issue in terms of the integration of "upstream" and "downstream" research. "Upstream" research is characterized as being partly basic, broadly general, and supportive; whereas, "downstream" research is characterized as being site specific, primarily adaptive, and useful without long delay for target groups of farmers (TAC, 1978; Gilbert et al., 1980; Harrington, 1980). Gilbert et al. (1980) described "upstream" research as finding

> ". . . out how to overcome major constraints common to a range of farming systems extended across one or more geographic zones. The partial or total removal of a constraint such as water availability in arid areas and soil fertility in the humid tropics can significantly expand the range of enterprises and techniques which can be potentially utilized by farmers. Such programs mainly contribute to the 'body of knowledge,' rather than develop practices specifically tailored to a local situation. Prototype solutions produced by 'upstream' FSR programs must be further adapted by 'downstream' FSR programs to specific local conditions. Further, 'upstream' programs may provide inputs into the establishment of research priorities for commodity improvement programs, since the 'upstream' perspective is broader in terms of commodities and disciplines than commodity improvement programs. And their geographic perspective tends to be broader than that of 'downstream' programs. Ultimately 'upstream' programs should rely on feedback from 'downstream' programs to sharpen their own research priorities or objectives. Extensive use of experiment station trials often characterizes 'upstream' programs" (Gilbert et al., 1980).

On the other hand:

> ". . . 'downstream' FSR programs begin with an understanding of existing farming systems and the identification of key constraints. However, in contrast to 'upstream' programs, 'downstream' FSR does not always seek to significantly alleviate key constraints. . ., but instead identifies areas of flexibility in the specific system through accommodating innovations to the reality of existing constraints. In so doing 'downstream' FSR, as emphasized earlier, depends primarily on existing research results for testing and incorporation directly—or with relatively minor modifications—into farming systems. On-farm trials and direct or firsthand interaction with farmers predominate while experiment station research tends to be minimal and restricted to adaptive rather than basic research" (Gilbert et al., 1980).

Gilbert et al. (1980) summarized their discussion about "upstream" and "downstream" research by saying that an FSR&D program should strive for some mixture of the two as determined by the availability of innovations that can be easily and rapidly integrated into existing farming systems. Where the pool of technologies is large, "downstream" programs can be effective in identifying and adapting the most promising approaches. Conversely, where basic or more general research is needed, an "upstream" approach may provide an appropriate mode for organizing research to cut across traditional disciplinary and commodity lines. At the minimum, a two-way flow of information is needed from farm level to research institution and back again in the form of appropriate technologies. To date, the IARCs have had the relative advantage in conducting "upstream" research, while the national programs have the advantage in conducting "downstream" research; however, a national program may engage in both types of research.

Another aspect of the type of research to favor concerns whether or not researchers should stress small farmers' welfare or their production. Sometimes increases in farmers' production are large enough to satisfy national objectives for larger output. When they are not, the FSR&D team should seek guidance from national policy makers.

3.4.5. A PARTIAL RESOLUTION OF THESE ISSUES

While the above discussion might suggest an excessive number of factors to consider in planning an appropriate strategy for a country's FSR&D program, some resolution of the issues is possible. Consider the following possibility. An FSR&D program might be initiated that emphasizes quick and modest changes in farmers' systems. Most small farmers will probably not want to risk much at first. In time and with adequate success, farmers may become interested in more substantive changes. Likewise, some early success will be needed to show FSR&D's effectivenss to those who fund the effort. Then, as researchers and farmers become better acquainted, more imaginative improvements can be planned.

Thus, the researchers could begin working to improve the yields of the farmers' subsistence crops. After these have been improved, part of the farmers' resources can be directed to higher valued crops or livestock with which the farmer is familiar. Finally, new crops and patterns can be introduced as farmers become more successful. This concept reflects the dynamics of a staged FSR&D approach. For example, such an approach is occurring in Ethiopia in which ILCA (1980) is experimenting with improved subsistence crops, with the idea of freeing land for forage crops and livestock improvement. Better livestock helps to improve traction and, in turn, crops.

3.5. SUMMARY

We began this chapter by describing the five major activities common to many FSR&D programs and the collaboration with experiment station and extension staff. Research and development results are fed back to earlier activities to improve research designs and add to the body of knowledge. Timing of FSR&D activities depends on the situation: for cropping systems in the tropics, rapid implementation of on-farm experiments is often possible; for livestock systems and other situations, more time is needed. FSR&D stresses on-farm research conducted by interdisciplinary field teams supported by disciplinary and commodity specialists. We closed the chapter by presenting alternative strategies concerning how much change, how soon to attempt change, where the ideas for change originate, and what type of change to consider.

In the course of gathering information for this book, we visited several national and international centers concerned with FSR&D. In Appendix 3-A, we provide a summary of some of these programs that reveals alternative strategies for FSR&D.

CITED REFERENCES

Byerlee, D., M.P. Collinson, R.K. Perrin, D.L. Winkelmann, S. Biggs, E.R. Moscardi, J.C. Martinez, L. Harrington, and A. Benjamin. 1980. Planning technologies appropriate to farmers: concepts and procedures. CIMMYT, El Batan, Mexico.

Collinson, M.P. 1979. Theme: Agrarian change, the challenge for agricultural economists. Micro-level accomplishment and challenges for the less developed world. A paper presented at the Int. Assoc. of Agri. Econ. 17th Conf. 3-12 Sept. 1979. Banff, Canada. CIMMYT, Nairobi, Kenya.

Delgado, C.L. 1978. The southern Fulani farming system in Upper Volta: a new old model for the integration of crop and livestock production in the West African savannah. Center for Res. on Econ. Dev. Univ. of Michigan, Ann Arbor, Mich.

Gilbert, E.H., D.W. Norman, and F.E. Winch. 1980. Farming systems research: a critical appraisal. MSU Rural Dev. Paper No. 6. Dep. of Agric. Econ., Michigan State Univ., East Lansing, Mich.

Harrington, L. 1980. Methodology issues facing social scientists in on-farm/farming systems research. A paper presented at a CIMMYT workshop on Method. Iss. Facing Soc. Sci. in On-Farm/Farming Sys. Res. 1-3 April 1980. El Batan, Mexico.

ILCA. 1980. ILCA the first years. ILCA, Addis Ababa, Ethiopia.

Technical Advisory Committee (TAC) Review Team of the Consultative Group on International Agricultural Research. 1978. Farming systems research at the international research centers. The World Bank, Washington, D.C.

Zandstra, H.G., E.C. Price, J.A. Litsinger, and R.A. Morris. 1981. A methodology for on-farm cropping systems research. IRRI, Los Banos, Philippines.

OTHER REFERENCES

Clyma, W., M.K. Lowdermilk, and G.L. Corey. 1977. A research-development process for improvement of on-farm water management. Water Mgmt. Tech. Rep. 47. Eng. Res. Center, Colorado State Univ., Fort Collins, Colo.

CRIA Cropping Systems Working Group. 1979. [Draft] Network methodology and cropping systems research in Indonesia, CRIA, Bogor, Indonesia.

Dalton, G.E. 1975. Study of agricultural systems. Applied Science Publishers Ltd., London.

Ford Foundation. 1977. Agenda papers, vol. 1, Middle East and Africa agricultural seminar. 1-3 Feb. 1977. Tunis, Tunisia. The Ford Foundation, New York.

Fumagalli, A., and R.K. Waugh. 1977. Agricultural research in Guatemala. Presented at the Bellagio Conf., Oct., 1977. Bellagio, Italy.

Hernandez X., E. (ed.) 1977. Agroecosistemas de México: contribuciones a la enseñanza, investigación y divulgación agrícola. Colegio de Postgraduados, Chapingo, and CIMMYT, El Batan, Mexico.

Hildebrand, P.E., and S. Ruano A. 1978. Integrating multidisciplinary technology generation for small, traditional farmers of Guatemala. A paper presented at the Ann. Meeting of the Soc. for App. Anthro. 2-9 April 1978. Merida, Mexico. ICTA, Guatemala.

IITA. [Undated]. The IITA farming systems program. IITA, Ibadan, Nigeria.

IRRI. 1976. International bibliography on cropping systems, 1973-1974. The Library and Documentation Center, IRRI, Los Banos, Philippines.

ISRA/GERDAT. 1977. Recherche et développement agricole: les unités expérimentales du Sénégal. CNRA Sem. 16-21 May 1977. Bambey, Senegal.

Laird, R. 1977. Agricultural research for development of traditional agriculture. Postgraduate College, National Agricultural School, Chapingo, Mexico.

Lawani, S.M., F.M. Alluri, and E.N. Adimorah. 1979. Farming systems in Africa, a working bibliography, 1930-1978. G.K. Hall & Co., Boston, Mass.

Mathema, S.M., and M.G. Van der Veen. [Undated]. Socio-economic research on farming systems in Nepal: a key informant survey in five cropping systems research sites. HMG, Min. of Food, Agric., and Irr. USAID/IADS, Nepal.

McIntosh, J.L., and S. Effendi. 1978. Cropping system research activities in Indonesia. A paper presented to the Cropping Sys. Working Group. 2-5 Oct. 1978. IRRI, Los Banos, Philippines.

PNIA. 1979. Guía metodológica para conducción de ensayos de finca. Secretaría de Recursos Naturales. PNIA, Comayagua, Honduras.

Salter, L.A., Jr. 1967. Scientific method and social science. p. 39-77. *In* A critical review of research in land economics. The University of Wisconsin Press, Madison, Wis.

Turrent, F.A. 1977. El agrosistema, un concepto útil dentro de la disciplina de productividad. p. 291–319. *In* Hernandez X., E. (ed.) Agroecosistemas de México: contribuciones a la enseñanza, investigación y divulgación agrícola. Colegio de Postgraduados, Chapingo, and CIMMYT, El Batan, Mexico.

World Bank. 1979. Senegal agricultural research project. The World Bank, Washington, D.C.

Zandstra, H.G., K. Swanberg, C. Zulberti, and B. Nestel. 1979. Caqueza: living rural development. IDRC, Ottawa, Canada.

Chapter 4

TARGET AND RESEARCH AREA SELECTION

Effective selection of target and research areas is one of the critical activities of the FSR&D process. Selection generally begins with high-level decision makers in a government deciding on one or more target areas as the focus of attention, and selection of research areas continues until improved technologies are diffused throughout the target area and possibly extrapolated into other areas.

The process of selecting areas for FSR&D activities is roughly as follows:

1) The decision makers select the target area or areas.
2) The FSR&D team divides the target area into subareas according to characteristics most important for the FSR&D effort.
3) The FSR&D team selects the research area within the target area.
4) The FSR&D team selects farms and farmers within the research area for conducting on-farm research.
5) When research results are promising, the FSR&D team and the extension service select multiple locations within the target area for validation of new technologies on a broader scale.
6) If the results of the multi-locational tests are satisfactory, the FSR&D team and governmental agencies select areas for pilot production programs to evaluate the new technologies on a more intensive scale.
7) After resolving any problems arising from the pilot production programs, the extension service and other relevant agencies implement the new technologics according to suitable subareas within the target area.
8) Finally, these governmental agencies may extrapolate relevant technologies to similar areas outside the target area.

We illustrate this sequence of activities in Fig. 4-1.

In this chapter, we discuss (1) the amount of data to collect for the various FSR&D activities, (2) the selection of a target area and subareas, (3) the selection of the research area, and (4) the early identification of opportunities for action. We close the chapter with a summary.

4.1. AMOUNT OF DATA TO COLLECT

The FSR&D team collects data during the different activities on similar topics, but at different levels of detail. The team begins with a broad, cursory overview of information relevant for target area and subarea selection. The overview will include physical, biological, and socioeconomic information. We represent this activity in Fig. 4-2 by a thin, continuous horizontal bar. While some of the material collected during target area and subarea selection is adequate for the research area selection, the team needs greater detail and other kinds of data.

The above requirements can be compared with still more intensive data requirements for problem identification and the research process. In Fig. 4-2 we show this need for increasingly greater detail on more specific topics by successively longer and narrower rectangles as we proceed from one activity to the next. Data gathering for FSR&D is an integrated and comprehensive process. We discuss the detailed information needed for FSR&D activities later in this book.

4.2. SELECTION OF TARGET AREA AND SUBAREAS

We now turn to a discussion of the first major activities in the FSR&D process—the selection of a target area and its division into subareas. We define our terms and then discuss the criteria and methods for selecting the areas, information needed, and selection of the areas.

4.2.1. DEFINITIONS

An FSR&D *target area,* may be selected for two basic reasons:

1) to meet the needs of the people living there
2) to take advantage of its agricultural potential.

In the second case, people may or may not already be living in the area.

When conditions in a target area are substantially different, it may be divided into smaller areas with similar physical, biological, socioeconomic, and farming systems characteristics. These smaller areas are the *subareas* for which improved technologies can then be developed. For example, Fig. 4-3 illustrates a hypothetical target area with four subareas. Subarea A consists of valleys and low plains with stony soil subject to flooding. Subarea B has the same type of land as A, but is settled by a different ethnic population with a different farming system. Subarea C is fertile plain. Subareas A, B, and C have good accesses to markets. The farmers in these three subareas

Figure 4-1. The sequence of locations of FSR&D work during the FSR&D process.

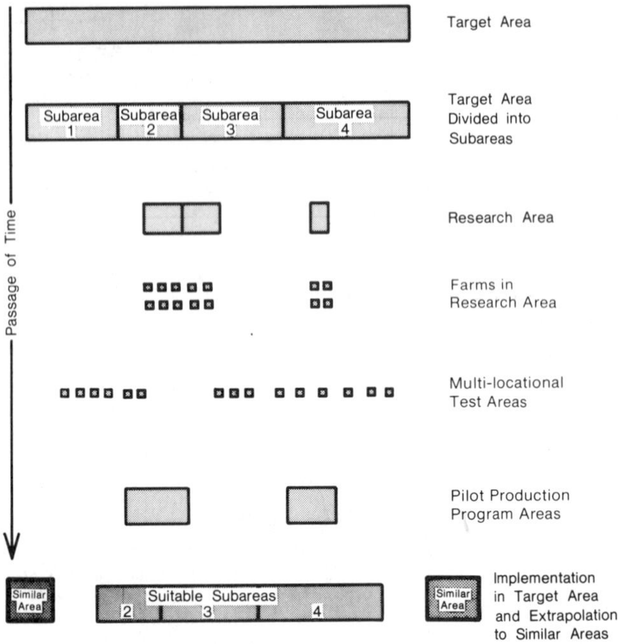

have mostly crop-based farming systems. Subarea D is erodible hill country with poor roads. The inhabitants have a livestock-based farming system and are nomadic during the dry season.

Some FSR&D practitioners use other means for dividing a target area. For CIMMYT's work, Byerlee et al. (1980) call the focus of the FSR&D effort the recommendation domain and define this term as "a group of roughly homogeneous farmers with similar circumstances for whom we can make more or less the same recommendation. Recommendation domains may be defined in terms of both natural factors—e.g., rainfall—and economic factors—e.g., farm size." In Appendix 4-A we present a tabulation of farming practices that CIMMYT used to aid in establishing recommendation domains.

ICTA in Guatemala searches for groups of farmers who are using similar methods for the same cropping patterns. If farmers grow the same crops in approximately the same manner, the researchers assume that the environment and other conditions are similar. ICTA's researchers assume an improved technology should apply to all members of the group, because the crops, means of production, and expected responses are similar.

In establishing the boundaries of a subarea, the FSR&D team does not look for an area with complete uniformity of all farmer circumstances. Such an area rarely, if ever, exists. Instead, the team seeks an area where large numbers of farmers are relatively homogeneous in their characteristics. Enough farmers should be involved to make the research effort worthwhile. By relatively homogeneous, we mean most farmers will respond to the new technologies in a similar way.

The first boundaries of a subarea may have to be tentative. Then, as the FSR&D process continues and the team learns more about the subarea, the boundaries may be revised.

Subareas are not necessarily contiguous. For example, in Fig. 4-3, three of the four subareas are divided. Furthermore, the team need not consider all subareas simultaneously. It might rank subareas according to some priority for deciding which ones to work in.

In Appendix 4-B we show the method Collinson (1979a) used for grouping farmers "into relatively homogeneous populations on the basis of their present farming system" in the Central Province of Zambia. "The key step in interpretation [of the collected information] is deciding the sources of variation which are critical in dictating resource allocation in farming systems of the area."

4.2.2. CRITERIA AND METHODS

The first step in selecting a target area is for the key personnel in the FSR&D program to identify potential criteria for the selection. Then those holding top policy and planning positions in the government discuss, select and order these criteria, and determine the information needed. FSR&D researchers collect and analyze the information and submit alternatives to the decision makers for their choice.

Compatibility with national and regional policies ranks high on the list of criteria for selecting target areas. For example, a national development policy in a country may favor assistance to poor farm families, providing food for urban populations, or increasing the country's foreign exchange balance. Table 4-1 based on an unpublished paper by Collinson (1979b) illustrates how selected characteristics can be used in comparing two potential target areas. If the government favors helping poor

Figure 4-2. The amount of data detail collected during FSR&D activities. The width of the rectangle indicates breadth of coverage; height of rectangle illustrates amount of detail.

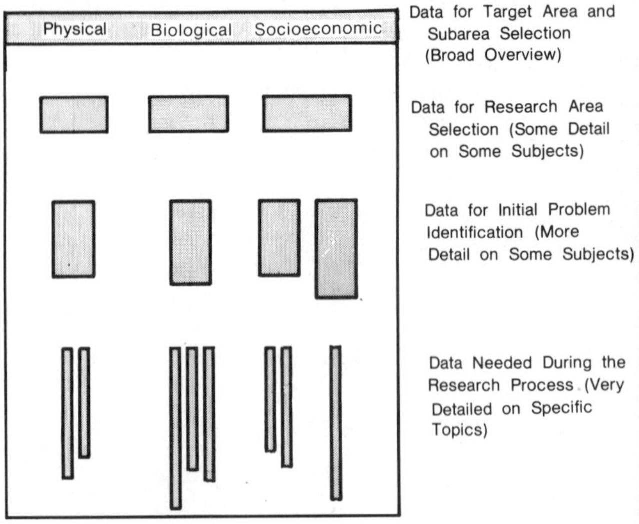

Figure 4-3. A hypothetical target area divided into four subareas showing a research area in three parts.

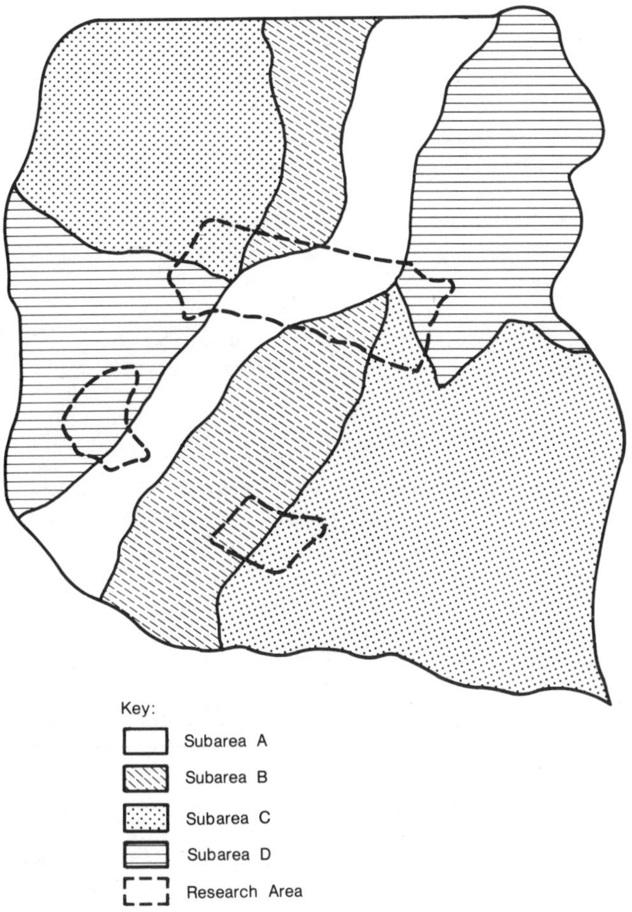

families, the decision makers would select target area A. Alternatively, if the government favors increasing urban food supplies or expanding exports, the decision makers would select target area B.

As another example, the national decision makers in one African country selected a sparsely populated region with improvement potential as a target area. They based their choice on the national policy to alleviate land pressure in its fertile, but overcrowded highlands.

Selection of a particular target area may satisfy more than one development goal. As an example, one Latin American country has the twin goals of improving the economic level of poor rural areas and of increasing political stability there. Farmers in one region of the country had traditionally supplemented their meager farm incomes with seasonal work in a neighboring country. When the neighboring country closed its borders to these migratory workers and threatened annexation of the region, the farmers' national government set up an FSR&D project to raise farm incomes and to solidify the farmers' loyalty.

As another alternative, the decision makers may select a target area on the basis of specific physical limitations or problems such as erodible slopes, flooding, soil

Table 4-1. Comparison of five characteristics of two potential target areas (Based on an unpublished paper by Collinson, 1979b).

Characteristics	*Target area A*	*Target area B*
Number of poor farm families	50,000	15,000
Number of poor farm people	300,000	100,000
Production of exports (US $)	---	1,000,000
Production of urban food supplies (US $)	50,000	200,000
Average family's cash income as percent of the national average	40%	150%

salinity, inadequate pasture, or animal disease. In Sumatra, *Imperata* grass overgrowing large areas presents a major problem to Indonesia's land settlement program. For this reason, the decision makers chose a target area that had large sections of *Imperata* infested land for an FSR&D program.

The following questions may help the FSR&D staff provide the information the decision makers need for determining broad policy options:

- Is the target area or subarea large and relatively similar in those environmental characteristics that have the most bearing on potential research results? If the answer is yes, the FSR&D team can apply the research results broadly and meaningfully within the target area.
- Is the target area similar to other areas? If so, some of the technology developed in the target area can be extrapolated to other areas.
- Does the area have the potential for rapid payoff from FSR&D? Factors that often determine the potential are (1) physical and biological conditions, (2) markets and infrastructure, (3) available technology, and (4) farmers' willingness to accept innovations.
- Do the area's environmental conditions facilitate application of technologies developed elsewhere? If so, the FSR&D process could be greatly shortened.
- What are some of the cost factors? For example, is the target area now served by existing governmental programs or private institutions? If so, cooperation with such programs or institutions could enable FSR&D to accomplish more, reduce costs, or save time. One example would be the help that an existing experiment station could offer. However, FSR&D may be needed more in areas where such links are not well established, and the costs are higher. Thus, high-level decision makers may have to make this choice.

In Appendix 4-C, we illustrate how environmental

suitability and other factors influenced the choice of target areas in Indonesia.

In Mali, several factors favored the Mali Sud region as the target area for a farming systems project. Regional characteristics (1) favored rapid development of the area and (2) offered a wide range of farms at different stages of modernization and a potential for integrating livestock and cropping enterprises. Rapid payoff was expected because of favorable agroclimatic conditions and good roads. The program costs were kept down because the area's supportive institutions were functioning well and because a research station had initiated farming systems work in the area.

4.2.3. INFORMATION NEEDED

As indicated in Fig. 4-2, the team uses a broad range of data for target area and subarea selection that includes information on national policy, farmers' characteristics, and the environment. Below, we (1) relate national policy with farmer characteristics, (2) discuss four aspects of the physical environment and provide an example, and (3) describe secondary and primary data sources.[1] In discussing this third item, we provide suggestions on gathering socioeconomic data, as well as other types of data.

National Policy and Farmer Characteristics

The criteria the decision makers and FSR&D team use in selecting the target areas will determine general data requirements. Consider national policy criteria. Suppose the primary objective of national policy is "to increase agricultural income on small farms." Then, for each potential target area the FSR&D team needs data such as the number of small farmers, the size of their farms,

[1]Secondary information means published or unpublished data collected for purposes other than the FSR&D project; primary information refers to data collected specifically for the FSR&D project.

agricultural income, cropping and livestock patterns, and management practices. In addition, the team should investigate the practices used by successful farmers and why others do not use these practices.

The exact wording of policy objectives is important. Suppose the word "agricultural" is left out before "income" in the above policy objective. Now the objective is "to increase the income on small farms." Thus, increases in income are not limited to agricultural income. Data collection must now include nonagricultural income on the farm such as spinning, weaving, or pottery making.

Physical Environment

In its work in Southeast Asia, IRRI developed an approach to classification of the physical resources of an area according to climate, topography, and soils. In the discussion that follows, we report on (1) IRRI's work on these three topics, (2) water, and (3) an example in Southeast Asia that integrates these factors.

Climate. In parts of Southeast Asia, IRRI researchers delineated four major climatic zones (with subdivisions) for rice, based entirely on rainfall patterns. See Fig. 4-4 (IRRI, 1974). These researchers explained "in general a fair amount of data on rainfall is available for most Southeast Asian countries. However, other macroclimatic parameters such as evaporation, temperature, relative humidity, wind speed, and particularly solar radiation are seldom recorded. Therefore, these data could not be used to evaluate evapotranspiration." In Appendix 4-D, we present IRRI's criteria for selecting the major climatic zones, the description of these zones, and their subdivisions.

For other regions of the world, for different crops, or for livestock, FSR&D researchers may need to consider other climatic factors.

Topography. According to IRRI (1974), "a convenient method for a broad-scale topographic classification of the rice lands of Southeast Asia is to divide the countries of the region into the physiographic units with which different forms of rice cultivation are closely associated." IRRI researchers used four subdivisions as a first approximation:

Figure 4-4. The rainfall zones for parts of Southeast Asia (Adapted from IRRI, 1974). Wet month means at least 200 mm rain. Pronounced dry season means at least 2-3 months with less than 100 mm rain.

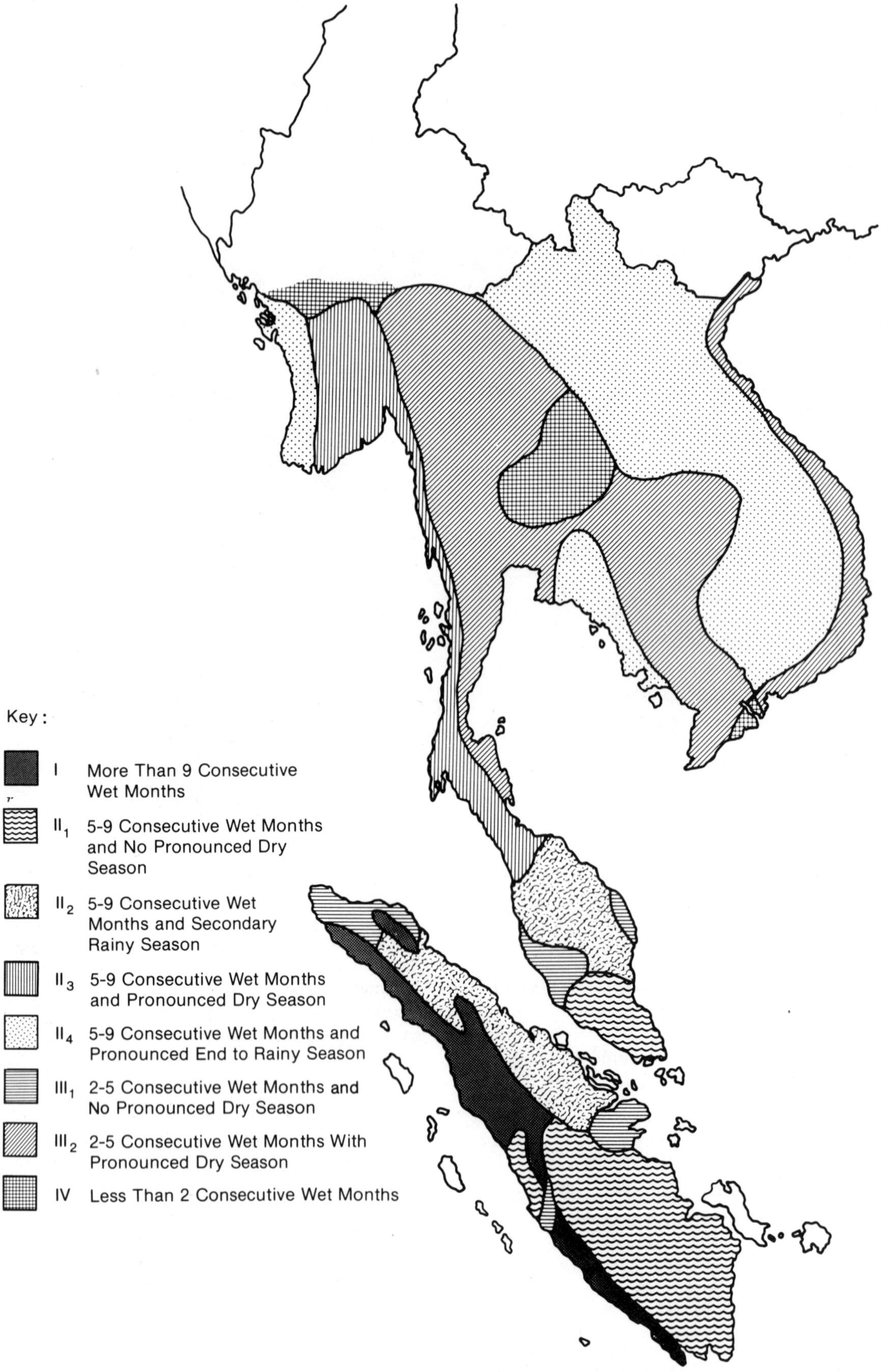

- coastal plains of marine, deltaic, and fluvial origin
- inland terraces and plains of ancient origin, usually well dissected
- inland terraces, fans, and valleys of recent origin
- sloping lands.

We show these four physiographic units in Fig. 4-5 and briefly describe them in Appendix 4-E.

Soils. IRRI (1974) lists the major soils in each physiographic unit that are important for rice growing (Fig. 4-5). In classifying these soils, IRRI relied on the soil taxonomy of the U.S. Department of Agriculture (USDA Soil Survey Staff, 1975). This taxonomy lists six soil orders—Inceptisols, Vertisols, Entisols, Ultisols, Alfisols, and Oxisols. Here we present IRRI's description of Ultisols as an example of soil information useful in target area description:

> "Ultisols. These are soils with a sandier topsoil, a more clayey subsoil, a low base saturation[2] and mainly 1:1 clays.[3] They occur in the older, but not the oldest surface in the tropics. In Southeast Asia, they are largely on the rolling and hilly lands. Probably much of the upland rice grown in shifting cultivation is on these soils, particularly in Sumatra, Kalimantan, and Thailand. Apart from this use for shifting agriculture, Ultisols form major areas for tree crops like rubber and oil palms, but the area under permanent annual cropping may be small. Major problems in annual cropping are erosion due to profile instability and high acidity with aluminum and manganese toxicity. Lime is needed to grow the more sensitive crops, e.g., corn and sorghum."

Water. With regard to water, especially for irrigation, such factors as the amount and period of availability, quality, and source are important. We discuss irrigation water in more detail in Sec. 5.4.1. and Appendix 6-A.

An Example in Southeast Asia. **Member countries of the Asian Cropping Systems Network are striving to increase small farm income by searching for ways to help farmers grow a second rice crop. Target areas are selected using data on the number of consecutive months of rainfall of more than 200 millimeters. Within this target area, data on topography, soil, and irrigation water guide the selection of subareas. In addition, researchers gather data on other topics, such as population density, agricultural production, prices, and cropping experiments.**

Secondary Information

Data on the broad physical, institutional, social, and economic factors are generally available from secondary sources. For example, census data provide information on population density, size of land holdings, land use, and crop and livestock production. Other sources include weather records, cadastral surveys, commodity studies, and reports of other programs.

Statistics and reports may be available at all levels—national, provincial, district, subdistrict, and village. Data sources include the offices of national planning, statistics, agriculture, community development, irrigation, research stations, and other agriculturally related offices, as well as those of private and public marketing organizations. For example, in an Egyptian project, statistics on agricultural production guided the selection of three target areas representing typical cropping patterns along the Nile River.

An FSR&D team might use satellite imagery for a general classification of potential target areas and aerial photography for more detailed analysis. The latter is especially useful in gathering and confirming information on land use. Such information helps in identifying soil and land types for cropping and livestock experiments. If available, current aerial photos can be compared with earlier aerial photos—preferably 20 years earlier—to identify trends (de Haan, personal communication).

Also, the team can prepare maps for FSR&D from sources such as satellite imagery, aerial photos, or government maps. Topographic maps on a 1:50,000 scale are often available for many areas of a country. Maps might show such items as population distribution, ethnic composition, land use, soil capability, vegetation, land tenure, ecozones, and percent of land cultivated. Fig. 4-6 is an example of one of the many types of maps that can be prepared as part of a reconnaissance soil survey (Thom, 1978).

Primary Information

Once relevant secondary data have been gathered and analyzed, the FSR&D team may need additional information to guide its selection of a target area. Primary sources may then be used. Formal and informal leaders, such as farmers, educators, local merchants, mayors, and religious leaders may be good informational sources. Also, technical specialists such as agricultural officers, agronomists, climatologists, and social scientists are useful. The team may obtain information from these sources through either informal or formal interviews. In Peru, a research team used informal interviews of merchants and others to select a highland region with easy access to the Lima market as a target area. In Appendix 4-B, we discuss how a research team gathered primary information for a province in Zambia, surveyed extension agents, and tabulated the data.

4.2.4. ACTUAL SELECTION

The selection of specific target areas depends on the key decision makers' values and on the country's decision-making structure. These values reflect the weights decision makers place on national and regional policy, as well as on socioeconomic and technical factors. Top-level decision makers such as the ministers of agriculture, planning, and finance are often involved. At other times, those

[2]Low base saturation refers to a soil with a small percentage of exchangeable metallic cations (i.e., Ca^{++}, Mg^{++}, K^{+}, NH_4^{+}).
[3]1:1 clays consist primarily of kaolinite minerals. Kaolinite is a non-swelling mineral of low cation exchange capacity.

Figure 4-5. The physiography of rice growing areas in parts of Southeast Asia (Adapted from IRRI, 1974).

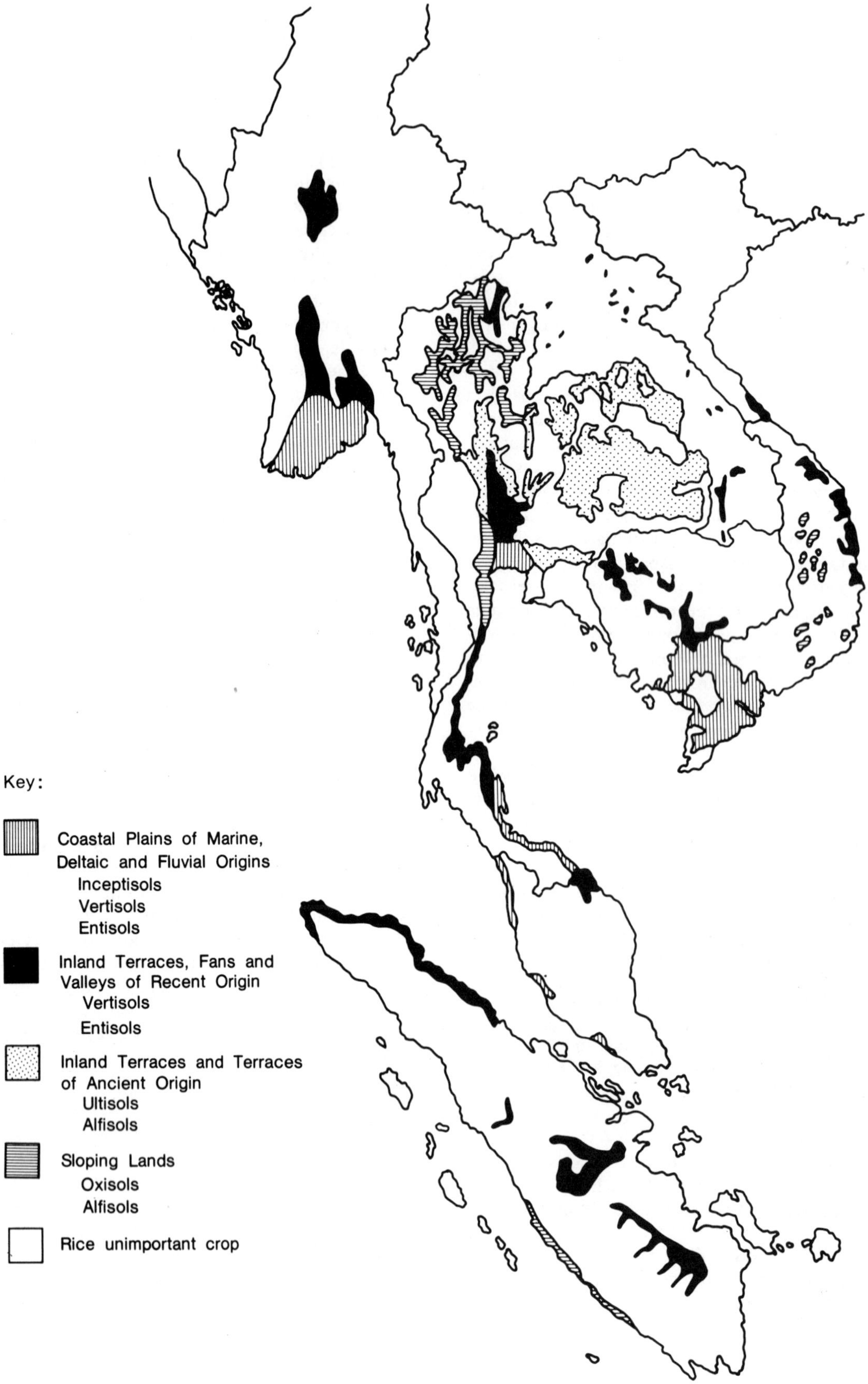

Figure 4-6. The percentage of land cultivated in an area of Kenya (Thom, 1978).

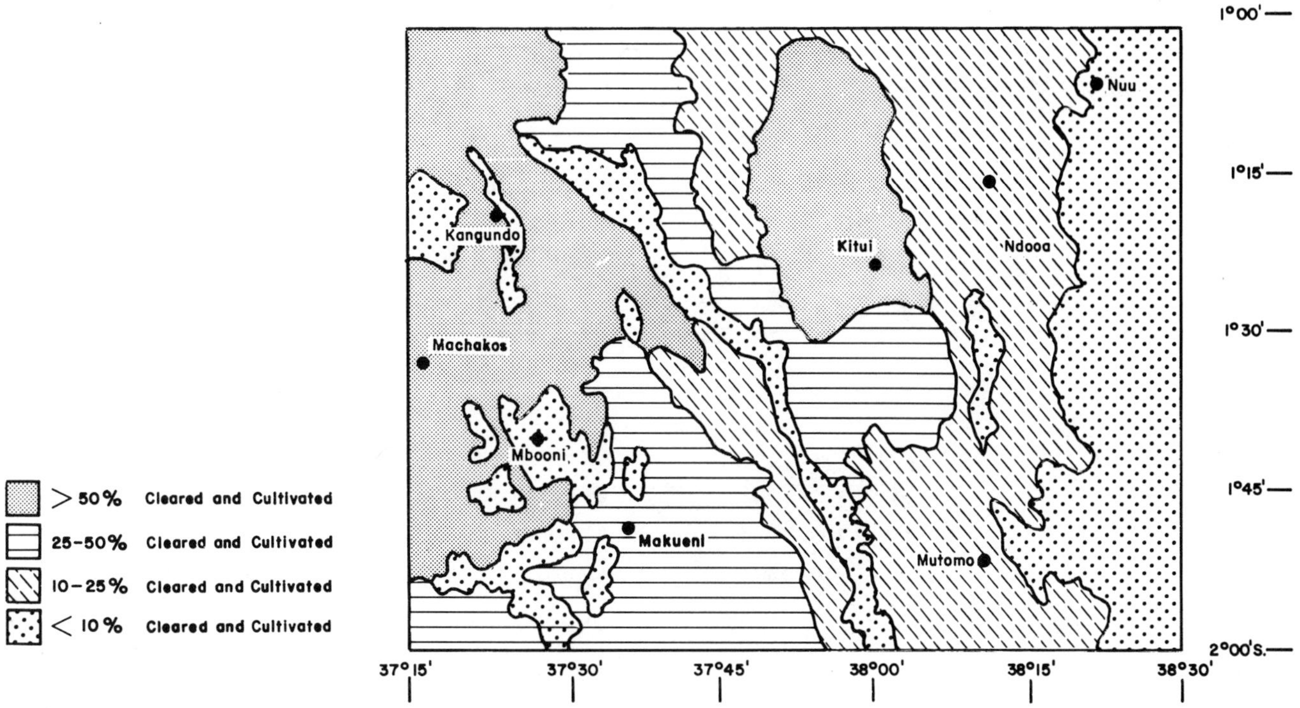

below the ministerial level make the decision. For example, the regional agricultural directors in one Central American country have considerable influence in such decisions, provided their decisions fall within broad policy guidelines. In another country, the director general of agriculture selects the target area.

Those responsible for deciding on the target area normally rely on their advisers to assist them in clarifying the alternatives and in providing background data. In this regard, FSR&D staff should be particularly helpful.

In summary, selecting a target area and subareas requires choosing criteria and then collecting and analyzing data pertinent to the criteria. Large amounts of detailed data should not be amassed. Generally, less than one month is enough time to select the target area.

4.3. SELECTION OF THE RESEARCH AREA

Once the target area and the subareas have been determined, the FSR&D team is ready to select the research area. In describing this process, we will cover (1) the definition of a research area, (2) criteria and methods for selecting the area, (3) staffing, (4) information needed, (5) means for collecting information, (6) selecting the area in stages when the target area and subareas are large, (7) actual selection, and (8) abandoning an area.

4.3.1. DEFINITION

A research area, where the FSR&D team develops improved technologies, may represent the whole target area or only some of the subareas. When the team properly selects a research area, the research results are transferable throughout the target area according to the subareas for which the technologies were designed.

The research area may, in turn, be a single area or comprise scattered areas. A single research area, reflecting the variations within the target area, usually provides advantages such as better logistics and team organization.

If the team cannot find a suitable single area, it might select several areas that when combined constitute the research area. Within this area, the team locates the trials and tests in a way that broadly samples the conditions within each subarea (Fig. 4-7). In Fig. 4-7, the research area is divided into three parts to cover minor variations within subareas.

A convenient location for the team's headquarters would be within or near the research area. Where the research area is fragmented and the parts separated by poor roads and communications, team operations could be difficult and costly. Where such conditions exist, the FSR&D team may need to establish two or more independent research areas.

4.3.2. CRITERIA AND METHODS

In this section, we provide illustrative suggestions that have helped others select research areas. Selection criteria relate to (1) the area's representativeness, (2) size, (3) accessibility, (4) closeness to an experiment station, and (5) cooperation with farmer contact agencies and leader support.

Figure 4-7. Hypothetical research area consisting of three parts. This figure is an enlargement of the 8,000 ha area shown in Figure 4-3. Selection of the location for the trials and tests is discussed in Chapter 7.

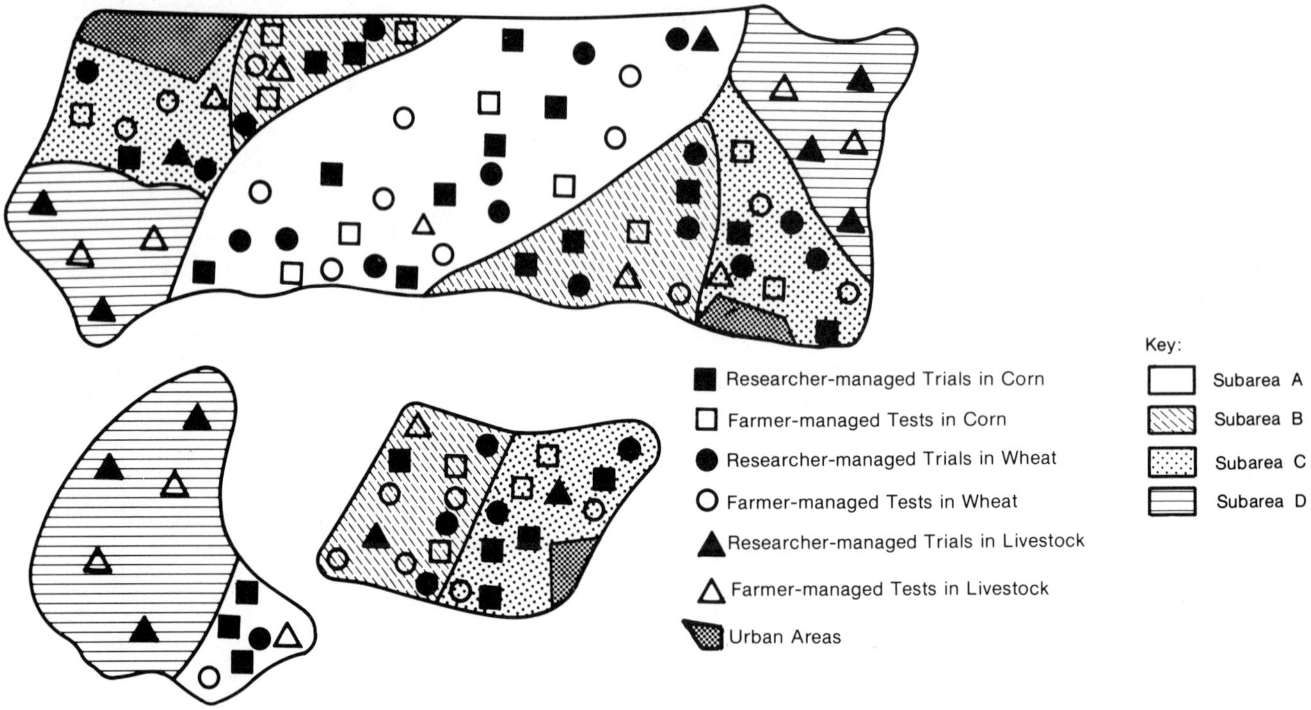

Representativeness of an Area

The overriding criterion for research area selection is its representativeness of the target area or subareas. This representativeness generally hinges on both environmental conditions—i.e., physical, biological, and socioeconomic—and the farming systems.

We suggest the FSR&D team determine the representativeness of a potential research area in this way:

1) Decide which of the target area's or subarea's characteristics such as rainfall, land type, distance to market, or ethnic composition of the farm population are the most important.
2) Take a sample of farm communities with these desired characteristics.
3) Collect data on these characteristics for each farm community in the sample.
4) Calculate the average value for the sample for each characteristic. These averages, taken together, describe the most representative farm community of the target area or subarea.
5) Label the farm community that deviates the least from these average values as the most representative of the area.

We illustrate how this method is being used in Indonesia in Appendix 4-C.

Size

The size of research areas varies. For example, the initial size of one of the research areas in Guatemala was 8,000 hectares. The area could be smaller in a densely settled region, such as in parts of India where considerable variability in farms and families occurs. The area could be much larger in a sparsely populated area when transportation is adequate. Most often, FSR&D teams use research areas large enough to include some variability in farmers' conditions. This variability allows the research team to test the adaptability of new technologies under different conditions and to make appropriate adjustments.

Accessibility

Easy access to all parts of the research area enhances cooperation between the research team and farmers, permits timely provision of inputs and marketing of outputs, and reduces the team's operating costs. Accessibility of the research area enables the support staff—scientists from the experiment station, government officers, and FSR&D researchers stationed in the nation's capital—to reach the research area quickly. Living facilities for the regional headquarters and field teams should be available nearby. Easy access is especially important during the early stages of FSR&D because meaningful results are often needed quickly. Later, when FSR&D becomes established, research in areas more difficult to reach may be undertaken.

Existence of a Nearby Agricultural Research Station

When research stations are close to the research area, the FSR&D team is more likely to receive effective cooperation from the experiment station staff. Experiment station staff can help the team with on-farm experiments and by conducting experiments on the experiment station.

Cooperation of Farmer Contact Agencies and Leader Support

Support from farmer-contact agencies such as extension, credit, and production groups is helpful to the FSR&D effort. Their provision of services such as information, credit, purchasable inputs, or marketing of outputs can be essential for the widespread adoption of FSR&D's results.

Help from formal and informal leaders at the national and local levels can also be critical to FSR&D's success.

4.3.3. STAFFING

In forming the FSR&D team to select the research area, the FSR&D leaders should select the organization's most experienced staff. These leaders may need help from specialists in agronomy, livestock production, agricultural economics, and farmer and social organizations.

Since livestock or crops are usually the focus of attention, at least one specialist in these fields is essential. An agricultural economist is needed to translate technical possibilities into economic practicality. A sociologist or someone with similar training may be needed for evaluating the sociocultural characteristics of farmers and the organizations affecting farmers. Other experienced staff may also be required, such as an agricultural engineer where irrigation or farm equipment is important, commodity specialists according to the most relevant crops, and specialists in pest management.

This team should integrate its efforts by operating in an interdisciplinary manner. For continuity, research and extension members, who will be working in the area, should be involved. Finally, women may need to be added if one or more would not otherwise be part of the group. Such involvement of female personnel is particularly relevant when women farmers manage many or all of the farming activities.

4.3.4. INFORMATION NEEDED

Keeping the criteria from Sec. 4.3.2. in mind, the FSR&D team now gathers information on alternative research areas. In Table 4-2, we list data categories for research area selection. In view of the limited budgets in most developing countries, the team needs to be selective in how much data to collect.

Table 4-2. Data categories for research area selection.

Physical environment
- Climate: e.g., rainfall, temperature, wind, sunny days
- Soil: e.g., physical, chemical, hydrological conditions
- Topography: e.g., slope, flood plain
- Irrigation: e.g., water source and quality, means and frequency of delivery, on-farm practices

Biological environment
- e.g., weeds, insects, diseases, birds, rodents, crop yields

Socioeconomic environment
- Resource availability: e.g., land, labor, cash, means of traction
- Infrastructure: e.g., supply of farm inputs, markets for farm output, transportation, electricity
- Market data: e.g., prices of farm inputs and commodities, traders
- Sociocultural characteristics: e.g., land tenure and inheritance systems, sexual division of labor in agriculture, religious beliefs concerning agriculture, openness to change
- Political and economic structures: e.g., national regulations, community groups, caste or clan systems, patron-client relationships, cooperatives

Production systems and land use
- e.g., major crops and livestock, cropping patterns, livestock characteristics, management practices

In Table 4-3, we show the type of data the Indonesian Cropping System Program uses for research area selection.

Table 4-3. Data required for systematic selection of rural communities as research areas in Indonesia (Adapted from CRIA, 1979).

Data	*Purpose*
Distance from main road (km)	To guarantee that the rural community is easily accessible
Area in each land use class (ha)	To permit the selection of rural communities with the largest area in the desired land use class
Relative area in each slope class (percent)	To avoid rural communities with atypical topography
Relative area in each soil texture (percent)	To avoid rural communities with atypical soils
Area planted to each crop, by month (percent)	To identify current production level
Population, by economic activity (number)	To determine importance of agricultural employment
Rainfall by month for past 10 years (mm)	To determine number of months 100 mm or more of rain and probability of less than 100 mm at beginning and end of cropping season
Participants (number) in the rice production program of the government	To determine the availability of credit and level of technology in the rural community
Months during which irrigation water is available (percent of area with less than 5, 6-7, 8-9, and 10-12 months of irrigation per year)	To identify areas according to irrigation regimes
Draft animal population (number)	To determine the availability of draft power
Tractor population (number)	To determine the availability of mechanical power

4.3.5. COLLECTION OF INFORMATION

Whenever possible, the FSR&D team uses secondary data for research area selection. Team members can usually obtain such data quickly, inexpensively, and simply. The team uses independent checks—other secondary sources, discussions with those knowledgeable in the field, spot surveys, and observations—to verify the reliability and accuracy of these data. In Table 4-4, we list the typical sources of data that researchers in Indonesia collect. Other countries may have different sources of secondary data such as tax offices or regional experiment stations. Often the latter are especially useful for soil surveys, rainfall records, and research results. Alternatively, the team may have to depend on data from national research centers and agricultural offices. Too frequently, data collected locally are sent to the national offices without copies being retained in the local offices; hence, local data resources may be limited.

When the FSR&D team finds secondary data insufficient for selecting a research area, direct observations and contacts with farmers may be necessary. In such cases, the team conducts reconnaissance surveys (sometimes called exploratory surveys).

Reconnaissance Surveys

Reconnaissance surveys of the research area include both aerial and ground surveys. For aerial surveys, the FSR&D team may use existing aerial photographs to obtain a quick and general impression of the obvious features such as terrain type, land use, transportation networks, and population centers. In addition, team members often fly over the area to observe conditions. For ground surveys, the team gathers information by observations and interviews. Besides the items listed under aerial surveys, the team observes such features as crops and livestock, field activities, power sources, soil types, and irrigation practices.

The team often interviews government administrators, agricultural researchers, personnel of farmer-contact agencies, businessmen, community leaders, and farmers and their families. Besides discussing and verifying their observations, the team members may ask economic and

Table 4-4. Typical sources of data for research area selection in Indonesia (Adapted from CRIA, 1979).

Extension office—district
Rainfall
Soil
Topography
Land use by type
Variety trials
Fertilizer trials
Pest surveillance
Demonstration plots

Extension office—subdistrict
Hectares in each crop
Planting and harvesting dates
Yields by crop

Extension office—field
Land ownership
Transportation
Support services
Markets
Power
Yield constraints
Constraints to crop intensification
Input subsidies

Village office
Land ownership
Tenure
Tenant-landlord relations
Landless labor
Support services
Irrigation system
Employment profile
Population
Transportation
Migration

Farmer-group interview
Current cropping pattern
Historical cropping pattern
Landless labor
Input availability
Input prices
Off-farm employment
Migration of agricultural labor
Water
Power
Input use
Yield constraints
Varieties
Planting decisions
Input levels
Constraints to cropping intensification

Input dealers
Input sales
Input availability
Input prices
Credit

Middlemen
Support services
Markets
Credit

Bank—field office
Credit

Irrigation office
Irrigation system

Statistician for district
Output prices
Agricultural production

social questions about farming systems, markets, and other unobservable factors. While farmer interviews may range from informal contacts to formal surveys, FSR&D practitioners generally favor quick, informal surveys for research area selection.

Experiments

At times, the reconnaissance survey and the available secondary data may still leave some doubt in the minds of the team members about the suitability of a proposed research area. If so, the FSR&D team may conduct exploratory experiments to learn more about the research area.

4.3.6. SELECTION BY STAGES

When the target area or subarea is large, the FSR&D team may select a research area in stages. First, the target area or subarea may be broken down into subdivisions. Where feasible, the team may use districts, subdistricts or other administrative units because secondary data are usually available for such entities. Then the team identifies the subdivision best suited for the research area. Within this subdivision, the team selects the research area.

The methods a team uses in research area selection and the emphasis on types of data vary from project to project and depend on FSR&D's mandate. For example, in selecting a subdistrict in Indonesia, the primary consideration was the number of hectares of a particular land-use type. In addition to the example in Appendix 4-C, we discuss a different method in Appendix 4-F—the one ICRISAT (Jodha et al., 1977) used in India to select the most representative subdistrict and research area.

4.3.7. ACTUAL SELECTION

Based on analyses of the foregoing data, the FSR&D team recommends a research area. The final selection, however, may involve governmental representatives at the national, regional, and district levels, and local leaders. Remember, target area selection is largely a policy deci-

sion that matches national objectives with practical possibilities, while research area selection is largely a technical decision that concerns how best to accomplish FSR&D objectives.

Some FSR&D practitioners consider one or two weeks sufficient to select a research area, if (1) the target area is fairly homogeneous in its natural and socioeconomic setting, (2) the selection is based primarily on physical and biological information, and (3) good secondary data are available. However, six weeks is more common. Even longer time is required if (1) the target area varies greatly such as in mountainous terrain with different climates and is in various stages of socioeconomic development, or (2) the team needs much socioeconomic, farming systems, or other primary information.

4.3.8. ABANDONING A RESEARCH AREA

Occasionally, after a research area has been selected and research started, national interests would be served better if the research area were abandoned. This may happen because of overlooked and unfavorable conditions such as farmers' unwillingness to cooperate, the appearance of a new obstacle such as a plant virus, or the discovery of a serious soil condition. In such a situation, the FSR&D leaders should have enough flexibility to shift to a new research area. The reasons should be explained to the affected government officials, local leaders, and farmers. Such actions do not preclude returning to the area later.

4.4. EARLY IDENTIFICATION OF OPPORTUNITIES FOR ACTION

In the process of collecting and analyzing data on the area, the FSR&D team may uncover opportunities for research or government action that can be started immediately. Solutions may appear so obvious and the consequences of failure so slight that waiting for subsequent FSR&D activities cannot be justified. These opportunities could relate to on-farm experiments, improvements to the infrastructure, or changes in governmental policy. However, we caution the team to be careful with such early actions, since more often than not the solutions are not as simple as they appear.

4.4.1. RESEARCH OPPORTUNITIES

A situation may dictate that the team initiate research as soon as practicable. For example, data collected during research area selection may indicate that drought during maize flowering regularly reduces yields. If early maturing maize varieties are locally available, they might be tested during the first available season. Test planting would not be costly and, if successful, could solve the problem quickly.

4.4.2. GOVERNMENT ACTION

Occasionally, by systematically reviewing the characteristics of an area, the team will be able to identify situations that call for quick government action. Such might be the case if government policy is not appropriate for an area, such as regulations for trucking or grain storage. Or from its experiences elsewhere, the team might know how to quickly improve a facility essential for the area's farmers. For example, the area may lack an all-weather access road, but may have all other production and marketing facilities for growing fresh vegetables to supply a nearby city during the rainy period. Starting a public dialogue on the advantages and costs of building the road may be in both the farmers' and the nation's interests. While the team will not have fully clarified the opportunities and problems of building the road, existing information might justify initiating the dialogue.

4.5. SUMMARY

The selection of the target area, subareas, and research area is normally the first activity confronting the FSR&D team. Proper choices at this stage can be crucial, since subsequent FSR&D activities and possibly successes are influenced by the suitability of the areas. Target area selection involves top-level decision makers in the government, FSR&D staff, and others who can help in matching national and regional goals with the potential of alternative areas. Identification of subareas and choosing the research area are largely technical matters that call on the expertise of the FSR&D team. Even so, advice and acceptance by national and local leaders are often needed before choosing the research area.

In choosing these areas, the FSR&D team gathers data on (1) national objectives and policies, (2) physical, biological, and socioeconomic environments, (3) farming systems, and (4) farmers' characteristics. Generally, the team begins this work by reviewing secondary sources of information such as published reports and records by local organizations, national ministries and agencies, and agricultural research stations. The team fills in missing data by direct means, such as reconnaissance surveys of the area. The total time for this activity may be as short as one or two weeks, but more often takes about six weeks — sometimes longer.

In addition, we covered other points in the chapter, such as the need for interdisciplinary teamwork, selecting research areas in stages when the target area or a subarea is large, justification for abandoning an area, and early identification of opportunities for action.

CITED REFERENCES

Byerlee, D., M.P. Collinson, R.K. Perrin, D.L. Winkelmann, S. Biggs, E.R. Moscardi, J.C. Martinez, L. Harrington, and A. Benjamin. 1980. Planning technologies appropriate to farmers: concepts and procedures. CIMMYT, El Batan, Mexico.

Collinson, M.P. 1979a. Understanding small farmers. A paper presented at a conf. on Rapid Rural Appraisal. 4-7 Dec. 1979. IDS, Univ. of Sussex, Brighton, UK.

______. 1979b. Farming systems research (FSR). CIMMYT, Nairobi, Kenya. (Unpublished).

CRIA Cropping Systems Working Group. 1979. [Draft] Network methodology and cropping systems research in Indonesia. CRIA, Bogor, Indonesia.

IRRI. 1974. An agro-climatic classification for evaluating cropping systems potentials in Southeast Asian rice growing regions. IRRI, Los Banos, Philippines.

Jodha, N.S., M. Asokan, and J.G. Ryan. 1977. Village study methodology and research endowments of the selected villages in ICRISAT's village level studies. Occ. Paper 16. Econ. Prog., ICRISAT, Hyderabad, India.

Thom, D.J. 1978. Human resources and social characteristics. *In* Kenya marginal/semi-arid lands pre-investment inventory. Rep. No. 6. Consortium for Int. Dev., Tucson, Ariz.

USDA Soil Survey Staff. 1975. Soil taxonomy: a basic system of soil classification for making and interpreting soil surveys. Agric. Handb. 436. USDA Soil Conser. Ser. U. S. Government Printing Office, Washington, D.C.

OTHER REFERENCES

Benchmark Soils Project. 1979. Development of the transfer model and soil taxonomic interpretations on a network of three soil families. *In* Benchmark Soils Project Progress Report 2, Jan. 1978-June 1979. Dep. of Agron. and Soil Sci., Univ. of Hawaii, Honolulu and Dep. of Agron. and Soils, Univ. of Puerto Rico.

Buol, S.W., F.D. Hole, and R.J. McCracken. 1973. Soil genesis and classification. Iowa State University Press, Ames, Iowa.

Calkins, P.H. 1978. Why farmers plant what they do: a study of vegetable production technology in Taiwan. p. 74-78. *In* AVRDC Tech. Bull. 8. Shanhua, Taiwan, Republic of China.

FAO. 1976. A framework for land evaluation. Soils Bull. 32. FAO, Rome.

______. 1974. Soil map of the world, Vol. 1. UNESCO, Paris.

Swindale, L.D. (ed.) 1978. Soil resource data for agricultural development. Hawaii Agric. Exp. Sta., College of Tropical Agric., Univ. of Hawaii, Honolulu.

Chapter 5

PROBLEM IDENTIFICATION AND DEVELOPMENT OF A RESEARCH BASE

After the target and research areas are selected, the FSR&D team identifies the specific problems farmers face and develops a research base. A clear identification of problems and opportunities and the development of a research base provide the team with information for subsequent research and development of improved technologies.

This chapter is divided into two parts. The first part deals with the identification of problems and opportunities. We refer to opportunities as well as to problems because they have different connotations. The second part concerns development of a research base and how to collect and manage data useful not only in problem identification but throughout the FSR&D process.

PART 1: IDENTIFICATION OF PROBLEMS AND OPPORTUNITIES

Identification of problems and opportunities is an iterative and dynamic process that continues throughout all FSR&D activities. The process is iterative because research and development results feed back to earlier activities to help improve subsequent research activities; the process is dynamic because accomplishments from one set of actions can be used to set new or revised targets and strategies for the next set of actions.

In the past, the problems of small farmers in the developing countries were often not clear to researchers, development specialists, or policy makers. These professionals failed to appreciate that existing farming systems best met farmers' needs by using farmers' resources and knowledge. This failure often led researchers to develop and others to attempt to extend technologies that were inappropriate for small farmers.

Farming systems research is more apt to design technologies that are appropriate and acceptable to small farmers because the FSR&D approach stresses an understanding of the farming system and the farmers' environment. The FSR&D approach encourages researchers to be open-minded and to revise their conceptualization of problems and opportunities as new information is gathered.

In the first part of this chapter, we will define problems and opportunities, and then discuss the general approach to problem identification, variations in problem focus, understanding the farmers' situation, analysis of problems and opportunities, and setting priorities for problems and opportunities. We close Part I with a summary.

5.1. DEFINITIONS OF PROBLEMS AND OPPORTUNITIES

Problems and opportunities are sometimes like two sides of a coin. Problems could be the result of constraints that prevent farmers from reaching their goals, and opportunities could be the potential for relieving the constraints. For example, farmers may have a problem in not being able to obtain yields comparable to other farmers in similar circumstances. Researchers cooperating with farmers might try to learn what prevents the farmers from doing as well as the other farmers. Once the problem's cause is identified, the opportunity becomes the potential for bringing the farmers' yields closer to those of the other farmers.

Sometimes farmers have problems they do not recognize and yet the opportunities for alleviating problems exist. These opportunities could be in the form of better use of underutilized or misused resources or the introduction of new technologies.

5.2. GENERAL APPROACH

The FSR&D team follows three basic steps as part of problem identification. The team:

- identifies existing farming systems and seeks to understand them and the environment
- identifies problems and opportunities for improving the system, the environment, or both
- sets priorities for research and implementation.

Fig. 5-1 sets out the various parts to the problem identification process. The first preliminary analysis (2.a. in Fig. 5-1) is based on information obtained during the selection of the target and research areas. Sometimes, a target area may be selected because of an overriding problem or opportunity. Such a situation was described in Sec. 4.2.2. where an area in Sumatra was selected because of a serious problem with *Imperata* grass. At times enough will be known about the problem to go directly to planning research activities. More often, the team will need to learn more about the farming systems and the environment of the area. The team usually begins with the collection (2.b. in Fig. 5-1) and analysis (2.c. in Fig. 5-1) of secondary data. Then, the team may know enough about the problems and opportunities to proceed directly to planning research ac-

Figure 5-1. A flow chart for the problem identification and development of a research base. See Figure 3-1 for more details on how the activities in this figure relate to other FSR&D activities, the experiment station, and extension.

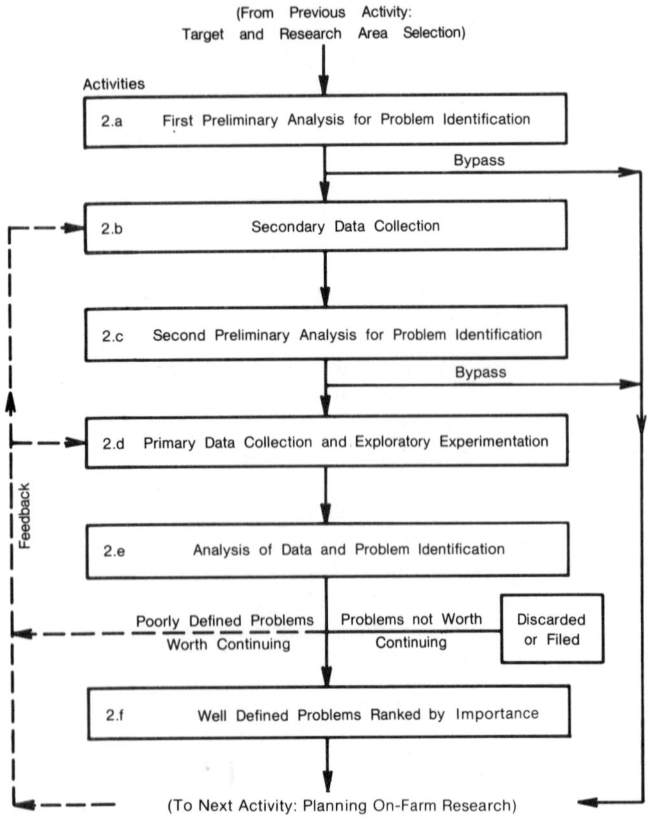

tivities; but, more likely, the team will find it necessary to collect primary data (2.d. in Fig. 5-1).

We offer a word of caution at this point. While bypassing part of the problem identification process is conceptually possible and, at times, can be justified, more often the team is ill-advised to do so. More likely, problems and opportunities that appear to be adequately understood will not be understood and the benefits of the FSR&D approach will be sharply reduced when a team bypasses the problem identification process.

A reconnaissance survey is commonly an initial step in collecting primary data. Some diagnostic analysis, such as soil pH, insect identification, or fertility tests, may be sufficient to identify the causes of some of the more apparent problems. When problems are complex, the team may initiate some experiments quickly to provide better data for identifying problems and opportunities. During the reconnaissance, the team seeks breadth of coverage so that major problems and opportunities will not be overlooked.

By analyzing the secondary and primary data (2.e. in Fig. 5-1) the team divides the problems and opportunities into those that are (1) well-defined, (2) poorly-defined, and (3) not presently worth pursuing. Well-defined problems and opportunities are those sufficiently clear and with benefits potentially great enough that they can serve as the basis for the next research activity. Poorly-defined problems and opportunities require further study and definition. Those not worth pursuing are problems with apparently no practical solution under present conditions or whose payoffs are substantially lower than those realized from other possibilities.

The first pass through the problem identification activity culminates with a set of well-defined problems ranked by criteria as discussed in Sec. 5.6. (2.f. in Fig. 5-1). During the first pass through the problem identification process, information from sources such as formal surveys, detailed studies, farm records, and on-farm experiments is usually not available. Data from these sources come later and materially add to a thorough understanding and penetrating analysis of the fundamental problems farmers encounter. In fact, problem identification, as an activity, continues throughout the FSR&D process.

In this chapter, we concentrate on this initial identification of problems and opportunities because the team often establishes the general approach to the research effort at this point. The methods for identifying problems and opportunities remain essentially the same throughout the FSR&D process, except for a narrowing of focus and a greater emphasis on detail and accuracy.

5.3. VARIATIONS IN PROBLEM FOCUS

While the approach to problem identification is often similar, the details differ. The FSR&D team may focus on any part of the whole farming system or it may concentrate on a predetermined subsystem. However, both approaches require some study of the whole farming system. The first approach requires an understanding of all subsystems and their interactions, while the second approach demands detailed knowledge only of the selected enterprise and its relationship with the rest of the system.

The approach an FSR&D team uses depends partly on its mandate. If the mandate is to increase productivity anywhere in the farming system, the first approach will probably be used. The FSR&D projects in Honduras and Guatemala follow such an approach. Where the mandate requires concentration on specific crops or livestock, the second approach will probably be used. Examples are FSR&D projects undertaken under the auspices of the International Maize and Wheat Improvement Center (CIMMYT) and the International Rice Research Institute (IRRI), in which the researchers focus on maize and wheat, and rice, respectively.

A more specific FSR&D approach is used in countries that have research organizations separated according to commodities. In Indonesia, for example, crop research and livestock research are administered by different governmental divisions. Thus, researchers would gain administrative approval more easily for separate crop or livestock projects than for combined projects.

5.4. UNDERSTANDING THE FARMERS' SITUATION

Effective FSR&D requires the selection of well-defined problems and opportunities that are applicable to a

significant number of farmers who operate in about the same way in response to similar environmental conditions. When such groups of farmers are identified, FSR&D results can be broadly applied. The key to identifying this farmer group is a good description of the farmers' environment and the farming system. For this, the FSR&D team distinguishes between

- environmental factors over which the farm household has little control
- farm management factors over which the farm household has considerable control.

While FSR&D implements change primarily through the latter, the team considers environmental factors too. Therefore, the team examines both to learn

- how the farming system interacts with the environment
- what flexibilities for change exist within the environment
- what farming strategies are likely to succeed
- how to influence local and national decision makers concerning support services and policies.

In learning about these factors, the team will need to use considerable judgment about the type and amount of data to collect. FSR&D budgets are normally limited and time and personnel are usually scarce. Furthermore, misguided approaches to collecting data frequently cause research programs to flounder. Thus, before collecting data, the team should have a reasonably clear understanding of how the data will be used; otherwise, the team should wait until the need manifests itself. An exception is when additional data can be obtained at little extra cost. Then, the researchers should remember that such data need not be analyzed just because they have been collected.

We now turn to a description of the farmers' environment and the farming system and then to analyses of the farming system.

5.4.1. DESCRIBING THE FARMERS' ENVIRONMENT

The farmers' environment can be divided into the physical, biological, economic, and social settings. While many facets of these environments can be used in FSR&D, some are more critical than others. The sections that follow cover what we have found to be the more important facets. Of course, individual situations and the experience of the FSR&D team will dictate which factors to emphasize.

Physical Setting

The more important physical factors are climate, water, and land. Under climate, the research team analyzes primarily temperature and rainfall as recorded at the nearest weather station. For temperature, the average monthly values, and the lows and highs during critical periods in the growing season are important. Rainfall analysis includes the study of monthly means and the time of the beginning and ending of the rainy season. Where irrigation is practiced, stream flow records should be analyzed. For these types of data, the longer the records, the better. The study of land takes location, slope, soil type, whether rain-fed or irrigated, and other factors into account. For additional information on physical factors, see Appendix 5-A on the physical resources of the research area affecting biological production, Appendix 5-B on land types and land evaluation, Appendix 6-A on climatic monitoring of a research area, Appendix 5-P for collecting data during reconnaissance surveys, and Appendixes 7-A and 7-B for data forms for characterizing and monitoring on-farm cropping and livestock experiments.

Biological Setting

The biological setting includes those factors that influence the health and vitality of plants and animals and the quality of harvested products. Most commonly the team identifies prevailing diseases and insects, according to the plants or animals being damaged or harmed. Other biological factors are considered too. Insects can be a serious destroyer of harvested grains. Weeds are particularly important in humid, tropical areas primarily as they interfere with plant growth; but, in some cases, weeds are allowed to grow and are used as animal fodder. Rodents and birds are other biological factors to consider. In some cases, birds—e.g., the African weaverbird—can devastate crops, especially if the crops are grown out of the normal sequence.

Economic Setting

Several aspects of the economic setting influence the farming system. Access to markets during critical periods is particularly important for increasing farmers' production. These critical periods correspond with the need to obtain credit, purchase inputs for production, and the marketing of crops. In marketing, pricing, storage, and reliable transportation are especially important when the farmers grow perishable crops. When investigating these factors, the FSR&D team should obtain information on any services that might be needed to support new technologies—for example, the roles of agricultural chemicals in small amounts or repair services for farmers' equipment. Other information that the team may need to gather concerns such items as processing facilities for farm products; channels of information; the performance of the extension service and cooperatives; seasonal wage rates; labor supplies; opportunities for off-farm employment; sources and costs of traction; and government regulations. In Appendix 5-C, we provide more information on economic factors as they relate to markets.

Social Setting

The specific nature of the social factors that the FSR&D team needs to consider is less subject to generalization than the above considerations because conditions vary so widely from location to location. Nevertheless, the team needs to keep a number of social factors in mind—particularly those social factors that influence farmers' acceptance of new technologies. These factors concern societal norms and customs related to land ownership and use, division of labor within society and the family, rights and obligations according to sex and age groups, descent and inheritance systems, and other community norms and customs as they support or restrict individual and cooperative efforts. We discuss the sociocultural environment further in Appendix 5-D.

5.4.2. DESCRIPTION OF THE FARMING SYSTEM

As noted in Sec. 2.3.5., the farming system is a unique and reasonably stable arrangement of farming enterprises that the farm household manages according to well-defined practices in response to the physical, biological, and socioeconomic environments and in accordance with the household's goals, preferences, and resources. These factors combine to influence farm outputs and methods of production. Having touched on environmental issues, we now turn to (1) a description of the household as an integrating factor, (2) the household's resources, and (3) farming enterprises. The FSR&D team needs to understand these aspects of the farming system before attempting to develop appropriate technologies for specific groups of farmers.

The team normally relies on secondary sources and reconnaissance surveys for initially understanding the characteristics of representative farms. These farms should be of sufficient number to justify the FSR&D effort. Usually the team bases its descriptions on the three general categories mentioned above. After problems and opportunities are identified and selected for further study, the team can then seek more detailed information using farm records, surveys, special studies, and other methods.

The general types of data within these three categories are described below.

The Household as an Integrating Unit

The farm household is a key element in FSR&D. Much of what FSR&D attempts to accomplish is the integration of the many factors influencing the farmers' choices of enterprises and methods of production, given the environment and the household's resources. In a sense, the farm family is the ultimate integrator; the FSR&D team is a helper in the process; and FSR&D methodology provides the framework for bringing the farmers and the team together. Consequently, the team needs to gain as much understanding as is practical about the farm family and the way it operates.

A starting point is for the team to distinguish between those families that are "nuclear" and those that are "extended." A nuclear family consists of the parents and their children. The extended family is a grouping together of two or more nuclear families. Types of extended families include not only parents and their married children, but also one man or woman with more than one spouse, members of several generations, or children of only one spouse when married siblings live in the same household. The extended family is common where cultivation practices require a labor force larger than the nuclear family. The extended family's solution to labor problems relies on a large permanent labor force, which must be fed, housed, and clothed over a long time. Such a unit is especially productive if the extended family has sufficient land and skills to maintain itself (Wolf, 1966).

With the family structure in mind, the team could then turn to a study of family decision making. Beal and Sibley (1967) propose a way of viewing decision making by farmers that may aid the FSR&D team in understanding why farmers do what they do. These authors distinguish among farmer characteristics, knowledge, beliefs, attitudes, behavior, and goals.

Characteristics. These are facts about farmers, such as their sex, age, education, literacy, and ethnic background.

Knowledge. This relates to what farmers know; for FSR&D, the team focuses on such items as the farmers' knowledge of alternative management practices, cropping patterns, sources of inputs, information, and markets.

Beliefs. These concern what farmers think is true, whether correct or not, based on the farmers' experiences and common knowledge. Beliefs, in turn, may influence the farmers' attitudes, behaviors, and goals. For example, even though farmers may be treated fairly when seeking credit, the anticipation of being treated unfairly may keep farmers from seeking credit.

Attitudes. These relate to farmers' feelings, emotions, and sentiments and may have a strong influence over farmers' decisions to accept or reject new technologies.

Behavior. Often farmers' past behavior may predict how farmers will react in the future. Behavior relevant to FSR&D concerns past actions such as farmers' work on and off the farm and visits to communities outside the region, marketing practices, use of credit, and adoption of new technologies.

Goals. These reflect what the farm family desires, i.e., what it is seeking. Goals are conditioned by the family's beliefs about what is attainable, as well as its sentiments, and are based, in part, on the norms of the community and farmers' general level of welfare.

Beal and Sibley (1967) used the above concepts for a sociological study of farmers in Guatemala to gain insight concerning which factors would be most important when introducing new technologies to farmers. This is one of the types of special studies that the FSR&D team might undertake itself or commission others to do. Alternatively, the FSR&D team might simply use the foregoing concepts as a checklist when studying the farm family.

We supplement the above description by providing additional appendixes on family decision making. These include (1) Appendix 5-E with additional details on the Beal and Sibley (1967) study, (2) Appendix 5-F with additional thoughts on decision making by small farm families, and (3) Appendix 5-G with a description of "decision trees" used as a method for learning about farmer decision making. This last appendix refers to the work of Gladwin (1979), which is an alternative approach to the statistical methods Beal and Sibley (1967) described.

The Household's Resources

Conceptually, the household's resources can be described as comprising land, labor, capital, and management. Which aspects of these four factors an FSR&D team considers depend on the situation. Below, we list some factors that FSR&D practitioners have found useful.

Land. Characteristics ascribed to the farmer's land, in the broad sense of the term, include

- size of holdings
- fragmentation of the holdings—e.g., whether the holdings are a single unit or are broken into pieces
- ownership—e.g., sole owner, joint husband and wife ownership, communal owner, long-standing tenancy, and short-term tenancy
- permanency of use—e.g., permanent, shifting, or nomadic
- landlord-tenant relationships—e.g., share of crops retained by the farmers, access to milk by the one who tends the herds, and division of inputs as between tenant and owner
- land quality—e.g., soil depth, texture, and presence of toxic substances
- terrain—e.g., slope, terraced or not, and concave or convex cross sections when on hillsides
- water availability—e.g., nearness of ponds or streams for livestock, irrigated or rain-fed farming, and dependability of supply
- location—e.g., access to markets and other services.

Labor. The family's labor includes members of the household who are capable of working and also the family's participation in cooperative efforts. Some relevant characteristics are

- number, age, and sex of the family members
- division of effort among the members according to preferences, individual profits, and customs
- general level of productivity and health
- division of time between farming and other activities—on the farm and off the farm
- extent and nature of cooperative efforts in terms of

obligations to others as well as help from others
- other responsibilities and factors that influence the way farm households allocate their time and effort.

Capital. This factor refers to physical and financial assets that include

- tools and equipment
- buildings and improvements to the land
- livestock and other assets capable of being sold to meet the farmers' needs or wishes
- cash from sale of crops, animal products, handicrafts, and from other sources
- access to credit.

Management. The skills of the household in organizing and carrying out the many farming tasks is a valuable resource. Management represents a considerable asset to the family and determines the household's efficiency in using its land, labor, and capital.

Farming Enterprises

The household integrates the foregoing resources in conducting various enterprises that relate to crops, livestock, and other activities such as processing of farm products, handicrafts, and off-farm employment. Although small farmers frequently engage in many enterprises to meet their varied needs, the FSR&D team usually concentrates on a limited number of enterprises—perhaps four or less. Even though this book emphasizes agricultural activities, those responsible for FSR&D may decide to focus on nonfarm activities. Some characteristics of enterprises to consider relate to

- general agricultural practices—e.g., methods of land clearing and preparation, means of traction, and pest management
- principal crops and cropping practices—e.g., rotations and combinations, varieties, yields, and schedules for such activities as seedbed preparation, planting, weeding, cultivation, and harvesting
- livestock and livestock practices—e.g., breeds, numbers, age and health, feed, yields, uses, and how tended, where, and by whom
- interactions between crops and livestock—e.g., complementarities such as crop residues used for

cattle feed, dairy cows used for traction, and competition among enterprises for land, labor, and capital
- overall cash, labor, and power requirements when the above enterprises are combined into the farmer's system with emphasis on periods of high requirements and restricted inputs
- purchased inputs—e.g., agricultural chemicals by types, amounts, uses, prices, sources, and dependability of supply
- disposal of production through family consumption, farm use, barter and sales, and marketing factors such as amounts sold, seasonal and annual price fluctuations, locations of sale, time spent selling, and means of transportation.

The Asian Cropping Systems Working Group developed a format for describing the existing cropping systems (Zandstra et al., 1981). The Group begins by listing the major crops and varieties produced on each land type within the research area, the time the crops are grown, and estimated yields. Next, the Group lists the major cropping patterns. Finally, the Group enumerates the principal cropping systems and the percentage of farms in the research area that follow each system. With such information, the Group is in a good position to understand the significance of cropping patterns and practices in the area and has the raw material with which to identify opportunities for improvement. We present further details on this approach in Appendix 5-H.

Furthermore, we include forms for recording and summarizing some of the information on the characteristics of farming systems in Appendix 5-I. These forms facilitate

- recording of labor required at various periods during the year, available family and non-family labor, and the costs of hired labor
- determining periods of cash scarcity and sources and costs of credit (Note: Careful attention needs to be paid to cash flows, since inopportune cash shortages may force farmers to sell their products at low prices at harvest time.)
- summarizing power and machinery available on the farm according to whether they are owned or rented, or the services are hired
- gaining an impression of the farmers' technical knowledge and the technical history of the area (Note: Technical knowledge and experience depend greatly on the area's stage of agricultural development.).

5.4.3. ANALYSES OF THE FARMING SYSTEM

Following initial data gathering, the FSR&D team will need to analyze the data it has collected. Possibilities for study of the farming system are to develop conceptual models of the system and hypothesize about ways to improve the system.

Conceptual Models

Study of the interrelationships among the components of a farming system are useful in two ways. One is to help classify the system, which is part of the selection of relatively homogeneous groups of farmers who are to be the focus of the team's efforts. The other is to help understand the system so that the team's efforts will be productive. Conceptual models in their simpler form identify the major components of the system and the links both among the components within the system and between the system and its environment. Later, additional data can be gathered to quantify these relationships and eventually to search for improvements in the system's functioning.

Below is a model in its simpler form developed by McDowell and Hildebrand (1980). We include a similar model with additional details in Appendix 5-J, along with other examples of conceptual models.

Fig. 5-2 shows the McDowell and Hildebrand model for a humid-upland farming system in Asia. This is one of eleven models they have developed for small farming systems in Asia, Latin America, and Africa (McDowell and Hildebrand, 1980). The authors' explanation follows:

> "The box identified as 'Market' represents all off-farm activities and resources (except land); hence it includes products sold or labor going off the farm as well as purchased inputs and household items. The 'Household' is the core of the farm unit. In preparing the models of the systems, labor use, sources of human food, household income, animal feed, and the roles of animals were the main focus. The solid arrows (——►) depict strong flows or linkages (e.g., more than 20 percent of total income arises from the sale of crops, animals, or household-processed products). Broken arrows (– – – –►) are used when sales of crops or animals contributed less than 20 percent of household income, the interchange among functions was intermittent, or there was no routine pattern identifiable . . . . Family labor applied on the farm was identified, but off-farm employment or the amount of hired labor was not quantified except generally and is indicated by broken or solid arrows.
>
> "For most products there is a direct relation to market, absent . . . when the household changes the characteristics of the product before sale (e.g., . . . milk to cheese . . .). Household modification is shown by solid arrows from crop or animal products to household to market."

In Appendix 5-K, we present a further discussion of this model of the humid-upland farming system in Asia. In this same appendix, we also include a discussion on the interactions between cropping and livestock systems that might help in designing research activities for these two enterprises.

While we believe conceptual modeling, as above, improves the understanding of the farming system, just viewing the farming system from a broad perspective helps

Figure 5-2. Humid-upland farming system in Asia, permanent cropping, moderate intergration of crops and animals (animals tethered or herded) [——→ depicts strong flows or linkages among the parts, while - - -→ depicts weaker flows or linkages] (McDowell and Hildebrand, 1980).

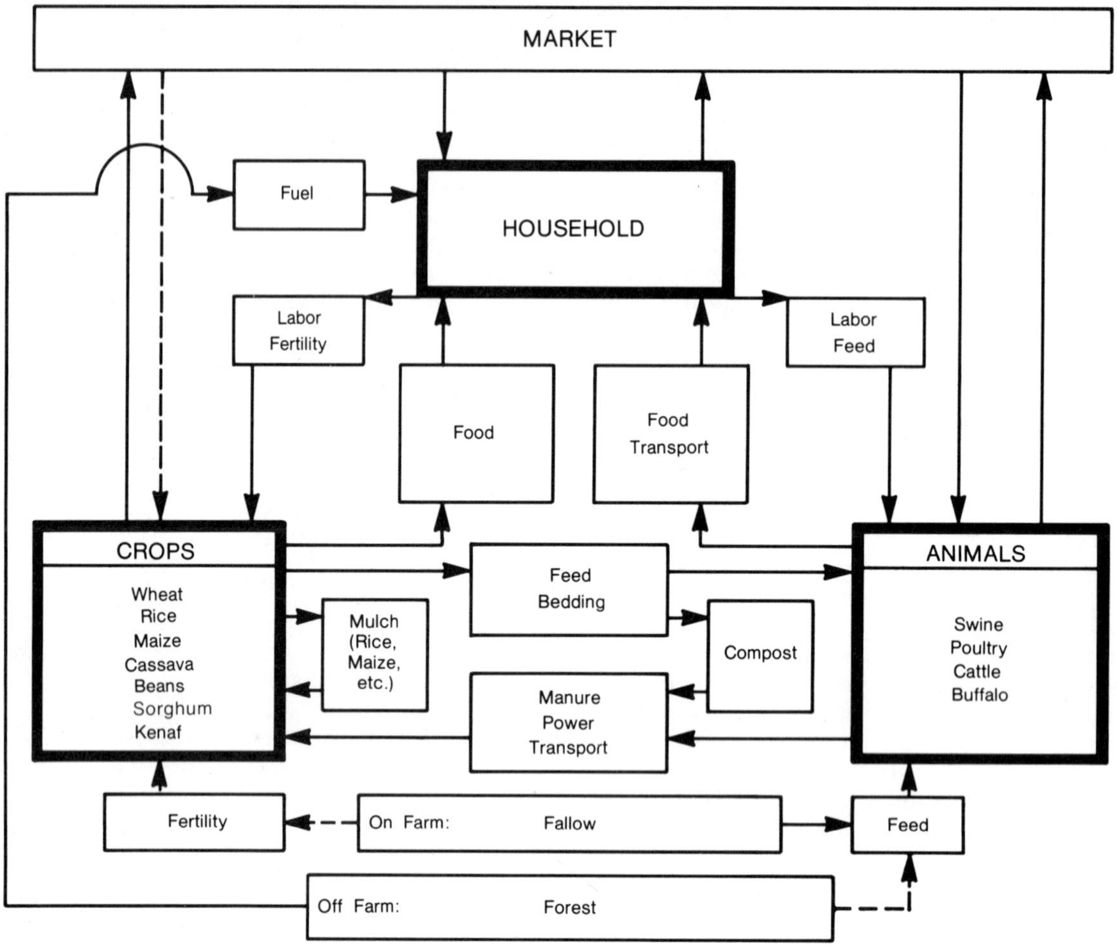

the FSR&D team. The following are examples of how a comprehensive view of the farming system led to the identification of interactions within the system and possibilities for improvement:

Farmers may combine two or more enterprises to make better use of scarce resources. For example, in a dry region of Northern Mexico, farmers are growing maize and castor beans together. Maize plants, with a shallow rooting system, use the moisture in the upper part of the soil more effectively while castor bean plants, with a deeper rooting system, use the moisture in the lower part of the soil more effectively.

Farmers in New Zealand often put both sheep and cattle into the same pasture, because the two species mostly use different parts of the pasture's growth. In addition, the cattle trample and break, and thereby help control the young fern, a weed, that sheep cannot adequately control.

In many cases, two or more farm enterprises use each other's products. For example, Egyptian farmers use the lower leaves of maize plants as feed for their cattle, and use cattle manure to help fertilize the maize.

Certain farmers in Central and South America grow maize and beans together. The maize stalks support the climbing beans and the bacteria on the bean's roots fix atmospheric nitrogen and make the nitrogen available to the maize plant.

Sometimes interrelationships between enterprises are quite intricate. For example, in Taiwan, farmers grow sweet potatoes for food or starch processing. The sweet potato peels are fed to swine, the swine manure is fed to fish, and the muck from the fish ponds is used to help fertilize the sweet potatoes.

Hypotheses for Improving the System

By working with conceptual models of the foregoing type, the FSR&D team should be in a better position to develop meaningful hypotheses for understanding and improving the farming system. These hypotheses will aid the team in sharpening its focus on the more important aspects of the farming system. The following example is hypothetical, but representative of situations described by McIntosh (1980):

An FSR&D team observed great variation in the number of rice crops farmers planted each year, the times and methods of planting, and the varieties used. The team tested several hypotheses regarding the relation of these variations to (1) availability of soil moisture according to rainfall, irrigated conditions, and type of soil, (2) variation in prices of rice sold and seed bought, and (3) the previous crop in the field. After analyzing available data, the team found that price fluctuations did not appear to influence farmers' rice growing methods. The other hypothesized relationships, however, proved meaningful. As illustrated in Fig. 5-3, in rain-fed fields with rather light textured soils, farmers raised only one crop of a transplanted, traditional rice variety. In rain-fed fields with clay soil, farmers planted two crops of early maturing varieties of rice. The first of these crops was dry seeded and the second transplanted. Where year-round irrigation was available, farmers produced three early maturing crops of transplanted rice.

Figure 5-3. The relation of rainfall distribution, availability of irrigation water, and soil type to the variety of rice and timing, method, and number of plantings (Adapted from McIntosh, 1980).

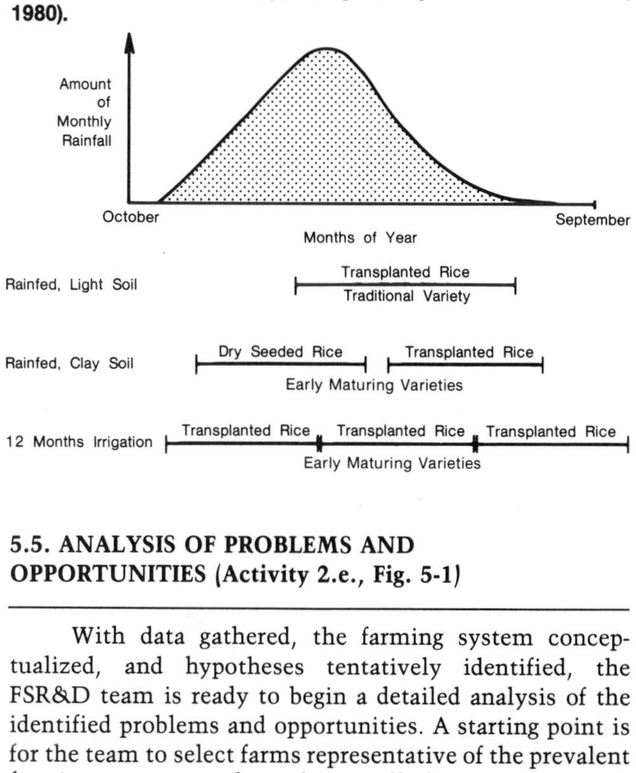

5.5. ANALYSIS OF PROBLEMS AND OPPORTUNITIES (Activity 2.e., Fig. 5-1)

With data gathered, the farming system conceptualized, and hypotheses tentatively identified, the FSR&D team is ready to begin a detailed analysis of the identified problems and opportunities. A starting point is for the team to select farms representative of the prevalent farming systems and see how well they compare with other farming systems. The team will have identified a problem if results from representative farmers are poorer than other farmers. These others may be the better farmers in the research area, good farmers in other areas, or even managers of experiment stations. The latter comparison is valid if the team allows for the better production environment such as better soils, water availability, information, and resources.

The team may identify a problem in the farmers' resource utilization. The farmers may be putting too many resources into one enterprise and too few into another. Suppose they are picking green beans daily, at a return of 25 cents per hour of labor, but are picking tomatoes only once a week, with a labor return of $5 per hour of labor. An opportunity for increased profits might be realized by harvesting tomatoes more often and green beans less often.

A low monetary return to the farmers' labor in an enterprise may, however, not be a problem at all; it may be the result of a purposeful decision by the farmers to mitigate a more serious problem. For example, farmers may be growing a low value, drought-resistant crop of cassava as insurance against the loss of their other, drought-susceptible food crops. The farmers' low return on labor from raising cassava should then be regarded as a cost of overcoming a serious problem—an excessive risk of drought loss. Alternatively, female farmers in areas with a high out-migration of males may plant cassava that, because of its long time to maturity, helps to spread out labor demands (Cloud, personal communication).

In identifying cropping and livestock problems, the team may use performance criteria, such as

- yield per unit of area, animal labor, or human labor
- value of product per farm or per unit area or animal
- net returns above variable cost
- net income to farm resources.

Among the other criteria that the team can use to measure the efficiency of specific resources are

- returns per unit of labor or cash
- amounts of carbohydrates or protein per millimeter of rain.

The team may measure intensity of land use by the indexes for land use intensity or multiple cropping; we show these calculations in Table 5-H-2 (Appendix 5-H). The team may use still other criteria to measure stability of returns over time or across locations. For an additional discussion of performance criteria, see Sec. 6.4.2. in Chapter 6, and Part 2 of Chapter 7.

In trying to identify farmers' problems, the team first looks for the factors that are limiting plant or animal growth and productivity. These factors may be part of the physical environment, such as frequent droughts or floods or poor soil fertility. Problems could also be of a biological nature such as weeds, insects, or diseases, or could be due to management factors. For example, farmers may be planting a wheat variety that tends to lodge when heavily fertilized; or they may be using a cropping pattern that requires more labor at critical periods than they have available.

Next, the team investigates why farmers have not solved these problems. It may find that farmers have good reasons for not using technology that, at first glance, appears to be better than what they are using now. For example, farmers may not use an insect spray that successfully controls a rice stem borer elsewhere in the country. Through investigation, the team might discover that the reasons for not using the spray include one or more of the following:

- The cost of the insecticide is greater for the farmers than the value of the rice they save by using it.

- Farmers do not know how to use the insecticide correctly or may be concerned about its effects on their health.
- Farmers cannot get sprayers when they need them.
- The insecticide is not available in the area or is sold in too large a package.
- The effectiveness of the insecticide has not been proven to the farmers' satisfaction.

Other reasons may explain why farmers cannot overcome their problems. For example, farmers may stay with a strain of chickens that produces just a few white-shelled eggs, instead of shifting to a much more productive strain that lays brown-shelled eggs because they think they have no market for eggs with brown shells.

Sometimes, an existing remedy for a problem is against farmers' beliefs or sentiments. For example, Hildebrand (1977) reported:

> "Among the indigenous farmers, young [maize] plants are treated as a child . . ., so they are almost never knowingly destroyed until they can provide a useful product. Hence, the farmers plant only a few seeds and then reseed if the number of plants drops too low in any hill. The net result is a less than optimum productive population. The usual technical solution is to plant a higher than necessary number of seeds and thin after germination to the desired number of plants per hill. But, for obvious reasons, this meets a tremendous cultural resistance on the part of these farmers, and will probably not be adopted on any large scale in this area."

A solution to one problem may create problems for another part of the farming system. For example, plowing a harvested maize field in the fall would allow earlier seeding in the spring and better utilization of early rains. However, in some places these harvested maize fields serve as cattle pastures in the fall, for which farmers have no substitute.

In another instance, Hildebrand (1979) reported,

> "In an irrigation project in a very dry area of eastern Guatemala, weeds are a severe problem and a definite limit to yields of the vegetables that are raised as the priority crop. However, this area was historically a dual-purpose cattle zone and the farmers still maintain their herds for production of meat and milk. During most of the year, forage is very scarce, so the farmers use the vegetable fields immediately following harvest as a source of feed. Hence, they tend to let the weeds grow to increase the feed supply, even though it knowingly reduces the yield of their principal crop and is counter to the recommendation made by agricultural technicians."

Farmers' abilities to see problems and opportunities in their farming system are often limited by the narrow range of their experiences. This is particularly so if they have little communication outside of their immediate area. In contrast, the FSR&D team tends to have experience in other regions of the country and may be familiar with literature on new agricultural technologies and research. Thus, the team may see opportunities for improving the farmers' system that the farmers do not. For example, farmers may only know the traditional way of farming in their area. They may produce one crop of rice per year, using slow methods of field preparation and a long-season variety of rice. The FSR&D team may be able to introduce new quick maturing rice varieties and herbicides that reduce the required time for field preparation. Together, these new technologies may enable farmers to produce two rice crops per year instead of one. The team may also know of other crops, such as soybeans or sorghum, that could be grown before or after the rice crop. Thus, the team may see opportunities for doubling or tripling the farmers' agricultural output, but perceiving such opportunities is beyond the limited experience of the farmers.

Winkelmann (personal communication) sums it up when he says that farmers have misconceptions. For example, in West Africa they believe that leaf mosaic on cassava is the natural way for the plant to grow. On the other hand farmers are "street wise," meaning they know how to survive in a hostile environment. In contrast the researchers, because of their education and training, can draw from a larger set of possibilities in solving problems. Integrating the capabilities of farmers and researchers is one of the major opportunities for FSR&D.

5.6. SETTING PRIORITIES FOR PROBLEMS AND OPPORTUNITIES (Activity 2.f., Fig. 5-1)

Setting priorities is the final activity in the initial pass through the identification of problems and opportunities. The FSR&D team might set its priorities on the following considerations:

- the seriousness of the problem as viewed by both the farmers and society
- the potential for solving the problem and gaining acceptance of the solution, and the ease of implementing the results
- the importance of the problem in some overall research strategy.

5.6.1. SERIOUSNESS OF A PROBLEM

The seriousness of a problem should be evaluated from the viewpoints of both the farmer and society. The farmer will usually be interested in the severity and frequency of the problem as related to the farmer's household and immediate associates. Decision makers who are responsible to society at large (farmers and others) will be concerned with the extent of the problem throughout the target area, and the long-run interests of the present and future generations.

Where a problem is important to both the farmers and to society at large, then both groups would agree that it should be included as part of the FSR&D team's research agenda. Where the interests of the farmer and society diverge, two possibilities exist. If the problem is in the interests of the farmers but not of society at large, the FSR&D team, being a representative of the government, is advised not to work on the problem. Alternatively, if the problem is in the interests of society and not of farmers, the government has the choice of whether to alter the farmers' environment, including incentives, to bring the farmers' interests in line with those of society or to leave conditions as they are.

These examples should clarify the above.

Commonality of Interests

If the government has concern for a specific group of farmers and the FSR&D team identifies a problem among them, such as damage from flooding or a disease affecting one of their major crops, then a commonality of interests exists. The FSR&D team could include such a problem in the priority list, provided the severity and frequency of the problem are great enough.

In Farmers' Interests but Not Society's Interests

Farmers may be pursuing a slash and burn pattern of farming on erodible hillsides and may be seeking help from the FSR&D team in ways to expand this activity. If the government sees this method as depleting the natural resource, the interests of society would be better served by not responding to the farmers' stated interests. A better solution would be to try to meet the farmers' needs in some other way.

In Society's Interests but Not in Farmers' Interests

The government may view unsettled lands as representing a major opportunity to increase agricultural production and to alleviate land fragmentation in overcrowded areas. If the farmers are unwilling to resettle, the government can upgrade the support services in the new area and offer incentives to individuals or groups of farmers in the hopes of persuading them to resettle and thereby bring the two viewpoints into agreement.

5.6.2. POTENTIAL FOR SOLVING THE PROBLEM

The FSR&D team can evaluate the potential for solving a problem, possible farmer acceptance of the solution, and ease of implementation according to these criteria:

- biological potential
- resource availability
- economic and financial feasibility
- sociocultural acceptability.

At this stage of the FSR&D process, the team tries to gain only a preliminary impression for these criteria. The team will make a more careful and detailed investigation of these factors when planning on-farm research (Sec. 6.2.4.) and during analysis of results (Part 2 of Chapter 7).

Under the heading of biological potential, the team investigates such questions as

- Do the physical and biological conditions in the research area provide opportunities to solve the problem?
- What information on potential solutions is available from experiment stations, farmers in the research area and in other areas, and the technical literature?
- Do the proposed technologies fit into the farmers' existing system?

Under the heading of resource availability, the team examines

- whether available resources are adequate to meet the resource requirements
- whether potential solutions reduce the employment of scarce resources
- whether the employment of under-utilized resources is increased
- whether farmers are able to apply the new technology.

Under economic and financial feasibility, the team tries to determine

- whether the benefits of potential improvements in the farmers' system offer sufficient incentives to interest family members (Note: incentives include coverage of the costs of purchases and any additional labor by the family and provide some crop or livestock surplus to offset the risk and effort of change)
- whether the potential solutions increase or decrease the stability of the farmers' production and earnings
- whether the farmer has sufficient cash or credit to pay for any increase in purchases and whether lenders differentiate between males and females in granting credit
- whether the potential solutions change the farmers' perception of risk through changes in the stability of production and requirements to obtain credit.

Under sociocultural acceptability, the team tries to determine

- whether the community's social and cultural values, norms, and customs help or hinder the acceptance of the proposed solutions.
- whether the farmers' perceptions, beliefs, knowledge, and attitudes facilitate or make more difficult the acceptance of the proposed solutions
- whether field team members have social or cultural values that hamper their working with certain groups or types of farmers
- whether farm family goals are served or altered if the proposed solutions are successful.

5.6.3. IMPORTANCE OF THE PROBLEM IN THE RESEARCH STRATEGY

After problem identification, the FSR&D team may think through a preliminary research strategy, which we call a development path. For example, ILCA (1980) planned a development path for small ruminant production in the Forest Zone of West Africa as shown in Fig. 5-4.

Figure 5-4. Generalized development path for small ruminant production in the forest zone [of West Africa] following completion of initial problem analysis and determination of market potential (ILCA, 1980).

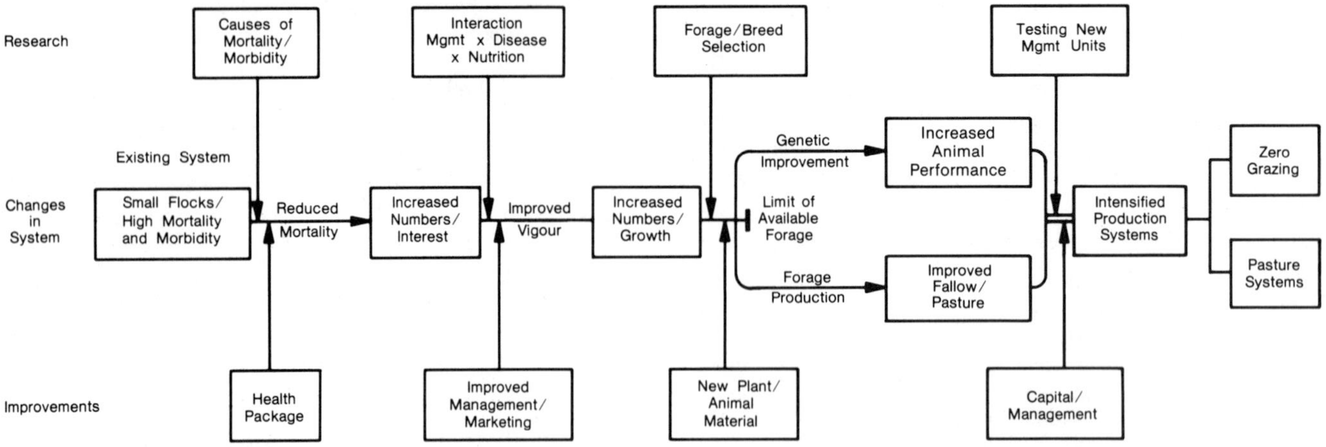

This development path calls for research to reduce disease incidence among the animals, followed by improvements in management, marketing, forage production and breeding stock, in that order.

The relative importance of the identified problems and opportunities in this development sequence is a useful criterion for establishing priorities and establishing the initial direction of the research program. While this activity is important, the FSR&D team will probably want to limit the time spent on such planning because the development path will change, possibly drastically, over time.

5.7. SUMMARY OF PROBLEM IDENTIFICATION

The identification of problems and opportunities serves an important purpose, coming as it does after the preliminary collection of secondary data and the completion of the reconnaissance survey. Such identification guides the FSR&D team in what research to initiate, and possibly what steps to take toward implementation. Since problem and opportunity identification is, at this point, based on preliminary and largely qualitative information, the team may not always start in exactly the right way, but hopefully in the right general direction. As more and better data are collected, the team, keeping an open mind and remaining flexible, should have little trouble in making the necessary adjustments in its research and implementation activities.

PART 2: DEVELOPMENT OF A RESEARCH BASE

The purpose of Part 2 on the development of a research base is to present methodologies for collecting information needed in describing and analyzing farming systems and in developing new technologies. While these methods are first used in identifying problems and opportunities, they are also employed during the rest of the FSR&D process. Our emphasis in this part is on data collected through surveys, observations, and record keeping, rather than from experiments. The latter will be discussed more fully in Chapters 6 and 7. The sections in this part cover general comments on collecting data, assembling secondary data, collecting primary data, combining data collection methods, data management, and a summary.

5.8. COLLECTING DATA: GENERAL COMMENTS

Data collection in a rural setting requires careful preparation and a special perspective. Often, farmers and their associates have never encountered researchers collecting data. Surveys may arouse both curiosity and suspicion—in some cultures researchers asking questions of and about women will arouse suspicion. We describe some of the more important points for collecting data in a rural setting in Appendix 5-L and for collecting data about women in Appendix 5-M.

The methods of data collection to be used in FSR&D are determined by

- the character of the FSR&D activities
- the stage in the FSR&D process at which data collection takes place
- the amount of detail and degree of accuracy considered necessary
- the size of the data base needed
- the resources—e.g., money, time, personnel, and equipment—available for data collection and analysis.

5.9. ASSEMBLING SECONDARY DATA (Activity 2.b., Fig. 5-1)

The FSR&D team can gather secondary data from several sources such as

- the national census—if the data are specific enough
- national archives
- reports of the village administration, local

marketing or credit offices, and the local extension service, irrigation or production program office

- reports on locally conducted experimental or extension demonstration plots
- agricultural publications and reports of university research and consultant studies—such reports often include excellent detail and good summaries of secondary data.

For some types of information, using secondary sources is the most efficient method of collecting data, and the team should explore this possibility before considering other methods.

The availability and quality of secondary data for FSR&D varies by subject matter and country. Secondary data may provide good information on physical environmental factors such as rainfall patterns, soil types, etc., but secondary data are often inadequate on the socioeconomic and biological aspects of the farming system.

The FSR&D team can determine the usefulness of secondary data by four criteria:

- The relevancy and specificity of the data for the particular study. For example, if the team needs weekly rainfall data, monthly rainfall records will not suffice; and biological data for a region may not apply to the research area.
- The clarity of defined terms. For example, a category of data labeled "returns" without further clarification is unsatisfactory. Do "returns" mean gross returns without any deductions or have some or all costs been subtracted?
- The accuracy and reliability of the data. Specifically, the team should check these data by comparing secondary data from different sources, by investigating the methods used in getting these data, and by conducting spot surveys.
- The recency of the data. As a general rule, the team should verify secondary data on socioeconomic topics that are more than 10 years old. This need for recency is much less critical for physical factors, for example, soil classification and rainfall patterns.

We present additional guidelines for judging secondary data in Appendix 5-N.

5.10. COLLECTING PRIMARY DATA (Activity 2.d., Fig. 5-1)

FSR&D teams commonly use several methods of primary data collection. These methods fall into two major groups—informal and formal. The informal methods discussed here include reconnaissance surveys, informal follow-up observations and interviews, and participant observation. Formal data collection includes single and frequent interview surveys, farm record keeping, monitoring, and case studies. The reconnaissance survey is the principal method for obtaining primary data during the initial identification of problems and opportunities. The other methods tend to follow the reconnaissance survey.

5.10.1. INFORMAL METHODS

Informal methods refer to surveys undertaken without questionnaires. This section begins by giving some of the advantages and disadvantages of informal methods, discusses approaches for conducting informal surveys, and closes with descriptions of reconnaissance surveys and participant observations. Reconnaissance surveys occupy a strategic position in FSR&D and include observations, discussions, and sometimes collection of physical data. Participant observations provide a means for gathering data over time, but, to date, have not been widely integrated into FSR&D methodology.

Advantages and Disadvantages of Informal Methods

During initial problem identification, the FSR&D team customarily relies heavily on informal methods to quickly gather broad-gauged data about farmers and local conditions. For example, informal interviews allow farmers and others to express their experiences without excessive structuring by the interviewer. This approach allows both the interviewer and the interviewee to pursue topics of interest freely and in depth. When interviews are conducted in a relaxed and friendly manner, the researchers and farmers will have a chance to become better acquainted.

Informal interviewing also gives the FSR&D team an opportunity to become acquainted with farmers' words, concepts, and ideas. This should lead to a much deeper understanding of farm families, their farming systems and environment, how they reason, and their decision-making process.

The team can use informal observations to check farmers' answers to questions administered in formal surveys. By verifying the accuracy of observable facts, such as pests and planted crops, the researcher can judge if the farmers understand the questions being asked and are accurately answering the questions.

Data gathered informally have some limitations because rigorous methodologies are not followed. For example, the farmers interviewed may have been selected purposively and not randomly. Thus, these data should not be subjected to statistical tests. Furthermore, researchers must be cautious in generalizing from informally collected data. Without a written questionnaire to work from, interviewers may not ask the same questions of all farmers, nor are they likely to ask questions in the same way. Thus, quantification, coding, computer analysis, and summarization become more difficult and the reliability of conclusions is more subject to question.

One way to overcome some of the disadvantages of informal surveys is to combine informal investigations with formal ones. For example, when administering a formal questionnaire, the interviewers may spend a few minutes on informal conversation and questioning about matters connected with the questionnaire. In general, the team usually sets the quantitative data gathered from formal surveys within the context of largely qualitative information obtained from informal surveys and observations. Another way is to carry out a content analysis of the results of informal interviews. This approach measures the substance of the interviews in a quantitative manner, which allows researchers to draw more meaningful information from the results. In Appendix 5-O, we include further details on content analysis.

To summarize, while informal methods have their disadvantages, they also have an important role for FSR&D. They aid the team in (1) quickly learning about farmers and farmers' conditions and (2) obtaining an early appraisal of researchable problems and opportunities. At times these problems and opportunities are so apparent that formal methods of data gathering are unnecessary and thus research can begin immediately and results can be obtained quickly.

Approaches for Informal Methods

Information gathered by informal observations and interviews will differ in both topic and detail, depending on the capabilities of those who do the gathering. Thus, informal data are best collected as a team effort, involving male and female staff from the physical, biological, and social sciences.

Some FSR&D practitioners believe that hypotheses and general guidelines should be developed before interviewing begins. They argue that this requirement forces interviewers to conceptualize the farming system and to take a systematic approach to farmers' problems. Consistency is thereby given to the research process, which prevents different teams from coming up with substantively different results for similar conditions. Others believe that a framework prepared before meeting farmers will predispose team members toward their own ideas, thereby blocking out opportunities to gain new insights. Such FSR&D practitioners prefer to go with a "blank mind." Which approach to take is a judgmental factor and depends on the nature of the team and the situation.

A difference of opinion also exists regarding the desirability of recording information in front of farmers. Some interviewers prefer writing down the information during the interview, if acceptable to farmers. Others believe that this practice restricts the spontaneity of farmers' reactions. Except for recording such items as quantities, names of products, and varieties, these interviewers prefer waiting to record answers and observations until they are out of the farmers' sight.

The Reconnaissance Survey

The reconnaissance survey, also generally known as a quick, informal, or exploratory survey, is a method of data collection that usually follows secondary data collection.

The reconnaissance survey has several distinguishing characteristics:

- It is conducted by the FSR&D team, assisted by commodity and disciplinary specialists, and extension agents.
- It emphasizes the collection of qualitative data.
- It uses informal, largely unstructured interviews combined with observation.
- It is carried out quickly. Sometimes farmer interviews are completed and the results written up in two weeks or less.

Objectives and Approach. The primary objectives of the reconnaissance survey are to provide orientation for the research, and to educate and develop teamwork among the FSR&D personnel. During the survey, the team will become aware of what additional data are needed to better understand present farming practices and to design useful and acceptable changes. The survey involves following up on the largely quantitative data gathered from secondary sources by obtaining information firsthand from farmers. This is the first time in the FSR&D process that the main attention is focused directly on what farmers do and why they do it.

The team members attempt to gain impressions of what enters into the farmers' decision-making with regard to their farming systems such as their knowledge and beliefs, their obligations, their goals, and their perceptions of risk. At this stage, the team formulates hypotheses to explain present farming practices.

The FSR&D team also undertakes reconnaissance surveys for other reasons. A survey early in the FSR&D process helps the team understand farmers' technologies, terms, units of measure, and explanations of why things happen. Besides being essential to on-farm experiments, such information helps the team develop material for subsequent surveys. Thus, the team will be better able to phrase questions in formal surveys so farmers can understand them.

Other objectives of the survey include helping the team to

- establish appropriate research objectives and methods
- decide on the type and size of the sample for formal surveys
- define more precisely the characteristics of homogeneous farmers for whom the research will be undertaken
- publicize forthcoming FSR&D activities
- locate collaborating farmers.

Interviewees. The team should consider interviewing a cross section of farmers such as

- Farmers who hold leadership positions. (They sometimes have useful perceptions about the reasons behind traditional practices and how these practices have changed over time.)
- Farmers identified by the extension service who will often have tried recommended technologies. (They will have information and opinions about problems

and potentials for these and other technologies.)

- Innovative farmers who have successfully developed improved technologies. (They will be valuable sources of information on potential technologies for other farmers in the area.)
- Women farmers who are both members and heads of households. (They can provide information on family decisions and resource allocation in areas where they have major responsibilities such as the care of small animals, garden crops, food processing, storage, and trading (Cloud, personal communication).
- Above all, farmers who are representative of major farming systems in the area.

The team may also want to continue interviews with people contacted previously during research area selection such as (1) extension agents, (2) bankers, (3) buyers of agricultural products, and (4) suppliers of inputs.

Observations. The two basic aspects of the reconnaissance survey method are (1) observation of farms and farm families, and (2) interviewing family members.

Team members, often working in pairs, observe farmers' fields and animals and notice such things as intercropping, spacing between rows, and condition of plants, animals, tools and equipment, and buildings. They may take plant and weed counts, study the rate of disease infestation, and grade pest damage. Nygaard (personal communication) reports that these types of observations and measurements by researchers and farmers were important to the FSR&D surveys undertaken in Syria by the International Center for Agricultural Research in the Dry Areas (ICARDA).

Tables 5-1 and 5-2 and Appendix 5-P illustrate the types of information that field teams have collected and found useful when observing and recording crop and livestock conditions during reconnaissance surveys. With suitable adjustments to meet local conditions, FSR&D researchers can use tables such as these as aids in quickly noting important characteristics about crops and animals.

Team members should also take advantage of their trips to the field to observe other aspects of the farming system. Possibilities include observations of household members and the type and quality of their food. In addition to observations, team members might take soil samples to measure depth, texture, and pH.

Team members' observations can be used for checking the validity of information obtained through farmer interviews. In this regard, the expertise of experienced researchers is particularly useful in identifying the underlying causes of observable and stated problems. Such information will often suggest follow-up questions to reconcile any inconsistencies between farmers' statements and the team's observations.

Interviews. At the beginning of the visit with members of the farm family, the FSR&D team may ask

Table 5-1. Summary of crop and soil observations from the reconnaissance survey.*

A . Crop ______
B. Cropping pattern ______
C. Stage of growth ______
D. Surface soil texture ______
E. Dates of observations ______
F. Observer ______
G. *Top growth characteristics* indicative of possible problems
  1. *Soil moisture relationships*: moisture stress______ excessive moisture ______ salinity stress ______
  2. *Nutritional relationships*: nutrient deficiency ______ soil pH______ soil salinity______ toxicity______
  3. *Pests* (e.g., weeds, diseases, insects)______
  4. *Cultural practices* (e.g., cultivation methods, weed control) ______
  5. *Other* (e.g., varietal and field uniformity, plant population) ______
H. *Root growth characteristics* indicative of possible problems
  1. *Soil moisture relationships* (e.g., color, distribution, depth) ______
  2. *Nutritional relationships* (e.g., toxicities, nodulation) ______
  3. *Pests* (e.g., cutworms, parasites) ______
  4. *Other* ______
I. *Soil characteristics* indicative of possible problems
  1. *Surface* (e.g., crusting, cracking, salts, structure) ______
  2. *Subsoil* (e.g., compact, layer, mottling, structure) ______
  3. *Other* ______
J. *Other* ______

*A table should be prepared for each crop in each cropping pattern.

Table 5-2. Summary of animal observations from the reconnaissance survey.*

A. Kind of animal ______
B. Animal use ______
C. Animal feed ______
D. Dates of observations ______
E. Observer ______
F. *Appearance of the animal*
  1. General condition (e.g., size, amount of fat, skin condition, liveliness) ______
  2. Symptoms of problems (e.g., swellings, growths, discolorations) ______
  3. Other ______
G. *Nutritional problems*
  1. Deficiencies (e.g., protein, minerals, vitamins) ______
  2. Excesses ______
  3. Toxicities ______
  4. Control measures ______
  5. Other ______
H. *Disease and parasite problems*
  1. Diseases or parasites ______
  2. Control measures ______
  3. Other ______
I. *Sanitation problems*
  1. Unsanitary conditions ______
  2. Control measures ______
  3. Other ______
J. *Other* ______

*A table should be prepared for each kind of animal.

wide-ranging questions such as

- How much of the family's harvest will be marketed and when?
- How much labor, credit, and other inputs are available to the family? Do women have difficulty obtaining these inputs? How are the inputs used?
- What are the family's food needs and preferences?
- What are the family's obligations to neighbors, relatives, friends, and religious institutions?

The team may then use the responses to decide what specific topics to emphasize during the remainder of the interview.

After each day's work, the team members gather and discuss such questions as

- What have they learned?
- Have they observed similar phenomena?
- Do they agree on their interpretations of what they have seen?
- What do they need to explore further?

During the reconnaissance survey, the team pays attention to the general practices of farmers in the area and to the variations in practices among the farmers. In each case, trying to understand why variations occur helps the team understand why farmers use certain practices. Variations among farmers should help identify environmental changes across the research area. Identification of trends—e.g., what traditional practices are being discarded and what new practices are becoming common—sheds light on farmer reactions to change such as increasing population or different market conditions.

Toward the end of the reconnaissance survey, the team might estimate the approximate frequencies—e.g., 0-10 percent, 11-25 percent, 26-50 percent, 51-75 percent, and 76-100 percent of farmers who use a particular practice (Byerlee et al., 1980). Finally, if all goes well, the team should be able to identify at least some of the more important constraints and opportunities in the farming system.

We include a description of the *sondeo*, the Spanish term for reconnaissance survey used by the Agricultural Science and Technology Institute (ICTA) in Guatemala, in Appendix 5-Q and we provide further guidance on the kinds of data to collect during reconnaissance surveys in Appendix 5-R.

Participant Observation

One type of informal observation and interview method is participant observation—a method from the social sciences. To use this method, a researcher lives with a farm family for several months, observing and recording what is going on and participating in the family's daily life to the extent possible. The process involves considerable informal interaction with the family.

Participant observation could be used in FSR&D for collecting information on farmer decision making and the factors that influence farmers' decisions such as social obligations, food preferences, and beliefs about plants. Participant observation would also identify interactions between different parts of the farming system and between the farming system and the environment—particularly socioeconomic aspects. By providing an understanding of the context into which changes are introduced, information from participant observation could serve as a basis for hypothesizing the possible effects of alternative actions by the farm family.

Participant observation could be particularly useful in situations where little is known beforehand about the farmers' culture. These observations could be a starting point for recording labor activities according to type, timing, and worker. Furthermore, these observations might provide the background for working with more complex systems.

Two disadvantages of participant observation are the time the method requires and its high cost. However, since participant observation provides an insight into farm families that is not easily obtained in other ways, the team may want to include an element of participant observation in its data gathering activities. Alternatively, a literature search of previous participant observations may provide information about farmers' past practices that helps to explain their present practices. Studies that relate farmer changes with environmental pressures could, in turn, aid the team in predicting farmers' future reactions to changes in the environment and to opportunities for adopting new technologies.

5.10.2. FORMAL METHODS

In FSR&D, formal surveys are generally undertaken to test and otherwise clarify the FSR&D team's reconnaissance and other findings and to follow up on important topics. Verification comes primarily through statistical procedures, but also through additional insights gained by experienced researchers. This section briefly covers some of the characteristics and issues of single and frequent interval surveys, questionnaire design, sampling, implementation of formal surveys, and three variations of formal surveys—farm record keeping, monitoring, and case studies. We provide additional details on data collection in Sec. 6.4.4.

Single Interview Surveys

The single interview survey often follows soon after the reconnaissance phase. A questionnaire is administered to farmers usually selected according to formal sampling procedures. The questionnaire may be used to

- verify and quantify findings of the reconnaissance survey
- collect information on topics of specific interest to the FSR&D team.

For example, in Syria an FSR&D team followed a general crop survey with a single interview survey designed specifically to learn about the cost of harvesting lentils.

The single interview survey is the least costly of the formal methods per unit of usable information and is best used for gathering data on phenomena that

- change slowly—e.g., land tenure or size of farm
- are one-time or infrequent occurrences—e.g., fertilizer purchases, date of planting, and similar farmer practices
- deal with such information as farmer knowledge, beliefs, attitudes, and goals.

Farmers can easily remember and describe this information. Also errors in farmer's statements for such information, called measurement errors, tend to be small.

On the other hand, the team is likely to obtain poor results when using the single survey method to collect data such as the amount of labor used daily for various farm operations during the growing season. Farmers do not remember these kinds of data, called continuous data, very well. In the single survey approach, measurement errors tend to be large for continuous data because the lack of a follow-up survey or observation does not permit direct verification of results.

Another error, called sampling error, occurs when the results of a survey are not representative of the group. Much variation may exist among farmers for any characteristic such as farm size. For a given situation, the larger the sample size, the less the sampling error. Thus, to improve the statistical reliability of results, researchers should take large samples when conducting single interviews. We discuss how to deal with measurement and sampling errors in Appendix 5-S.

Moreover, a single interview is a poor instrument for gathering sensitive information such as farm income, because the interviewers have little chance to build rapport with the farmers. Also, the FSR&D team must be cautious in making cause and effect statements based on data from single interviews. Often, several explanations can be offered for the information obtained from a single interview survey. Finally, in societies in which members do not pool their incomes, income data obtained from heads of households can be incomplete (Simmons, 1976). Husbands and wives may not know the income of the other, nor wish to divulge their own income (Staudt, personal communication).

In the single interview method, the team may take an hour and a half, or even more, of a farmer's time. However, over the life of the project, the single interview approach takes less of a farmer's time than any other formal data collection method. This is an advantage of the method, because team members need to be sensitive to their use of the farmer's time, both in gaining the farmer's cooperation and in not being a burden during those times of the year when the farmer is particularly busy.

Frequent Interview Surveys

The frequent interview method involves collecting data from a limited number of farms on a repetitive basis. This may continue for a year or more. For example, periodic visits to farmers by the National Agricultural Technology Center (CENTA) in El Salvador continued for more than a year. Other frequent interview surveys may cover shorter periods such as the growth period of an annual crop.

The frequency of visits depends on the topics, the degree of accuracy required, and the funds available. For example, a researcher with experience in farm management surveys in Africa prefers twice-weekly interviews even though they are costly and require much time to summarize. He found that daily interviews pestered farmers too much while more than three days between visits resulted in loss of accuracy (Friedrich, 1976).

The frequent interview survey can be used to show progress, trends, and fluctuations over time, and to gather information on specific aspects of the farming system at specific times or places. This type of survey is well-suited for collecting continuous data such as labor and cash flows and food consumption. Because data are continuously recorded, errors due to faulty farmer recall are minimized, as are errors in measurement and observation on the part of the interviewer. This method, because of the frequent contact, has the potential for establishing close rapport with the farmer. Furthermore, when carried out for any length of time, the method provides a mechanism for transmitting information from the researcher to the farmer and back again.

Frequent interviewing does, however, have disadvantages. The more serious ones concern the time to obtain results, staff requirements, and validity of the data. Interviewers have to be in the field continually for long periods, ranging from several months to a year. In addition, data collected by means of frequent interviewing are open to sampling error, since the high cost of the method permits only a small number of farmers to be included in the sample. Since the frequent interview method requires much time to collect the data, it cannot be used when the information is needed quickly.

The team must be aware of the threats to the validity of the data during interviews. For example, farmers may unknowingly try to help the team by responding with the socially correct answers. Or farmers may feel obliged to answer questions about topics for which they have little feeling or experience. In such cases, farmers' responses may not reflect the real conditions on the farm or in the community. Therefore, the FSR&D team should consider the validity issue when designing surveys and in interpreting the results. In Appendix 5-T, we provide some suggestions on how to do this.

Questionnaire Design

A good questionnaire is important to the FSR&D team when it makes a formal survey. The questionnaire links the FSR&D team with the farmers. A sequence for the development of multidisciplinary questionnaires is shown in Fig. 5-5. Each stage is discussed below.

The team begins by reviewing what is known from secondary sources and the reconnaissance survey, and develops a list of additional data needs. Keeping in mind the characteristics of both farmers and interviewers, the team next decides on the type of questions, for example, multiple choice or open ended—i.e., respondents are not forced to select from a set of predetermined answers. All questions need to be carefully and clearly worded, so as to communicate the intended meaning of the inquiry. This is helped when the team uses appropriate local terminology and units of measure.

The questions should be arranged in a logical progression from the farmers' standpoint, starting with simple, more general questions and proceeding to the more specific, difficult, and sensitive areas. Sometimes, breaking the logic or sequence of questioning may be desirable to keep from leading farmers to what they believe are the expected answers. A way to check on the validity of the farmers' responses is to ask the same question in more than one way.

Next the team has to decide on the appropriate layout and length of the questionnaire and to take time to train interviewers. These interviewers then assist in

Figure 5-5. The steps in designing a questionnaire (Adapted from Marketing Research by T.C. Kinnear and J.R. Taylor). Copyright © 1979 McGraw-Hill Book Company. Used with the permission of McGraw-Hill Book Company.

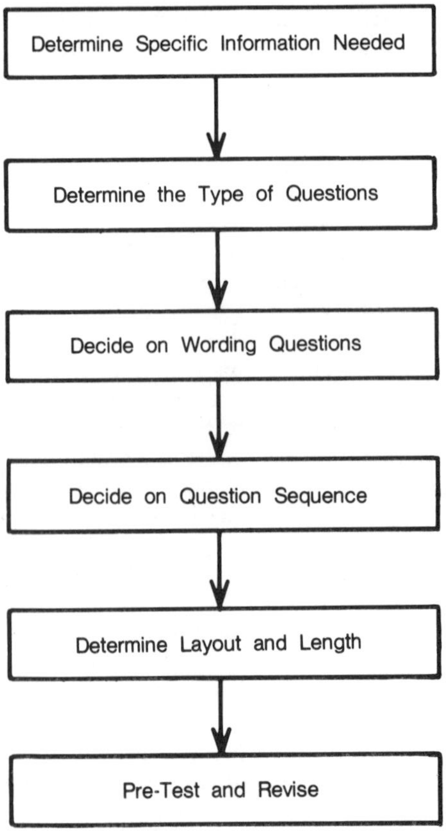

pretesting the questionnaire. Pretesting is a trial run during which the questionnaire is administered to a limited number of farmers who are then asked if they had problems understanding the questions. During this trial run the team also has an opportunity to test its data analysis techniques. The team then revises the questionnaire, in cooperation with the interviewers and perhaps the help of a farmer advisory group. We provide additional details on questionnaire design and examples of poorly versus clearly worded questions in Appendix 5-U.

Designing a questionnaire should not only take into account the best way in which to ask questions, but should also consider ways to increase the efficiency of data collection and processing. In Sec. 5.12., we discuss data management further.

Sampling

If the essential characteristics of farm families and their farming systems in an area were the same, the FSR&D team would not have to be concerned with sampling. The team would only need to select one farm family to find out about the rest of the families. Farm families, however, vary; no one is exactly like any other. Thus, by sampling, the team selects a small group of farm families to represent a larger group. The team uses information from this small group to generalize its findings to all farm families of a particular type in the research area and, therefore, does not have to interview all of the farm families. Sampling saves considerable staff time and money as compared with taking a census and usually produces results adequate for FSR&D purposes. In Appendix 5-V, we discuss sampling procedures at greater length.

Implementation of Formal Surveys

With a questionnaire developed and pretested and a sample drawn, the formal survey is ready to be implemented. Successful completion of the interviews calls for a cadre of trained interviewers and close supervision by the FSR&D team. We elaborate on these requirements and others in Appendix 5-W.

Variations of Formal Surveys

Researchers may normally think of sample selection, questionnaires, and frequency of contact when hearing about formal surveys, but other approaches to formal surveys are possible. One of these, farm record keeping,

has played a key role in some FSR&D projects. Two other approaches are monitoring and case studies.

Farm Record Keeping. A variation of the frequent interview method involves farmers keeping regular, often daily, records of specific farming activities. The FSR&D team initiates farm record keeping once it knows the types of farmers and situations it will be studying. These records generally focus on inputs and outputs associated with specific crops and livestock activities. To keep these records simple enough for farmers and technicians to manage, data on household items such as family income and personal expenditures are generally left out. Market prices and technical descriptions of inputs can usually be obtained as part of special studies or other data gathering procedures—not from farm records.

When records are kept by farmers participating in the on-farm experiments as well as by other farmers in the research area, the records can form the base for evaluating FSR&D's effectiveness. Such information, supported by studies and surveys, also assists the team in interpreting its research results and in planning research for subsequent seasons. If farm records are started early in the project's life, they can be substituted for baseline surveys. When kept for several years, these records reveal the effects of technological changes in the farming systems.

Farm records are especially useful for recording the kinds of data that farmers and their families soon forget. Examples of data recorded include how much work the farm family performs daily on each crop or animal type, when and how long the family hand weeds, and when it irrigates certain fields. In Appendix 5-X, we provide additional details on the types of data to collect and the approach taken to record keeping by ICTA in Guatemala.

If well managed, farm record keeping is less expensive and more accurate than the other methods of collecting data over a long period of time. However, obtaining reliable data from farm records requires the FSR&D team, usually the technical assistants, to regularly and carefully check these records. These assistants generally live in the research area during the life of the project. While making their visits, the assistants should observe the farming system and report their findings to the rest of the team. In Guatemala, technical assistants with ICTA have been able to oversee from 40 to 50 farm records per season.

Farm record keeping normally requires some literacy. However, two methods have been devised to include illiterate farmers in this process. One is to have literate members of the farm household, usually young children, help in keeping the records. The other method involves forms with pictures and symbols. Hatch (1980) designed such a form which we also include as part of Appendix 5-X.

Monitoring. Another variation of formal surveys is monitoring. In this section we will report on three types of monitoring useful to FSR&D. These three relate to obtaining data on climate, recording data from on-farm experiments, and gathering information on livestock systems.

Climatic data need to be monitored as part of planning on-farm experiments and interpreting the results. Such data are used to locate experiments according to different climatic conditions. Also, such data are needed for judging whether the climatic conditions prevalent during the experiments represent typical or atypical conditions.

The FSR&D team generally collects data on rainfall, temperature, wind, and sometimes solar radiation. This information should be summarized weekly and monthly by the team's technical assistants. Then senior members of the field team should review these data periodically. If the assistants are having difficulty preparing the weekly summaries, the team might consider omitting some of the items, or reducing the number of locations where measurements are taken. In Appendixes 5-A and 6-A, we provide more information on climatic monitoring.

Monitoring of cropping and livestock experiments includes two types of observations—those about crop and livestock performance and those about farmers' management practices. This monitoring requires field team members to visit the experiments, particularly farmer-managed tests, to record farmers' activities and plant growth or animal performance during the experiments. Such monitoring is particularly important for farmer-managed tests, because the team needs to keep track of what the farmers do and what results they obtain. For researcher-managed and superimposed trials, the team obtains the necessary data as a natural part of its experimental activities.

Observations of crops and animals, for example, include those that indicate symptoms of stress, deficiencies, toxicities, pest and disease infestations, plant stands, and changes in animal health. The researcher records farmers' activities as they relate to such factors as planting dates and animal feeding rates. The results from such monitoring help the team to (1) identify production problems and (2) understand farmers' management practices.

ILCA uses monitoring techniques to observe how livestock production systems change with the introduction of production-oriented projects. This method of data collection complements ILCA's other work such as identifying constraints and carrying out detailed analyses. ILCA (1978) comments on the purpose of its monitoring activities as follows:

> "It will also be important to examine the response of traditional systems to development processes. Indeed, the monitoring of ongoing development programmes needs to receive a high priority, since these programmes represent unique experiments which can never be reproduced in the confines of a research station. If not given early attention, a great volume of information crucial to future livestock development will be lost. At first these studies are likely to be mainly in eastern Africa, where existing development programmes already affect a wide range of pastoral societies, though they would be selected also for their wider relevance to Africa as a whole."

In Appendix 5-Y we provide additional information

on ILCA's approach to the monitoring of livestock systems.

Case Studies. The case study approach, a special form of the frequent interview survey, involves an in-depth analysis of a small number of farms selected because of their representativeness of farming systems in the research area. Formal interviews, combined with observation and informal discussions with the farm household are repeated regularly, sometimes for an entire year.

The case study method is particularly suited for investigating the whole farming system and the interrelationships among its parts and between the system and its environment. One output of such a study is an integrated model of the farming system, showing, in quantitative terms, flows of money, materials, energy, and information (Hart, 1979). Such information is useful in predicting probable repercussions of changes in any part of the system.

The primary advantage of the case study is the accuracy and detail of the data. This in-depth view can be most helpful in interpreting the data gathered on many farms during reconnaissance and single interview surveys. However, the method is costly and time consuming. In addition, as with the frequent interview survey, sampling errors can occur, since few farms are usually studied. In Appendixes 2-A and 5-J we discuss Hart's (1980) case study work in Honduras.

5.11. COMBINING DATA COLLECTION METHODS

FSR&D teams use several methods to gather data. Each method is limited in the quality and type of information it can provide. Achieving the right combination of methods involves a careful study of research needs and resources. As different kinds of data are often needed at different stages of FSR&D, the methodologies used should be reevaluated periodically. A carefully developed combination of data collection methods can take advantage of the best points and minimize the drawbacks of each.

Following are some possibilities:

- A reconnaissance survey is almost always followed by one or more of the other data collection methods.
- Informal observation and casual interviewing should accompany formal data collection methods.
- The single interview survey is often combined to good advantage with frequent interview surveys, farm record keeping, or case studies.

Concerning this last point, a single interview survey at the outset minimizes the delay in moving from the descriptive phase to the problem solving stages of FSR&D. Much of the information, however, may not be detailed enough for later research. Following the single interview survey, one or more of the other methods can provide the required detail. At times, these in-depth studies uncover data that contradict the information gathered during the reconnaissance or single interview surveys. The FSR&D team should take this information into account.

To overcome the criticisms of the single and frequent interview methods, Norman (1976) suggested two levels of sampling. The first sample would be a large one involving the collection of data that change slowly or represent infrequent occurrences. Such a sample, which could be used at infrequent intervals, would minimize sampling errors and would not involve large measurement errors. The second level of sample would be a much smaller one, in which the team concentrates more on such continuous data as labor or fertilizer input for each field. Measurement errors would be minimized in this second sample through frequent interviewing and direct measurements. However, because of its small size, the second sample may contain a high degree of sampling error. In Appendix 5-S, we present the discussion of Norman (1976) on this method of two levels of sampling in greater detail and how he used it in a study in Northern Nigeria.

We also include another example in Appendix 5-S, in which Guceliogiu (1976) combined record keeping and frequent interviews in a household income and expenditure study in Turkey.

5.12. DATA MANAGEMENT

The objective of data managment is to put collected data as quickly as possible into forms useful for analysis, while at the same time minimizing the chance for errors. Too often, collecting data is considered an end in itself and questionnaires are designed without sufficient attention paid to how the data are to be analyzed. Because the FSR&D approach emphasizes quick results, increasing the efficiency of data collection and analysis is important.

Norman (personal communication) mentions that ICRISAT is using a form in Upper Volta that illustrates the advantages that can be gained from improving the linkage between data collection and data processing. For example, instead of collecting labor data for different fields on one form and seed inputs for those fields on another form, all data for each field are recorded on one form. Data are placed on different parts of the form depending on the operation undertaken. In addition to increasing the efficiency of data analysis, this method provides a good check on whether each operation has been fully carried out.

In Appendix 5-Z, we present some general methods for collating and tabulating data. This appendix summarizes the advantages and disadvantages of three common methods for data preparation—i.e., tabular sheets, sorting strips, and computers. Also included is a section on programmable calculators.

Data analysis is preferably done by the field team. This team is usually in a better position to evaluate the implications of the data than researchers in a distant headquarters. If, however, the data have to be sent away, the field team should take precautions against the data's loss. Thus, the team might want to keep the original data in the field office and send copies for the analysis.

5.13. SUMMARY OF RESEARCH BASE DEVELOPMENT

In the development of a research base and in carrying out the FSR&D process, both secondary and primary data and informal as well as formal methods of data collection are important. The role of each of these depends on such factors as the specific country, the character of FSR&D, the stage in the FSR&D process, and the available resources. In the early research stages, informal methods are generally preferred. They are more effective in (1) establishing rapport between the FSR&D team and farmers, (2) developing team cooperation, and (3) providing the initial orientation for problem and opportunity identification, research, and policy implementation. Gathering data in a more systematic and quantifiable way—i.e., by formal methods—tends to become more important in the later stages of research and implementation.

CITED REFERENCES

Beal, G.M., and D.N. Sibley. 1967. Adoption of agricultural technology by the Indians of Guatemala. Rur. Soc. Rep. No. 62. Dep. of Soc. and Anthro. Iowa State Univ., Ames, Iowa.

Byerlee, D., M.P. Collinson, R.K. Perrin, D.L. Winkelmann, S. Biggs, E.R. Moscardi, J.C. Martinez, L. Harrington, and A. Benjamin. 1980. Planning technologies appropriate to farmers: concepts and procedures. CIMMYT, El Batan, Mexico.

Friedrich, K.H. 1976. "Farming type areas" useful in sampling. p. 40. *In* B. Kearl (ed.) Field data collection in the social sciences: experiences in Africa and the Middle East. Agricultural Development Council, New York.

Gladwin, C.H. 1979. Cognitive strategies and adoption decisions: a case study of nonadoption of an agronomic recommendation. *In* Economic Development and Cultural Change 28:1:155-173.

Guceliogiu, O. 1976. Be prepared to yield to the weather. p. 146. *In* B. Kearl (ed.) Field data collection in the social sciences: experiences in Africa and the Middle East. Agricultural Development Council, New York.

Hart, R.D. 1980. One farm system in Honduras: a case study. *In* Activities at Turrialba 8:1:3-8. CATIE, Turrialba, Costa Rica.

______. 1979. An ecological systems conceptual framework for agricultural research and development. Presented at an Iowa State Univ.—CATIE—IICA Sem. on Agric. Prod. Sys. Res. 19 Feb. 1979. CATIE, Turrialba, Costa Rica.

Hatch, J.K. 1980. A record keeping system for rural households. MSU Rural Dev. Paper No. 9. Dep. of Agric. Econ., Michigan State Univ., East Lansing, Mich.

Hildebrand, P.E. 1979. Generating technology for traditional farmers—the Guatemalan experience. Presented at the 9th Int. Cong. of Plant Prot. 5-11 Aug. 1979. Washington, D.C. Socioeconomía Rural, Sector Público Agrícola, ICTA, Guatemala.

______. 1977. Socioeconomic considerations in multiple cropping systems. Presented at the 16th Ann. Reunion of the Board of Dir., IICA. 18 May 1977. Santo Domingo, Dominican Republic. Socioeconomía Rural, Sector Público Agrícola, ICTA, Guatemala.

ILCA. 1980. ILCA the first years. ILCA, Addis Ababa, Ethiopia.

______. 1978. The monitoring programme: report to ILCA's programme committee. ILCA. Addis Ababa, Ethiopia.

Kinnear, T.C., and J.R. Taylor. 1979. Marketing research: an applied approach. McGraw-Hill Book Co., Inc., New York.

McDowell, R.E., and P.E. Hildebrand. 1980. Integrating crop and animal production: making the most of resources available to small farms in developing countries. Presented at Bellagio Conf. 18-23 Oct. 1978. The Rockefeller Foundation, New York.

McIntosh, J.L. 1980. Cropping systems and soil classification for agrotechnology development and transfer. *In* Proc. Agrotech. Transfer Workshop. 7-12 July 1980. Soils Res. Inst., AARD, Bogor, Indonesia and Univ. of Hawaii, Honolulu.

Norman, D.H. 1976. Factors that influence recall. p. 67-68. *In* B. Kearl (ed.) Field data collection in the social sciences: experiences in Africa and the Middle East. Agricultural Development Council, New York.

Simmons, E. 1976. Economic research on women in rural development in northern Nigeria. Overseas Liason Committee, American Council on Education, Washington, D.C.

Wolf, E.R. 1966. Peasants. Prentice-Hall, Englewood Cliffs, N.J.

Zandstra, H.G., E.C. Price, J.A. Litsinger, and R.A. Morris. 1981. A methodology for on-farm cropping systems research. IRRI, Los Banos, Philippines.

OTHER REFERENCES

DeBoer, A.J., and A. Weisblat. 1978. Livestock component of small-farm systems in South and Southeast Asia. Presented at the Bellagio Conference. 18-23 Oct. 1978. The Rockefeller Foundation, New York.

De Tray, D.N. 1977. Household studies workshop. *In* A/D/C Seminar Report No. 13. The Agricultural Development Council, Inc., New York.

Dillon, J.L., and J.B. Hardaker. 1980. Farm management research for small farmer development. FAO Agric. Ser. Bull. 41. FAO, Rome.

Duff, B. 1978. The potential for mechanization in small farm production systems. Presented at the Bellagio Conf. 18-23 Oct. 1978. The Rockefeller Foundation, New York.

Epstein, T.S., and D.H. Penny. 1972. Opportunity and response: case studies in economic development. C. Hurst & Company, London.

FAO. 1978. Report on the agro-ecological zones project, volume 1: methodology and results for Africa. World Soil Res. Rep. 48. FAO, Rome.

______. 1977. Land evaluation standards for rainfed agriculture. World Soil Res. Rep. 49. FAO, Rome.

______. 1976. A framework for land evaluation. Soils Bull. 32. FAO, Rome.

Friedrich, K.H. 1977. Farm management data collection and analysis. FAO Agric. Ser. Bull. 34. FAO, Rome.

Gilbert, E.H., D.W. Norman, and F.E. Winch. 1980. Farming systems research: a critical appraisal. MSU Rural Dev. Paper No. 6. Dep. of Agric. Econ., Michigan State Univ., East Lansing, Mich.

Hart, R.D. 1979. Agroecosistemas: conceptos básicos. CATIE, Turrialba, Costa Rica.

Lowdermilk, M.K., W.T. Franklin, J.J. Layton, G.E. Radosevich, G.V. Skogerboe, E.W. Sparling and W.G. Stewart. 1980. Development process for improving irrigation water management on farms: problem identification manual. Water Mgmt. Tech. Rep. 65B. Water Mgmt. Res. Proj., Eng. Res. Center, Colorado State Univ., Fort Collins, Colo.

Odum, E.P. 1971. Fundamentals of ecology. Saunders, Washington, D.C.

Payne, S.L. 1951. The art of asking questions. Princeton University Press, Princeton, N.J.

Yang, W.Y. 1965. Methods of farm management investigation. FAO Dev. Paper No. 80. (revised). FAO, Rome.

Chapter 6
PLANNING ON-FARM RESEARCH

After having identified the more relevant problems and opportunities confronting specific groups of small farmers in the research area, the FSR&D team is then ready to plan the research program. This effort should produce a work plan for research leading to improvements for the identified groups of farmers. Simply, the team uses its knowledge of these farmers and their farming systems, the environment, and available technologies for planning a suitable research program for the farmers' and society's needs. The potential for improvement becomes the basis for setting research objectives, selecting research activities and methods, coordinating the efforts of experiment station staff and other supporting organizations, and outlining the FSR&D team's tasks and responsibilities.

The FSR&D team may wish to follow the activities listed below in establishing its research program. This approach is general and would be followed most closely when first initiating research in an area. Other approaches are possible, since planning is highly personalized and depends on local conditions. Moreover, as the team gains experience in an area, the team may give some elements only cursory attention or even skip them.

The planning activities we propose involve

1) laying the groundwork for on-farm research
2) making preliminary analyses of on-farm experiments
3) considering alternative research activities and methods
4) finalizing plans for on-farm experiments.

FSR&D leaders can facilitate this process by holding one or more regional workshops prior to the season's or year's activities. At these workshops, FSR&D members discuss and agree on the research objectives, approaches, and details. The workshops should be held soon enough to allow the field staff to be selected and the field teams to be organized before the season starts. Attendees to the workshop include FSR&D staff from the national and regional headquarters, the FSR&D field teams in the region, specialists from other organizations, and representatives of local groups.

Before discussing these activities, we need to stress that this chapter is a companion to Chapter 7 because research is first planned and then executed and analyzed. We use "on-farm research" in the titles of these two chapters to emphasize on-farm research procedures. Procedures for other types of research, including those on the experiment station, are well-established and generally understood. As Fig. 6-1 shows, the sequence of these two chapters starts with planning on-farm research (Activity 3), moves to on-farm research (Activity 4.a.) and ends with analysis of research results (Activity 4.b.). The feedback lines indicate the iterative nature of the approach. That is, the analysis and extension of results feed back to planning on-farm research and other FSR&D activities. With each looping of activities, more information is gained, problems and opportunities are better defined, and increasingly more specific research is undertaken. For clarity in Fig. 6-1, we do not show the linkages with the experiment station or extension. These linkages were shown in Fig. 3-1.

6.1. LAYING THE GROUNDWORK

In arriving at a suitable plan of on-farm research for a region, FSR&D teams need to undertake several activities. These include

1) reviewing priority problems and opportunities
2) appraising the organization's capabilities and resources
3) appraising present technologies
4) setting assumptions about near-term conditions
5) categorizing and setting research priorities
6) developing hypotheses for testing
7) establishing research collaboration.

The conclusions the teams draw from these activities are essential for the next activity, which is for the teams to prepare preliminary analyses of alternative on-farm experiments.

6.1.1. REVIEWING PRIORITY PROBLEMS AND OPPORTUNITIES

The FSR&D team begins by carefully reviewing the priority problems and opportunities selected during Activity 2. This review will probably produce a modified—and smaller—set of problems and opportunities from those selected during problem identification. By considering fewer possiblities and having more time for investigation—than during the reconnaissance phase—the team sharpens its analysis by focusing on the most promising possibilities.

Figure 6-1. A flow chart for planning and executing on-farm research and analysis. See Figure 3-1 for more detail on how the activities in this figure relate to other FSR&D activities, the experiment station, and extension.

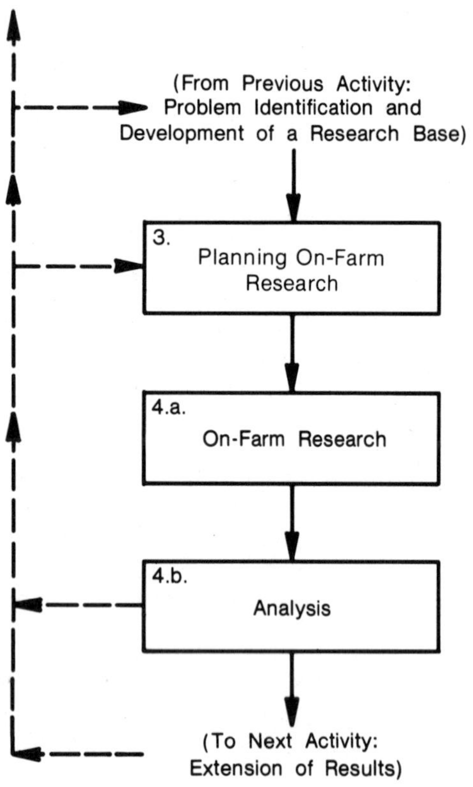

In making this review, the team should seek help from appropriate specialists whenever it lacks the appropriate experience. However, by participating in target and research area selection and by its initial identification of farmers' problems and opportunities, the team should be reasonably prepared for this planning task. The team will have gained an understanding of the farmers' objectives, the prevalent farming systems in the area, the environmental characteristics, and any generally supported research strategy for the area's farmers.

6.1.2. APPRAISING THE ORGANIZATION'S CAPABILITIES AND RESOURCES

FSR&D team leaders need to consider their capabilities and resources in relation to the tasks confronting them. Specifically, they need to consider the organization's strengths and weaknesses as related to the problems and opportunities to be investigated. When the leaders perceive that the needs for research are substantially above their capabilities, they will probably want to consider approaching the government for additional support. Should additional support not be given, the FSR&D leaders will need to pay particular attention to effectively planning the FSR&D program and the teams' activities.

6.1.3. APPRAISING PRESENT TECHNOLOGIES

In deciding among the alternative possibilities for research and development, the FSR&D team needs to know what technologies are available within the area. Present technologies serve as a good starting point for solving some of the more pressing problems and beginning on-farm experimentation. The team should be able to identify potentially relevant technologies by reviewing reports and data from local experiment stations and from regional, national, and even international research organizations. The team members might also talk with those knowledgeable about the area's research and farming practices. The ideas generated can then be screened through researcher-managed and superimposed trials. In this way, the team seeks to match technological possibilities with the identified problems and opportunities.

At times, the team may be uncertain about farmers' conditions and the suitability of available technology. When this happens, the team could begin with a few farmer-managed tests to learn how farmers react to new situations. When working with farmers in this manner, the team should not subject farmers to much risk and should be sure farmers know the exploratory nature of the experiments.

6.1.4. SETTING ASSUMPTIONS ABOUT NEAR-TERM CONDITIONS

To help in the selection and design of research activities, the FSR&D team needs to make reasoned assumptions about the environment over the next five or six years. Often, the most reasonable assumption is that the future will repeat the past—this assumption is especially true where similar farming techniques have been practiced for centuries.

Occasionally, the team will be able to identify changes that are currently taking place or are reasonably certain to influence the farmers' environment by the time improved technologies are developed and diffused. When changes are occurring or pending, prudence suggests the team estimate, as best it can, the situation most likely to prevail. When the team cannot do this with sufficient certainty, the team must make alternative assumptions. Then, the preferred technologies might be those that apply to the broadest range of conditions.

An alternative approach when encountering uncertainty is to focus the team's research on those technologies applicable to the most certain conditions and delay research on other technologies until conditions can be ascertained more clearly.

6.1.5. CATEGORIZING AND SETTING RESEARCH PRIORITIES

The selected problems and opportunities can next be divided into whether the farmer does or does not have control over the factors necessary for making improvements. We generally consider that the farm household has reasonably complete control, within the limitations set by society, over the family's resources and activities pertaining to the farmers' system. Similarly, we consider that the farm household has little effective control over the environment, except occasionally through organized groups of farmers acting on their own or by influencing the government.

Those planning FSR&D will probably find that most of the practical research opportunities will involve factors over which the farmers have control. The reason for this is that environmental conditions are not easily altered. Consequently, much of the FSR&D effort will be directed to changes in cropping and livestock patterns and farm management by means of on-farm experimentation. Nevertheless, the team should not overlook opportunities to conduct research on the farmers' environment through special studies and by other means.

6.1.6. DEVELOPING HYPOTHESES FOR TESTING

With the foregoing accomplished, the team should be able to state the hypotheses for research on problems and opportunities. Combined, these hypotheses will establish the general direction and the nature of the overall FSR&D effort. For example, a major objective might be to add a second or third crop to the farmers' cropping pattern.

Researchers should form these hypotheses in such a way that (1) the tests of the hypotheses will yield meaningful results for the overall approach and (2) the technologies will be within the farmers' and agencies' capacities to implement. For example, the hypothesis "an earlier planting date for a second crop will increase the farmers' overall output per unit of scarce land" would not be useful if the farmers cannot prepare the land in time. On the other hand, if a means has been demonstrated for quicker land preparation, then the hypothesis about an earlier planting date could lead to useful experiments and conclusions.

6.1.7. ESTABLISHING RESEARCH COLLABORATION

By integrating the various hypotheses about priority problems and opportunities, the FSR&D team should be able to formulate a coordinated research progam. More than likely, the team will need assistance from qualified organizations and individuals in implementing this program. For example, help might be sought in clarifying an animal disease, learning how a water users' organization functions, trying to secure more favorable credit, or finding a solution to a soil salinity problem. Once the key elements of the research program are identified, responsibilities can be assigned to the field teams for on-farm research, to experiment station staff for work on and off the station, and to other research collaborators for studies supportive of the FSR&D effort. The way the FSR&D team goes about obtaining such collaboration depends on its position and authority within the governmental hierarchy.

Before finalizing the research program, these teams will need to gather additional data and subject their findings to preliminary analyses to confirm the reasonableness of the approach. This analysis of on-farm experiments, which we discuss next, is sometimes called an *ex ante* or prefeasibility study.

6.2. MAKING PRELIMINARY ANALYSES OF ON-FARM EXPERIMENTS

An analytical framework can aid the FSR&D team in designing on-farm experiments. This framework guides the team in making preliminary estimates of overall feasibility and includes consideration of physical, biological, economic, financial, and sociocultural factors. The requirements of these analyses help the team focus on collecting the most relevant types of information.

This section introduces the preliminary analyses of on-farm experiments. To begin, the FSR&D team should try to predict the implications of the experiments. By considering a range of possibilities, the team will be in a better position to decide whether to go ahead with the experiments and, if so, how to design them. In the remainder of this section, we explain an approach to preliminary analyses, which also applies to subsequent analyses. The principal distinction between the preliminary and subsequent analyses is that the former is based on generalized estimates of values and the latter is based on experimental results and more precise values obtained from farmers' records, monitoring, and special studies. Details on the methods of analyses are covered in Part 2 of Chapter 7.

6.2.1. ALTERNATIVE SOLUTIONS

Before deciding on a particular approach, the team should consider the alternatives. Failure to consider the better alternatives will produce inferior results no matter how well the experiment is designed. Searching for alternatives takes an open mind, imagination, and considerable judgment and experience. Categories of alternatives to consider include

- increasing output from a given level of resources such as substituting a new technology for an old one—e.g., introduction of a new variety—or introducing better management practices—e.g., planting densities or on-farm conservation of water and soil
- increasing resources—e.g., more land, credit, and cooperative labor—coupled with increasing output enough to justify the increase in inputs
- reducing farmers' risk through more reliable inputs, more uniform outputs, or more stable prices
- reducing the inputs to produce a given output
- increasing farmers' satisfaction in other ways than the above such as increasing family health through better nutrition or working conditions.

6.2.2. FARMERS' CONDITIONS

The farmers' potential acceptance of technological change requires improvements over what the farmers' conditions would be were the changes not introduced. Because most farmers' conditions are stable, this means

studying the farmers' current situations. Such study means understanding, as best possible, the farmers' environment, objectives, resources, enterprises, and management practices.

In addition, some estimate is needed of the degree of improvement sufficient to interest farmers in change. When yield increases are the objective, some researchers use a 30 percent increase as the minimum amount farmers can easily discern and, therefore, are willing to accept. However, this percentage is simply a "rule of thumb" to use until the FSR&D team makes its own estimates. In making these estimates the team should consider the household's relative affluence. Households that are better off and accustomed to change may accept values less than 30 percent. On the other hand, households operating near subsistence will be constrained by shortages of cash and credit and will generally be most concerned about producing a stable food supply and other family requisites. Households near subsistence levels may require both yield increases of more than 30 percent and assurance that the possibility of losses is not great. Alternatively, farmers may think in terms of an increase in profitability (see Minimum Acceptable Return in Sec. 7.7.4.).

With such knowledge about farmers' conditions, the team should be able to predict the acceptability of change based on reasoned estimates about the differences in alternatives. Where the farmers' conditions are changing, the team will need to estimate, as best it can, what future conditions would be like in the absence of any improvements it might propose.

6.2.3. PERSPECTIVES

The team must understand the farmer's perspective because farmers can accept or reject the proposed changes in their enterprises and management practices. In addition, the team should ascertain whether its proposed changes will be in society's immediate and future interests. Simply, the team has this responsibility by virtue of its being a government entity.

When dealing with these two perspectives, the team will usually find that considering the farmer's viewpoint calls for specific information about the technology and the farmer. As the team gains experience by working with farmers in the farmers' fields, it should be able to anticipate with some accuracy farmers' reactions to new technologies. On the other hand, taking society's interests into account is more general and, therefore, more subject to alternative interpretations. For example, the team might judge that the increased yield from using pesticides is large enough to interest certain farmers in the research area. In contrast, the team could encounter difficulty in judging the long-run effects on the environment from the widespread use of these pesticides. For the latter situation, the team might well seek expert advice.

6.2.4. TECHNICALLY VIABLE DESIGNS

In planning experiments, the FSR&D team is advised to work toward technically viable designs rather than toward optimal designs. Optimality does not have much operational meaning within the complexity of farmers' circumstances. On the other hand, technically viable designs can be prepared to raise the farmers' benefits enough to gain their interest. By technically viable designs, we mean those that are responsive to conditions likely to prevail when the technologies are broadly introduced to farmers. To accomplish this, the team will need to look into a range of conditions covering the experiment, farmers' conditions, and the environment.

Physical Conditions

Soil, topographic, water, and climatic conditions should be representative of conditions encountered by the groups of farmers for whom the technologies are being designed. Consider the following illustrative suggestions. If shallow, stony soils prevail, then the team should select these types of soils for the on-farm experiments. If most of the farmers' land is on hillsides, that is where the team should place the experiments. If the farmers plant under rain-fed conditions, the team should do likewise.

Biological Conditions

The team will need to consider the farmers' existing cropping and livestock patterns as the starting point for introducing changes in patterns and in mangement practices. Then, the team should study the biological characteristics of alternatives to learn how they can be incorporated into the farmers' existing system. For example, the team might consider whether the farmers have enough time after harvesting their traditional crops to plant any new short-season crops. As another example, the team should be aware of any livestock's feed requirements if it plans to change plant species in which the amount and quality of fodder is significantly changed. Also, the team needs to consider the farmers' particular pest problems.

Economic Conditions

The team should be reasonably sure that future economic conditions will support the change in technology. For example, the technology's requirements for labor, supplies, and services need to be available, and the output should be acceptable for the family's use or be in demand by others. Also, the potential improvements should interest the farmers.

Financial Conditions

The team will need to compare the monetary requirements for any proposed changes against the farmers' financial resources. In checking on the financial needs of such purchases as supplies and equipment, the team should consider the farmers' reaction to credit, dealing with lenders, and the time required for these transactions.

Sociocultural Conditions

A number of farmer characteristics, which we categorize as sociocultural factors, are implied in the foregoing considerations. The team needs to be sensitive to the influence of the community and prevailing customs on farmers' decisions. Where the community and customs do not exert strong control over the farmer, the team might find considerable flexibility in design possibilities and should take advantage of this flexibility. At other times, the social system will restrict the types of technologies than can be introduced. In either case, we advise the team to be aware of the farmers' sociocultural setting.

6.2.5. ESTIMATING VALUES

The foregoing considerations require the team to estimate both quantities and prices for inputs and outputs. In making these estimates, many analysts use "conservative" estimates. By conservative we mean increasing costs or reducing benefits over what analysts believe are the most likely possibilities. Their purpose is to be more certain that suggested changes will be equal or better than their calculations indicate. In this way, they feel they are protecting themselves from criticism and the farmers from loss. The analysts' inclinations are understandable, but changing costs and benefits in this way obscures the analyses. Analysts are not able to demonstrate the potential of proposals in their clearest light. Consequently, we recommend that the team select values that are as representative as possible of the values they believe will prevail if the technologies are introduced.

After obtaining an unbiased estimate of a new technology's value, the team can apply analytical techniques designed specifically for taking uncertainty into account. For this, the team needs to gather data on alternative quantities and prices over a range of possible outcomes. The team can then use these additional estimates to evaluate how a proposed technology looks under varying assumptions. We present ways for making these calculations under the headings of Risk and Sensitivity Analysis in Sec. 7.7.4.

Finally, for subsistence farmers in particular, many inputs and outputs do not involve cash transactions. Consequently, the reader may wonder how to place values on these items. The solution is straightforward. The team estimates the "opportunity cost" of the input and output and uses this as the market-based value. The opportunity cost is the value of an item in its best alternative use. We will say more about this subject under Net Benefits in Sec. 7.7.3.

6.2.6. EVENTUAL CONSEQUENCES

Finally, the FSR&D team should take the precaution of trying to anticipate the eventual consequences of the changes it proposes. The consequences apply to both farmers and the environment. While precision in predicting the full range of possible effects is unrealistic, the team should try to estimate how specific small farmer groups might be influenced. When the government is unable to protect the interests of the farmers for whom the research is intended, the team may want to work on other technologies that do not threaten farmers' welfare.

We offer two examples to illustrate the point. First, consider the introduction of a new variety that increases yields through the application of agricultural chemicals. If small farmers frequently cannot obtain these chemicals while the larger farmers consistently can, the net result could make the small farmers worse off. They would be worse off should their output remain the same and prices fall because of increased production in the area. The second example, comes from Kusum Nair (personal communication). Nair reported a situation in India in which an irrigation project brought water to the farmers' fields for the first time and thereby made the land more valuable. For one reason or another, the wealthier farmers acquired this land because of its higher value, and those who originally occupied the land were forced to look elsewhere for their livelihood.

6.3. CONSIDERING ALTERNATIVE RESEARCH ACTIVITIES AND METHODS

The FSR&D team can call upon a variety of research activities and methods when implementing an FSR&D approach. An important part of the planning process is to identify which activities to undertake and which methods to apply. Table 6-1 provides a matrix of possibilities. The "X"s indicate the more common application of methods to activities, with the strength of the relationships increasing as the number of "X"s increases. As the table indicates, the strongest relationships are the application of researcher-managed trials for the development of technology, farmer-managed tests for adaptation of technology to farmers' conditions, and surveys and record keeping to learn about the farmers' systems. In this section, we provide additional information on this subject.

6.3.1. RESEARCH ACTIVITIES

We have divided FSR&D into the following activities, which are particularly relevant for cropping systems research: technology development, farmer adaptation, management of the farming system, climatic analysis, and special studies. We also include management of natural resources because farmers' activities are sometimes strongly influenced by the way these resources are managed. Examples include supply of irrigation water, conservation of soil and water on steep slopes, and management of rangelands. Where an FSR&D team considers livestock, special studies might include modeling of livestock systems and monitoring of changes resulting from development programs.

Technology Development

Research on technology development is undertaken to (1) better understand the individual and combination of technical factors affecting plant or livestock production and (2) develop new and improved technologies. This research relates to cropping and livestock patterns, mixed crop and intercrop combinations, cropping pattern management—e.g., use of higher yielding varieties, pest control, fertilizer response, and planting dates—and livestock management—e.g., animal nutrition, pest control, feeding trials, and use of crop residues as animal feed. Technology development may also apply to on-farm water and soil erosion control in rain-fed areas and to various moisture conservation practices for irrigated areas. Some of the literature on FSR&D refers to research on these items as the development of component technologies (TAC, 1978; and Zandstra et al., 1981).

Farmer Adaptation

The FSR&D team uses its understanding of the farming system, background information on technology development, and farmers' suggestions to propose improvements in the farmers' systems. These improvements include such possibilities as introducing new or additional crops or animals into existing patterns, redesigning farmers' management of existing patterns, finding better

Table 6-1. Matrix of FSR&D activities and methods.*

Research Methods	*Research Activities*					
	Technology development	*Farmer adaptation*	*Management of the farming system*	*Climatic analysis*	*Special studies*	*Management of natural resources*
Researcher-managed trials	XXX				X	X
Farmer-managed tests		XXX				X
Superimposed trials	XX	X			X	X
Surveys		X	XXX		X	
Record keeping		XX	XXX		X	X
Monitoring		XX		XX		X
Research station support	XX				XX	

*The "X"s indicate the common application of a research method to a research activity, with the multiple "X"s indicating a higher frequency of application. The absence of an "X", however, does not preclude the use of a method for a particular activity.

methods for storing crops, and so forth. The approach to farmer adaptation of technology is for the researchers to learn how farmers react to introduced change—to what extent the introductions are accepted, modified, or rejected—and to learn why farmers act the way they do.

Management of the Farming System

The FSR&D team usually conducts surveys and sets up record keeping on specific aspects of farming systems to augment the information it gains from problem identification and on-farm experiments. For example, for a given farming system, the team may want to find out about the farmers' year-round use of farm labor, the periodic value of sales of crops and livestock, seasonal feed requirements of livestock, timing of field operations, periodic expenditures for crop and livestock production, harvesting and post-harvesting losses, planting and harvesting dates, and areas devoted to specific crops. The teams collect such information by sampling and recording data across the area and, thereby, obtaining more complete and representative descriptions of farmers' management practices.

Climatic Analysis

Analyzing climatic data is a routine activity for evaluating results from crop and animal experiments. At least one weather station usually is established in the research area to obtain information on rainfall, maximum and minimum temperatures, and relative humidity. Additional rain gauges may be placed at different topographical locations, sometimes on the farmers', schools', or others' properties. Whenever possible, the team obtains detailed information on climate from nearby stations. We present more details on climatic monitoring in Appendix 6-A.

Special Studies

The FSR&D team may decide that special studies are needed when the cause of a production problem is difficult to identify, when a new crop or animal of unknown performance in the area is being considered, or when a management practice is new to the area. These studies may involve biological experiments at the local experiment station to determine the agroclimatic adaptability of new plant species or to learn how to control disease or insect infestations affecting plants and animals. Or the team may need to determine the market's acceptance of a new crop. When a new technology substantially changes the household's activities, a sociological study may be needed to understand how the family and society are likely to react. Where nitrogen fertilization is urgently needed, but fertilizer is too expensive or not available, the team may decide to study alternative possibilities such as the introduction of legumes or better residue management. Sometimes the need for these types of studies comes only after proposed technologies prove unsuccessful or after more has been learned about a particular problem.

Management of Natural Resources

Activities concerning management of natural resources go beyond the single farm to the consideration of the farmers' general environment. The FSR&D team, either itself or with the help of others, seeks to improve environmental conditions for farmers in the target area. For example, (1) improvements may be sought in watershed management, such as soil and water conservation and maintenance of rangelands; (2) improvements may be made to the area's irrigation system; or (3) research may be directed toward solving problems involving a high water table or salt-affected soils.

6.3.2. RESEARCH METHODOLOGIES

A variety of research methods have been taken from various origins and adapted to FSR&D's specific needs. Particularly useful are researcher-managed trials, farmer-managed tests, superimposed trials, surveys, record keeping, monitoring, and experiment station support. We discussed surveys, record keeping, and monitoring as part of formal methods in Sec. 5.10.2. We will now discuss the other four methods.

Researcher-Managed Trials

Researchers manage experiments on farmers' fields to develop appropriate technologies for specific groups of farmers. These trials help the team in a variety of ways. For example, they:

- provide a means for screening available technologies according to their suitability for different types of farmers and conditions
- help the team define the characteristics of the research area more precisely
- may be used to partition the research area according to physical gradients that cannot be recognized visually, such as a change in moisture availability with distance from an irrigation source or across rainfall gradients
- assist the team in recognizing the gap between current and potential yields
- provide an opportunity for the team to work with and learn from the farmers
- give the team the opportunity to experiment with riskier treatments because the farmers' welfare is not at stake
- allow the team time to identify some of the more difficult and less successful experiments before proceeding to farmer-managed tests.

The team uses methods and techniques similar to those at experiment stations. However, by moving off the station, the team is better able to take farmers' conditions into account. The researchers try to simulate farmers' conditions to the extent possible. This can be facilitated through questioning, observing, and having farmers perform much of the work. Nonexperimental variables are generally set to represent farmers' conditions, but may sometimes be set at the level recommended by the extension service. Researchers will often pay farmers for their labor and the use of their land so that farmers do not suffer losses from poor experimental results.

Among the more common types of researcher-managed trials are those designed to (1) investigate cropping patterns; (2) develop better management technology related to such factors as improved plants or animals, pest and parasite control, soil fertility, animal nutrition, planting dates, and crop-animal interactions; and (3) evaluate alternative management practices. As noted above, these trials can also be used to study cropping and livestock responses to alternative environmental conditions. For example, the researchers may wish to learn if soil conditions, such as low pH or excessive salinity, induce nutrient toxicities or deficiencies in plants and animals.

Norman (personal communication) points out that some researcher-managed trials are not part of an FSR&D effort. This occurs when the experiment station staff (1) wishes to learn how a technology responds to a physical or biological environment different from that encountered at the station and (2) is not primarily concerned about how the technology fits specific farmers' conditions.

Farmer-Managed Tests

Farmer-managed tests provide the FSR&D team with an excellent means for evaluating how new technologies fit into the farmers' system and how farmers react to the proposed changes. The changes relate to a wide range of possibilities covering new cropping and livestock patterns, management practices, and changes in resource use. For best results, farmers need to manage these tests using resources normally available to them and without excessive interference from the team. Once the farmers understand the purpose of the tests and the essential elements for conducting them, the team members serve mainly as advisers. In fact, the way farmers alter the tests, together with their reasons, are important test results. Such information can be the basis for modifying the technologies and identifying opportunities for further research.

The team should design the tests in cooperation with the farmers. Plots need to be large enough to permit accurate measurement of the farmers' activities, particularly the use of family labor. In this way, the tests will receive adequate attention from the farmer and the team will be able to observe how the farmers allocate their labor and other resources. The team judges the acceptability of the proposed technologies through farmers' records, observations of farmers' practices, discussions with farmers, measurements of results, and calculations of profitability.

Because these tests essentially belong to the farmers, the team should not introduce patterns and management practices until the team believes the changes will perform at least as well as the farmers' normal practices. Stated differently, these tests are not for research in the early stages of development where the details have not been worked out and outcomes are uncertain. An exception is when farmer-managed tests are undertaken primarily to learn how the farmer and the system respond to change. On these occasions, the farmer needs to know the nature of the experiment and the possibilities of loss.

These tests also have their drawbacks:

- Experimental conditions are difficult to control.
- Tests must be replicated on different fields.
- Resulting coefficients of variation are high.
- The number and complexity of treatments are limited.

Nevertheless, these tests show the researchers how farmers are likely to react to new technologies should they be introduced on a broad scale. Such tests are superior to other methods in revealing how farmers employ their resources. The team learns how adaptable and stable the technologies will be within the research area. In addition, because the farmers sometimes manage the experiments under the supervision of extension workers, the number of experiments can be increased relative to researcher-managed trials.

Superimposed Trials

Superimposed trials combine the methods of researcher-managed trials and farmer-managed tests to examine technologies for a range of conditions. These trials tend to be single factor experiments—e.g., fertilizer treatments—that are superimposed on farmers' ongoing activities or on farmer-managed tests. Because they are single factor experiments and can be worked into these other experiments, superimposed trials are generally less expensive than other experiments.

The trials have a variety of uses, as the following examples indicate:

The trials effectively evaluate the suitability of farmers' cultural practices within and across land and soil types. An example is the evaluation of the effectiveness of a chemical over the range of soil textures within the research area.

The trials can efficiently evaluate appropriate *levels* of technology, for example, the *amount* of fertilizer to be applied rather than whether it should be applied. Such trials apply to various practices—insect and disease control for plants and animals, weed control, soil application of fertilizers, and micronutrient foliar sprays.

The trials are exceptionally well-suited for insect control studies that require large plots or many animals. These conditions are needed to accommodate the drifting of insecticide sprays and to provide a reasonable opportunity for infestation of mobile pests that seldom attack crops and animals uniformly.

The trials are very useful for studying biological response to a new technology over time. This may be important for research on plant or animal insect control. If, for example, the host crop is planted over six weeks, the insecticide treatment should be spread over six weeks so that the growth stage of the plant most susceptible to insect attack will occur during the various growth stages of the pest. Similarly, insects that attack animals can be researched over time to determine the influences of weather and the insects' growth cycle on infestation. Such research can be important in determining the extent of infestation and effectiveness of the control methods.

The trials are good for evaluating simple procedures, such as application of micronutrient foliar sprays, that are unfamiliar to farmers. Because such trials are easily implemented, they interfere little with the farmers' other field operations.

Finally, researchers can use the trials to obtain the farmers' opinions of labor-consuming innovations and possible modifications or adjustments the farmers would make in adopting the practices. This is especially important for the design of farmer-managed tests that involve time-consuming operations. In some trials, researchers may want to determine the time required for the farmer to use the new technology over a relatively large area. Researchers can time alternative farmer practices over a large enough plot to give reliable results and then discuss these results with the farmers. In this way, the researchers should obtain a reasonably good estimate of various factors—e.g., labor costs, timing of activities, and equipment needs—and how the technology ought to be changed. With this information, the team should be in a better position for designing the farmer-managed tests.

Experiment Station Support

Staff and facilities of the experiment station can contribute much to the FSR&D effort. As explained in Sec. 3.4., agricultural research strategies use various combinations of on-farm and experiment station activities. The experiment station is both a source of potentially adaptable technologies for the small farmer and also a place to seek help once problems and opportunities have been identified on the farm.

The stations—whether regional, national, or international—serve a variety of purposes. By using methodologies that are more precise and where nonexperimental variables can be better defined, the stations can provide more statistically reliable and broadly applicable results than on-farm research. For example, stations frequently provide basic information on genetically improved plants and animals, soil fertility, multiple cropping, animal nutrition, and pest control. The national centers offer a point of transfer of improved germ plasm and other information from the international centers to individual countries. Improved germ plasm received from the international centers is usually used in local breeding programs, and in some cases, eventually multiplied for local use. The stations usually have weather monitoring equipment and sometimes evapotranspiration experiments. In addition, they may be convenient locations for housing FSR&D staff, training, holding seminars, and providing other support services.

While FSR&D helps to integrate on-station and on-farm research, FSR&D researchers normally undertake few experiments on the station. These experiments are normally left to those researchers engaged in disciplinary and commodity research as part of the experiment station's program. A reason for this division of effort is that on-farm research usually takes so much time that the field team has little opportunity to do much else.

6.3.3. APPLYING METHODS TO ACTIVITIES

Especially in the early stages of FSR&D, the team will need to consider (1) an effective division of effort among the various research activities, and (2) which research methods to apply. During problem identification, the team will be analyzing the current level of technology in the area. If little technology is suitable for the selected group of farmers, the initial emphasis should be on technology development. As Table 6-1 indicates, researcher-managed and superimposed trials and research station support are the most appropriate methods. But where a large body of technology is suitable for introduction at the farm level, more attention can be given early to farmer adaptation research. In this case, farmer-managed testing is the principal method.

We advise FSR&D's management to secure the best available staff when planning the initial research effort. This is the point where experienced staff with sound judgment helps set FSR&D on the right path.

The proportion of the total FSR&D effort devoted to each activity varies with FSR&D goals, available technology, stage of agricultural development, and the amount of FSR&D experience. As an example, in the first year about 40 percent of the initial effort might focus on technology development research and then the emphasis might gradually shift to farmer adaptation testing over the next two years (Fig. 6-2). In the first year, climatic analyses take more time than later as the team sets up weather equipment and outlines procedures for data gathering, compilation, and calculations. Other activities – i.e., those involving management of the farming system, special studies, and management of natural resources – account for the rest of the FSR&D team's effort.

Figure 6-2. An example of distribution of effort among four research activities for an FSR&D project over a three-year period.

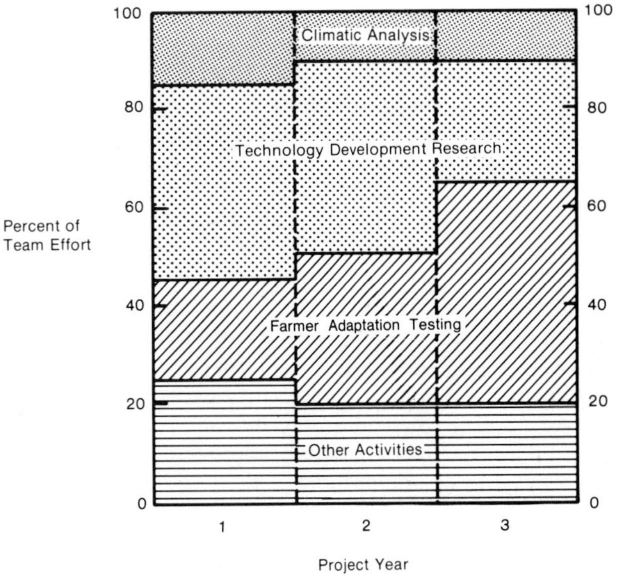

6.4. FINALIZING PLANS FOR ON-FARM EXPERIMENTS

In preparing for on-farm experiments, the FSR&D team needs to (1) decide on the design conditions under which the experiments are to be conducted, (2) search for improvements, (3) establish design standards for the experiments, and (4) gather additional data to finalize the research designs.

6.4.1. DECIDING ON THE DESIGN CONDITIONS

In setting up the experiments, the team needs to decide what the design conditions will be. Important considerations include choices of cropping or livestock patterns, resource availability, and management practices. McIntosh (1980) describes four conditions that the Central Research Institute for Agriculture (CRIA) in Indonesia uses as the basis for designing and conducting experiments with alternative cropping patterns. These conditions and the reasons for choosing each one are listed below:

- farmers' present cropping patterns and management practices – to provide baseline information for comparison with other patterns and practices
- farmers' present cropping patterns with input and market constraints removed – to evaluate how farmers would alter their management practices were input and market conditions more favorable
- new cropping patterns with low levels of inputs – when trying to influence farmers to accept new patterns
- new cropping patterns without input and market constraints and with new technical assistance provided – to estimate the potential for improving farmers' production.

In addition to these four conditions, the FSR&D team should consider a fifth possibility. This would be the condition whereby farmers' resources and cropping patterns remain unchanged, but management is improved. An FSR&D team can add this fifth alternative to evaluate how much improvement can be accomplished with a minimum of change to farmers' conditions.

Setting forth these conditions governs the way improvements in technology are introduced, and consequently, the way experiments are designed and hypotheses are tested. Selectively choosing from among these five design conditions should allow the team to (1) identify reasons for the spread in farmers' productivity and (2) learn how to introduce change under varying conditions.

6.4.2. SEARCHING FOR IMPROVEMENTS

This section describes the FSR&D team's search for possible improvements. The significance of this activity can, perhaps, be appreciated by recalling our discussion in earlier sections. During problem identification, the team found alternatives worth exploring. For the more interesting ones, the team agreed upon a set of hypotheses to be tested through on-farm experiments. The team might have hypothesized that lowland maize would respond economically to increased fertilizer application. The team then needs to search for the best fertilizer choices. That is, the team might have looked at alternative fertilizer types, methods of application, and quantities to be applied for each of the five conditions described in Sec. 6.4.1.

In its search for possible improvements, the team might emphasize one or more of the following:

- comparisons within the research area
- comparisons outside the research area
- farmers' uses of resources
- productivity criteria.

Comparisons Within the Research Area

The team can sometimes identify opportunities for improvement by studying farmers' systems within the research area. Opportunities arise when yields from crops and livestock are low on some farms and considerably higher on others. When these differences occur on nearby farms, the team should find it easier to identify the causes of the differences. Possibilities include differences in management, farmers' resources, access to animal traction, land tenure, and so on.

When proper attention is given to the reasons for differences in production, the team should be able to identify research possibilities that are responsive to farmers' interests and within the farmers' capabilities to change. Nevertheless, we urge the team to recognize that farmers with low yields may not have the resources—e.g., quality land—or environmental conditions—e.g., easy access to markets—for replicating their neighbors' successes.

Comparisons Outside the Research Area

Where yields within a research area are uniformly lower than in other areas, the team may not be able to identify the causes of the differences as easily because the reasons for the low yields are less apparent. The reasons could come from a range of farmer characteristics or environmental conditions. The team could begin its examination at several points. Possibilities include using the approach described in Sec. 4.2.3. to reexamine the subareas' boundaries. This approach included separating the farmers according to such factors as population characteristics; farming systems including the distinction between irrigated and rain-fed agriculture; the quality of support services; and the physical environment—climate, soil, and topography. Then, for common conditions, the team attempts to isolate the distinguishing factors that could lead to the differences in output. As an example, if the amount and distribution of rainfall in the area seriously constrains farmers' production, the team might study how others outside the area have dealt with this problem.

Another approach would be for experienced researchers to compare potential yields with existing yields and then judge to what extent these potentials might be realized. Even though the gap between farmers' output and the biological potential may be difficult to bridge, simply knowing whether the gap is large or small can be useful in searching for ways to improve farmers' output. That is, if the potential is found to be very close to the farmers' output given the physical conditions, then the team would probably take a different approach than if the gap were large. For example, DeDatta et al. (1978) reported on a study of gaps in rice yields.

During the planning workshop, the teams may wish to decide how to estimate the production potential of the more important crops within the research area. Such estimates provide guidelines for cropping pattern research and identification and clarification of factors possibly limiting growth. Considerable experience and expertise is required, however, to make reliable estimates of potential yields for a particular environment. Three guides may be followed. First, the researchers might use available survey data to search for farmers with higher yields than other farmers. These could be innovative farmers or those with above average resources. Second, the researchers could estimate the top production potential from published estimates for similar agroclimatic zones. One of the sources of such information is the World Soil Resources Report 48 (FAO, 1978). Third, production potential based on experiment station results might be used.

Farmers' Use of Resources

Another approach is for the team to focus on individual farmers' use of their resources. Through such analyses the team seeks to identify (1) ways for the farmers to use existing resources more efficiently and (2) those resources most limiting the farmers' production. In applying this approach, the team might follow the steps listed below:

1) List the resources farmers are currently using such as labor, animal power, equipment, crop and animal residues, purchased inputs, cash, and credit.
2) List the present level of use for each of these resources.
3) Identify and rank the most limiting resources confronting the farmers.
4) Estimate the increase in farmers' benefits from the better use of currently available resources.
5) Estimate the increase in farmers' benefits from modest increases in the most limiting resources (Zandstra et al., 1981).

The fourth step, involving improvements at the current level of resources, would be applicable when the team does not believe the farmers will be able to acquire additional resources. Concerning the fifth step, modest increases in the most limiting resources could come about by having farmers reallocate their use of existing resources, or by finding ways to increase farmers' total resources as through additional credit. By increasing resources incrementally, the team obtains a good indication of the significance of individual restrictions. Also, moderate increases in the most limiting resources probably represents both the amount of change a farmer could adequately handle and the level of support that could be provided initially.

Productivity Criteria

The foregoing approach can be applied from a slightly different angle—one involving a more generalized consideration of productivity. Such productivity criteria measure the efficiency of resource use, which is the amount of output per unit of input—e.g., crop yield per hectare.

The experiences of the team and participating specialists applied to specific situations under study provide the basis for identifying opportunities for improvement. An example of such an opportunity might arise from a situation in which the household's labor is fully employed during parts of the year and underemployed at other times. Identifying technologies or management practices that will permit the farmer to smooth out the household's labor requirements should increase output.

By comparing farmers' objectives with their resources, the FSR&D team can develop a set of criteria for improvement. Some of the farmers' objectives could relate to yields of subsistence crops, production of animal feed, number of livestock, total or net cash income, land ownership, calories or protein consumed, and stability of production. Resources, for example, could be land, the household, head of the household's labor, cash or credit, and irrigation water. Dividing an item in the first set by one from the second produces indexes such as tons of rice per hectare of irrigated land, kilograms of cotton per hour of family labor, total or net income per cash expenditure, or average calories consumed per household member.

With cropping pattern tests, the team will often look to biological productivity of individual crops as well as to the productivity of the pattern as a whole. For patterns with the same crop in sequence, the team will find it relatively simple to sum the yields for each sequence. However, when the pattern contains different crops, the team needs to consider some common term. One way is for the team to convert individual outputs to their monetary, caloric, or protein equivalents. The monetary measure is especially good when much of the farmers' crops are sold; the caloric and protein measures are good when farmers who are near the subsistence level have nutritional deficiencies. Alternatively, when similar crops such as maize and sorghum are grown in sequence, the measure of productivity may sometimes be made in terms of the total grain produced.

Where two or more dissimilar crops are grown in the same field at the same time, the team could measure the efficiency of land use by the land equivalent ratio (LER). As Harwood (1979) explains, the LER is the "area needed under sole cropping to give as much produce as 1 hectare of intercropping or mixed cropping at the same management level, expressed as a ratio. LER is the sum of the ratios or fractions of the yield of the intercrops relative to their sole-crop yields." We discuss the LER further in Appendix 6-B.

Which of these productivity measures to choose depends (1) on which ones best capture the farmers' situation—e.g., objectives, resources, and constraints; (2) on the ability of the team to identify the most limiting factors; and (3) on the best opportunities for improvement. In using these measures, the team may also focus on issues such as the risks associated with planting before the rains start, the influence of harvesting dates on pest damage and prices received in the market, and the turnaround time between crops. These issues are part of the technically viable designs discussed in Sec. 6.2.4. Several excellent references that may be used for analyzing biological productivity are Alvin (1977), Mitchell (1970), Milthorpe and Moorby (1974), and FAO Report 49 (1977).

6.4.3. SETTING DESIGN STANDARDS

After gaining an idea about the possibilities for improvements, the team should be ready to set standards for the experiments. These are general standards that guide the team in (1) setting up the experiments so that useful results will be obtained, (2) avoiding unnecessary detail and complexity, and (3) gaining uniformity among experiments across areas and over time. This third factor allows for differences in experimental conditions due to climate, economic factors, and other uncontrollable variables, as well as for identifying trends over time. Coming up with clearly defined standards is especially important when new researchers are being trained and when the rate of staff turnover is high. By formalizing research methods and establishing standards that carry over from year to year, the FSR&D team will lessen the problems a new or inexperienced staff might cause.

Many of these standards are unique to the country's program and the preferences of those in charge of FSR&D's technical program. Consequently, our discussion will be on some of the more general issues pertaining to researcher-managed and superimposed trials and farmer-managed tests. The issues we will cover are (1) types of farmers, (2) locations of experiments, (3) number of experiments, (4) design complexity, (5) experimental design characteristics, (6) methods of analyzing research results, and (7) methods for handling incomplete experiments.

Types of Farmers

The team will need to decide something about the characteristics of the farmers to be chosen as cooperators in the experiments. Even for farmers described as being relatively homogeneous in terms of their resources, cropping and livestock patterns, and management practices, the FSR&D team will find differences in attitudes and the willingness to cooperate. The team must decide whether to select those farmers who show the greatest willingness to cooperate or to select farmers who are more representative of the area's average farmers. Rather than recommend one or the other, we will simply state the advantages of each approach for now and discuss the subject further under Research on Crops in Sec. 7.3.1.

If the more cooperative farmers are selected to participate in the program, the team will probably find its work easier and experiments completed more rapidly. If other farmers seek advice and help from the cooperative farmers, then these cooperative farmers would be an asset to the team and the extension service when diffusing the improved technologies. On the other hand, cooperative farmers may not be representative of other farmers, partly because of their attitude, and the experimental results might not be applicable to other farmers. Also, those who appear to be the most cooperative at the outset, may not be the most effective cooperators in the long run.

In some cases, a farmer's background may influence the success of the experiments. Thom (personal communication) reports an instance in which an area's farmers were generally uncooperative because the research team first contacted two farmers who had little standing in the community. One of the farmers had an unacceptable family background and the other was a newcomer to the village.

Locations of Experiments

Once the farmers have been selected, the team will need to decide how to choose among alternative fields and locations within the fields. Replicated experiments conducted within the same farmer's field give the greatest uniformity of conditions for the experiments; and additional experiments across the research area give greater understanding of the area. Where experimental plots are small, as with researcher-managed trials, the team can take the foregoing into account when locating the experiments. On the other hand, when farmer-managed tests are being designed and farmers' fields are small, the team may have no other choice than to spread the experiments throughout the research area.

Another issue concerns the location of the experiments within the farmers' fields. Our contacts with FSR&D practitioners lead us to recommend that experiments, for the most part, be located in the center of the farmers' fields. By doing so, the farmers are more apt to apply their normal practices to the experimental area. This advantage would be particularly important for farmer-managed tests. Also, researchers will be able to observe on-farm conditions more closely as they go to and from the experimental plots. In some cases, however, experiments can be conducted near the edge of the farmers' fields without disrupting the farmers' normal procedures. This would be appropriate for some of the superimposed trials.

In Appendix 6-C, we present additional guidelines for locating experiments on farmers' fields.

Number of Experiments

Important to the complexity and thoroughness of the research program will be the number of experiments the team attempts during the year. This decision depends partly on the team size, members' experience, the nature of the research program, and the size of the area. Waugh (personal communication) suggests that 20 researcher-managed trials per researcher are too many. If the number is about 15, the researcher has sufficient time to visit with the farmers and analyze the results adequately. In contrast, more farmer-managed tests can be handled by a researcher; and with the help of extension, the number could be considerably greater than 20. But the team must work out the number of trials and tests for itself.

Design Complexity

The team's effectiveness can be enhanced if its members can adequately judge the appropriate level of experimental complexity—both for the team and for the farmers. While researcher-managed trials can be complex and involve a large number of treatments, FSR&D practitioners generally agree that the number of treatments should be kept small. This slower, step-by-step approach to research (1) fits in with the more difficult conditions under which on-farm research is conducted, (2) adapts better to the lesser experience of the field team when compared with experiment station staff, and (3) reflects the need to introduce technical change relatively slowly through farmer-managed tests.

Finally, when planning the experiments, the team should look forward to the eventual diffusion of the resulting technologies. These technologies should be within the extension service's capabilities to implement. Of course, these capabilities can be enhanced in various ways, including training and the extension service's involvement in the FSR&D process.

Experimental Design Characteristics

Table 6-2 contains information on the characteristics of on-farm experiments that the FSR&D teams may wish to use in setting their design standards. This information was compiled from the experiences of FSR&D practitioners. Plot size, number of treatments and replications, and field design are responsive to the methods and objectives of each experiment. To illustrate, researcher-managed trials are biologically oriented, are on small plots, have more treatments and replications per field, embody more complex designs, and are conducted by research staff in ways that have many similarities with experiment station research. In contrast, farmer-managed tests have socioeconomic as well as biological objectives, are on large plots, have fewer treatments and replications per field, use simpler designs, and are conducted by farmers. Superimposed trials have characteristics intermediate between researcher-managed trials and farmer-managed tests.

Not included in this table, but worth mentioning are the requirements for livestock and cropping pattern tests. For on-farm tests with large animals, a minimum of 20 to 30 animals is recommended, as we discuss further in Sec. 7.1.2. For cropping pattern tests, probably 40 to 50 test fields are required, as we discuss under Field Design of Experiments in Sec. 7.3.1.

Table 6-2. Comparison of researcher-managed trials, superimposed trials, and farmer-managed tests for cropping systems experiments.

*Characteristics of trials and tests**	*Researcher-managed trial*	*Superimposed trial*	*Farmer-managed test*
Plot size	Generally small — on the order of 75 square meters	Both large and small	Generally large — on the order of 1,000 square meters
Number of treatments	5-20	4-6	2-4
Number of replications per field	1-5†	1-2	1-2
Total replications across farms, per land type	4-5†	4-10	4-25
Field design‡	Completely randomized, randomized complete block, randomized incomplete block, split plot	Completely randomized, randomized complete block, randomized incomplete block	Completely randomized, randomized incomplete block, paired treatments
Sensitivity to treatment differences	Medium to high	Medium to high	Low to medium
Types of data collected	Physical and biological	Predominantly physical and biological, but some socioeconomic	Physical, biological, socioeconomic

*These characteristics will vary with experimental objectives, type of treatment, farm size, and cooperating farmers.

†Usually all replications will be placed on one farm field to give the complete experiment. However, if the field is small and only one or two replications on a field are possible, additional replicates will be placed on other fields of the same land type to give a total of four or five replications for the experiment.

‡See Appendix 6-D for description of field designs.

Finally, we provide further information on alternative field designs and appropriate statistical procedures in Appendix 6-D.

Methods of Analyzing Research Results

During this planning stage the team needs to agree on how it will analyze the research's results. This will influence the types of data to be collected, the way the experiments are conducted, and how decisions about the technologies are reached. Following are some questions that the team may want to consider

- Which statistical procedures should be used in analyzing the results of the biological experiments?
- Will whole farm analysis be attempted, or will partial budget analysis be used (see Sec. 7.6.3.)?
- What minimum yield increases and changes in risk are acceptable to farmers?
- What type and rate of profitability will be considered acceptable to both farmers and society?
- What coefficients of variation (C.V.'s) will be considered satisfactory for the different types of experiments?

Methods for Handling Incomplete Experiments

Experiments can be voided for a number of reasons, including natural phenomena, farmers' actions or inaction, or social unrest. Some examples include insect infestations, flooding, droughts, farmers' harvesting of the plot before yields have been measured, livestock's destruction of crops, and labor shortages during critical periods. Regardless of the reason for the disruption, the FSR&D team should plan on appropriate action for handling these instances. As a minimum, researchers whose experiments were voided should record the circumstances and explain the reasons for the voidance. The team should keep these reports as part of the research results. In some cases, conditions causing experimental failure may be regularly occurring events that need to be considered when evaluating the effectiveness of proposed changes in farmers' practices.

Of particular importance to the FSR&D team is to learn why farmer-managed tests are not completed. The team needs to be alert to the reasons for the incompletions, as when the farmer becomes involved in other farm activities and is unable to hand weed the experimental plot. While identifying such possibilities beforehand is seldom possible, the planning workshop should outline general guidelines on how to proceed when such situations arise. For example, the team should generally agree when to (1) abandon an experiment, (2) modify and continue a farmer-managed test, and (3) convert a farmer-managed test into a researcher-managed trial.

6.4.4. GATHERING ADDITIONAL DATA

Gathering data on the research area is a continuing process. It begins when FSR&D teams identify the target area and continues as long as a team is in the research area. When moving into an area for the first time, the team should gather and study secondary data followed by one or more reconnaissance surveys. Should the planning workshop follow shortly thereafter, the team will probably need additional information before finalizing the seasonal or annual research program. Below, we provide suggestions on collecting and dividing data according to physical and biological conditions and economic and sociocultural conditions. Such data could be collected as part of this planning phase.

Physical and Biological Conditions

When comparing yields in different locations, the physical and biological conditions often help explain a good part of the differences. Such an analysis centers on land type, climate, and growth characteristics of relevant crops and animals. Data on land type and climate were collected during problem identification (see Physical Setting in Sec. 5.4.1.). If data are missing, the team may have to extrapolate or interpolate using data from similar locations. Where such estimates of data are not practical, preliminary experimentation may be the best way to learn how plants and animals respond to the physical environment.

In analyzing the biological data, the team should gather and analyze information from available literature on the growth periods and requirements of the proposed crops and animals. For example, the growth periods for plants include emergence, vegetation, reproduction, and ripening. Data to collect on plant requirements concern such topics as nutrients, climate, and water. From available information, the team can describe the plant in terms of (1) the types of root and top growth; (2) nutritional requirements; (3) sensitivity to soil acidity, salinity, and depth, and to extremes in climatic conditions according to growth stage; and (4) means of harvesting. The growth characteristics of the crops proposed for the cropping patterns can then be matched with prevailing physical conditions.

Important soil properties for the team to consider are drainage, slope, depth, acidity, fertility, and texture. As an example, some plants that are particularly sensitive to soil acidity can be grown successfully in strongly acidic soils only after liming. Not only may this be expensive, but secondary nutritional deficiencies may occur. While alternative crops more adapted to high soil acidity may be more appropriate biologically, farmers may not want to grow them.

Important climatic properties are rainfall and its seasonal distribution, seasonal air temperature, humidity, day length, and solar radiation. Cereals that are sensitive to high temperature or to certain temperature and humidity combinations during flowering may not produce a full seed head. This information would be significant when designing a pattern if shifts in the time of the cropping pattern results in flowering during an unfavorable climatic period for grain development. With cereals that are sensitive to high temperatures during flowering, yields could be greatly reduced under such conditions.

In limited rainfall areas, the team should consider the time of rains in relation to soil texture, slope, and the plant's growing season. If the rains coincide with the growth period for a certain crop, reasonably good yields of drought-resisting varieties can be obtained in medium-textured soils on level to gently undulating topography. The moisture intake rate will be high, runoff will be minimal, and soil and water resources will be used effectively. Conversely, crop production will be less successful if the crop does not fit the rainfall pattern, if the soils have a low water holding capacity, if surface runoff is high, or if the crop lacks drought-resisting properties.

Where a new variety is being introduced, its expected yield and influence on other crops may be difficult to estimate. Growth and management requirements of the variety in similar agroclimatic zones would be a preliminary guide. If the agroclimatic zone and land type of the research area appear suitable, then the team should estimate the farmers' capability for growing the new variety. If the expected yields are sufficiently high, the new variety or cropping pattern is suitable for the research program. Introducing or changing intercropping patterns can be considered similarly. For these considerations—e.g., new introductions and new cropping combinations—the best approach may be for the team to conduct researcher-managed experiments.

In Appendix 6-E, we provide an example that illustrates the procedures for designing new cropping patterns of the type discussed above. We do not provide an example for livestock, but the general procedures are similar.

Economic and Sociocultural Conditions

During the reconnaissance survey, the team collected information to help its members understand the farmers' decision-making process within the area's economic and sociocultural setting. During the planning phase, this knowledge is augmented by further examinations to gain a better idea about how farmers make decisions. The team looks further into those factors described in Sec. 5.4.2., namely, farmer characteristics, knowledge, beliefs, attitudes, behavior, and goals.

Illustrative of the significance of the farmers' goals, the team might find that farmers engage in activities indicating their preference for off-farm income. With this knowledge, the team could logically look for ways to reduce on-farm labor requirements. The team might do this by planning experiments involving chemical weed controls or draft animals.

When planning the experiments, the team incorporates the farmers' ideas and reactions about change. For example, the team may believe that a shorter season rice variety will allow the farmer to grow two crops a year instead of one. In discussing these possibilities with the farmers, the team needs to look at factors such as

- the family's food preferences
- how the family would dispose of production increases
- what family and community obligations might arise from the new technology
- what demands would be placed on the household's labor and how responsibilities would be divided among its members
- farmers' requirements for accepting the new technology and their reactions to risks associated with change.

Most of the above questions relate to farmers as they currently are, rather than assuming farmers will change. Often such a stance is appropriate for the team during the short run. On the other hand, as farmers become exposed to the team and the successes of new technologies, they acquire new knowledge, which, in turn, changes their beliefs and attitudes and eventually their behavior and goals. Keen observation and help from specialists in the behavioral sciences should aid the team to integrate such possibilities for change into its research plans. Because of this possibility, the team needs to ascertain flexibilities for change within the sociocultural setting and the farming system.

As an example of such flexibility, the Kofyar of Northern Nigeria enlarged their household work groups when environmental conditions forced a shift from extensive to intensive cultivation. The institution of the extended family made it possible for the Kofyar to enlarge their household work group through increased polygyny, which was supplemented by hired labor (Netting, 1965). These changes made the Kofyar receptive to several labor intensive practices.

In contrast, sometimes the farm households are not willing to change. When this occurs, the team should consider whether it has taken farmers' conditions and reactions adequately into account. When the team's appraisal is reasonably accurate, rejections could be due to farmer misconceptions. For example, farmers may not be using fertilizer because they erroneously believe it "burns the soil." In this case, the team could explore ways to help the farmers see for themselves whether fertilizers are damaging. One approach is for the team to discuss the concepts with the farmers and then conduct a superimposed trial in which the farmers risk little on the experiment. After seeing the results, the farmers' knowledge and beliefs might change.

To sum up, during the planning phase, the team seeks to understand the farm household, the circumstances under which it operates, the flexibilities for change, and which changes are most likely to be accepted. When the team takes these factors into account, it increases the possibilities that technologies acceptable biologically, economically, and financially will also be acceptable socially and culturally. For example, by following these procedures, the team should not find itself conducting experiments on crops that farmers will not eat or cannot sell in the market.

6.5. CONDUCTING REGIONAL PLANNING WORKSHOPS

Some FSR&D teams use regional workshops to aid in planning the research program. These workshops provide an interdisciplinary setting for those who will participate most directly. In this section, we provide a brief description of some of the more important features of these workshops. Included are comments on the nature and purpose of the workshop, those attending, and some of the activities.

6.5.1. NATURE AND PURPOSE

A scheme for initiating activities in the research area for the first time is to hold a workshop there to explore alternative ways to carry out the research program. This workshop could be followed by one or two months in which FSR&D team members gather data in preparation for a second workshop. The second workshop would be the point at which the FSR&D team finalizes its plans for the coming season or year. For ongoing programs, a single, annual workshop may suffice. In such a case, the general approach to the research program will have been set and the regional headquarters team will be able to prepare for the workshop while performing its other activities. Such preparation will normally be during slack periods in the on-farm activities. Where two distinct cropping seasons occur annually and where time permits, workshops may be held prior to each season.

Whichever the case, the workshop should be completed about one month before the cropping season or livestock activity begins. This schedule should provide the team with sufficient time to prepare for the on-farm experiments.

6.5.2. ATTENDANCE

Generally, the leader of the regional headquarters team heads the workshops. This leader should draw heavily on the field teams and members of the regional headquarters staff in preparing for these meetings. Also, in setting the workshop agenda, the leader will probably want to consult closely with the national headquarters team. In addition, the leader will normally invite others to attend such as representatives from the experiment station (if separate from the regional or national headquarters teams), the extension service, farmers' groups, and other organizations concerned with the FSR&D program. If the experiment station is part of the FSR&D program, the station's specialists would normally participate in the workshop as members of the regional headquarters or field teams. Specialists from the other groups can provide information about technologies available outside the research areas, and farmers' representatives are valuable for evaluating proposed management alternatives and new technologies. Should the regional team find the need for specialized studies, those who can help determine the study's scope and terms of reference should be invited to the workshop.

6.5.3. ACTIVITIES

The workshop's activities encompass the range of topics described in the previous sections of this chapter. The regional team can use these topics as a checklist in deciding which topics are suitable for discussion during the workshop. Since meetings are not always easy to keep in focus, the leader should take care so that those topics proposed for the workshop are of the type calling for group consideration and decision. If two workshops make up the research planning activity, then the time between workshops can be used for more narrowly focused activities. The following are some topics that cut across disciplines and are, therefore, appropriate for consideration during the workshop.

Preparing for the Workshop

To aid in effectively using staff time, the regional team should prepare for the workshop in as much detail as practical. And each field team should prepare an initial work plan for its research area. Since each field team will have participated in the reconnaissance survey and in other on-farm activities, it should be knowledgeable about the physical, biological, and socioeconomic environments and alternative practices likely to be acceptable to farmers.

The work plans prepared by the field teams will be instrumental in shaping the direction of the overall plan of work, particularly when:

- The field teams are experienced and knowledgeable about their research areas.
- The regional and national headquarters teams have been effective in conveying the overall objectives of the FSR&D effort to the field teams.
- The work plans of individual field teams can be accommodated by supporting disciplinary and commodity specialists.

The workshop is further aided if the regional teams prepare data summaries, resource and activity maps, and other compilations of their work.

The leader of the workshop can facilitate its smooth functioning by identifying those who might serve on interdisciplinary task groups. Persons and assignments can then be confirmed as one of the workshop's first items of business. Assigned topics could be any one of the activities described earlier in this chapter such as the review of problems and opportunities, proposing alternative research activities and methods, and setting design standards. During the workshop, each group could be asked to present a summary of its findings for discussion and action. By the end of the workshop, the activities of these various groups should be pulled together in a form that becomes the basis for the season's or year's plan of work for the region.

Deciding on Data Requirements

Because data can be costly to gather and analyze and because data requirements often cut across several activities and disciplines, the workshop is a convenient place to agree on what data to collect, who is to collect the data, and in what form. The team should consider such needs as those associated with on-farm experiments, farm records, climatic monitoring, and surveys. If this topic is assigned to a task group, the group could direct the discussion and eventually come up with forms and instructions for gathering and analyzing the data. Where field teams have been functioning in the research area, the task group could also (1) review the field team's work in data collection, (2) evaluate suggestions for the field team's improvement, and (3) propose solutions. For example, if the task group finds that the team is taking more data than needed or cannot keep up with its assignments, adjustments would be in order.

Drawing up the Work Plan

As noted, the output of the regional workshop is a plan of work to guide the regional headquarters and field teams. This plan should contain

- the objectives to be accomplished by the forthcoming research effort
- descriptions of activities and responsibilities
- a timetable for accomplishments by individuals and groups
- the basis for evaluating accomplishments
- plans for taking corrective action, when needed.

In preparing the timetable, target dates should be carefully indicated. For example, in a cropping pattern experiment, tentative dates should be given for all field operations from planting through harvest for each crop. When the experiments pertain to researcher-managed or superimposed trials, the timetable would apply to the researcher; when the experiments pertain to farmer-managed tests, the timetable highlights critical times for observing farmers' activities. Time should be allotted for data preparation and analysis. Because of the short turnaround between harvesting and land preparation for follow-up crops in some parts of the world, the field team needs to organize itself carefully for the short time between crops so the data can be analyzed and the results from the previous season's experiments applied to the next season's experiments. Similar schedules should be prepared for all on-farm cropping and livestock experiments, surveys, special studies, climatic monitoring, and record keeping.

When the work plan and time schedules for the on-farm research are completed, they are distributed to the relevant field and headquarters personnel. This material

serves as a primary guide for the day-to-day activities of each field team. The regional leader should also be prepared to adjust these schedules as necessary. Should a field team encounter major difficulties in fulfilling its work plan, the regional leader should consider calling a special meeting or workshop for help in preparing corrective action.

A convenient way to monitor team progress is for relevant regional headquarters staff to meet monthly with each field team. At these meetings summaries of work, including experimental results, can be presented and discussed, and future work can be planned. Account can also be taken of additional staffing, logistic, and financial requirements. In this way, preparation for subsequent regional workshops will not fall as an excessive burden just before the workshop convenes.

Setting Staff Assignments

As the workshop members develop the research activities and draw up the plan of work, staffing requirements will become apparent. Tasks according to activity and person should be prepared. Those responsible for carrying out each task should help draft the assignment. In this way, the person will understand his or her assignment, will know how it relates to the team effort, and can be held responsible for the results. Assignments should be prepared for each member of the team. We provide suggestions for the assignments of the field team leaders, field team researchers, technical assistants, and disciplinary and commodity specialists in Appendix 6-F.

Regional leaders will also need to provide the field team with a means for acquiring additional labor during peak periods of activity such as during planting, weed control, and harvest. Such labor may be recruited from participating farmers or from other local sources.

Reporting of Results

Because reporting of research is so important, we wish to provide a few suggestions on this topic before closing this chapter. As noted above, field team members are responsible for preparing reports on their research activities. These reports are funneled to field team leaders, who use them for internal evaluation of the team's accomplishments. At the end of each research period, the field team carefully reviews and discusses these reports and prepares summaries to be used in subsequent workshops. These reports:

- contain recommendations on those technologies that appear suitable for multi-locational testing
- point out problems and opportunities that have been identified or for which a better understanding has been obtained
- propose a plan for research during the next season.

In this way, FSR&D is both continuing and iterative. Results from the planning workshop flow into on-farm research and analysis, proceed to extension of results, or feed back to problem identification and development of a research base—as illustrated in Fig. 6-1.

6.6. SUMMARY

We have divided this chapter into five parts. The first concerned the initial steps in the planning process whereby FSR&D teams review the list of priority problems and opportunities, appraise their organization's capabilities and resources, search for relevant technologies, make assumptions about near-term conditions, categorize and set research priorities, develop hypotheses for testing, and establish collaboration with other groups.

Next, we described some of the more important factors to consider when making preliminary analyses of on-farm experiments. The team makes these analyses to gain an initial understanding of possible physical, biological, economic, financial, and sociocultural responses to the proposed changes. Factors for the teams to consider in such an analysis include alternative solutions, farmers' conditions, farmers' and society's perspectives, development of technically viable designs, estimates of values, and eventual consequences from introducing changes to the farmers' systems.

The third part reviewed alternative research activities and methods. The activities include technology development, farmer adaptation, management of the farming system, climatic analysis, special studies, and management of natural resources. Methods for conducting these activities include researcher-managed trials, farmer-managed tests, superimposed trials, surveys, record keeping, monitoring, and research station support.

In the fourth part, we discussed design conditions, the search for improvements, establishing standards for the experiments, and gathering additional data before finalizing the research plan.

The last part concerned regional planning workshops that the FSR&D teams can use to implement the foregoing activities. This discussion centered on the nature and purpose of the workshops, attendance, and activities. By the time the workshop is completed, the regional headquarters and field teams should be prepared for the next FSR&D activity, namely, on-farm research and analysis.

CITED REFERENCES

Alvin, P.T. 1977. Ecophysiology of tropical crops. Academic Press, New York.

DeDatta, S.K., K.A. Gomez, R.W. Herdt, and R. Barker. 1978. A handbook on the methodology for an integrated experiment: survey on rice yield constraints. IRRI, Los Banos, Philippines.

FAO. 1978. Report on the agro-ecological zones project, volume 1: methodology and results for Africa. World Soil Res. Rep. 48. FAO, Rome.

______. 1977. Land evaluation standards for rainfed agriculture. World Soil Res. Rep. 49. FAO, Rome.

Harwood, R.R. 1979. Small farm development: understanding and improving farming systems in the humid tropics. Westview Press, Boulder, Colo.

McIntosh, J.L. 1980. Cropping systems and soil classification for agrotechnology development and transfer. *In* Proc.

Agrotech. Transfer Workshop. 7-12 July 1980. Soils Res. Inst., AARD, Bogor, Indonesia and Univ. of Hawaii, Honolulu.

Milthorpe, F.L., and J. Moorby. 1974. An introduction to crop physiology. Cambridge University Press, London.

Mitchell, R. 1970. Crop growth and culture. Iowa State University Press, Ames, Iowa.

Netting, R. 1965. Household organization and intensive agriculture: the Kofyar case. Africa. 35:422-429.

Technical Advisory Committee (TAC). Review Team of the Consultative Group on International Agricultural Research. 1978. Farming Systems research at the international agricultural research centers. The World Bank, Washington, D.C.

Zandstra, H.G., E.C. Price, J.A. Litsinger, and R.A. Morris. 1981. A methodology for on-farm cropping systems research. IRRI, Los Banos, Philippines.

OTHER REFERENCES

Andrews, C.O. and P.E. Hildebrand. 1976. Planning and conducting applied research. MSS Information Corp., New York.

Collinson, M.P. 1979. Understanding small farmers. A paper presented at a conf. on Rapid Rural Appraisal. 4-7 Dec. 1979. IDS, Univ. of Sussex, Brighton, UK.

Gilbert, E.H., D.W. Norman, and F.E. Winch. 1980. Farming systems research: a critical appraisal. MSU Rural Dev. Paper No. 6. Dep. of Agri. Econ., Michigan State Univ., East Lansing, Mich.

Grant, E.L., W.G. Ireson, and R.S. Leavenworth. 1976. Principles of engineering economy. The Ronald Press Co. New York.

Hildebrand, P.E. 1979. The ICTA farm record project with small farmers: four years experience. ICTA, Guatemala.

______. 1977. Generating small farm technology: an integrated multidisciplinary system. A paper presented at the 12th West Indian Agric. Econ. Conf., Caribbean Agro-Economic Soc., 24-30 April 1977. Antigua, WI.

IRRI. 1979. Report of the 8th cropping systems working group meeting. 28-31 May 1979. Nepal.

Nair, Kusum. 1962. Blossoms in the Dust. Praeger, New York.

Perrin, R.K., D.L. Winkelmann, E.R. Moscardi, and J.R. Anderson. 1976. From agronomic data to farmer recommendations: an economics training manual. Inf. Bull. 27. CIMMYT, El Batan, Mexico.

Chapter 7
ON-FARM RESEARCH AND ANALYSIS

With the on-farm research planned as described in Chapter 6, the FSR&D teams are then ready to implement the on-farm research. The regional and field teams will have (1) agreed on the objectives, (2) outlined the experiments and other research activities, (3) developed suitable work forms, (4) assigned initial tasks to team members, (5) recruited technical assistants from the research area, and (6) sought assistance from specialists. The team's efforts will be accomplished primarily through researcher-managed and superimposed trials and farmer-managed tests that are complemented by farm record keeping, monitoring, experiment station research, surveys, and special studies.

In Part I of this chapter, we concentrate on suggestions to the field team and others who will implement cropping and livestock experiments on farmers' fields. Our suggestions concern such activities as on-farm planning, field design of experiments, dealings with farmers, and measuring results. Since many of the suggestions apply equally well to different experiments, we will comment on these topics the most fully when we first mention them.

We discuss both cropping and livestock experiments under the headings of researcher-managed and superimposed trials and farmer-managed tests. We also include a discussion on organizing the team's efforts in implementing the experiments. In Part 2, we concentrate on analytical procedures primarily as they apply to cropping experiments. Each part closes with a summary.

PART 1: ON-FARM RESEARCH

In this part, we provide sections on researcher-managed and superimposed trials, farmer-managed tests, and team organization. The section on farmer-managed tests is substantially longer than the others because of its relevance to FSR&D and the need to provide detailed instructions. The topics within these sections include our suggestions for the FSR&D teams as they organize, implement, and report on the on-farm research program.

Some of the simpler and more common statistical procedures for analyzing the biological results from these tests are detailed in Appendix 6-D. For more complex procedures, the field teams should consult an agricultural statistician. If that is not possible, team members with a background in statistics might refer to standard texts on agricultural experimentation, such as that by Little and Hills (1978).

7.1. RESEARCHER-MANAGED TRIALS

As noted in Sec. 6.3.2., researchers undertake researcher-managed trials under farmers' conditions primarily to develop new technologies acceptable to farmers. Even though experiments are conducted on farmers' fields or with farmers' livestock, many of the experimental conditions and procedures are typical of those encountered at the experiment station. For example, experimental designs, plot size or numbers of animals, replications, measuring crop or animal production, and statistical procedures are similar. These similarities occur because the experiments are under the researchers' control—both the experimental variables and some nonexperimental variables. The latter are set to represent conditions under which the new technologies will be applied, such as methods of plowing or time of planting. By concentrating on controlled experimental conditions, researcher-managed trials do not generate much information about how farmers will respond to the new technologies being developed. The best information on farmers' reactions comes from the farmer-managed tests.

Staff experienced in experiment station procedures will, therefore, not find the approach to researcher-managed trials much different from what they are accustomed. Consequently, we will use this section to point out some of the factors that depart from conditions typical of the experiment station.

7.1.1. RESEARCH ON CROPS

In this section, we offer suggestions for cropping experiments that pertain to field selection, field design of experiments, cultural practices and data collection, monitoring progress, and analysis and reporting of results. The analysis in this last item is brief, since a detailed discussion of analysis is the subject of Part 2 of this chapter.

Field Selection

During the planning of on-farm research, Chapter 6, we talked about selecting fields representative of the physical environments and cropping patterns to be modeled. This usually means placing experiments on fields within well-defined land types or with well-defined cropping patterns. When this is done, the FSR&D team is in a position to extend its research results to similar conditions within and outside the target area.

The land for this research may be rented or borrowed from the farmers. Normally, the team can easily make arrangements with the farmers when they understand the team's goals. Rental payments may be a fixed cash sum, the produce from the cropping trials, or some other equitable arrangement.

Should a large tract be offered without charge, the FSR&D team should be wary of the donor's motives. Such donors may be more interested in personal or political gain than in the experiment.

Finally, before renting large tracts of land, the team should be certain that tenants will not be displaced from the land.

Field Design of Experiments

Because researchers manage the experiments, the designs for researcher-managed trials can be more complex than for farmer-managed tests. In Appendix 6-D we provide a discussion of alternative field designs. FSR&D leaders will need to weigh the merits of alternative approaches and the research abilities of their field teams. While small plots are normally chosen, some large plots may be needed for studies of such things as soil and water resources. The team should finalize these and other issues such as the number of tests and the data to be collected during the regional planning workshops.

Replications of experiments can be either contiguous as in most experiment station work or dispersed on different farms, depending on the experiment and field sizes. Contiguous replicates are used for studies when one field represents the entire experimental area. Dispersed replicates are used when farms are small or when the team wishes to measure the variability of a factor across the research area. Here, we summarize the work of Gines and Zandstra (1977) on dispersed researcher-managed experiments:

> Their study was designed to evaluate further the results of farmer-managed tests from the previous year, when 17 trials of dry-seeded, early maturing rice were scattered over the study area. Rice yields were highly variable and ranged from total failure to more than five metric tons per hectare. The availability of the iron and zinc micronutrients were thought to be largely responsible for the differences. To obtain more accurate information on this, the team designed a researcher-managed trial with five rice varieties of known tolerance to iron and zinc deficiency. The trial consisted of two replications of each variety placed in small plots in each field where the farmer-managed tests were conducted the previous year. The team stratified the data in various ways and found that distance from the irrigation canal accounted for the most variation. Using this information together with other observations, the team deduced that the iron and zinc availability decreased as the result of an increase in soil pH near the canal. Calcium in the irrigation water apparently was precipitated as it left the canal to form a calcareous surface soil that increased the pH.

Cultural Practices and Data Collection

Field teams should normally follow farmers' cultural practices for nonexperimental variables at the level recommended by the local extension service if this level represents a reasonably attainable future level for the farmers being studied. Farmers' cultural practices may apply to such activities as land preparation, seed selection, crop establishment, weed control, and fertilization. When farmers representative of the group being studied do much of the nonexperimental work, they will be able to apply their regular practices. When the researchers directly or otherwise do much of this work, they will need to make certain that they understand and know how to apply farmers' practices. Decisions on which experimental variables to select, such as pest control, should have been made during the planning stage. However, some flexibility is needed to alter these earlier decisions after the team enters the field.

The following experiment illustrates how the researchers reacted to unfavorable research results before an experiment was completed:

> Nelson et al. (1980) tested the hypothesis that improved irrigation practices and an increase in plant stand would significantly increase the yield of cotton. The farmer's practice of basin irrigation and low seeding rates gave stands of only 22,000 plants per hectare because of crusting and poor seed germination. Using improved practices, the cotton was planted at a higher seeding rate on both single-row and double-row beds. Also, the crop was furrow irrigated instead of basin irrigated. Although the farmer's plant stand was much below the 84,000 plants per hectare of the improved practice, the yields for the improved practice were no better than for the farmer's practice.
>
> The cotton plants were about 2.5 meters high for both the farmer's practice and the improved practice. With the improved practice, chemical control of attacking insects was so poor that yields were below the potential. The combination of a high plant population and tall plants made walking through the field and operating the backpack sprayer so difficult that insects could not be controlled. The poor stand from the farmer's practice, however, allowed the farmer to walk through his field easily with the same sprayer and to maintain good insect control. Although stands were poor, cotton production per plant was high.
>
> About mid-season, when the difficulties of insect control became apparent, the researchers removed every third row of plants from the plot with the improved practice. The result permitted easier access for insect control. Although this correction was made late in the season, yields increased 15 percent over the traditional practice. This experience clearly demonstrates the need to test new technology under farmers' conditions. By observing the farmer's management practices, the field team was able to correct the experiment and analyze the results.

Also, in some cases, researchers may wish to depart from farmers' practices when they believe the departures will not invalidate the results. For example, tractors may sometimes be used for land preparation even though most of the farmers use animals or manual labor. However, such departures can mask important differences between researchers' and farmers' conditons and should be used cautiously.

Adhering to local practices would include accepting the same distribution schedules of irrigation water, rather than a more favorable one, even though the irrigation authorities might allow such a schedule to "assist" the trial. The same would be true of planting time. The planting time for a research trial can be set somewhat arbitrarily. Unless the date of planting is an experimental variable, the most useful planting times for developing new technologies would be close to those of the local farmers.

When measuring crop yields, sampling techniques similar to experiment station procedures will have been set during the planning workshop. The workshop will also have provided the team with general guidelines for developing details on data collection. These include the types of information, how to record the information, and frequency of collection. Data should describe the field plots, weather conditions, field operations, and crop performance. The research approach and data forms should be finalized before implementing the experiments. We show typical forms for collecting data on cropping experiments in Appendix 7-A.

Monitoring Progress

The field team needs to monitor the seasonal progress of the trials to help understand the results. Technical assistants should visit the fields at least twice a week. Researchers need to observe the crop's progress weekly and to note any unusual growing conditions or growth responses. The team should

- record when the crops reach various growth stages such as germination, flowering, grain filling, and ripening
- record when each field practice is completed
- record and explain significant delays or omissions in planned field operations.

Analysis and Reporting of Results

With the harvest completed and yields measured, the field team then analyzes the data and reports on the results. Specifically, the team:

1) summarizes the quantitative data for each of the cropping tests; ordinary statistical procedures are normally used to estimate variability and establish confidence intervals for comparing new practices with farmers' practices
2) summarizes other information and integrates it with the statistical findings to arrive at an overall evaluation of the experiment
3) prepares a report on the findings that includes suggestions for further action
4) uses the report in preparing for the analysis workshops (Sec. 7.11.).

For small plot experiments typical of researcher-

managed trials, the overall coefficient of variation is often low so that standard analysis of variance procedures can be used to study treatment differences. The results usually indicate the confidence interval for each mean.

When dispersed replicates are used, as in the Gines and Zandstra (1977) example, the overall coefficient of variation may be high. In such cases, the team should try to identify the principal reasons for the actual or potential variation of results. Data from completed experiments can be helpful in searching for causes of variance and in designing new experiments. Also, more sophisticated statistical procedures, such as blocking techniques, may be the best approach for designing complex experiments. For these designs, the team will probably need to seek the advice of a statistician or an agronomist experienced in designing such experiments.

7.1.2. RESEARCH ON LIVESTOCK

The general methodological approaches to livestock and cropping trials have much in common, such as identifying farmers, planning experiments, monitoring progress, and analyzing and reporting results. The approaches also have their differences. Livestock trials tend to last longer than most cropping trials. For larger animals, the trials may last from 2 to 4 years to account for the animals' life cycles. When experiments are even longer than this, they are often more effectively integrated into a regional or national research center's outreach program.

The nature of livestock experiments and the procedures for conducting them are influenced by the types of animals being studied. For large animals, such as cattle, the team can seldom take over complete management of the experiment. Farm households use these animals in a number of ways—e.g., traction, transportation, milk, and meat—and therefore cannot let outsiders control them.

Nonexperimental variables may vary greatly. Thus, the team may concentrate on monitoring and farm record keeping, at least in the first and second years of the experiment.

In FSR&D, livestock experiments must generally be replicated on many farms. The reasons are that (1) many farmers have only one or two large animals and (2) the genetic variability of the animals and the management variability between farms are great. Thus, 20 to 30 animals may often be a minimum sample size for each treatment. Even then, coefficients of variation tend to be high and require complex statistical methods of design and analysis. Therefore, the field team will probably need help from an agricultural statistician.

Experiments with small animals can be less complicated than with large animals. For example, more chickens and ducks than cattle will be on small farms so the researcher may be able to place several treatments on a single farm. Also, fewer farmers will be involved, which reduces management variation and improves estimates of treatment effects and differences.

If the FSR&D team identifies the need for improved animal breeds, the team should seek help from regional or national experiment stations. Developing improved breeds tends to require more time and specialized knowledge than FSR&D teams generally have. Alternatively, a coordinated program in which experiment station staff begins research by introducing new animals into the farmers' system may be practical. The FSR&D team then conducts these applied researcher-managed trials to learn about the animal's performance under farmers' conditions. When livestock is closely integrated into the farmers' systems, the team may want to consider mixed farming experiments.

In Appendix 7-B, we provide illustrative data forms for characterizing and monitoring livestock experiments. Also, the reader can find additional information on ways to analyze livestock performance in ILCA (1980), CATIE (1978), and other reports by these and possibly other organizations dealing with livestock systems research.

7.2. SUPERIMPOSED TRIALS

Superimposed trials are a low-cost means of preliminary evaluation of relatively simple changes in technology. Experience is sufficient for us to offer several methodological suggestions for cropping systems. In contrast, we have uncovered little on which to base methodological recommendations for superimposed livestock trials. Consequently, we will limit our discussion on livestock in Sec. 7.2.2. to an example of a superimposed livestock trial in Colombia.

7.2.1. RESEARCH ON CROPS

This section covers field selection, field design of experiments, and implementation of experiments. Other aspects of superimposed trials for crops are similar to those concerning researcher-managed cropping trials as discussed in Sec. 7.1.1.

Field Selection

While many features of field selection are similar to those discussed earlier, some are different. Superimposed cropping trials may be located on fields that represent "targets of opportunity." These involve fields that are not identified for trials until a farmer's crop is already growing. Then, fields with reasonably uniform conditions are selected according to their suitability for testing specific management practices, such as farmers' dates of planting, application of irrigation water, intercropping, and fertilization. An advantage of this approach is the reduced variability caused by differences in farm management. As a result, the trials provide a better estimate of treatments or practices than other methods for the specific conditions selected.

Field Design of Experiments

The field team has considerable flexibility in designing field experiments for superimposed cropping trials. Where fields are large enough, the team can use randomized complete blocks in which all treatments for a given block are on a single field. Where fields are small, the team can use randomized incomplete blocks with two or three treatments per farm. In either case, treatments are replicated on other farms.

Plot size can vary more with superimposed trials than with either the generally small plots of researcher-managed trials or the generally large plots of farmer-managed tests. The team's choice of plot size depends on the treatment and the size and shape of the field.

With chemical applications of fertilizers or herbicides, which have small border effects, small plots can be placed in two rows along one side of the field as shown in alternative "a" of Fig. 7-1. Placing small plots along the side of a field rather than in the middle minimizes interference with farmers' regular operations. However, the researchers should make the plots wide enough so that border effects are not a serious factor.

If the team wishes to compare farmers' traditional practices of hand weeding with herbicide controls, the small plots with chemical treatments can be placed in one part of the field, with the rest of the field being the farmers' treatment. However, the researcher needs to select a section of the farmer's field comparable with the test plot for making comparisons and analyses.

Alternative "b" in Fig. 7-1 represents large plots, which are needed for treatments involving such activities as estimates of farmers' labor inputs or studies of insect and disease control measures.

Plot configuration depends on the size and shape of the field. When treatments are placed on row crops, strips parallel to the rows should be used as in alternative "b" of Fig. 7-1. Plots to study insect or disease control in a large field can easily be made by placing strips lengthwise in a rectangular field as in alternative "b," by quartering a reasonably large field as in alternative "c," or by placing strips across the field as in alternative "d."

Finally, in placing plots on farmers' fields, the team should take care that the number and types of treatments

Figure 7-1. Field designs of possible plot arrangements for superimposed trials. T_1, T_2, and so on indicate treatments. Fields a and b represent row crops and c and d represent closely planted crops. "7-1a"—Experimental chemical treatments for weed control (T_1-T_4) with T_5 the farmers' treatment (hand weeding). "7-1b, c, d"—Large plots representing insecticide or fertility treatments differing in size and arrangement because of different field shapes. One treatment is the farmers' treatment, e.g., T_1. The small rectangles without numbers represent randomly selected areas for measurement of yields.

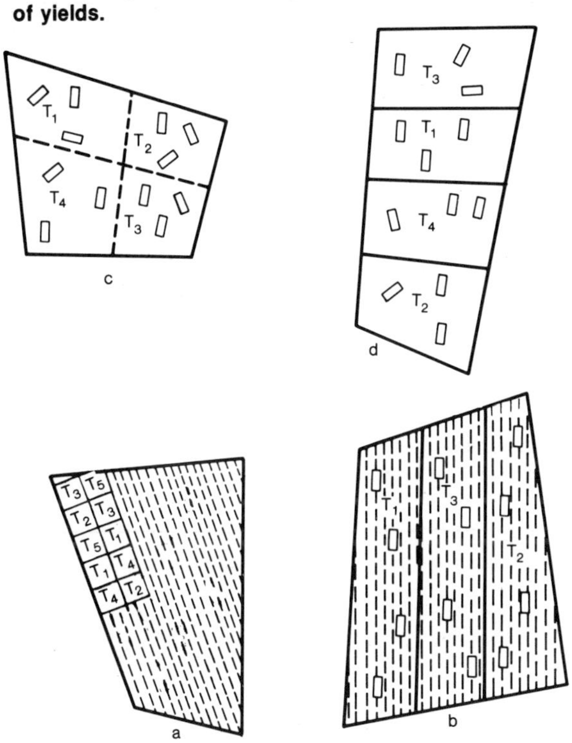

do not interfere with the farmers' regular activities. Otherwise, farmers' practices will be changed, thereby negating the concept of the superimposed trial.

Implementation

Here, we consider the topics of conducting the experiments and monitoring the results. Except for the treatments applied by the researchers, farmers complete all activities on their plots at the regular times. The researchers supervise and assist farmers with only that part of the farmers' activities involving the treatments. The researchers normally measure crop yields from larger plots by taking several samples from each of the plots. We give additional guidelines for sampling large plots in Sec. 7.3.1. on Measuring Crop Yields. For small plots, researchers harvest the crops as they would on the experiment station.

Those responsible for the superimposed trials need to visit the experimental plots regularly to note progress and any extraordinary events. Data monitoring should concentrate on the nature and timing of farmers' activities affecting the experiment. For "targets of opportunity," the researchers will have to rely on farmer interviews to learn what happened before the experiment began. When trials are superimposed on farmer-managed tests, the team has the opportunity to integrate the data requirements of both types of experiments.

7.2.2. RESEARCH ON LIVESTOCK

An example of a superimposed livestock trial is one in which an FSR&D team tests a new livestock technology on farmers' herds. Except for changes introduced by the researchers, farmers manage their herds in their customary way.

Below, we describe such a trial conducted jointly by the Colombian Agricultural Institute (ICA), the *Caja Agraria* (agrarian bank), and CIAT. In this trial, the effects of early weaning of beef calves—a new technology for the area—was compared with local cattle raisers' customary practice of not weaning at all (CIAT, 1974 and 1975).

An experiment station project was already underway in the plains of eastern Colombia to determine the effects of improved versus native management of beef cattle. As the project progressed, researchers found that one of the major factors influencing reproductive rates was lactational stress. That is, cows whose calves were weaned after only two or three months of lactation rebred almost immediately with a calving rate of 99 percent. In contrast, the researchers found that cows whose calves were weaned after nine months of lactation have a calving rate of only 64 percent (Stonaker et al., 1979).

In the light of these findings, a group from ICA, the *Caja Agraria*, and CIAT looked into the situation further and found that (1) without weaning, the cows might not kick their calves off and dry up naturally until the calves were from 10 to 14 months old; (2) although a market did not currently exist for 2 to 3 month old calves, the group felt that such a market could be developed if an adequate supply of calves were offered; (3) a number of farmers from the study area had cattle loans from the *Caja Agraria*; (4) *Caja Agraria's* technical staff visited the farmers frequently; and (5) technical assitance was part of loan procedures.

With this background information, the following experiment was conducted

1) The researchers carried out a preliminary survey to find 10 farmers interested in cooperating in a test involving early weaning of calves.
2) The research used a table of random numbers to select 10 cows from each farm. These cows had calves that were 2 to 3 months old.
3) The researchers randomly weaned 5 calves from each group of 10, leaving the other 5 to be weaned by the farmers' usual practice. This gave two treatments per farm.
4) CIAT agreed to purchase the 2 to 3 month old calves at prices that would give farmers an overall income comparable to that from selling their 10 to 14 month old calves.
5) The farmers kept the experimental herd with the rest of their animals and managed them all in the same way.

Six months after the early weaning, an experienced technician examined these cows for pregnancy. The overall differences in pregnancy rates for the on-farm experiment were even greater than the results obtained from the experiment station project. Moreover, cooperating farmers supported these findings by observing that cows with early-weaned calves rebred soon after lactation stopped.

7.3. FARMER-MANAGED TESTS

Farmer-managed tests are particularly important to the FSR&D process because they allow the farmers to participate in testing new technologies. In doing so, farmers reveal to researchers their reactions to these technologies. Furthermore, we emphasize the discussion of farmer-managed tests because such procedures are less well-known. Our discussion centers on crop research, since most of the experience has been here.

7.3.1. RESEARCH ON CROPS

We begin this section with a fairly lengthy discussion about the farmers who will cooperate in these experiments. This initial portion covers farmer selection, incentives to cooperating farmers, agreements with farmers, and farmer-researcher relationships. After that, with the exception of a section on measuring crop yields, the sections are the same as for researcher-managed trials with crops: namely, field selection, field design of experiments, cultural practices and data collection, monitoring progress, and analysis and reporting of results.

Farmer Selection

Because of the key role that farmers play in farmer-managed tests, the field team needs to pay particular attention to the selection of cooperating farmers. The team should take the whole farm household into account when selecting the farmers, since willingness and ability to cooperate depend on more than the head of the household. FSR&D practitioners have generally found that (1) farmers are willing to cooperate in these experiments and (2) they will do what they can to conduct the tests properly, but (3) sometimes they have limited capacity to cooperate. These limitations relate to the household's resources and the uncertainties of the farmers' environment. Thus, the team needs to do more than to find farmers who are representative and willing to cooperate. The team also needs to design experiments that the farmers are able to conduct, given their limited resources and the prevailing environmental conditions.

Incentives to Cooperating Farmers

As a general rule, FSR&D practitioners urge that incentives *not* be provided farmers in farmer-managed tests. By incentives, we mean some form of encouragement to cooperate in the experiment, such as outright payment in cash or materials, reduction in the cost of inputs, or reimbursement for losses. The reason is that these tests are intended to show how farmers will react to the new technologies when applied to their conditions. Should the tests be "fouled" by creating unrealistic conditions, then experimental results will be misleading. For instance, if fertilizers were part of a farmer-managed test and were of-

fered to the farmers without charge, the farmers would probably not value the fertilizers as highly as if they were to pay for them. This difference in attitude could change the farmers' methods of fertilizer use, and the results of the tests.

Consequently, we stress that FSR&D teams exercise considerable caution when providing cooperating farmers with incentives. Having said this, we now turn to some of the instances when the team might wish to consider providing farmers with some form of incentive. Following are some possibilities:

- When new materials such as seed and chemicals are introduced to a region, chances are that they are not yet available locally. Or if they are available, the conditions of sale may not be those that would prevail should the technology become common to the area. For instance, (1) current prices could be high because of low demand, (2) packages may be too large for small farmers' use, and (3) credit may not be available. Should the technology prove successful, then as experience has shown elsewhere, conditions could become more favorable.

  In these instances, the team would be justified in subsidizing the high cost of the inputs to the point that costs to the farmers represent those expected to prevail should the technology be implemented broadly. Where credit is a problem, the team might assist the farmers in obtaining it under conditions appropriate for such farmers. Another possibility is for the team to supply the farmers with these experimental materials. Farmers would be expected to pay for these inputs at harvest time.
- Team error is another instance in which cooperating farmers might be compensated for losses. Since this situation would not be recognized before the farmers undertake the experiment, compensation agreements would normally come after the fact. Except for clear-cut errors that can be attributed directly to the team, creating the impression that compensation will be provided farmers in case of their loss—relative to the farmers' normal yields—should probably not be implied by the team. Otherwise, the team might be forced to defend itself against imagined, as well as real shortcomings on its part.

  An example is the experiment described in Field Design of Experiments (Sec. 7.1.1.). During initial experimentation with a newly released rice variety (IR28), the effects of a calcareous soil condition on available iron and zinc for dry-seeded rice were not adequately evaluated. Because of iron deficiency much of the rice grew poorly and could not compete with the weeds. Several cooperators put a major effort into hand weeding, which was a losing effort. The crops were lost, sometimes too late for replanting. However, no compensation was offered the farmers. Notwithstanding, most of the farmers cooperated in the following year's program.
- Where the team destroys or keeps the harvested sample, farmers are justified in receiving an amount equivalent to that destroyed or what they would have gained using their traditional methods. But this is hardly an incentive. Rather, it is reimbursement.

To close this discussion on incentives, normally the risks of farmers' loss during farmer-managed tests are not great. The FSR&D process is set up to guard against such possibilities. The team is not expected to initiate farmer-managed tests until it knows, with reasonable certainty, that the proposed technologies are suitable for the area. Also, participating farmers are accustomed to the risks of farming. The example above demonstrates farmers' willingness to accept some risk of loss when participating in an experiment.

Ultimately, the decision on incentives is a management decision that hinges on an evaluation of (1) the farmers' welfare, (2) the manageability of implementing incentives, and (3) maintaining the integrity of the experiments. Moreover, when incentives are given for farmer-managed tests, the team should also attempt to provide comparable incentives to other cooperators in the FSR&D program. Finally, any incentive program should conform wth government procedures and fall within the FSR&D budget.

Agreements with Farmers

Besides identifying cooperating farmers and setting incentives, the FSR&D team may have to decide which farmers are to participate in the different experiments. Were all the trials of the same type, this would not be an issue. However, frequently some experiments are much more profitable than others. In these situations, the team should be sensitive in deciding how the experiments are divided among the interested group. One way to settle this issue is for the FSR&D team to meet with participating farmers as a group. During the meeting, the objectives of the program and experiments can be outlined, the continuing nature of the FSR&D activities can be stressed, and the means for equitably distributing the experiments among interested farmers can be established.

During such meetings farming practices for each test can be reviewed and agreements reached. Also, this gives the team an opportunity to finalize plans for its monitoring responsibilities, what support farmers are to receive, how incentives and contingency plans are to be implementd, and how to help ensure that the farmers and team members will keep their part of the agreement.

Farmer-Researcher Relationships

A critical factor in farmer-managed tests is the relationship between farmers and researchers. This relationship hinges on the extent to which researchers attempt to direct farmers' activities. Fundamental to the effectiveness of farmer-managed tests is the researchers' awareness that these tests belong to the farmers. Farmers must be allowed enough freedom so that test results reflect their responses to the suggested changes.

The difficulty arises from the need for farmers to be interested enough and to understand the tests well enough to give the new technologies a fair test. Researchers can aid this process in several ways:

- by exposing farmers to the concepts of the new technologies before starting the farmer-managed tests, possibly by conducting researcher-managed or superimposed trials in the area beforehand
- by learning farmers' management and environmental conditions so that the tests (1) do not contain elements foreign or disagreeable to the farmers, (2) contain elements that the farmers favor, and (3) introduce change that is within the farmer's ability to grasp
- by preparing an outline that conveys the essential steps of the test to the farmers, such as the sequence of farm operations and the inputs required to perform them (we provide an example of such a schedule in Appendix 7-C)
- by monitoring and otherwise being available to provide the farmer with technical information and demonstrations about the tests.

The danger in farmer-managed tests is that the researchers may (1) become excessively interested in the results, (2) feel pressured to demonstrate an area's high production potential, or (3) think that the purpose is to instruct the farmers about improved farming practices. These problems can be intensified when a farmer's fields are separated so the researchers cannot see how the tests are integrated with the farmer's other activities. When researchers respond in one of these ways, they may be applying pressure on farmers to follow procedures that researchers would follow. Alternatively, the researchers might help the farmers by providing resources and services that are normally not available to the farmers, such as tractors from an experiment station to prepare the farmer's land so that the seedbed can be completed on time and the experiment remains on schedule. But these acts defeat the purpose of the tests. When the researchers are not ready to let the farmers manage the tests using their resources in their way, chances are good that the test is premature and the experiment should really be conducted as a researcher-managed or superimposed trial.

Researchers should realize that the results of farmer-managed tests will be different from those of researcher-managed trials. Because farmers carry out their activities to fit their available time, they may omit some operations. These adjustments allow the tests to become fully incorporated into the farmers' total effort. Or, farmers may change cropping patterns once the tests are underway. For example, an initial design may have involved 24 farms, three patterns, and eight replications. One of the patterns, used as the control, should be the farmers' traditional pattern. Table 7-1 is such a possibility.

When the season was over, the farmers might have changed the 24 patterns and replications as shown in Table 7-2.

Thus, three farmers returned to their traditional pattern, two changed to a mono-cropping pattern, and 11 rather than 16 stayed with the new patterns. The shifts in patterns could represent farmers' responses to climatic and socioeconomic conditions, or to their reappraisal of the value of the test.

Table 7-1. Planned cropping-pattern experiments.

Replicated patterns	*Cropping patterns*
8 (control)	rice-sweet potatoes
8 (new design)	rice-rice-mung beans
8 (new design)	green corn-rice-cowpeas
24 total	

Table 7-2. Actual cropping-pattern experiments.

Replicated patterns	*Cropping patterns*
11 (control)	rice-sweet potatoes
6 (new design)	rice-rice-mung beans
5 (new design)	green corn-rice-cowpeas
2 (other)	rice
24 total	

An analysis of farmers' changes, whether in patterns or practices, is as important to the FSR&D team as the crop responses. Such analyses (1) become the means for identifying and measuring farmers' constraints in adapting the changes to their conditions, and (2) help identify problems and opportunities for additional research.

Field Selection

When selecting fields for farmer-managed tests, the field team needs to be concerned about how the tests on these fields fit into the rest of the farmers' activities. This consideration is particularly important when the timing of operations for the experimental plot is substantially different from the farmers' other activities. Such a situation could arise when planting a crop earlier than normal so as to fit a second crop into the growing season.

Other suggestions include

- The experimental field should not block or otherwise interfere with farmers' regular activities—e.g., the normal movement of livestock or equipment to and from the fields.
- For irrigated crops, the various plots should all be located in areas with relatively similar irrigation conditions.
- For nonirrigated crops grown experimentally in areas where the bulk of the farmers' crops are flooded rice, researchers should be sure the water table represents the conditions they wish to study.

Field Design of Experiments

Some factors for the FSR&D team to consider in the field design of treatments include the number and replication of treatments, environmental variation, and plot size.

The number of treatments on a farmer's field is often

restricted because (1) farmers' fields are customarily small, (2) the size of the treatment plots must be large enough for the farmers to work as they are accustomed, and (3) farmers have limited experience in managing different treatments in the same field. Therefore, the field team usually limits the treatments on a farmer's field to one or two and then places additional treatments and replications on other farmers' fields. The control, which is the farmer's usual practice, can be in the same field as another treatment when the field is large. When the field is small, the control can be on a nearby field with similar characteristics as the field with the new treatment or treatments.

Where the testing involves alternative management practices for a given crop on a single land type, the number of replications of a treatment should probably be at least four. In Appendix 7-D, we provide an example of the field design for a farmer-managed cropping test. The test involves four treatments for one crop on each of three land types and five replications on each land type, giving a total of 60 test fields.

Cropping pattern testing requires a more complex field design than simply testing alternative management practices on the farmers' traditional patterns. In a completely randomized design, usually two or three different patterns are tested with four or five replications on each of several land types. Zandstra et al. (1981) recommend that 40 to 50 cropping pattern test fields in a research area are needed to obtain a reliable evaluation of the patterns. In Appendix 7-E, we present an example of the field design of a farmer-managed cropping pattern test based on Zandstra et al. (1981).

Farmer-managed tests may result in coefficients of variation that range from 30 to 50 percent—considerably larger than the coefficients for researcher-managed trials. These large variations frequently result from (1) micro-variations in farmers' fields, such as the plot being on a ridge or in a trough; (2) variations in farmers' access to resources or in management practices, such as the use of inputs and timing of activities; and (3) flexible farmer-researcher relationships that allow for mid-seasonal changes in cropping patterns or management practices.

While the magnitude of the coefficients of variation can seldom be brought down to the level of the researcher-managed trials, field teams can do several things to keep the coefficients within acceptable levels. For example, the team can

- select experimental blocks by land types or other distinguishing characteristics of the farmers' environment and management
- attempt, within these general conditions, to eliminate as much of the micro-variation as practical
- plan more tests
- standardize farmer-researcher relationships.

We describe possibilities for alternative field designs in Appendix 6-D.

Plots should be large enough so that the tests give a reliable measure of the farmers' normal operations as related to labor practices, cash and credit requirements, application of seed and agricultural chemicals, timing of activities, and other aspects. This is the reason we gave 1,000 square meters as an appropriate size of plots (Table 6-2). Thus, a test plot may sometimes be an entire field.

Cultural Practices and Data Collection

Once the researchers and farmers reach an understanding about how the test should proceed, farmers should be ready to begin the test. This understanding should be reached well in advance of the normal planting dates, so that the tests do not disrupt the farmers' routine. Because farmers customarily plant over an extended time and because of the importance of planting dates, researchers should closely observe this aspect of the farmers' operations. Where planting dates range over several months, yields may vary by more than 50 percent. Norman (personal communication) suggests that for such wide-ranging planting dates, the team may want to consider that the farmers are really using different cropping technologies. While the crop may be the same, its management in terms of labor and other inputs and its yield and value could be significantly different.

For accurate evaluation of cropping tests across an area, the farmer-managed tests should all be planted at about the same time. However, farmers in their desire to cooperate with the FSR&D team might plant earlier than they normally do. Then, should rains be advanced, early planting might give the farmers a chance to plant a second crop during the same year. Thus, the test result would be more favorable. By advancing the planting date, farmers alter the original design.

In such a situation, the researchers may want to keep the intent of the original design by adding fields to maintain the planned spread in planting dates. This not only gives researchers information on the planned patterns, but also provides information about how different patterns respond to weather variability.

In preparation for its monitoring of the farmer-managed tests, the field team needs to finalize its data collection forms before the season begins. These forms should agree with the general approach established during the planning workshop. As indicated for researcher-managed trials, the data should include information about the weather, the plot, farming operations, and crop performance such as its progress through the growth stages, yields, and the influence of pests. But additional data are needed on such items as labor inputs by household members, detailed cash expenditures by crop, other aspects of the family's activities, and the general socioeconomic setting as related to the experiment. In Appendix 7-A, we provide forms typically used for collecting some of these data.

Monitoring Progress

Much of the team's monitoring of farmer-managed cropping tests is similar to that for researcher-managed trials. In addition, the team needs to pay attention to socioeconomic matters, and to how farmers manage their time and make decisions. For instance:

- Did market or sociocultural conditions have any unusual influence on farmers' actions?
- Which member of the family is responsible for which activities?
- How do farmers react to unplanned events, such as unusually late rains or a sudden pest attack?
- How do the families occupy themselves during slack farming periods? Do they maintain the farm, seek off-farm employment, engage in handicrafts, or rest?

Insight into items such as the above can help researchers understand farm households and over time develop an improved approach to identifying suitable technologies. While some of this information is sought as part of special studies, observing farmers' responses to given situations is frequently more enlightening than their responses to survey questions.

Some of this information can be worked into the farm records and other data sheets. In other cases, the information is most easily acquired by having the team members keep daily logs of their activities and observations. These logs can subsequently be of considerable use to those who keep them and to the whole field team—particularly during the planning and analysis workshops (Sec. 6.5. and 7.11., respectively).

Measuring Crop Yields

Accurate measurement of yields is fundamental to evaluating the acceptability of alternative cropping patterns and practices. For example, grain production may be measured as the yield of grain for human consumption and forage for livestock. Where significant losses occur during or after harvesting—e.g., losses from grain shattering or from pests—researchers should measure or estimate these losses as well.

The best estimate of yields is to harvest the entire plot. This approach poses no problem for small plots, but does for the larger farmer-managed plots. A satisfactory alternative for large plots is to sample within the plot—sometimes called subsampling. *Subsampling* reduces the time and cost of measuring yields, but introduces the danger of biasing the results. The potential for bias is most serious when plant growth in a field is uneven and the sampler has some reason for wanting the experiment to look better or worse than it really is. The team can reduce sampling bias by specifying on a map beforehand the areas to be sampled, such as the small rectangles shown in Fig. 7-1.

A common technique in measuring yields is to make three to five crop cuts—subsamples—in each plot using a grid system to randomly select the coordinates for each cut. With sampling procedures for multiple cuts standardized, the team can compare sampling errors for different seasons. As the team gains experience, it can modify its sampling procedures and thereby reduce sampling errors.

Depending on the crop, researchers should sample from 5 to 10 square meters for closely planted crops and forages and from 5 to 10 linear meters per pair of randomly selected rows for cultivated row crops. Because yield measurements vary with the sampling method, the team should follow the same procedures for each type of crop and select comparable procedures for different types of crops.

After researchers weigh the samples, they usually give them back to the farmers. When samples are destroyed during testing, such as in evaluating a crop's nutritional value, farmers should be reimbursed according to prior agreement.

We do not recommend a form of subsampling called panicle sampling, which has sometimes been used for measuring cereal yields. In this approach, the sampler goes through the field and selects panicles of grain. A drawback to the method is the team has little basis for assuring that the sampler will not intentionally or unintentionally bias the sample by the way the panicles are selected.

Analysis and Reporting of Results

Data analysis and reporting procedures for farmer-managed tests are similar to those described for researcher-managed trials under Analysis and Reporting of Results in Sec. 7.1.1. The team generally uses farmers' traditional patterns and practices as the statistical control and follows statistical procedures suitable for dispersed replications. For these tests, researchers need to interpret results in the light of climatic and other variable conditions prevailing during the tests, and to note how these conditions compare with historical data.

Another topic requiring researchers' attention is how to handle incomplete tests. While experiments can be abandoned for various reasons during researcher-managed and superimposed trials, this topic is particularly relevant for farmer-managed tests. When farmers abandon tests, the team should follow the general guidelines set out during the planning workshops (see Methods for Handling Incomplete Experiments in Sec. 6.4.3.). Following these general instructions, the team first analyzes why the farmer changed the tests, whether the tests can be salvaged, and what corrective measures to take for subsequent tests.

For example, if hand weeding was part of the technology, a farmer may not have had access to additional labor. Alternatively, after starting to weed, the farmer may have concluded that this was a waste of time and stopped weeding. Other cooperating farmers who did weed for the same test may have secured additional labor or concluded the procedures were worth following even though difficult. In this example, the team needs to learn why some farmers weeded while others did not. If the reasons include farmers' available resources, then the farmers may have been inappropriately stratified when designing the experiment. On the other hand, if farmers' conditions are about the same, then the individual decisions to weed or not to weed represent the natural dispersion of acceptance that occurs when dealing with individuals.

As another example, should fertilizers not be applied because farmers could not obtain them, researchers need to investigate why the fertilizers were not available. If the fertilizers were unavailable because of a persistent national shortage, then the experimental design was wrong and the team should learn from this experience. In contrast, if the fertilizers were temporarily unavailable because the local supplier happened to close the store—for example, to attend a burial—then the closing of the store is simply one of the random events that causes experiments to depart from the plan. In this case, the experiment is valid and the results should be included in the analysis.

7.3.2. RESEARCH ON LIVESTOCK

Farmer-managed tests with livestock offer a wealth of interesting possibilities, particularly those involving mixed systems—i.e., the combination of cropping systems and livestock systems. Unfortunately, procedures tend to be complex methodologically and we have uncovered little that serves as practical guidelines for this type of research. Therefore, we will simply indicate research possibilities, suggest ways for developing suitable methodologies, and offer some suggestions for data collection and analysis.

Research Possibilities

Because of the diversity and self-sufficiency of many small farmers, research to improve farmers' use of livestock is particularly interesting. Some possibilities for farmer-managed tests include treatments to evaluate

- control of animal diseases, insects, and internal parasites
- benefits of increased nutritional or caloric intake
- effects of changes in the numbers and types of animals
- alternative uses of crop residues and animal manure
- alternative forms of animal traction
- alternative allocations of family land and labor among subsistence food production, pasture production, and care of livestock.

As complements to these experiments, the field team may wish to seek regional or national help in supportive studies. Such studies help in designing the experiments and in interpreting the results. Possibilities for supportive studies concern (1) the effectiveness of livestock as a store of wealth, (2) the social customs surrounding the acquisition, care, and use of livestock, and (3) a family's nutritional benefits from keeping animals. These studies supportive of farmer-managed tests help in evaluating the livestock experiments within the whole farm setting (ILCA, 1978).

Development of Methodologies

Historically, most livestock researchers have concentrated on cattle experiments at experiment stations. The result has been that methodologies for farmer-managed livestock tests are not well-defined. Consequently, we can offer only rudimentary suggestions for such tests.

To start, an FSR&D team might use the approach for farmer-managed cropping tests as a guide. As testing proceeds, researchers should allocate enough time for studying the differences between the crops and livestock approaches. With this knowledge, the researchers could then produce livestock procedures specific to the interests of those in the research area.

Farmer-managed livestock experiments should involve the farmers' animals within the farmers' setting. Because of the value of large animals to the farmers and because of the occasional risk to the animals' health, the FSR&D team will need to offer farmers some guarantee in case such an animal should become incapacitated or die. While being prepared for this eventuality, the likelihood of such occurrences should be slight. As with cropping experiments, much of the risk to farmers should be removed before new technologies are brought to them for testing. Tests with small animals usually do not involve risks to the same degree as with large animals.

Where the team places its livestock on the farmers' fields, researchers will need to ascertain that farmers have a genuine interest in the animals and know how to care for them. This will be particularly important for expensive animals such as cattle. In fact, the team may not have an adequate budget to place many animals on farmers' fields in this way. Thus, FSR&D management may want to approach donors or lenders—both domestic and international—for assistance in financing livestock purchases.

Experiments involving young livestock should begin with birth and follow through the critical periods to maturity or to some predetermined time. Thus, when farmers agree to cooperate in livestock experiments, they are committing themselves to feed, water, watch over, and otherwise care for the animals for an extended time. Farmers should be aware of this when they agree to the experiments. In turn, researchers and their assistants need to monitor, at regular intervals, the animal's well-being in terms of weight gain and loss, appearance, and general health.

For reasons given earlier, treatments are generally limited to one or two per farmer with the traditional practice serving as the control. The treatments are replicated on other farms.

Data Collection and Analysis

Because of the breadth of activities involving livestock research, data requirements are substantial—so much so that one of the first year's activities might be a survey of the farmers' management of the livestock system and how this system integrates with cropping systems. Surveys dealing with mixed systems should include the family labor pattern, the contribution of manure to cropping and other uses, cash flows associated with livestock activities, and the farmers' attitudes toward adoption of new livestock technologies. Delgado (1978) conducted such a survey as the first step in defining the Fulani mixed system in Upper Volta. Because of the need to collect such information about the farmers' livestock system, cropping experiments that are part of mixed-system testing may have to wait for the team to collect this background information and to analyze it before designing experiments on integrated cropping-livestock systems. In Appendix 7-F, we describe a mixed-system experiment in the Ethiopian

highlands; and in Appendix 7-B we provide tables that may be useful to the team in gathering data on livestock.

Many of the data collection methods for livestock are similar to those described for cropping tests. Generally, climatic data gathered for crops will meet the needs of livestock experiments. Just as for crops, temperature extremes and extended periods of rain or drought influence animal growth and well-being. Besides temperatures, researchers need to follow other environmental conditions to determine the impact on experimental results. For instance, nonexperimental variables having an influence on results include animal diseases and pests, grazing conditions, and availability of drinking water. Of course, some of these conditions may be related to rainfall and temperature. The team should summarize, analyze, and report these and other data.

7.4. TEAM ORGANIZATION

In this section, we cover briefly some of the ways a field team might organize its efforts for implementing on-farm research. The topics are the team leader's activities, assignment of resources, review sessions, and integration with local organizations.

7.4.1. TEAM LEADER'S ACTIVITIES

An initial step in preparing for field experiments is for the leader to review the team's objectives and responsibilities as set during the planning workshop. Next, the leader, with help from team members, might draft an integrated set of responsibilities that identifies the general tasks for each member and the team. The staff members are normally assigned tasks based on team responsibilities rather than on a member's discipline. For some objectives, several researchers may need to work jointly on an activity under the direction of a task leader. At other times, assignments may be so narrowly defined that a single team member can handle the assignment. In this case, the assignment would logically go to the person with the most appropriate discipline and experience. Technical assistants are also tentatively assigned to groups or individuals for direction and on-the-job training.

The leader might call a meeting to present the proposed plan to the team for further discussion, modification, and approval. After reviewing the materials, team members could draft their job descriptions as they envision them and pay close attention to how each position interacts with other positions. These job descriptions could then be discussed and rewritten as necessary. When finalized these descriptions become the basis for work assignments and evaluation of a team member's performance. Such group action, in establishing member responsibilities to the FSR&D effort, helps to develop a team approach (see Sec. 10.6.1. for further discussion of interdisciplinary teamwork). As the work progresses, the leader can adjust assignments in the same way as they were set initially.

7.4.2. ASSIGNMENT OF RESOURCES

Once assignments have been agreed upon, the team can identify its resource requirements. Each group or individual within the team should compile a list of requirements for carrying out the assignments. Items to consider include such things as portable field equipment for planting, spraying, and harvesting; field testing equipment for measuring soil pH, soil and water salinity, and water flows; ordinary hand tools; supplies of seeds, agricultural chemicals, and veterinary items; and office supplies. Particularly important is adequate transportation so that members will not be delayed during critical periods.

The leader matches resource requirements with availability. Where serious shortfalls occur, the leader can call another meeting to gain team input and support for the best way to modify the team's activities and individual responsibilities. Sometimes experiments are delayed because special equipment, materials, or other inputs are unavailable. Then, the team should replace the experiments by other experiments that can be conducted with available inputs and are within the team's overall research objectives.

7.4.3. REVIEW SESSIONS

The team's leader should periodically hold sessions to review the team's activities. These meetings can be used to

- coordinate the team's activities
- set targets for accomplishments and assess team progress
- identify solutions to individual and team problems
- report on individual and group experiments
- convey findings of general interest to the team
- propose topics for future discussions
- prepare the team for the planning and analysis workshops.

While the output from these meetings contributes directly to the team's operational responsibilities, the meetings also help build effective teamwork.

7.4.4. INTEGRATION WITH LOCAL ORGANIZATIONS

The team will function most effectively in the research area when it is able to work with other groups concerned with improving farmers' conditions. Possible groups include the extension service, irrigation districts, and farmers' organizations.

In addition to the extension staff participating as part of the FSR&D teams, other extension personnel should be encouraged to learn about the field teams' activities. Extension service personnel may be the most aware of farmers' problems, how farmers use their resources, and which improvements interest farmers most. When extension personnel are integrated into the FSR&D effort, they will become better informed about FSR&D's goals and activities and will be more effective in diffusing the results.

Engineers and other staff attached to local irrigation districts are important when the field team works with farmers who irrigate. The field team should become acquainted with this irrigation staff and invite them to participate in the field team's activities as appropriate. For example, irrigation staff might be asked to serve as consultants or to accept part-time assignments with teams to assist in water measurements and related activities concerning on-farm water use.

Others with whom the team might develop contacts include representatives of local farm cooperatives, lending institutions, water-users' organizations, public officials, and private groups serving farmers in the research area. The more that local farmers and farm service organizations cooperate in the FSR&D effort, the greater the opportunity for the team's success. Moreover, because the field teams are at the "grassroots" level, they have an excellent opportunity to develop strong local support for the FSR&D approach.

7.5. SUMMARY

This part contains suggestions for helping FSR&D teams conduct their experiments. We have written sections on researcher-managed and superimposed trials and farmer-managed tests. Within these sections is material common to all three types of experiments, such as field selection, field design of experiments, cultural practices and data collection, monitoring progress, and analysis and reporting of results. We devote considerable space to farmer-managed tests, with additional sections on farmer selection, incentives to cooperating farmers, agreements with farmers, farmer-researcher relationships, and measuring crop yields. Throughout Part 1, we gave more emphasis to experiments with crops than with livestock. We closed this part with suggestions for team organization.

PART 2: ANALYSIS

We present this part as a guide for FSR&D teams in planning their research activities and analyzing the results. Researchers should understand analysis procedures if they are to adequately set up their experiments and plan supporting studies and data collection. Because many of the analyses apply to on-farm experiments, we have written this part with the field teams in mind. They will not need more than a rudimentary background in economics or statistics before receiving instructions on the material in this part. Such instructions should, most appropriately, come from a national FSR&D training program of the type described in Sec. 11.2.1.

In Part 2, we integrate many of the physical, biological, economic, financial, and sociocultural aspects of FSR&D that we brought up in earlier sections of this book. Furthermore, we emphasize economic procedures, because many experimental results require economic interpretation once biological results have been analyzed. Our approach draws heavily on CIMMYT's manual, *From Agronomic Data to Farmer Recommendations* (Perrin et al., 1976). That manual presents analytical concepts in a way that field teams should be able to learn and apply quickly. We do not include details of statistical analyses, because we discuss these in Appendix 6-D.

We have divided this part into sections on (1) concepts of analysis, (2) illustrative designs and analysis procedures, (3) acceptability of new technologies, (4) sociocultural feasibility, (5) other analysis procedures, and (6) analysis workshops.

7.6. CONCEPTS OF ANALYSIS

In this section we discuss concepts of analysis as they relate to (1) an integrative approach, (2) prediction versus acceptance of new technologies, and (3) partial budget analysis versus whole farm analysis.

7.6.1. AN INTEGRATIVE APPROACH

Analysis in FSR&D is an integrating activity in which the team seeks to simulate the breadth of activities and considerations that farmers do intuitively. The more researchers are able to assimilate farmers' decision-making processes, the more accurately they will be able to anticipate farmers' decisions.

The team begins the analysis by seeking to understand the household and the farming environment. After the on-farm experiments are completed, the team takes the results as a basis for judging whether a technical change represents a biological improvement. That is, the team wants to know if a new technology produces more from a given set of resources, meets farmers' requirements with less resources, or helps stabilize inputs and outputs. The team also wants to know whether the experimental results represent real improvements for farmers or whether the results could be simply due to chance. To do this, the team needs to consider the environmental setting in which the experiments are conducted. For example, the team should decide whether conditions such as rainfall, temperature, or labor supplies were sufficiently representative of "typical" conditions. If yes, then the team can reasonably conclude that statistically significant results should have general validity.

When experimental results are acceptable biologically, the team must still decide whether farmers will be interested and have the resources and capabilities to implement the changes. To help in deciding about the farmers' interests, the team can place monetary values on farmers' inputs and outputs for the current and new technologies. Results based on monetary values can then be compared with farmers' preferences concerning profit, risk, and other factors.

Our use of monetary values as a basis for comparing alternatives does not imply that farmers are only interested in money—e.g., that only those outputs sold for cash or those inputs purchased with cash count in the analysis. Rather, we use monetary values because this measure is the most convenient for comparing many diverse factors. We will pursue this topic further in the section on net benefits (Sec. 7.7.3.).

The teams should not confine their analyses to only those experiments that meet the biological scientists' criterion of a 0.01 or 0.05 significance level. Results with different levels of significance, such as 0.10, should also be analyzed. The reason is that biological researchers tend to be more rigorous in requiring that the technologies they propose have little chance of being recommended erroneously. Farmers, on the other hand, may be seriously in need of help. They may be willing to consider new technologies even when researchers are not convinced of the technologies' validity. This is especially true when the cost of failure is not great and when researchers might be delayed in conducting additional experiments to reach firmer conclusions about a technology's validity.

When biological and economic results are both satisfactory, the team still needs to check on financial feasibility. A technology is financially feasible when farmers are able to secure the cash resources for implementing the change and will subsequently be able to repay any borrowed money according to the terms of the agreement.

Finally, the team needs to observe how farmers react to alternatives that the team judges acceptable biologically, economically, and financially. Farmers' reactions can be obtained (1) from farmer-managed tests, (2) by noting acceptance rates of new technologies, and (3) through discussions with farmers. Where all indications are that a new technology should interest farmers but it does not, researchers need to study the situation further. The team should review its calculations. If all is in order, the team should examine its knowledge and assumptions about the farm household and the sociocultural environment. Then, the team might seek help from social scientists with appropriate experience and orientation.

7.6.2. PREDICTION VERSUS ACCEPTANCE

The foregoing discussion contains two approaches to establishing the validity of new technologies for farmers. One is the attempt to predict how farmers will react once exposed to the technology. The other is to let farmers have the technology and observe how they respond.

The Predictive Approach

When done well, the predictive approach is valuable. It forces the FSR&D team to systematically and logically evaluate a technology according to farmers' conditions. The team can try to predict how farmers might react should the new technology be made available to them. Clearly, this is the only type of appraisal that can be made for the planning of experiments, as discussed in Sec. 6.2. The approach is useful for analyzing the results of researcher-managed and superimposed trials and when deciding how to proceed. Even though researchers have the biological results of the trials, the analyses are still predictive. This is because the trials are under the researchers' control, are on small plots, and are based on estimates rather than measurements of farmers' responses.

Measurement of Acceptability

The other approach is to test the new technologies with farmers. In this way, researchers can tell through measurements, observations, and discussions with farmers how the new technology fares. Measurements of acceptability come primarily through farmer-managed tests and from recording farmers' actions and comments following the tests.

The field team influences farmer-managed tests in that farmers agree to a set of procedures and receive guidance from the team. However, farmers still have considerable management flexibility in implementing the tests. Because the plots are large and farmers manage them, the team has a good opportunity to obtain sound data on farmers' management practices. These practices include timing of activities and the allocation of the family's land, labor, and other resources. By obtaining actual data on yields, output prices, and types and costs of inputs, the team will be in a good position to calculate the profitability of the alternative technologies. We will take up the subject of farmer-managed tests as a measure of acceptability in Sec. 7.8.1.

The team will have further evidence of a technology's acceptability by recording farmers' acceptance of the technology following farmer-managed tests. The advantages of these measurements is that the influences of the researchers will have been removed so that the farmers will be acting on their own. We will say more about acceptability in Sec. 7.8.

Complete reliance on acceptability of test results has the disadvantage that the team obtains information on only the conditions prevailing at the time of the evaluation. When climatic or other conditions are "typical" and the farmers are representative, researchers can feel reasonably confident that the results can be applied to similar situations. But the team will find difficulty in explaining why farmers act the way they do, or how to interpret the results when conditions are not typical. For this, predictive studies are needed. Thus, both approaches to analysis contribute to an adequate understanding of proposed changes to farmers' systems.

7.6.3. PARTIAL BUDGET ANALYSIS VERSUS WHOLE FARM ANALYSIS

Different situations call for different approaches to analysis. Where the emphasis is on incremental change such as with seasonal crops, partial budget analysis is an appropriate approach. In contrast, research on more complex systems over longer periods of time generally calls for more comprehensive measures. Whole farm analysis is appropriate for these types of situations. We discuss both approaches below.

Partial Budget Analysis

Partial budget analysis is one of the simplest methods for analyzing the acceptability of relatively minor changes to ongoing operations. Brown (1979) reported,

> "It is a form of marginal [incremental] analysis designed to show, not profit or loss for the farm as a whole, but the net increase or decrease in farm income resulting from the proposed changes."

Such an approach is particularly well-suited to FSR&D when small changes are made to the farmers' systems. For many of the proposed treatments, farmers will make small, incremental changes to the way they farm. These changes apply to only a limited number of activities; the rest of the farmers' activities remain the same. Partial budget analysis is not only appropriate under these conditions, it is the preferred approach. The reason for the preference is that the analyst needs concentrate on only the *changes* to the farmer's system, not the whole farm. By lessening the workload, the researchers are able to focus their attention on a relatively few factors, which helps improve the quality and timeliness of the analysis. These changes refer to the differences to the farmers' system with and without the proposed change in technology.

Whole Farm Analysis

Should the introduction of new technologies force farmers to reorganize substantial portions of their activities, either all at once or over time, then partial budget analysis is inappropriate. The whole farm must be studied. Whole farm analysis can be conducted by broadening the scope of analysis by observing, recording, and otherwise estimating all of the farmers' activities, including the family's nonfarm enterprises.

The need to broaden the perspective to the whole farm may occur when considering livestock and crops as a mixed system. ILCA favors whole farm analysis for much of its work, as the following quotation from de Haan (personal communication) indicates:

> ". . . we find whole farm tests sometimes absolutely necessary. For example, the traditional system in the Ethiopian Highland programme is characterized by a small farm size (± 2 ha) [in which 70 percent of the area is] used for subsistence grain cropping and with a serious deficit in animal nutrition for livestock with a low genetic capability for traction and milk production. Introducing forage at the level of 1000 M^2 plots per farm would not prove anything, as it would not contribute enough to the total fodder supply. Therefore in this case it was necessary to (a) increase yields per ha of subsistence crops, (b) introduce forages in the space so created, and (c) use this forage through genetically improved livestock. Only a whole farm approach was therefore possible here."

When such studies are conducted informally, the team will be able to understand reasonably well the interrelationships among the parts. Through comparative analyses, the team can consider a broad range of factors and generally conclude whether improvements over the farmers' existing system have been found.

However, searching through a range of feasible solutions is generally complex and time consuming. Instead, some form of modeling procedure is usually required. For example, IRRI (Jayasuriya, 1979) uses linear programming as part of its approach and ILCA (1980) uses a variety of systems research techniques. The linear programming studies search for optimum solutions through consideration of farmers' objectives, values of output, production relationships, and available resources to arrive at improved cropping and livestock patterns and management practices. Experiments, farm records, monitoring, surveys, and special studies provide inputs to such optimization studies. We emphasize, however, that these are relatively sophisticated studies that should not divert the attention of field teams. Rather, such studies are long-range approaches that, when integrated with field teams' on-farm research, help in setting the directions of future research.

7.7. ILLUSTRATIVE DESIGNS AND ANALYSIS PROCEDURES

We begin this section with an illustration of alternative designs. Then, we follow with (1) a discussion of biological results and (2) an application of partial budget analysis and other techniques for evaluating economic and financial feasibility.

7.7.1. ILLUSTRATIVE DESIGNS

Perrin et al. (1976) illustrated the application of economic analysis to the biological results from three cropping experiments. Two of these are what they call "Yes-No" experiments and the other is a "How Much" experiment. A Yes-No experiment aids the researcher in deciding whether to recommend a new technology—e.g., the introduction of a new crop variety or the use of fertilizers. A How Much experiment provides the researcher with information useful in recommending the level at which a technology should be applied—e.g., the *amount* of fertilizer to apply or the seeding rate.

The Yes-No experiments can be farmer-managed tests when kept simple enough for the farmers to manage. In one experiment concerning maize production, Perrin et al. (1976) compared farmers' traditional technology with a high-level technology comprising nitrogen and phosphate fertilizers, soil and foliar insecticides, and herbicides. These two treatments were conducted at 26 locations within the research area. While such an experiment would normally be too complicated for most farmers to manage, some farmers who have experience with experiments might be able to conduct such tests. Until then, however, the experiments should probably be conducted as researcher-managed tests.

A second experiment compared three wheat varieties with and without fertilizer application. Of the six treatments—i.e., 3 varieties × 2 fertilizer levels—one represented farmers' conditions, namely, the traditional variety without fertilizer—i.e., zero level of application. Following our suggestion in Sec. 6.4.3., this should still be a researcher-managed or superimposed trial because the number of treatments is greater than four. This is another example of a Yes-No experiment in which researchers search for the best variety with and without fertilizers. The fertilizer level for the experiment could come from the researcher's prior knowledge, or from earlier researcher-managed trials.

The third experiment described in Perrin et al. (1976) was designed to identify the appropriate level of application of nitrogen and phosphate. This How-Much experiment involved 12 treatments and eight replications and is appropriate for researcher-managed trials. The experiment considered 12 combinations of these two fertilizers applied to the farmers' variety. The applications started with the lowest level of control—i.e., no nitrogen or phosphate—and gradually increased to a maximum application of 150 kg/ha of N and 50 kg/ha of P_2O_5. Because of its detail, we will use this example in subsequent sections to illustrate the techniques of partial budget analysis.

7.7.2. BIOLOGICAL RESULTS

Table 7-3 provides data on the 12 fertilizer treatments, including yields for each of eight trials—i.e., replications—and the averages for each treatment and trial. We will begin by looking at the averages of these treatments and later note the variability of results.

Fig. 7-2 depicts the expected increases in yields with increasing levels of fertilizer application at three levels of phosphate. The rates of nitrogen application represent the relevant range when nitrogen is applied without phosphate—i.e., little additional yield is gained from additional applications of N. However, the slopes of the two curves where 25 kg/ha and 50 kg/ha of P_2O_5 are applied with nitrogen indicate that yields have not begun to peak out. However, as we shall see shortly, higher levels of P_2O_5 applications are not needed to draw relevant conclusions from this experiment. The reason is that the value of the additional yield from applying extra fertilizer is not great enough to offset the additional costs of materials and labor. To show how this works, we must first discuss the concepts associated with net benefits and farmers' preferences. However, before turning to this subject, we wish to comment on the setting within which biological results like these take place.

To put biological results such as these into perspective, the FSR&D team needs to consider the environmental conditions prevailing during the experiment. For example, if weather conditions are considered "typical" for the area, the team can proceed with the analysis of results. By typical we mean those weather conditions that lead to crop yields or livestock production and health that are likely to prevail more often than not. If, however, weather conditions were not "typical," the team should exercise caution when drawing conclusions from the results. Yield results for atypical weather conditions can be quite different from those occurring when weather conditions are close to the average.

Even when the weather appears to have been typical during the growing season, the team should investigate

Table 7-3. Maize yields by fertilizer treatment, eight trials (Adapted from Perrin et al., 1976).

	Fertilizer treatment (kg/ha)												
N:	0	50	100	150	0	50	100	150	0	50	100	150	
Trial P_2O_5:	0	0	0	0	25	25	25	25	50	50	50	50	Avg.
	tons/ha												
1	0.40	1.24	3.63	3.76	0.79	2.58	4.23	4.72	1.67	2.51	3.28	3.66	2.71
2	1.53	2.60	5.14	5.32	1.67	3.79	5.10	6.83	1.41	4.13	5.89	6.27	4.14
3	4.15	4.86	4.80	4.87	4.44	5.00	4.97	5.28	5.12	5.66	6.36	6.62	5.18
4	2.42	3.82	5.23	4.48	2.36	4.54	6.26	7.17	1.61	4.41	5.38	6.58	4.52
5	1.64	1.92	2.08	2.19	2.04	3.21	3.12	2.93	1.44	3.44	3.32	3.62	2.58
6	1.61	2.94	4.14	4.34	1.81	3.92	3.61	3.81	1.18	3.89	5.38	4.92	3.46
7	4.74	5.41	4.29	4.92	4.91	5.22	5.38	5.14	5.10	4.88	4.54	5.28	4.98
8	1.21	2.33	1.97	2.23	1.53	2.78	2.49	2.80	1.37	3.51	3.75	4.35	2.53
Avg.	2.21	3.14	3.91	4.01	2.44	3.88	4.40	4.84	2.36	4.05	4.74	5.16	3.76

Figure 7-2. Average yield response to nitrogen [at three levels of phosphate] (Perrin et al., 1976).

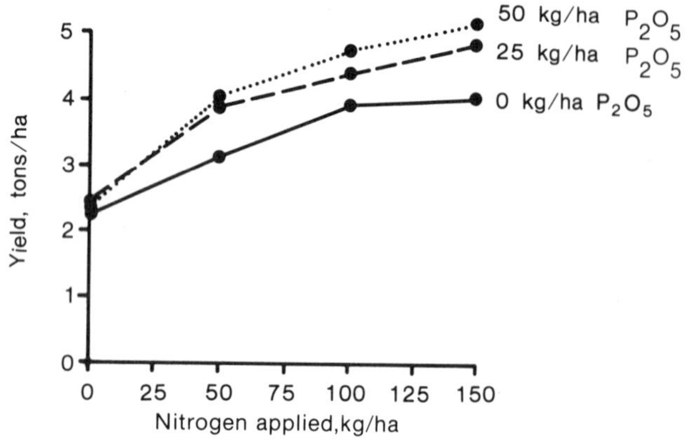

the effects of weather during each phase of the plant's growth. For example:

- Was the crop establishment slowed or enhanced by too little or very favorable rainfall?
- Is a new crop variety needed that is more adaptive to variations in rainfall conditions?
- What was the possible effect on the crops of winds or very low or high temperatures?
- Did farmers change the timing or type of operations or make other adjustments to compensate for variations in weather and moisture availability?

We have provided Appendix 6-A on climate, Appendix 7-A on crops, and Appendix 7-B on livestock as guidelines for collecting data that should be useful to the team in answering the above types of questions.

Such an analysis will help the team understand the variability that normally occurs even when "typical" weather conditions prevail. Because biological performance varies from year to year, data from several seasons may be needed before recommending a new technology to farmers—especially when that technology constitutes a substantial change in the farmers' present system, as with new varieties or cropping patterns. In other cases, new technologies can be recommended more quickly. For example, recommendations for the removal of a severe nutrient deficiency or toxic soil conditions as with soil salinity can generally proceed without delay.

The team should also raise the following types of questions when conditions are not typical. For example, how do farmers respond to unique events such as typhoons, floods, droughts, or volcanic activity; and how often do these incidents occur?

The team can obtain much of the information needed to evaluate these conditions from observations of farmers' management practices. Some of the farm records, monitoring, and other data gathering can help fill in the gaps in the team's knowledge of on-farm conditions. Although many of the team's conclusions about interactions between crop performance and environmental conditions must, of necessity, be only impressions, such evaluations are important. Team members need to reflect on the meaning of the experiments and the transferability of experimental results over time and across areas.

7.7.3. NET BENEFITS

Farmers will undertake change when they perceive that the benefits will outweigh the costs. The difference between the two are the net benefits. For partial budget analysis, net benefits represent the increases in benefits over costs, when compared with the farmer's existing farming enterprises. For example, the benefits might be an increase in maize yields that require additional inputs of labor and fertilizer. When the net benefits are great enough, we can expect farmers to accept the change.

Because such changes involve different items—e.g., labor, materials, services, and farmers' outputs, we need some means for expressing them in common units. As we indicated earlier, monetary values serve this purpose. When items are traded for money, these transactions can serve as the value to the farmer; but when items are not bought and sold, some other reference of value is needed. For this we introduce the concept of opportunity cost. The opportunity cost of an input to, or of an output of, the farmers' system is its value in the best alternative use. A few examples should help clarify this concept:

If a new maize technology requires the farmer to spend more time fertilizing the field and if the farmer would ordinarily work in town during this period for 20 pesos/day, then a day's work on the farmer's field has an opportunity cost of 20 pesos, less the daily cost of going to and from town.

Or, if the family normally purchases maize for consumption, then the additional maize output from the improved technology might replace part or all of the family's maize purchases. The benefit of the additional maize production that substitutes for purchases could be measured using the price of maize in the market place, adjusted by (1) adding the benefit of not having to transport maize from the market to the farm and (2) subtracting the cost of having to harvest the maize. The common point of comparison between the maize the farmers produce with the new technology and the maize they would otherwise have to purchase is maize available for consumption at the farm.

Alternatively, let us suppose that the family already produces its minimum maize requirements, but decides to consume more maize from the increased yield of the improved technology. Now, the benefit of the maize production is measured using the field price of maize. When farmers consume the maize instead of selling it, we can assume that they value it at least as much as the price they would receive from its sale.

In the first example, the cost of the farmers' labor in applying the fertilizers is estimated by the money the farmers give up by not working in town. In the next two examples, benefits of increased maize output are estimated differently depending on whether farmers normally sell or purchase maize. In Appendix 7-G we provide additional details on estimating farmers' net benefits from alternative treatments.

By estimating the value of money transactions and

the opportunity costs, the team can approximate the net benefits of technology changes to farmers. Of course, some values are difficult to estimate in these terms—e.g., the value ranchers attach to their livestock or to their children being able to attend school. These values should be noted and used as additional factors in predicting farmers' reactions. Fortunately, many changes do not involve such issues and the net benefit calculations are sufficient to produce good results. Risk is another factor to consider when predicting farmers' reactions. Later, we will introduce the approach Perrin et al. (1976) suggested for dealing with risk.

In estimating benefits and costs, the team should strive to identify the most important ones. This includes estimates of the types, amounts, and values for each factor important to the analysis. While some of these factors may be difficult to estimate, we advise the team to make "educated guesses" rather than to ignore them simply because they cannot be estimated precisely.

Where values are likely to change, the team will probably need help from regional or national headquarters economists in projecting future values. Also, instead of relying on single-valued estimates for the more uncertain and important factors, the team can use sensitivity analysis in considering alternative values.

Sensitivity analysis, which we will illustrate later, is an approach whereby alternative values are varied individually to learn to what extent they would change the conclusions about the proposed technologies. For example, the team can make estimates over a range of values for a commodity to learn how these changes influence the economic attractiveness of the new technology. The range of values tested should represent conditions reasonably likely to occur—perhaps 8 or 9 chances out of 10. If the change in values produces a switch in the farmers' choice, then the outcome is sensitive to this factor. Factors found to be sensitive generally require further analysis, whereas factors that are not sensitive do not require further analysis.

Estimating Benefits

For partial budget analysis, benefits are generally increases in output, decreases in inputs, or reductions in risk. We will discuss risk shortly. For now, let us consider the benefits from increased yields resulting from the application of fertilizers. Table 7-4, which is based on the average maize yields from the treatments presented in Table 7-3, shows the gross field benefit for each of the 12 treatments in dollars per hectare. In this case, the dollars simply represent some national currency, not the currency of any particular country.

Gross field benefits are obtained by multiplying the net yield per hectare times the field price. The net yield is that average output of the crop per hectare, less harvesting losses and any storage losses. For maize and wheat, Perrin et al. (1976) suggested that these losses could be about 10 percent, or possibly more. In Table 7-4, assuming losses of 10 percent, they estimated net yield as follows: e.g., for N = O and P_2O_5 = O, 90 percent × 2.21 tons/ha = 1.99 tons/ha. Henceforth, we shall refer to the treatments by the amounts of fertilizers applied—e.g., (0,0) represents the above case, (50,0) represents 50 kg/ha of N and no P_2O_5, (0,25) represents no N and 25 kg/ha of P_2O_5, and so on.

The field price is called the money field price when farmers sell their crops in the market or the opportunity field price when they consume the crops. The money field price is the market price less costs that vary with yield such as harvesting, transporting, storing, and other such costs incidental to converting the crop from the field to the point of transfer to the buyer. In Table 7-4 the gross field benefit of $1,000/ton of maize assumes a market price of $1,200/ton and $200/ton for the costs that vary with yields. Should the farmer sell the crop to a buyer who picks up the harvested grain at the farmer's field, the market price would be lower and the farmers would not incur the costs of transportation.

The opportunity field price is the amount of money the farmer either (1) forgoes by consuming the grain rather than selling it, or (2) saves by not having to purchase the grain in the market.

When harvests are drawn out, or the farmer stores the grain and sells it over an extended period, the team should estimate the market values over the period of the sales. Since prices may vary considerably over such a period, the team should use some weighted average of the sales prices. Where the farmer stores the grain before selling it, the associated costs of storage, such as labor for drying and handling should be included. The costs of the storage structures should be estimated as well. When farmers' storage structures are already in place, these are fixed costs and should not be counted; when such storage structures must be built to accommodate the increase in yields, these costs become variable costs and should be estimated. We discuss long-term investments in Sec. 7.10.2.

When a crop yields more than one item of value, such as straw from wheat and rice or stalks and leaves from maize, the team should value these items in the same way as described for grain. The field price would be based on either what the farmer or others would pay to obtain these items, or on the eventual increase in output from using them. For instance, if maize stalks and leaves are fed to cattle, their value is the increase in yield from cattle products for having consumed these items less the increased costs such as labor associated with the feeding.

These benefits accrue to farmers in full when they own the land and provide all inputs. When someone else owns the land, farmers generally share the crop or pay a fixed rent. When the crop is shared, this naturally reduces the gross field benefits the farmer receives. These benefits are reduced according to the terms of the agreement. When farmers pay a fixed rent regardless of intensity of land use, this is no longer a variable cost and does not enter into the partial budget calculations. We provide additional comments on fixed and variable costs in the next section.

Estimating Variable Costs

The partial budget approach is based on the distinction between variable and fixed costs. Variable costs are

Table 7-4. Partial budget of averaged data from fertilizer trials, per hectare basis (Perrin et al., 1976).

Item		Fertilizer treatment (kg/ha)											
	N:	0	50	100	150	0	50	100	150	0	50	100	150
	P_2O_5:	0	0	0	0	25	25	25	25	50	50	50	50
(1) Average yield (ton/ha)		2.21	3.14	3.91	4.01	2.44	3.88	4.40	4.84	2.36	4.05	4.74	5.16
(2) Net yield (ton/ha)		1.99	2.83	3.52	3.61	2.20	3.49	3.96	4.36	2.12	3.64	4.27	4.64
(3) Gross field benefit ($/ha at $1000/ton)		1990	2830	3520	3610	2200	3490	3960	4360	2120	3640	4270	4640
Variable money costs:													
(4) Nitrogen ($8/kg N)		0	400	800	1200	0	400	800	1200	0	400	800	1200
(5) Phosphate ($10/kg P_2O_5)		0	0	0	0	250	250	250	250	500	500	500	500
(6) Variable money costs ($/ha)		0	400	800	1200	250	650	1050	1450	500	900	1300	1700
Variable opportunity costs:													
(7) Number of applications		0	1	2	2	1	1	2	2	1	1	2	2
(8) Cost per application (2 days at $25)		50	50	50	50	50	50	50	50	50	50	50	50
(9) Opportunity cost ($/ha)		0	50	100	100	50	50	100	100	50	50	100	100
(10) Total variable costs ($/ha)		0	450	900	1300	300	700	1150	1550	550	950	1400	1800
(11) Net benefit ($/ha)		1990	2380	2620	2310	1900	2790	2810	2810	1570	2690	2870	2840

simply those costs that vary with the level of output; fixed costs remain constant with changes in the level of output. The principle on which partial budget analysis rests is that the farmer is already producing at a satisfactory level and will continue to do so without changes in the system. Thus a change that produces positive net benefits represents a more profitable position for the farmers. Whether the farmer will change depends on the farmer's attitude toward profit, risk, and other factors. Fixed costs are already being incurred by the farmer and will remain at the same level whether the farmer accepts the change or not.

As with benefits, variable costs can be divided into those that represent variable money costs and variable opportunity costs. The concepts are the same as with benefits. A variable money cost such as for nitrogen and phosphate shown in Table 7-4 is the cost of purchasing these fertilizers in the market. If additional labor is hired for delivering fertilizers or for their application (not shown in Table 7-4), then these labor costs would be included as a variable money cost.

Variable opportunity costs are variable costs not incurred through money transactions. The most important of these is the value of family labor. The value to be placed on such labor depends on the opportunities open to the farmers and the farmers' preferences. Below are possible situations:

Busy Periods When Hired Labor is in Great Demand

During busy periods, farmers will usually be looking for hired labor to help them with their crops or animals; they would not be interested in working for others during these times. The reason is that their labor produces more value for them when applied to their farms than from the income earned off their farms. This means that the opportunity cost of their labor is higher than the market wage. Perrin et al. (1976) suggested that the opportunity cost of the farmers' labor might be 125 percent of the cost of hired labor. That is, if the wage for hired labor is \$20/day, the opportunity cost of the farmers' labor would be \$25/day (i.e., \$20/day × 125 percent). The premium of 25 percent—i.e., 125 percent − 100 percent—over the going wage rate represents the higher value of the farmers' labor from working on their own farms rather than from working off their farms.

Slack Periods When Hired Labor is Not in Great Demand.

Were the farmers not to accept the new technology, which requires extra labor, they might have engaged in other productive activities. For example, they might have worked in their fields, repaired their equipment, or engaged in handicrafts. The value of these other activities is usually less than the going wage rate during slack periods. The team might use the value of such increased production to estimate the value of the farmers' labor. Thus, if farmers spend more time weeding a crop, this extra effort should lead to a higher output for that crop. The value of this greater output can then be used for estimating the opportunity costs of the farmers' labor. Alternatively, Perrin et al. (1976) suggested a value ranging somewhere between 50 and 75 percent of the going wage rate. Part of the reduction is due to farmers' preferences for working on their own farms. When off-farm employment opportunities coincide with the farmers' slack period, the percentage might be higher than 75. Another alternative to working on the new technology would be for the farmers to spend their time in leisure. Since leisure is worth something to the farmers, researchers are seldom justified in assigning a zero opportunity cost to the farmers' labor.

With these concepts in mind, the team should be able to estimate the variable costs of the different treatments. For this, the team needs to understand both the technology being tested and the farmers' management methods. In the example in Table 7-4, the treatment (50,0), is estimated to cost \$450/ha—i.e., \$400/ha for the nitrogen at \$8/kg N and \$50 for application of the fertilizer. The \$50/ha is the opportunity cost of the farmers' labor. Had the farmers hired labor to apply the fertilizer, this labor cost would have shown up as a money cost—unless the farmers paid the laborers with grain or in some other way.

For the treatment (50,0), the value of net benefit shown in Table 7-4 is \$2,380/ha, in which the gross field benefit is \$2,830/ha and variable costs are \$450/ha. Out of this \$2,380/ha, farmers (1) must pay their fixed money costs, such as any rent, taxes, and seed, and (2) hope to receive a satisfactory return for the use of their fixed resources—i.e., land, labor, capital, and management. Remember that these costs are considered fixed in this case because they do not change with the purchase and application of fertilizers or the other activities. Had the treatment involved changing seed or weeding practices, then the costs of seed and changes in weeding would have become variable costs instead of fixed costs.

One final comment on costs. Had the value of additional fodder from the increases in grain yield been calculated, these benefits could simply have been added to the gross field benefit of the grain; and the net benefit at the bottom of the table would have increased. Any additional costs of handling the fodder would also have been added to the variable costs. The net benefit at the bottom of the table would then have incorporated these additional values. Normally, little would be gained by trying to allocate—i.e. assign—costs to the grain and fodder activities.

7.7.4. ECONOMIC CRITERIA

Economically feasible alternatives for farmers can be found by considering four factors:

- dominance among alternatives
- the marginal rate of return
- the farmer's minimum acceptable return
- allowances for risk.

Dominance

By graphing the results of the net benefit calculations from Table 7-4, we can gain a clearer understanding of their meaning. Fig. 7-3 shows the relationship among the net benefits and the level of variable costs. The solid line connects the most attractive values for the different amounts of variable costs. The dotted line simply indicates what Perrin et al. (1976) considered a more logical progression from lower to higher levels of fertilizer ap-

Figure 7-3. Net benefit curve for the fertilizer trials. Numbers in parentheses represent kg/ha of N and P_2O_5 respectively (Perrin et al., 1976).

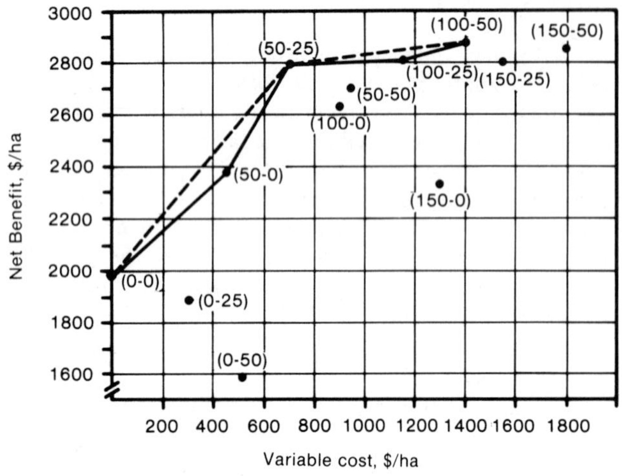

plication. The shape of the dotted line—and even of the solid line—is a reasonably good approximation to the expected yield response from increased applications of fertilizer.

By studying Fig. 7-3, we can conclude that points on the solid line are superior to those below the line. Why? Because net benefits are greatest for any given level of variable cost. For example, (50,0) at a variable cost of $450/ha produces $2,380/ha in net benefits, whereas (0,50) at a variable cost of $550/ha produces only $1,570/ha in net benefits. The latter treatment costs more and produces less and therefore is a poor choice.

This type of analysis leads to the identification of those alternatives that are undominated by the others—i.e., are superior to the others. Table 7-5 shows those alternatives that are undominated and therefore remain as viable alternatives to consider in predicting which of the treatments the farmers might prefer.

Table 7-5. Dominance analysis of fertilizer response data (Perrin et al., 1976).

Net benefit ($/ha)	*Fertilizer treatment (kg/ha)* N	*Fertilizer treatment (kg/ha)* P_2O_5	*Variable cost ($/ha)*
2870	100	50	1400
2840	*150*	*50*	*1800**
2810	100	25	1150
2810	*150*	*25*	*1550**
2790	50	25	700
2690	*50*	*50*	*950**
2620	*100*	*0*	*900**
2380	50	0	450
2310	*150*	*0*	*1300**
1990	0	0	0

*Dominated alternatives.

Marginal Rate of Return

The reader might be inclined to think that (100,50) would be the best alternative, since this treatment produces the highest net benefits—i.e., $2,870/ha. But this conclusion is not necessarily correct. To show why, we need to consider how the farmer feels about spending additional money and effort in variable costs to obtain extra net benefits. For this, we introduce the concept of the farmer's marginal rate of return, which is the incremental increase in net benefit divided by the corresponding increase in variable cost.

The initial calculation of the marginal rate of return should be made by comparing the incremental net benefit and the incremental variable cost for the undominated alternatives with the two lowest variable costs. For example, using values from Table 7-5, we can calculate the marginal rate of return between (0,0) and (50,0) as follows:

$$\frac{\text{Incremental net benefit}}{\text{Incremental variable cost}} = \frac{2380 - 1990}{450 - 0} = \frac{390}{450} = 0.87$$

This value of 0.87 is usually represented as a percent and means that for every dollar spent as a variable cost, the farmers recover that dollar plus 87 cents.

Should this level of profit satisfy the farmers and if they have the resources, we could expect them to abandon the old practice (0,0) in favor of the new one (50,0).

This procedure continues by comparing pairs of undominated alternatives and moving progressively to higher levels of variable cost. After each marginal return is calculated, it is compared with the farmers' minimum acceptable return. A farmer's minimum acceptable return is the minimum percentage that would interest the farmer in spending extra money. If the marginal rate of return is above the minimum acceptable return, then the higher level of expenditure is acceptable; otherwise it is not.

If the higher level is accepted, then this new level is tested against the next higher level of variable cost of an undominated treatment. In this case, we compare (50,0) with (50,25)—using values from Table 7-5—and the marginal rate of return is

$$\frac{\text{Incremental net benefit}}{\text{Incremental variable cost}} = \frac{2790 - 2380}{700 - 450} = \frac{410}{250} = 1.64$$

Assuming that the earlier rate of return, 87 percent, is acceptable, we can be reasonably confident that the 164 percent rate is also acceptable.

The analysis proceeds in this way until the marginal rate of return falls below the minimum acceptable return. When this occurs, the team compares the previously acceptable treatment with the undominated treatment with the next higher variable cost. Let us suppose the team had evidence that farmers would reject the increment yielding an 87 percent rate of return. In that case, the next comparison would be between the traditional practice (0,0) and (50,25). The marginal rate of return would then have been

$$\frac{\text{Incremental net benefit}}{\text{Incremental variable cost}} = \frac{2790 - 1990}{700 - 0} = \frac{800}{700} = 1.14$$

The team must now decide if farmers would accept or reject a marginal rate of return of 114 percent.

Proceeding in this way, the team progressively compares all relevant alternatives—i.e., those that are undominated. The alternative remaining is the one the farmers should find economically most attractive, provided the team has estimated the farmers' reactions correctly and no major differences in risk are associated with the alternatives.

The calculation of the marginal rates of return for the five undominated alternatives is shown in Table 7-6. Intuitively, we would expect the farmers to be interested in increasing variable costs up to the (50,25) level. Beyond that level the farmers' marginal rates of return are considerably less and could very well be unaccceptable. Consequently, we tentatively conclude that the treatment most likely to be of interest to farmers is neither the one producing the highest yield (150,50) nor the one producing the highest net benefit (100,50), but (50,25). However, we cannot be confident in this conclusion without first considering in more detail the farmers' minimum acceptable return.

Minimum Acceptable Return

Very high levels of marginal returns will generally interest farmers in change. As these levels decrease, farmers become less interested. At some point, they would no longer be willing to incur additional variable costs. We call this point the farmers' minimum acceptable return. Although risk can be handled in different ways, we follow the suggestion of Perrin et al. (1976) and include a risk premium to the cost of capital to estimate the minimum acceptable return. The risk premium is an amount added to the cost of capital to compensate the farmers for the uncertainties of change and any increased variability in outcomes. Farmers react differently to risks depending on the degree of change, the uncertainty of the outcomes, the farmer's financial position, and similar factors. Until the FSR&D team gains a feeling for the farmers in its areas, we suggest accepting the 20 percent risk premium per cropping season as Perrin et al. (1976) recommended.

The cost of capital can be either the effective rate charged by lenders or the opportunity cost of the farmers' own funds. The effective rate is what borrowed money actually costs farmers. Because we will be dealing with seasonal rates in the illustrations that follow, the effective interest rate will be computed for the season and not for a full year. The effective seasonal rate includes the stated interest rate and any additional costs such as a loan servicing charge and insurance. Although we did not do so, we could have added to the effective interest rate the value of the farmers' time in acquiring and repaying the loan.

Table 7-6. Marginal analysis of the undominated fertilizer response data, per ha (Adapted from Perrin et al., 1976).

	Fertilizer treatment				Change from next highest benefit		
Row	N	P_2O_5	Net benefit	Variable cost	Marginal increase in net benefit	Marginal increase in variable cost	Marginal rate of return
(1)	(2)	(3)	(4)	(5)	(6)	(7)	(8)
(a)	100 kg	50 kg	$2870	$1400	$ 60	$250	24%
(b)	100	25	2810	1150	20	450	4
(c)	50	25	2790	700	410	250	164
(d)	50	0	2380	450	390	450	87
(e)	0	0	1990	0			

Example of calculations: the top amount in column 6 ($60) is the difference between the amounts for lines a and b in column 4; the top amount in column 7 ($250) is the difference between the amounts for lines a and b in column 5; and the top amount in column 8 (24%) is the top amount in column 6 divided by the top amount in column 7.

Based on an example from Perrin et al. (1976) we can show how an annual interest rate of 12 percent for a fertilizer loan ends up being 21 percent on a seasonal (six months) basis. The calculations are as follows:

$1000	cost of fertilizer
× 0.12	annual interest rate of 12 percent
120	annual interest
× 0.5	fraction of the year (6 months)
$ 60	interest charge
$1000	amount borrowed
60	interest charge
50	service fee
100	insurance premium
$1210	Total

The farmer receives the $1,000 loan to buy fertilizer and must pay back $1,210 for an effective six month's rate of 21 percent. Combining this effective interest rate with a risk premium of 20 percent, we end up with a miniumum acceptable return of 41 percent that the farmer must receive to be interested in the fertilizer treatments. A marginal rate of return of 41 percent gives the farmer 21 percent to repay the bank loan and leaves 20 percent compensation for the risk of making the change. Because these are approximate values, we will henceforth use a rounded value of 40 percent to represent the minimum acceptable return.

Earlier, in Sec. 6.2.2., we proposed a minimum yield increase of 30 percent as a possible lower limit to interest farmers in a new technology, whereas here we propose a minimum return of 40 percent. These two limits, which come from different sources and are only approximations, are not necessarily in conflict. First, farmers may simultaneously have a yield increase of 30 percent and a minimum acceptable return of 40 percent. For example:

1) Net yield increase from 2.0 tons/ha to 2.6 tons/ha = 0.6 tons/ha (30% increase)
2) Gross field benefit at $1,000/ton gives an increase from $2,000/ha to $2,600/ha = $600/ha (30% increase)
3) Increase in total variable costs from $800/ha to $1,229/ha = $429/ha
4) Net benefit from $1,200/ha (i.e., 2,000 - 800) to $1,371/ha (i.e., 2,600 - 1,229) = $171/ha
5) Minimum acceptable return =

$$\frac{\text{Incremental net benefit}}{\text{Incremental variable cost}} = \frac{1371-1200}{1229-800} = \frac{171}{429} = 0.40$$

Second, the same farmers may require different rates of return under different conditions. For example, Zuberti et al. (1979) points out that many farmers require higher rates of return the greater the money inputs associated with the improved technology.

Another comment on the 40 percent minimum is that this percentage includes the cost of interest and other charges paid to the bank. This is the reason loan costs were not added to the production costs of fertilizer and labor in the section on Estimating Variable Costs (Sec. 7.7.3.).

Returning to Table 7-6, we can now see that the marginal rates of return of 87 percent and 164 percent would be adequate to interest a farmer who requires a minimum return of 40 percent. In like fashion, the farmer would not be interested in incurring additional variable costs to implement treatments (100,25) or (100,50). The marginal rate of return between (50,25) and (100,25) is only 4 percent. Since (100,25) is unacceptable, we must now compare (50,25) with (100,50). The marginal rate of return for this comparison is only 11 percent, as follows:

$$\frac{\text{Incremental net benefit}}{\text{Incremental variable cost}} = \frac{2870-2790}{1400-700} = \frac{80}{700} = 0.11$$

Using values from Table 7-6, the most economically attractive alternative appears to be our original choice of alternative (50,25).

Before closing this section, we need to mention that farmers will sometimes use their cash reserves to purchase inputs such as fertilizers. Even though using these funds does not require the farmer to pay an interest charge to the bank, their use to purchase fertilizer also has an opportunity cost. Had the farmer not used the funds to purchase fertilizer, the funds could have been loaned or applied to the farmer's other activities. To keep from complicating the discussion further, we can assume that they are loaned at approximately the same rate as the bank. When this is not the case, the team will need to estimate the farmers' opportunity costs of capital if the team expects to predict farmers' reactions reasonably well.

Risk

Besides the risk premium, which is largely to compensate farmers for the uncertainties of change, farmers are also concerned with the variability in outcomes of the alternative treatments. So far we have suppressed information about variability by dealing only with average yields. Perrin et al. (1976) suggested considering the lower 25 percent of the yields as a way to account for the variability of outcomes. When selecting the lowest 25 percent of the yields, the team should not eliminate complete failures as long as they represent conditions farmers would face—even though infrequently.

Perrin et al. (1976) suggested the lowest 25 percent of the values, instead of the single lowest value, since farmers are not likely to base their decisions on a single worst outcome. Stated differently, farmers are accustomed to uncertainty and although they prefer to reduce it, they probably would not reject a good opportunity because of the chance that a single outcome would be unfavorable.

With this approach in mind, we can return to Table

7-3 and identify the individual yields for each trial of each treatment. Then, we can calculate individual net benefits. The method for calculating the net benefit is the same as presented in Table 7-4. Thus, for trial 1 of (0,0) in which the yield was 0.40 tons/ha, the net benefit is \$360/ha (i.e., 0.40 tons/ha × 90% to convert to net yield × \$1,000/ton = \$360/ha net benefits). Net benefits for each of the yields are calculated in the same way and shown in Table 7-7.

Using the worst 25 percent of the trials means looking at the average of the worst two of the eight trials in our example. These values are shown in Table 7-8. From this, we can see that the choice of (50,25) is nearly the best when considering the average of the worst two trials. This average for (50,25) is \$1,710/ha, which is only \$20/ha less than \$1,730/ha for (50,50). If the farmers were seriously concerned about picking the alternative with the highest value in Table 7-8, they would pick (50,50). On the other hand, the overall average of the eight trials for (50,50), as shown in Table 7-7, is \$2,690/ha, which is \$100/ha less than \$2,790/ha for (50,25). If the farmers prefer (50,50) over (50,25), they would be saying that they are willing to give up \$100/ha as an overall average to save \$20/ha for the average of the worst two trials. We doubt that farmers would make this choice, but this is a question that the team should put to the farmers for their reaction.

This example does not present us with a problem of choice, since the treatment with the highest acceptable net benefit is also nearly the best for the worst 25 percent of the trials. Should this situation not occur, the team would have to probe more deeply to learn how farmers feel about the trade-offs between average net benefits and stability of results. One possibility is for the team to apply a still higher risk premium when considering treatments with high average yields yet low minimum yields.

Sensitivity Analysis

Besides the above, the team can also test the stability of results by looking into alternative possibilities for some of the other values affecting net benefits. Examples include the price of maize, fertilizer costs, and the opportunity cost of labor. For instance, what would happen to the choice between the (50,25) and (100,50) should the price of maize increase (see Fig. 7-3)? One way to approach this issue is to find the price of maize that would cause the farmer to change in favor of the larger fertilizer application.

With this particular case, the break-even point occurs when the extra variable cost of the more expensive alternative yields a 40 percent rate of return. Calculations for the break-even point—i.e., the value of maize that makes the farmers indifferent between (50,25) and (100,50)—are shown below. For this calculation, we must separate the gross field benefits into the net yield of maize and the field price of maize. We obtain the former from Table 7-4 and let P represent the latter. Then, using the same type of calculations as before:

$$\frac{\text{Incremental net benefit}}{\text{Incremental variable cost}} = 0.40$$

$$\frac{(4.27P - 1400) - (3.49P - 700)}{1400 - 700} = 0.40$$

$$\frac{0.78P - 700}{700} = 0.40$$

Solving for P, we have $0.78P = 980$

and $P = 1256$

Checking on this result, we have

Net benefit:

(100,50)	$4.27 \times 1256 - 1400 =$	3963
(50,25)	$3.49 \times 1256 - 700 =$	3683
	Incremental net benefit =	280
	Incremental variable cost = $1400 - 700 =$	700

$$\frac{\text{Incremental net benefit}}{\text{Incremental variable cost}} = \frac{280}{700} = 0.40$$

Thus, at a field price of \$1,256/ton, (100,50) yields \$280/ha more in net benefits and costs \$700/ha more in variable costs than does (50,25). The marginal rate of return for the incremental variable costs of (100,50) over (50,25) is 40 percent. This confirms \$1,256/ton as the break-even value of P.

The reader may recall that the value of P was reduced by \$200/ha to convert the market price to a field price. By adding \$200 back again, the comparable market price would be \$1,456/ha. Now that the team has this value, it may need to consult those who study price trends in the region to learn the extent to which a price increase for maize of some 21 percent—i.e., \$1,456 ÷ \$1,200—is likely. If this represents a strong possibility, then the team may want to reevaluate its estimate of a market price of \$1,200/ha. If the possibilities are not great, the team could hold to its original prediction that (50,25) is the treatment farmers are most likely to accept.

A similar type of break-even analysis could be made for changes in the cost of fertilizer or the opportunity cost of labor. But in this case, the team should test the recommendation against the undominated treatment with the next lower level of variable cost, i.e., (50,0). The reason for going in this direction is that as variable costs increase, alternatives with lower variable costs become relatively more attractive.

By completing calculations, such as the above, the team gains considerable knowledge about the nature of the various alternatives. This information can be used to question and otherwise observe how farmers react to a range of possibilities. In this way, the team learns about farmers' potential response to treatments with different variable costs, rates of profitability, and stability of net benefits.

Table 7-7. Net benefits to fertilizer treatments by trial (Adapted from Perrin et al., 1976).

Trial		Fertilizer treatment (kg/ha)											
	N:	0	50	100	150	0	50	100	150	0	50	100	150
	P_2O_5:	0	0	0	0	25	25	25	25	50	50	50	50
		$/ha											
1		360	670	2370	2080	410	1620	2660	2700	950	1310	1550	1490
2		1380	1890	3730	3490	1200	2710	3440	4600	720	2770	3900	3840
3		3740	3920	3420	3080	3700	3800	3320	3200	4060	4140	4320	4160
4		2180	2990	3810	2730	1820	3390	4480	4900	900	3020	3440	4120
5		1480	1280	970	670	1540	2190	1660	1090	750	2150	1590	1460
6		1450	2200	2830	2610	1330	2830	2100	1880	510	2550	3440	2630
7		4270	4420	2960	3130	2120	4000	3690	3080	4040	3440	2690	2960
8		1090	1650	870	710	1080	1800	1090	970	680	2210	1980	2120
Avg.		1990	2380	2620	2310	1900	2790	2810	2810	1570	2690	2870	2840

Table 7-8. Minimum net benefits from eight fertilizer trials (Adapted from Perrin et al., 1976).

		Fertilizer treatment (kg/ha)											
	N:	0	50	100	150	0	50	100	150	0	50	100	150
Net benefit	P_2O_5:	0	0	0	0	25	25	25	25	50	50	50	50
		$/ha											
Worst		360	670	870	670	410	1620	1090	970	510	1310	1550	1460
Second worst		1090	1280	970	710	1080	1800	1660	1090	680	2150	1590	1490
Average of worst two		725	975	920	690	745	1710	1375	1030	595	1730	1570	1475

7.7.5. FINANCIAL FEASIBILITY

Even though the team identifies the alternative that is economically the best, this does not mean that the team can recommend it. The team must first check whether the alternative is financially feasible. Financial feasibility means that farmers are able to obtain the money required to implement the technology and will have enough money to repay any loans associated with the technology.

We can take the average of the lowest 25 percent of the trials in the maize example and compare the money generated by the treatment against the treatment's money requirements. We suggest the lowest 25 percent of the trials because the team will need to check on the household's ability to repay any financial obligations during times when yields are low. If these money obligations can be met in times of low yields, they are even more likely to be met when yields are high.

Since the foregoing analysis suggests treatment (50,25) as the best choice, we will concentrate our attention on the financial feasibility of this alternative. The team will need to estimate (1) what portion of the output will be sold, or used to substitute for maize purchases; and (2) what portion of the variable costs will require money payments. To illustrate for (50,25), if

- all of the outputs were sold at $1,000/ton, the money the family receives for the average of the worst two years would be $2,412/ha, as follows: From Table 7-3, the two lowest yields are 2.58 tons/ha and 2.78 tons/ha, which gives an average of 2.68 tons/ha. Then, 2.68 × 90% (to obtain net yields) × $1,000/ton = $2,412/ha
- the household borrows $650/ha to pay for the fertilizers—i.e., variable money costs from Table 7-4—and agrees to repay the loan in six months at an effective six months' interest rate of 20 percent
- the household does not have to pay money for the fixed costs of growing maize,

then we can conclude that treatment (50,25) is financially feasible. The ratio of money income to money obligation is 3.1:1 and the excess of money after loan repayment is $1,632/ha as calculated below:

Money income from the treatment =	$2,412/ha
Loan obligation:	
$650/ha × 1.20 (loan + interest) =	780/ha
Net	$1,632/ha

Ratio, i.e., (2412 ÷ 780):1 = 3.1:1

With ratios of this magnitude, the farmers should seldom have difficulty repaying their loans. Even for the worst 25 percent of the times, the farmers will have three times the money needed to pay off the debt. In contrast, as the ratios approach 1:1, the farmers can expect that in some years, the new technology would not produce enough money to repay the loans.

Other assumptions would give different results, such as

- the household increases its consumption of maize and thereby does not receive or save as much money
- the household must use part of the money revenues to pay for money obligations associated with the fixed costs of maize production, or possibly other household debts.

But these are individual household situations that explain why some farmers accept a new technology and others do not. The team can only generalize about these situations.

The team might also make another calculation that compares the financial implications of the treatment (50,25) with the farmers' present practice (0,0). If we assume that farmers are presently selling all of their maize at $1,000/ton, then their present money position for the average of the lowest 25 percent of the trials would be a net money inflow of $725/ha. We obtained this result in the same way as for (50,25). This comparison leads us to conclude that the farmers' money position would be improved by treatment (50,25). The net amount of $1,632/ha from the foregoing paragraph versus $725 also favors treatment (50,25) since fixed costs of maize production or other household obligations are the same for both treatments. Thus, treatment (50,25) is financially attractive on this basis as well. The major drawback to (50,25) is that it requires farmers to purchase more than they are accustomed for (0,0), which in some cases could be a deciding factor.

The team may find the test for financial feasibility even more important for longer-term investments, such as those for pumps, land clearing, or livestock purchases. In these cases, especially, the economic feasibility may be attractive, but the farmers may have difficulty either acquiring investment funds or repaying loans. We provide an example of the financial implications of long-term loans in Sec. 7.10.2.

Finally, the effective cost of borrowed money to the farmer was used in estimating the farmer's minimum acceptable return. Should the team find that the actual cost of borrowing money is substantially different from the value used in estimating this minimum acceptable return, then the economic calculations will probably have to be redone using the revised values.

7.8. ACCEPTABILITY OF NEW TECHNOLOGIES

Besides the predictive approach, the FSR&D team can also expose farmers to the new technologies and observe the farmers' reactions. Common ways for doing this in FSR&D are through farmer-managed tests and estimates of acceptability. These two approaches will now be discussed.

7.8.1. ANALYSIS OF FARMER-MANAGED TESTS

Since we introduced farmer-managed tests in earlier sections of this book, we will add only a few comments about evaluating farmers' reactions. These comments center on farmers' changes in the experimental designs—for both cropping and livestock patterns and for management practices.

Evaluating the farmers' tests is a subjective matter that begins when the field team compares actual results with those anticipated during the planning workshop. Farmers' alterations of the tests do not invalidate the tests. When the reasons for the changes are understood, the results may provide valuable guidelines for designing future experiments. For example, if a farmer shifts from single cropping to double cropping, or if a second crop is lost because the farmer plants late, the team should still consider these results. One possibility is for the team to treat the results as though they were for a new cropping pattern design. After such adjustments, the team should describe the new patterns and compare the results with those anticipated for the original design, or with other experiments that were completed as designed.

We illustrate the concept using data from Tables 7-1 and 7-2 on the number of planned and actual replications (see Table 7-9). The researcher should use the actual number of replications for any numerical analysis involving incomplete replications.

In examining the differences between planned and actual patterns, the team should seek to learn why farmers made the changes and estimate the probability of the changes happening again. If a change was due to climatic fluctuations, then the team should review any long-term climatic records, estimate the probability of such conditions occurring again, and then reconsider the results in the light of these probabilities.

Likewise, if the shift in cropping patterns was due to a single infrequent event, the chances of such an event happening again should be appraised and the experiment judged in this light. Such infrequent events need not be restricted to physical conditions such as flooding, drought, or pest invasions, but could also include economic, social, or political disruptions. Thus, socioeconomic conditions in Sri Lanka during 1977 restricted tractor fuel supplies in the northern part of the country when tractors were needed to prepare the land for the next rice crop. Because of this disruption in land preparation, experimental results were not typical of the normally rapid crop establishment (IRRI, 1978).

Similarly, shifts in the timing of farming activities may result because of family problems such as illness or death. When these occur, the tests are often inappropriate for recommending technological change. However, the results provide the team with information about how farmers react to unexpected circumstances. For this reason, the results should be analyzed and kept for future reference, but they should probably not be analyzed as part of the experimental data.

Depending on the results of the farmer-managed tests, the team can make several recommendations. For example, if the tests turn out well, the team might recommend proceeding to multi-locational testing. The team might prefer to do further testing with slight modifications to the research design. If farmers encountered numerous problems with these tests or if farmers made significant changes, the test might be redesigned. If the test brought new information to light, the team might want to go back to researcher-managed trials or even to the experiment station where researcher control would be greater. Although not likely at this stage, the team might want to abandon the test if other research opportunities appear more promising. We provide an example of the analysis of cropping pattern experiments in Indonesia in Appendix 7-H.

To correctly evaluate the results of farmer-managed tests, the team should note how farmers allocate their time to the experiments. Should farmers be too supportive of the experiments, they might spend a disproportionate amount of time on the experiment. In that case, the experiment would yield better results than would normally be the case. Alternatively, farmers might wish to show that their customary practices are superior to anything the team might have to offer. Consequently, the team needs to be alert to such possibilities and ask such questions as, "Does the farmer spend relatively more or less time on the test plot than on the same crop on another part of the farm? Does the farmer require extra time to grow the crop according to the improved technology than the indigenous technology?" In reviewing answers to such questions, the team should seek to learn how flexible farmers are in altering their time and other resources. By its nature, much of this analysis is subjective.

The team also needs to note any substantial delays or omissions from the planned experiment. Such changes can usually be identified by comparing the outline of planned farm operations with actual practices, as il-

Table 7-9. Comparison of planned and actual cropping-pattern experiments.

	Replicated patterns			
	Number		*Actual as a percent of*	
Cropping patterns	*Planned*	*Actual*	*Planned*	*Total*
rice-sweet potatoes	8	11	138%	46%
rice-rice-mung beans	8	6	75	25
green corn-rice-cowpeas	8	5	63	21
rice	0	2	--	8
Totals	24	24	--	100%

lustrated in Appendix 7-C. These comparisons show when specific operations such as planting or harvesting take place, turnaround times between crops, and the extent to which farmers adjust their cropping patterns. When extended delays or omissions occur, the team should focus on evaluating farmers' priorities and why farmers believe their actions are justified. If farmers' reasons for the changes appear justified to the team, this may become an identified problem for subsequent study. One such possibility would be that the team overestimated the resources available to the farmers or underestimated resource requirements.

7.8.2. ACCEPTABILITY INDEX

Besides observing and recording farmers' activities during the farmer-managed tests, ICTA has developed another method for estimating the acceptability of new technology to farmers in the research area. ICTA staff use an acceptability index, which they derive from data on the rate of acceptance of new technologies introduced into an area. The staff has calculated indexes based on data both from farmers participating in the farmer-managed tests and from farmers participating in record keeping.

The index is obtained by (1) multiplying the percentage of farmers who adopt the new technology by the percentage of the crops on their farms so affected and (2) dividing the product by 100. Thus, if 60 percent of the farmers accept the technology on 50 percent of their crops, the index is 30—i.e., $60 \times 50 \div 100 = 30$. According to Hildebrand (1979), ICTA considered an index of 25 as being large enough to justify its recommending the technology to the extension service.

Even index values less than 25 have significance. For example, an index of 9 in which 90 percent of the farmers apply the new technologies to 10 percent of their fields could mean that the technology is widely, but cautiously accepted. Further testing would seem in order. Or, 10 percent of the farmers applying the new technology to 90 percent of their fields suggests that some farmers are greatly impressed with the technology. Those conducting the studies should be challenged to find out the characteristics of the farmers who have adopted the technology. Once learned, the technology could be recommended to the extension service for this specific group. As the above examples indicate, both percentages making up the index need to be considered individually (Waugh, personal communication).

ICTA has found that farmers are selective in their acceptance of change (Hildebrand, 1979). Data collected over three years from farmer-managed tests in La Maquina, Guatemala, show acceptable indexes for improved seed, planting distances, and foliar insect controls, but unacceptable indexes for five other technologies. These results led ICTA to reduce the number of technical changes offered to farmers for testing at any one time. ICTA also found that the acceptability index improved as their teams learned more about farmers' conditions and perfected their approach to farmer-managed tests.

As for the calculation of acceptability indexes using farm records, Hildebrand and Ruano (1978) report that these records:

> "provide information which is used for longer run evaluation on changes in practices and yields, and comprise a more representative sample than of only those farmers who participated in Farmers' Tests [i.e., farmer-managed tests]. Ultimately, a completely randomized sample of all target farmers will need to be conducted to determine adoption of technologies, but this has not been undertaken in any area to date."

An advantage of ICTA's plans to measure acceptability of the technologies among the area's farmers over time is that the team will learn about long-run acceptability. Some good technologies may take time to be accepted. This might be the case when the change is a substantial departure from the farmers' normal practice and time is needed for farmers to accept the change. Also, some technologies are not adopted readily because of inadequate services or supplies, as when agricultural chemicals are not in the right form or quantities to interest the farmers. In other cases, a new technology may find favor during particular climatic conditions, but then fall out of favor should results be poor during unfavorable conditions.

In any case, the combination of farmer-managed tests, estimates of acceptability, and related studies help to give the team a greater understanding about how farmers actually react to the new technologies.

7.9. SOCIOCULTURAL FEASIBILITY

The FSR&D process provides its teams with procedures for understanding farmers and the farmers' environment, for designing appropriate studies and experiments, for predicting the acceptability of new technologies, and for observing farmers' reactions through farmer-managed tests and other means. Ideas will be generated for improving the experiments, such as removing farmers' constraints in labor, materials, credit, and the like. However, at times the team may encounter situations not explained technically. That is, not all farmer rejections can be explained away by the lack of labor, resources, or a chemical not being offered in the proper-sized package.

Getting to know the farm household takes time. Below, are three examples in which farmers followed different practices on different land and crops:

- A farmer applied many more resources to one plot than another. Because one plot was much more productive than the other, the farmer devoted his major effort to the more productive plot. He used the poorer plot for whatever time and resources were not needed on the better plot. Consequently, the same farmer used considerably different levels of management on the two plots.
- Farmers working both communal lands and their private lands were reported to use the communal

lands for growing subsistence crops and their private lands for growing cash crops.

- Hildebrand (personal communication) reports that Guatemalan farmers are more inclined to experiment with cash crops than with subsistence crops.

These three examples dramatize the need to understand farmers—e.g., their goals, attitudes, and beliefs—and the circumstances under which they farm.

Sometimes farmers' understanding is at fault, such as the belief that herbicides "kill the crop." At other times, the team simply may not know, without further investigation, why farmers reject some technologies and accept others. When farmers reject a technology for "no apparent reason," the team may have misunderstood the farmers and their sociocultural setting. In such cases, the team should consider obtaining help from specialists in the social sciences, such as sociology, applied anthropology, geography, and political science. Those trained in these areas are more likely to identify characteristics of farmers, their families, and their communities that others do not see. For example, the "decision trees" as described by Gladwin (Appendix 5-G) are designed to help the researcher understand *why* farmers do what they do, not just *what* they do.

By following the analytical procedures described thus far in Part 2 of this chapter, the team will learn much about the acceptability of a new technology from the farmers' point of view. When further sociocultural studies are still needed, the team should be able to focus the studies on a relatively narrow set of topics. Also, where the team's researchers and the area's farmers are of different ethnic groups, the team may want to learn more about the farmers using the type of study Beal and Sibley undertook in Guatemala (see Appendix 5-E).

7.10. OTHER ANALYSIS PROCEDURES

In much of Part 2, we have concentrated on the acceptability of technologies to farmers. We will now discuss three other topics that should interest FSR&D teams: (1) further data analysis, (2) long-term investments, and (3) analyses from society's point of view.

7.10.1. FURTHER DATA ANALYSIS

FSR&D teams can use experimental and supportive data for purposes other than learning about the acceptability of proposed technical changes. These data can also help the teams learn more about the technologies themselves, the farmers' systems, and the environment. Researchers at the regional and national headquarters, at times assisted by consultants, might use this information to conduct multiple regression and other types of more complex analyses as a basis for learning more about the research area. Experiments and the data they produce provide a basis for identifying functional relationships between inputs and outputs, such as the yield responses from fertilizers applied under varying conditions.

Some of the relationships between inputs and outputs of farming systems relate to factors, such as

- planting dates and densities
- dates of first substantial rainfall after the dry season, and total rainfall during the cropping season
- stored soil moisture at time of planting
- timing and availability of irrigation water
- number of irrigations and methods of application
- seasonal distribution of forages for animals
- fertility of soil and rates of fertilization
- availability of animal manures for crops and crop residues for animal feed
- timing and degree of pest damage and effectiveness of control measures
- excessive salinity or acidity in soil conditions
- availability of labor during peak seasonal demands
- availability of traction power during critical times.

These relationships between output and growth controlling inputs can be used to help set priorities for research based on those relationships that have the greatest influence on production. For example, an analysis of these relationships could indicate whether additional crop establishment should take priority over weeding or application of insecticides or fertilizers; or that threshing and marketing of the first crop should take priority over second crop establishment.

7.10.2. LONG-TERM INVESTMENTS

Farmers may also want to make technical changes requiring investments of longer duration than those associated with seasonal inputs such as fertilizers. Examples include purchases of livestock and equipment, planting of trees for wood and crops, clearing land, and making on-farm improvements to receive irrigation water. In none of these cases is the value of the investment used up in a single season or year.

Analyzing the net benefits of such investments calls for more complex procedures than those for analyzing seasonal expenditures. Rather than go into the details of such analyses, we will simply list a number of factors for the teams to consider and suggest that they receive instructions on the subject. Suitable references include *Economic Analysis of Agricultural Projects* by Gittinger (1972) and *Principles of Engineering Economy* by Grant et al. (1976). Economists at regional or national headquarters should be able to assist the team in learning how to analyze these types of investments.

Some factors for the team to consider when designing and analyzing long-range investments include

- the estimated life of the investment and how the needs of the farmer will change over time
- the chances that better technologies will be forthcoming soon
- any salvage value, should the investment have a short life
- the need for periodic servicing, repairs, and replacements

- the estimated annual benefits and operating costs over the life of the investment
- possible variations in these estimates
- alternative investment possibilities available to farmers
- sources of finance for the investment and terms for repayment.

Concerning the sources of finance, we mentioned in Sec. 7.7.5. that financial constraints may be more severe for long-term investments than for short-term investments. The reason is that lenders may not wish to offer loans for more than a few years. In contrast, the life of the improvement may be much longer. We offer a simple example to illustrate the point. A farmer might wish to invest in a pump that has a useful life of seven or eight years. Assume that the benefits of this pump will build up as the farmer gains experience with irrigated farming. Thus, money earnings during the early years may be small. Now if the seller of the pump or some lending institution demands repayment within two or three years, the farmer may, very likely, not have enough income or money reserves to pay off the loan in such a short time. In this sense, the pump could be quite attractive economically and yet be financially infeasible because the farmer cannot meet the loan repayment schedule. In such a situation the farmer may have to wait to accumulate the money before purchasing the pump, or else seek other sources of finance that do not have to be repaid so quickly.

7.10.3. ANALYSES FROM SOCIETY'S POINT OF VIEW

Because FSR&D is of national concern and nationally organized, FSR&D teams may need to justify some of their expenditures and recommendations from the national—i.e., society's—point of view. Means for evaluating the effectiveness of research programs are not standardized and may be difficult to make, as we discuss in Sec. 10.8. On the other hand, sometimes the FSR&D teams may recommend that the government, or private organizations, invest in support of farmers' activities. Alternatively, the teams may have to present analyses that compare the interests of the target area's present farmers with the rest of society. When these matters arise, the team might resort to a social benefit-cost analysis of the proposed investments.

Below, we list some of the factors to consider in analyses involving the national interest:

- efficiency in the use of the country's resources—both private and public
- conservation of the country's nonrenewable resources
- improvement in the distribution of income among the country's population
- accomplishment of other national objectives
- contribution to any development strategy for the country's low-income farmers and to agricultural development in general.

Analyses of this type call for training in economic development of a type generally not available to the regular members of FSR&D teams. Most of such expertise lies within the ministry of planning, and the planning units of other ministries. Consequently, when such issues arise, we suggest that the FSR&D team seek help from these sources within the country.

General references on the topic of social benefit-cost analysis include those by Gittinger (1972), Little and Mirrlees (1974), Sassone and Schaffer (1978), and United Nations Industrial Development Organization (1972). In addition, the World Bank's Economic Development Institute regularly offers courses in project preparation and analysis suitable for the developing countries.

7.11. ANALYSIS WORKSHOPS

After the season's experiments are completed, the regional and national headquarters teams may wish to hold an analysis workshop. This workshop could review the results of the on-farm experiments, surveys and special studies, and data from monitoring and record keeping. Such a workshop might be held during (1) a slack period in the field teams' activities, (2) the regional planning workshop, or (3) some other convenient time.

The group attending the meeting could help the field teams in applying analytical procedures, help interpret the meaning of the results, and recommend how to proceed. The teams can also receive help in (1) identifying and refining problems and opportunities, (2) suggesting new technologies for testing, (3) reaching a better understanding of the farmers' systems for improving on-farm experiments, and (4) stratifying the research area prior to transferring research findings to other areas.

Participants in the workshop should include the field teams and the regional headquarters team, as well as members of the national headquarters team, specialists from the experiment station and the extension service, and others who can analyze and interpret the results. To make good use of the workshop's time, the field teams should have completed at least preliminary analyses of their experiments and have prepared reports (1) summarizing the experiments' objectives, research methods, major findings, and conclusions and (2) offering recommendations for future action. Field team leaders can play a key role in these workshops by (1) aiding the team members in preparing their reports, (2) seeing that team reports are clearly written and duplicated for those attending the workshop, and (3) being ready to explain their team's work when called upon. The regional headquarters team should probably host the meeting, arrange for inviting participants, and otherwise help organize activities.

7.12. SUMMARY

While we have commented on analysis procedures in earlier sections of this book, this part contains much more detail on the subject. We (1) emphasized the integration of physical, biological, economic, financial, and sociocultural aspects, (2) distinguished between prediction and measurement of acceptability as a means for judging farmer acceptance of new technologies, and (3) gave our reasons for concentrating more on partial budget analysis than whole farm analysis. After that, we used an example from Perrin et al. (1976) to illustrate analysis procedures for cropping experiments. The illustration began with biological results and then proceeded to measurement of net benefits and economic and financial feasibility. This was followed by sections on (1) acceptability of new technologies to farmers, (2) sociocultural feasibility, and (3) other analysis procedures. Finally, we discussed the use of analysis workshops as an aid to field teams in analyzing the results of their experiments and in preparing for next season's research.

CITED REFERENCES

Brown, M.L. 1979. Farm budgets. World Bank Staff Occ. Paper No. 29. World Bank, Washington, D.C.

CATIE. 1978. Milk and beef production systems for the small farmers using crop derivatives. *In* 1978 Progress Report. CATIE, Turrialba, Costa Rica.

CIAT. 1975. Annual Report. CIAT, Cali, Colombia.

______. 1974. Annual Report. CIAT, Cali, Colombia.

Delgado, C.L. 1978. The southern Fulani farming system in Upper Volta: a new old model for the integration of crop and livestock production in the West African savannah. Center for Res. on Econ. Dev., Univ. of Michigan, Ann Arbor, Mich.

Gines, H.C., and H.G. Zandstra. 1977. Performance of five rice varieties under dry seeded management in marginal soils. Presented at the 8th Ann. Meeting of the Crop Sci. Soc. of the Philippines. 5-7 May 1977. Mountain State Agricultural College, La Trinidad, Benguet, Philippines.

Gittinger, J.P. 1972. Economic analysis of agricultural projects. The Johns Hopkins University Press, Baltimore, Md.

Grant, E.L., W.G. Ireson, and R.S. Leavenworth. 1976. Principles of engineering economy. The Ronald Press Co., New York.

Hildebrand, P.E. 1979. Incorporating the social sciences into agricultural research: the formation of a national farm systems research institute. ICTA, Guatemala, and The Rockefeller Foundation, New York.

______, and S. Ruano. 1978. Integrated multidisciplinary technology generation for small, traditional farmers of Guatemala. A paper presented at the Ann. Meeting of the Soc. for App. Anthro. 2-9 April 1978. Merida, Mexico. ICTA, Guatemala.

ILCA. 1980. ILCA the first years. ILCA, Addis Ababa, Ethiopia.

______. 1978. Animal production systems in the high potential highlands of tropical Africa. ILCA, Addis Ababa, Ethiopia.

IRRI. 1978. Report of the 7th cropping systems working group meeting. 2-5 Oct. 1978. Los Banos, Philippines.

Jayasuriya, S. 1979. New cropping patterns for Iloilo and Pangasinan farmers: a whole farm analysis. IRRI Sat. Sem. 21 July 1979. IRRI, Los Banos, Philippines.

Little, I.M.D., and J.A. Mirrlees. 1974. Project appraisal and planning for the developing countries. Heinemann Educational Books, London.

Little, T.M., and F.J. Hills. 1978. Agricultural experimentation. John Wiley and Sons, Inc., New York.

Nelson, L., A. Koreen, and C.J. deMooy. 1980. A field study of traditional and improved methods of cotton culture. p. 63-94. *In* Improving water management on farms. Water Mgmt. Res. Proj. Eng. Res. Center, Colorado State Univ., Fort Collins, Colo.

Perrin, R.K., D.L. Winkelmann, E.R. Moscardi, and J.R. Anderson. 1976. From agronomic data to farmer recommendations: an economics training manual. Inf. Bull. 27. CIMMYT, El Batan, Mexico.

Sassone, P.G., and W.A. Schaffer. 1978. Cost-benefit analysis: a handbook. Academic Press, New York.

Stonaker, H.H., J. Gomez, and M.C. Amezquita. 1979. Calf production in the Colombian *llanos* as influenced by early weaning, mineral and urea supplementation and pastures. Abstr. 71st Ann. Meeting, Am. Soc. An. Sci. Univ. Arizona, Tucson, Ariz. 28 July -1 Aug. 1979. American Society of Animal Sciences, Corvallis, Oregon.

United Nations Industrial Development Organization. 1972. Guidelines for project evaluation. United Nations, New York.

Zandstra, H.G., E.C. Price, J.A. Litsinger, and R.A. Morris. 1981. A methodology for on-farm cropping systems research. IRRI, Los Banos, Philippines.

Zuberti, C.A., K.G. Swanberg, and H.G. Zandstra. 1979. Technology adaptation in a Colombian rural development project. *In* A. Valdez, G.M. Scobie, and J.L. Dillon (ed.) Economics and the design of small-farmer technology. Iowa State University Press, Ames, Iowa.

OTHER REFERENCES

Byerlee, D., M.P. Collinson, R.K. Perrin, D.L. Winkelmann, S. Biggs, E.R. Moscardi, J.C. Martinez, L. Harrington, and A.

Benjamin. 1980. Planning technologies appropriate to farmers: concepts and procedures. CIMMYT, El Batan, Mexico.

Collinson, M.P. 1972. Farm management in peasant agriculture: a handbook for rural development planning in Africa. Praeger Publishers, New York.

Harwood, R.R. 1979. Small farm development: understanding and improving farming systems in the humid tropics. Westview Press, Boulder, Colo.

IRRI. 1977. Proceedings, Symposium on Cropping Systems Research and Development for the Asian Rice Farmer. IRRI, Los Banos, Philippines.

______. 1975. Annual report. IRRI, Los Banos, Philippines.

Jolly, A.L. 1952. Unit farms. p. 172-179. *In* Tropical Agriculture, Vol. 29, January - December 1952. Port-of-Spain, Trinidad.

McDowell, R.E., and P.E. Hildebrand. 1980. Integrated crop and animal production: making the most of resources available to small farms in developing countries. Presented at Bellagio Conf. 18-23 Oct. 1978. The Rockefeller Foundation, New York.

Reuss, J. 1980. Analysis of sampling variation within and among wheat fields in Punjab. p. 411-474. *In* Improving water management on farms. Water Mgmt. Res. Proj., Eng. Res. Center, Colorado State Univ., Fort Collins, Colo.

Roumasset, J.A. 1977. Risk and uncertainty in agricultural development. A/D/C/ Sem. Rep. No. 15. The Agricultural Development Council, Inc., New York.

Shaner, W.W. 1979. Project planning for developing economies. Praeger Publishers, New York.

Sparling, E.W., W.D. Kemper, J.E. Hautalouma, M.K. Lowdermilk, G.V. Skogerboe, and W.G. Stewart. 1980. Development process for improving irrigation water management on farms: development of solutions. Water Mgmt. Tech. Rep. 65C. Water Mgmt. Res. Proj., Eng. Res. Center, Colorado State Univ., Fort Collins, Colo.

Spencer, D.S.C. 1972. Micro-level farm management and production economics research among traditional African farmers: lessons from Sierra Leone. Rural Employment Paper No. 3. Dep. of Agric. Econ., Michigan State Univ., East Lansing, Mich.

Steel, R.G.D., and J.H. Torrie. 1980. Principles and procedures of statistics. McGraw-Hill, Inc., New York.

Valdes, A., G.M. Scobie, and J.L. Dillon (eds.) 1979. Economics and the design of small-farmer technology. Iowa State University Press, Ames, Iowa.

Virmani, S.M., M.V.K. Siva Kumar, and S.J. Reddy. 1978. Rainfall probability estimates for selected locations of semi-arid India. ICRISAT, Hyderabad, India.

Wharton, C.R., Jr. 1971. Risk, uncertainty, and the subsistence farmer. p. 566-574. *In* G. Dalton (ed.) Economic development and social change: the modernization of village communities. The Natural History Press, Garden City, New York.

Zandstra, H.G. 1978. Cropping systems research for the Asian rice farmer. Cropping Systems Program, IRRI, Los Banos, Philippines.

Chapter 8
EXTENSION OF RESULTS

In FSR&D, research develops new technologies and proves their worth to a relatively small number of farmers, whereas the extension service or a similar agency diffuses the new technologies to as many farmers as practical. Not only must these complementary roles be recognized and fulfilled within the FSR&D process, but the transition between research and extension must be effective. This requires continual communication and cooperation between researchers and extension workers. Both should recognize the other's contributions throughout the FSR&D process.

In this book, we assume that most developing countries have institutions responsible for educational programs in agricultural production. The term extension designates those agencies whose role is the diffusion of agricultural information. In FSR&D, extension's primary role is cooperating with research and diffusing new and tested technologies to farmers in the target area.

In the sections that follow, we (1) explore how extension is integrated into each of the FSR&D activities, (2) discuss the staffing and organizational means for accomplishing this integration, (3) provide details on extension's involvement in FSR&D, (4) describe two implementing activities in transferring new technologies—i.e., multi-locational testing and pilot production programs, (5) present some problems and solutions in extension, and (6) close with conclusions.

8.1. INTEGRATION OF EXTENSION INTO FSR&D

Since FSR&D is a relatively new concept and the effort, to date, has concentrated on research methodology, the role of extension in FSR&D has not been fully established. However, we are not without some insight into how FSR&D research and extension can be integrated. This insight comes from (1) FSR&D practitioners who have had experience in bringing about effective cooperation between the two activities and (2) what appears, to us, as the realistic needs of FSR&D.

In Fig. 8-1, we provide an example of the division of effort between extension, research, and others concerned with improving small farmers' conditions. The "others" include national decision makers, production program personnel, farmers, and representatives of farm organizations.

In our opinion the most significant aspect of the process illustrated in Fig. 8-1 is the involvement of extension with research and others throughout the FSR&D process. Extension's participation increases from about 20 percent of the total activity during target area selection to a peak of about 60 percent during multi-locational testing. Extension is less involved during the planning and implementing of on-farm experiments, which are primarily the responsibilities of researchers. In the pilot production programs, many other institutions become involved and extension's share of the total effort declines to about 25 percent.

Fig. 8-1 ends with pilot production programs, since the scope of this book stops at the point where new technologies are transferred to extension and other production-oriented groups. At this point, new technologies will have been tested multi-locationally and through pilot production programs. Then, suitable technologies will be ready for diffusion regionally and possibly nationally. During such diffusion, extension will normally be called upon to take the major responsibility for education and promotion at the farm level and for coordinating the activities of participating institutions.

8.2. STAFFING AND ORGANIZING FOR EXTENSION'S INVOLVEMENT

Extension's lack of experience with FSR&D also limits the firmness of our recommendations about how to organize an extension staff for FSR&D. Consequently, the material that follows represents our judgment about the best way to proceed. These suggestions will undoubtedly need refinement as additional experience is gained from national FSR&D activities currently underway and those soon to be implemented. In this section, we introduce the concept of an extension specialist in farming systems (ESFS) and then elaborate on how the extension service might be organized to accommodate FSR&D.

8.2.1. EXTENSION SPECIALIST IN FARMING SYSTEMS

A potentially key position in FSR&D is that of the ESFS. This position bridges the gap between the generation of technology by FSR&D's researchers and the broad-scale diffusion of improved technologies by extension. Responsibilities of the ESFS include

- learning FSR&D procedures by working closely with the researchers
- familiarizing researchers with extension's capabilities, needs, and viewpoints

Figure 8-1. The division of effort between extension, research, and "others" for each activity in the FSR&D process. These are theoretically useful divisions of effort; the actual divisions of effort depend on the specific conditions in each country. "Others" include those such as national decision makers, production program personnel, farmers, and representatives of farm organizations.

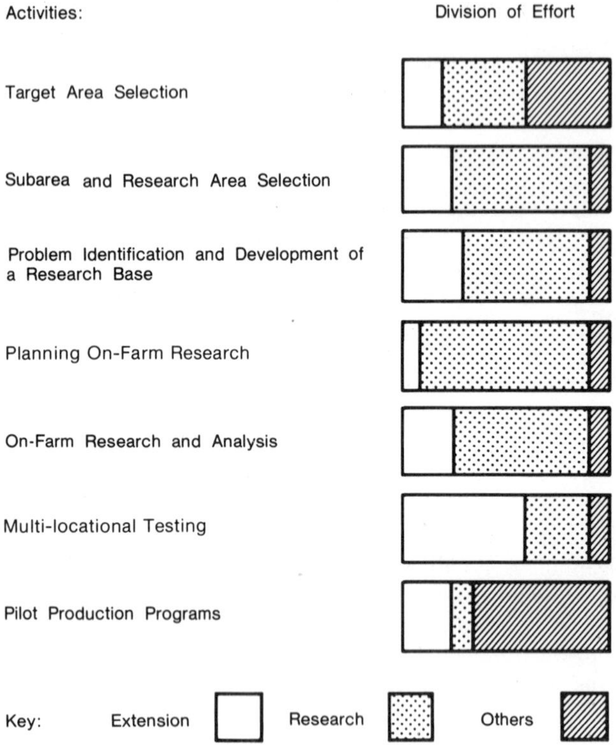

- educating extension personnel in FSR&D philosophy and methods
- training extension personnel in the application of new technologies
- generally coordinating the activities between research and extension, especially during multi-locational testing and the pilot production programs.

To our knowledge, this position has not been formally implemented; however, an ESFS position has been included as part of the regional FSR&D team in a project being planned for Zambia (Norman, personal communication). ESFSs can also be effective as members of both the field teams and the national headquarters team.

The ESFS ought to be specialized in a recognized discipline and be well-grounded in FSR&D philosophy and methodology. While an ESFS might function adequately with an educational degree at the diploma level—i.e., one year less than the bachelor's degree—close contact with research scientists will normally be enhanced when the ESFS has a bachelor's or master's degree. To help integrate ESFSs quickly into the FSR&D teams, they should be among the first groups to be trained in FSR&D.

8.2.2. PROPOSED ORGANIZATION OF EXTENSION AT THREE LEVELS

In this section, we propose a three-level organizational scheme for extension that should help implement FSR&D on a broad scale. To facilitate our discussion, we will divide extension's activities related to FSR&D into the same categories as in Sec. 3.3.—namely, field, regional, and national teams (see Fig. 8-2). In larger countries, FSR&D practitioners often divide regions into subregions.

At the field level, the extension agent is the principal extension position. The extension agent usually lives in the rural community to which he or she is assigned. If the workload is large, extension assistants may help by working under the close supervision of the extension agent. Depending on the number and quality of educated personnel in the country, extension assistants may or may not have graduated from high school. They frequently work directly with individual farmers or farmers' groups, or they may instruct local farmer leaders who, in turn, work with farmer groups.

The regional extension officer is the chief supervising officer at the regional extension level. The subject matter specialists advise and train the extension agents. These specialists are experts in such fields as agronomy, plant protection, livestock, farm management, water management, and farm implements. As ESFSs become trained, they will become the key extension specialist for FSR&D.

We recommend that a country consider making ESFSs regular members of the FSR&D teams at all levels. Under this arrangement, the ESFS would still maintain formal ties with the extension service. If the ESFSs cannot

Figure 8-2. Suggested extension organization for FSR&D.

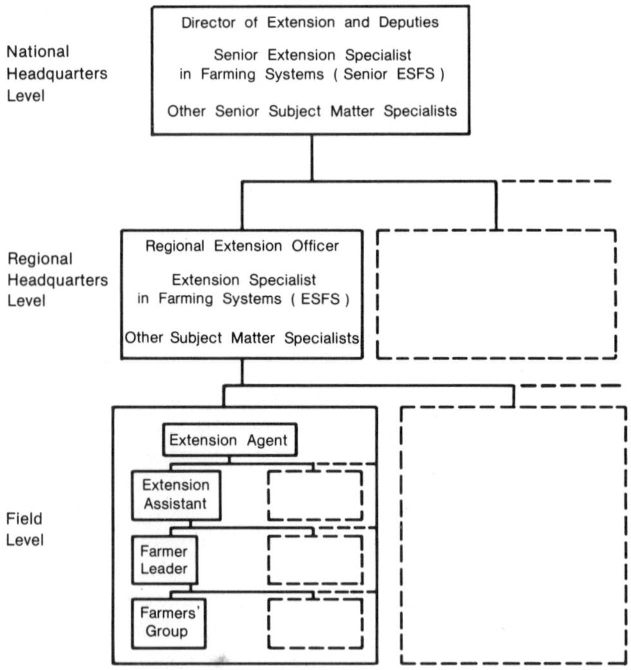

be integrated into the FSR&D teams, then ESFS positions should, at least, be set up in extension's regional headquarters. From there, the ESFS could establish effective working relationships with the FSR&D teams and also with extension staff at national headquarters and extension agents and extension assistants in the field.

At the national headquarters level, the director of extension and the director's deputies develop policy and make key decisions on extension's role in FSR&D. When FSR&D projects are operating in several regions, a senior ESFS stationed at national headquarters is desirable. This job includes supporting the regional ESFSs, strengthening extension-research ties in FSR&D throughout the country, keeping informed of FSR&D activities abroad, and helping develop FSR&D training programs. Senior extension specialists in other fields such as education, training, communications, publications, and social sciences also support extension's involvement in FSR&D at the national headquarters.

In Fig. 8-3, we show how a new technology might be quickly diffused to large numbers of farmers. Here, the ESFS instructs and supervises extension agents in the new technology; each agent transfers the technology to 10 local farmer leaders; and each one of the local farmer leaders diffuses the technology to four groups of 25 farmers. In this scheme one ESFS, with the help of 10 extension agents and 100 local farmer leaders, can reach 10,000 farmers. Variations of this transfer system are possible. For example, instead of acting alone, the ESFS may be supported by the regional extension officer or by members of the FSR&D field teams; instead of local farmer leaders, extension assistants could be used; and so on.

Figure 8-3. Diffusion of a new technology from one ESFS to 10,000 farmers (Adapted from Waugh, undated).

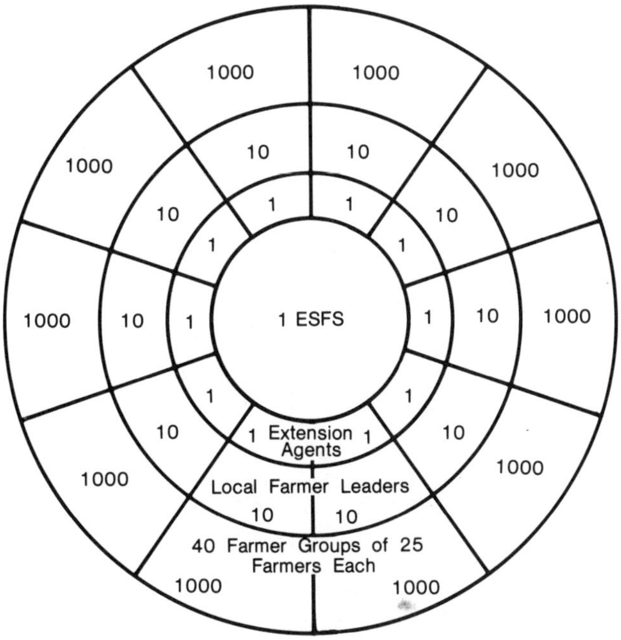

8.3. DETAILS OF EXTENSION'S INVOLVEMENT IN FSR&D

In Table 8-1, we list some of the ways extension participates in each of the FSR&D activities. The specific items listed illustrate what might be done. Naturally, actual situations will be different both in terms of the types of activities shown in Table 8-1 and the degree of involvement indicated in Fig. 8-1. Extension's actual participation depends on such factors as

- the FSR&D strategy
- extension's experience in FSR&D and its leadership
- organizational agreements between extension and research
- the capabilities of extension's personnel
- extension's budget and other responsibilities.

Below, we discuss extension's involvement in each of the FSR&D activities.

8.3.1. TARGET AREA SELECTION

Extension's input to target area selection comes primarily from its national headquarters staff. At this level, the staff is familiar with national objectives and policies and is generally knowledgeable about the various regions of the country. The extension staff at the regional level can also help by providing information specific to its area.

8.3.2. SUBAREA AND RESEARCH AREA SELECTION

The regional staff will make extension's major contribution to subarea and research area selection. The regional staff know what major differences exist within the target area and the field level staff can assemble additional information as required. Beginning at this point and extending through the pilot production programs, the ESFS becomes the primary link between extension and research.

8.3.3. PROBLEM IDENTIFICATION AND DEVELOPMENT OF A RESEARCH BASE

Extension's field staff and the ESFS assist researchers during problem identification and development of a research base. The extension agents and their assistants are familiar with local conditions and have close contacts with farmers and others living in the research area. Thus, they can help familiarize the researchers with the area during the reconnaissance surveys and at other times. The ESFS can help train extension's field staff in the techniques of surveys, monitoring, and in other ways that support FSR&D activities.

8.3.4. PLANNING ON-FARM RESEARCH

Although planning on-farm research is mainly the

Table 8-1. Extension's participation in each of the FSR&D activities.

Activity	*Extension's participation*
Target Area Selection	• Suggest relevant criteria for target area selection. • Cooperate in assembling and analyzing secondary and primary data for target area selection.
Subarea and Research Area Selection	• Cooperate in choosing the relevant criteria for subarea and research area selection. • Cooperate in assembling and analyzing secondary data and in making preliminary surveys, especially in selecting and locating farmers and other interviewees.
Problem Identification and Development of a Research Base	• Help researchers become familiar with local conditions and establish contacts with farmers and others. • Cooperate in assembling secondary data and in making the reconnaissance survey. • Participate in problem identification. • Provide comprehensive outlook of farming and community systems.
Planning On-Farm Research	• Contribute knowledge of current farmer practices and farmer's environment. • Help researchers select farmers for trials and tests. • Help in making farmer surveys. • Provide feedback from farmers to researchers and vice versa.
On-Farm Research and Analysis	• Assist in supervision of farmer-managed tests. • Check on farmer acceptance of new technology. • Provide feedback from farmers to researchers and vice versa. • Help in making surveys and special studies, supervising farm record keeping and climatic monitoring.
Extension of Results	• Organize field days on trials and tests in farmers' fields.
Multi-Locational Testing	• Assist in selection of farms. • Help in supervising tests. • Cooperate in adapting new technologies to different conditions. • Provide feedback from farmers to researchers and vice versa. • Help in preliminary packaging of the new technology for diffusion, and in developing preliminary transfer methods.
Pilot Production Programs	• Help determine feasibility of new technology on intensive scale. • Assist in bringing about needed changes in support systems. • Assist in defining and coordinating tasks of cooperating institutions. • Help finalize packaging of new technology and transfer methods for widespread diffusion.

researchers' job, the extension staff can help here as well. Field and regional staff from extension will know the local conditions where research activities are to take place and can help in identifying collaborating farmers. Consequently, key representatives from extension should be asked to participate in the planning workshops.

8.3.5. ON-FARM RESEARCH AND ANALYSIS

With the ESFS as the prime contact, extension field staff can assist with on-farm research, interpretation of results, and follow-up investigations after farmers have completed the farmer-managed tests. When trained in FSR&D, such staff can greatly expand the capability of the FSR&D team to conduct farmer-managed tests throughout the research area. Also, the field staff can contact farmers the season after their participation in farmer-managed tests to learn the extent to which these farmers used or abandoned the technologies introduced by the tests. Thus, the field staff helps in providing data on the acceptability of the new technologies to the farmers under study. In addition, the field staff can help the FSR&D team and farmers acquire inputs such as agricultural chemicals and seeds in time for the experiments.

8.3.6. EXTENSION OF RESULTS

When the ESFS and other members of the extension service are integrated into the FSR&D effort, the extension of results is greatly facilitated. Extension will have contributed to the research design and will be able to anticipate research results. In this way, diffusion becomes easier and more effective. Extension's main efforts will be with broad-scale diffusion using the extension agents and their assistants. But before reaching that point, the ESFSs can help in the transfer process through their active involvement in multi-locational testing and pilot production programs. Because of the significance of these two activities, we will discuss them separately in the next two sections.

8.4. MULTI-LOCATIONAL TESTING

In the first part of this section, we describe the nature of multi-locational testing and how the FSR&D team and extension staff participate. In the second part, we provide an example from Southeast Asia based on Zandstra et al. (1981).

8.4.1. NATURE AND PARTICIPATION

As pointed out in Sec. 4.2.1., the target area and even the subareas contain some variability in farmers' conditions. During multi-locational testing, a technology developed within the research area is tested broadly in subareas within the target area. To aid in testing new technologies throughout a subarea, the research area should contain part of the subarea within its boundaries.

Thus, in multi-locational testing the FSR&D team adapts technologies to the varying conditions in the subareas. When the subareas are well defined, the amount of adaptation is small. In the process, the team:

- associates differences in the performance of new technologies with the factors causing the differences
- seeks to find out what adjustments in the new technologies are needed for adapting them to conditions somewhat different from those encountered in the research area
- verifies, or revises if necessary, the subarea boundaries for the recommendations associated with each new technology
- assists in preparing instructions for the extension agents' use in diffusing the new technologies throughout the target area
- may eventually consider extrapolating the results to similar areas and groups outside the target area (Zandstra et al., 1981, and Pantastico et al., 1980).

The FSR&D team usually manages the multi-locational tests. Extension personnel assist in supervising such functions as timely application of chemicals for fertilization and for pest, disease, and weed control. Extension staff thus becomes more familiar with the new technology. This will help extension workers to do a better job when they are called upon to widely diffuse this technology. They also can inform the FSR&D team about problems they observe during testing. The farmers, themselves, perform the routine farm operations such as preparing the land, weeding, and any irrigating. During multi-locational testing, the team adheres to the management methods developed for the new technology in the research area. This allows meaningful comparisons of technology performance across the various test locations.

Sometimes the FSR&D team and extension personnel may want to continue multi-locational testing of the new technology for two or three seasons. In that way, they can observe the technology's performance under various conditions. Such precautionary procedures are important when the introduced technologies mean substantial changes in farmers' practices and when local conditions contain considerable variability.

In preparing for multi-locational tests, the ESFS informs the extension field staff about what kinds of farms are needed. The staff then assists the researchers in finding suitable locations. While this staff has more responsibility for the multi-locational testing than for the farmer-managed tests, final responsibility for conducting multi-locational tests still rests with the researchers because the technologies are still being developed. The extension field staff and the ESFS assist the researchers in adapting the new technology to different conditions in the target area and provide feedback from farmers to researchers.

With the help of other subject matter specialists and the FSR&D team, the ESFS prepares instructions for implementing the new technologies. In these instructions, the ESFS describes the individual steps and resource requirements in detail.

8.4.2. AN EXAMPLE FROM SOUTHEAST ASIA

The Asian Cropping Systems Network has begun to implement the concept of multi-locational testing in some of its member countries. The procedures, which Zandstra et al. (1981) described, help to illustrate how cropping patterns and land types are selected for testing.

In this example, the FSR&D team developed three technologies for the research area shown in Fig. 8-4. These technologies relate to experimental cropping patterns labeled A, B, and C for land types I, II, and III. (In Appendix 5-B, we discuss land types). After experimenting, the team excluded land type III from further consideration because all three cropping patterns performed poorly on it. Cropping pattern A was only applicable for land type I and cropping pattern C was only applicable for land type II, whereas cropping pattern B was applicable for both land types I and II.

The team selected three expansion areas within the target area for further testing of the new cropping patterns. The team's criterion for selecting these expansion areas was the amount of land types I and II that the areas contained. At least one of the two land types had to be present

Figure 8-4. Expansion area for multi-locational tests (Adapted from Zandstra et al., 1981).

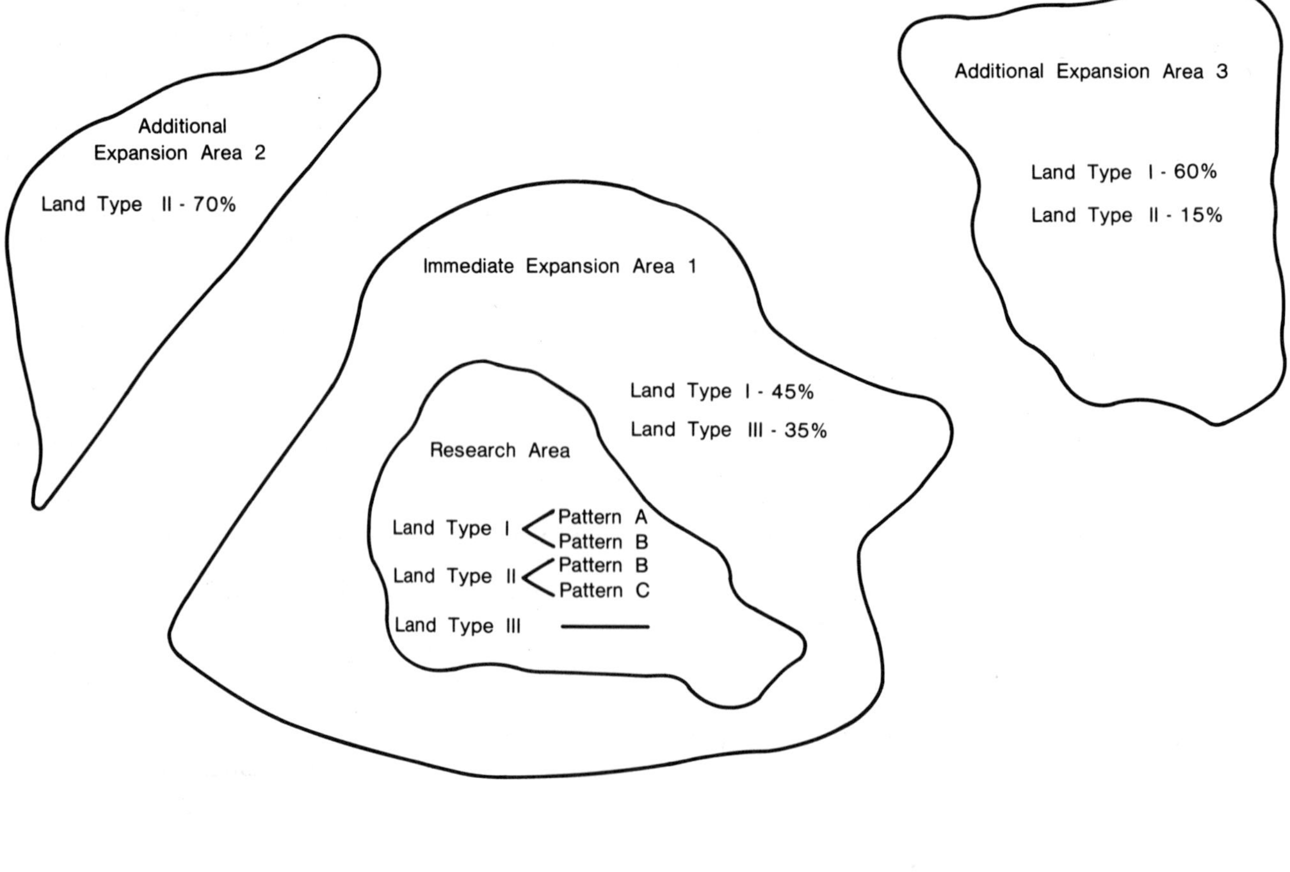

in a large enough area to make testing of the patterns worthwhile. One of the three expansion areas surrounds the research area.

In Table 8-2, we present the team's multi-locational testing design for the three expansion areas. In expansion areas 1 and 2, the team tested each applicable cropping pattern on each suitable land type. From above, we noted that cropping patterns A and B were applicable for land type I and cropping patterns B and C were applicable for land type II. However, the team did not test patterns B and C in land type II in expansion area 3 because the amount of land was too small. The results from these cropping pattern tests provided the information needed for adapting research results to other locations in the target area.

Table 8-2. Multi-locational testing design for three expansion areas.

	Expansion area					
	1		*2*	*3*		
Cropping pattern	*Land type I*	*III*	*Land type II*	*Land type I*	*II*	
	(Number of multi-locational tests)					
A	6	-	-	6	-	
B	6	-	6	6	-	
C	-	-	6	-	-	

8.5. PILOT PRODUCTION PROGRAM

Once the new technologies have been tested multi-locationally and modified as necessary, pilot production programs are initiated. Such programs test both how the support systems function—e.g., suppliers of inputs—and the environment reacts—e.g., commodity prices—when new technologies are introduced in the area on a relatively large scale. A new technology suitable for a single farm may not be suitable when many farmers simultaneously adopt it. This might occur when inputs are in short supply or outputs flood the market. A pilot program is designed to gain information on the adequacy of such factors as local commodity markets, credit, labor, agricultural chemicals, equipment, transportation, and information systems.

The size of the pilot production program needs to be large enough to test the various supporting organizations and activities. Therefore, the actual size will vary according to (1) the types of organizations and activities being tested, (2) farming densities within the testing area, (3) the current organizational capabilities of FSR&D and extension, and (4) the resources of the supporting groups. For example, Haws and Dilag (1980) suggested that pilot production testing in the Philippines be applied to several hundred hectares.

A pilot production program should define the roles of participating institutions—e.g., the banks, suppliers of agricultural inputs, marketing cooperatives, and any irrigation organizations. The program also provides a final evaluation of new technologies' benefits and costs before introducing changes on a broad scale (Zandstra et al., 1981).

During a pilot production program, the ESFS helps coordinate the activities of those concerned. These may include extension staff at all levels, researchers, and other personnel. Activities relate to selecting the location, laying the groundwork, and implementing the program.

At this stage in FSR&D, the ESFS finalizes instructions for applying the new technologies. In this, the ESFS receives help from the FSR&D team and from extension specialists in education, training, publications, communications, promotion, and other relevant disciplines. Earlier, the ESFS and others prepared these instructions as part of multi-locational testing, but on a preliminary basis. The ESFS and other collaborators also train the extension staff in implementing the new technologies and in promotional strategies and techniques. This training is preparatory for the widespread diffusion of the improved technologies.

Haws and Dilag (1980) described a pilot production program in the Philippines, as follows:

> "A report of the results of the multi-location tests is prepared which is then presented to the regional director of the Bureau of Agricultural Extension (BAEX) for his evaluation and approval of the proposed pilot production program. An economic analysis is also presented. If the regional director approves the proposal, a second meeting is held with the provincial governor. All other agencies whose inputs are needed to support the farmer in the establishment of the new technology in the pilot production program area are invited. Some of the agencies involved are: a) the governor of the province who presides at the meeting; b) the municipal mayor of the target area; c) the regional directors of the various agricultural agencies, i.e., BAEX, [Bureau of Plant Industry, Philippine National Bank], Bureau of Soils, [National Grains Authority], etc.; d) the local pesticide and fertilizer dealers; and e) the rural banks.
>
> "If the committee approves the proposal, a 'Memorandum of Agreement' . . . is drawn and signed by all parties concerned, stating what contributions each will make to the overall plan, i.e., IRRI will provide technical information; [National Grain Authority] will purchase all the rice harvested; banks will release loans to farmers in the program, etc. This agreement is a vital part of the organization of the program because it has a tendency to bind all parties to the program and thus make it a living document. It then becomes their program.
>
> "It is also at this time that a name is given to the proposed program. A name gives identity to all members in the pilot area and makes legitimate the activities of the committee in the eyes of the farmers in the program. The name MASAGANA 99 proved to

be a magic word for rice production in the Philippines. It can do the same for any program if the name is carefully selected."

In Appendix 8-A, we provide the 1976 memorandum of agreement among groups participating in this pilot production program in the Philippines.

8.6. PROBLEMS IN EXTENSION

Certain problems confronting extension in developing countries may reduce that organization's effectiveness in FSR&D. Problems typically concern extension-research ties, training, orientation, organization, and budgets. We will now discuss these problems and propose some solutions, including ways that an FSR&D approach contributes to the solutions.

8.6.1. TIES BETWEEN EXTENSION AND RESEARCH

Research and extension in developing countries frequently do not coordinate their efforts to help small farmers, especially in those situations in which the concepts of FSR&D are not applied. This failure weakens the effectiveness of both activities. Without a continual flow of new technologies and tested recommendations from research, extension can run out of technologies to extend; and without close links with extension to provide feedback from farmers, researchers' work can lose much of its relevance. Researchers compound the problem when they believe (1) farmers cannot provide information useful for improving technologies and (2) extension workers are professionally below them and not worth listening to. Researchers need to appreciate that their work has little merit unless farmers widely adopt the technologies they produce.

To help solve this problem, research and extension should meet and try to understand each other's objectives, interests, and capabilities. When such meetings are successful, one group becomes aware of its dependence on the other group and, therefore, has an incentive for developing stronger ties. One of FSR&D's aims is to bring about this type of cooperation.

An example of an attempt to develop closer ties between research and extension is a 1978 letter of understanding between ICTA (research) and DIGESA (extension) in Guatemala. ICTA was created without formally making extension part of its activities. After several years of operation, both organizations realized the need for closer cooperation. Therefore, they drew up an agreement to "integrate efforts and contribute resources to have a greater number of production alternatives that have been validated under proper ecological and social conditions, and to make more effective their transfer to and acceptance by the farmers of the country" (ICTA and DIGESA, 1978 as translated by the Consortium for International Development). This agreement includes these essential points:

- National development programs create the need for coordination and mutual support by those organizations holding complementary objectives.
- Agricultural research, promotion, and training should be merged into a single effort leading to technologies that farmers will adopt.
- DIGESA will convey to ICTA problems arising during technology transfer; and ICTA will provide DIGESA with technologies suitable for farmers' adoption.
- To accomplish the above, both organizations will integrate their efforts and contribute resources to that effect.
- Each organization will appoint a coordinator to represent its interests and to select projects in their mutual and individual interests.

In Appendix 8-B, we provide our translation of this Letter of Understanding.

Another organizational approach for linking extension's and research's efforts in FSR&D is based on Benor and Harrison (1977). In Fig. 8-5, we illustrate how extension and research might be linked at the national, regional, and field levels. At the national headquarters level, a committee with membership consisting of key staff from both extension and research could be established. This committee would develop policies and general procedures for cooperation between the two organizations. The chairman of this national committee should probably be the head of research, with the chairman's deputy being the head of extension. The senior ESFS could play a major role in the committee—for example, by being the permanent secretary and by developing alternative recommendations and presenting them to the committee for consideration. This committee might be an alternative to or operate as a subcommittee of an organization with broad membership and responsibilities, such as the national advisory committee mentioned under Institutional Linkages in Sec. 10.4.1.

Figure 8-5. Linkages between extension and research at the national, regional, and field levels.

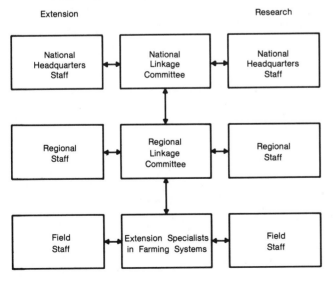

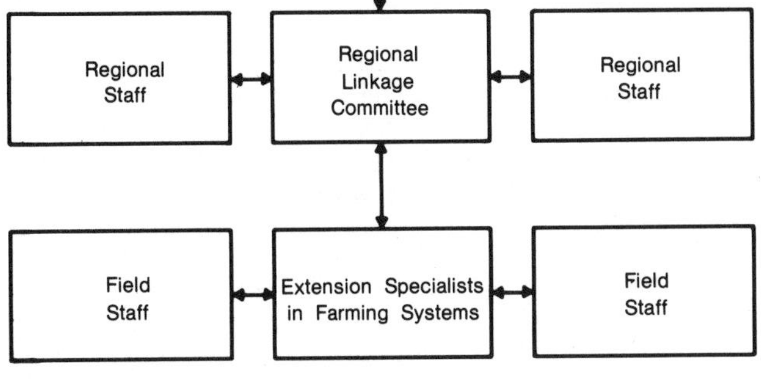

The regional committee linking extension and research is similar in structure and function to the national committee, except that the personnel and issues of cooperation are regional. The regional ESFS could have important functions at both the regional and field levels. At the regional level, the ESFS might operate within the regional committee as the senior ESFS operates at the national level; and at the field level, the ESFS could help coordinate the activities of extension's and research's field staffs. In performing these functions, the regional ESFS works within the policies and procedures set by the national and regional linkage committees.

8.6.2. TRAINING

The training that extension's field personnel receives in developing countries is often inadequate. Such training may consist only of preservice instruction in the classroom with little opportunity for practical fieldwork, and scarcely any provision for follow-up or refresher courses. Extension agents are told to tell farmers what to do. They are not told to listen to farmers, nor are they told to inform researchers about farmers' needs. The lack of thorough, up-to-date training contributes to the field staff's low productivity and morale. Furthermore, the field staff's training does not adequately prepare it for FSR&D tasks, such as identifying farmers' problems, supervising farmer-managed tests, and participating in multilocational testing and the pilot production programs.

If an FSR&D program is contemplated, field staff performance often needs to be improved substantially and quickly. Accomplishing this through in-service training is covered in Sec. 11.2.2.

8.6.3. ORIENTATION

Extension agents in developing countries often tend to be single commodity oriented—for example, they may be rice, vegetable, wheat, or livestock agents. They are not trained to see the farm as an integrated system. They may observe the immediately favorable impact of a new practice on a crop or animal, but they do not relate these changes to the whole farming system. Because FSR&D

focuses on the whole farm, extension agents who are assigned to FSR&D will normally need reorientation.

8.6.4. ORGANIZATION

Many extension organizations in developing countries are highly centralized. Decisions are made at the top, with little input by the field staff. Too frequently, extension agents simply do what they are told. Thus, superiors often overlook the farmers' needs and extension agents' comments.

In FSR&D, the field staff needs more freedom of action. For example, within the guidelines of the regional plan of work, an extension agent trained in FSR&D should (1) play a part in selecting farmers for farmer-managed tests and for keeping farm records and (2) be heard when expressing ideas about problems and opportunities arising from the agent's experiences. Some extension organizations associated with FSR&D, as in Honduras and the Philippines, have become more decentralized. With this setup, more decision-making authority has been shifted to local and regional groups.

8.6.5. BUDGETS

Considering their potential workload, the extension organizations in many developing countries have inadequate budgets. Low budgets lead to inadequate staff pay and insufficient funds for operations. The effect can cause low staff morale, which in turn leads to extension's low credibility with farmers, researchers, and others.

Extension's participation in FSR&D should change this situation. We hope the material we have presented will demonstrate to extension's decision makers how FSR&D serves extension's interests. When convinced of FSR&D's effectiveness, these decision makers should be willing to support the new program by allocating appropriate staff and funds. By receiving and diffusing better technologies, extension becomes more effective and should then be able to support its claim for a larger share of the national budget. This, in turn, gives extension the opportunity to further strengthen its organization and program.

8.7. CONCLUSIONS

Extension's role in FSR&D is supportive during the earlier activities and then takes on major responsibilities during multi-locational testing, pilot production programs, and widespread diffusion. The importance of extension is highlighted when researchers realize that the ultimate payoff is broad farmer acceptance of the new technologies they produce.

To improve their organization, decision makers in extension need to analyze the ability of their organizations to perform their assigned responsibilities in FSR&D. Difficult decisions may have to be made in restructuring and reorienting the present organization to make it fit FSR&D's needs.

CITED REFERENCES

Benor, D. and J.Q. Harrison. 1977. Agricultural extension: the training and visit system. The World Bank, Washington, D.C.

Haws, L.D., and R.T. Dilag, Jr. 1980. Development and implementation of pilot production programs. Presented at the Cropping Sys. Conf. 3-7 March 1980. IRRI, Los Banos, Philippines.

ICTA and DIGESA. 1978. Carta de entendimiento entre El Instituto de Ciencia y Tecnología Agrícolas (ICTA) y La Dirección General de Servicios Agrícolas (DIGESA). ICTA, Guatemala.

Pantastico, E.B., R.T. Dilag, Jr., and L.D. Haws. 1980. Utilization of site related research in multi-location testing. Presented at the Cropping Sys. Conf. 3-7 March 1980. IRRI, Los Banos, Philippines.

Waugh, R.K. [undated]. El caso del ICTA en Guatemala como institución dedicada a la generacíon y validación de technología para pequeños agricultores. ICTA, Guatemala.

Zandstra, H.G., E.C. Price, J.A. Litsinger, and R.A. Morris. 1981. A methodology for on-farm cropping systems research. IRRI, Los Banos, Philippines.

OTHER REFERENCES

Axinn, G.H., and S. Thorat. 1972. Modernizing world agriculture: a comparative study of agricultural extension education systems. Praeger Publishers, New York.

University of Minnesota Project in Chile. 1971. An experiment in production education. Univ. of Minnesota, St. Paul, Minn. (Also in Spanish: Una experiencia en educación en producción.)

Chapter 9

DECIDING ON AN FSR&D APPROACH

The previous chapters gave information about the general nature of FSR&D, how the process operates, and how to conduct FSR&D activities. With this background, the reader should be able to consider whether or not an FSR&D approach will interest a country's leaders. These leaders and their advisers will probably want to consider a number of issues that relate to the country's progress in its research and development efforts for small farmers. Therefore, in this chapter, we present issues on development policies, farmers' needs, personnel and organizational capacities, FSR&D's comparative costs, and related issues.

9.1. BASIC ISSUES IN RESEARCH AND DEVELOPMENT

At a policy-making level, a range of issues must be addressed in deciding whether to use an FSR&D approach. The basic issue concerns identifying which conditions are most conducive for organizing research and development along FSR&D lines.

Most policy makers are interested in increasing the returns from their investments in agricultural research. Therefore, the question arises as to whether satisfactory returns are being secured from current investments in research and development. This is particularly important in most developing countries, where shortages of scientific personnel and insufficient research budgets restrict agricultural research to only modest efforts. Within this context, policy makers must be continuously alert to ways in which research policies and programs can respond better to national needs.

A review of the conditions in a number of developing countries suggests that agricultural research and development are undergoing important changes as governments seek to enhance the income and productivity of small farmers. The character and origin of these changes can be brought out by asking questions about society's goals and small farmers' needs.

9.1.1. ARE THE ACTIVITIES OF THE RESEARCH AND DEVELOPMENT PROCESS CONSISTENT WITH THE NATIONAL DEVELOPMENT GOALS?

Frequently research priorities are not consistent with national goals. At times, research and development activities are haphazard, with program leaders paying little attention to each other or to broad developmental goals. Under such conditions, policy makers have difficulty understanding research's purpose. When reviewers do not know the priority or direction of research, they will have difficulty in evaluating whether the research effort is fulfilling its intended purposes.

Even when research results are proved applicable, policy makers must still ask if the results are being disseminated quickly and broadly. Frequently research produces a backlog of new technologies, but few channels are available for rapidly diffusing these technologies. This condition often occurs when researchers and the extension service are isolated from each other. Policy makers need to seek ways to bridge this gap so that research and extension can collaborate more effectively.

9.1.2. IS THE RESEARCH AND DEVELOPMENT PROCESS PRODUCING RESULTS THAT ARE RELEVANT TO SMALL FARMERS' NEEDS?

Many policy makers are asking whether research for agricultural development is producing scientific innovations that relate directly to production problems and opportunities encountered by small farmers. In many instances, research activities are not addressing these important concerns. Research at experiment stations and within the commodity programs may be quite intensive, but alone these activities may do little to increase small farmers' production.

A related question is, "Do small farmers provide extension agents and researchers with feedback on new technologies?" Policy makers are beginning to assess research and development organizations' performance in terms of whether small farmers have an opportunity to (1) influence which research is undertaken and (2) evaluate the resulting technologies. Where a communication gap exists between researchers, extension agents, and the farmers, new technologies may not be relevant.

9.1.3. SOME ANSWERS

Because of their historical concern with exports, many developing countries directed their agricultural research and extension efforts toward commercial farmers who produced commodities for export. Agricultural

research and extension often served these commercial farmers quite well, but virtually ignored traditional farmers. With the growing interest in total food production and the welfare of the poor, the lack of attention toward traditional farmers is being reconsidered. FSR&D has emerged from these needs.

Opportunities for applying an FSR&D approach become apparent when examining individual situations. For example, the reader might (1) compare the availability of agricultural technology in a country with those technologies being used by the country's small farmers, (2) then observe the various levels of agricultural development. Where agriculture productivity is low, considerable potential probably exists for applying FSR&D concepts. These concepts concern how to select technologies that fit the farmers' situations, which technologies fit where, and how to integrate the technologies into the farmers' system. *How to use available technologies* is more important than *generating new technologies*. FSR&D is particularly well-suited for this situation (Harwood, personal communication). On the other hand, highly developed agriculture will often be constrained from further growth by a lack of improved technology—e.g., varieties, mechanization, and pest control—leaving more scope for traditional research.

Policy makers in developing countries who seek more effective research and development processes are beginning to recognize the advantages of FSR&D for helping small farmers. For one thing, the FSR&D approach makes it possible to consider the highly integrated activities of subsistence farming. For another, FSR&D, along with some other approaches, offers opportunities for more control over research's direction by explicitly choosing geographical areas and specific groups of farmers. In this way, FSR&D procedures make agricultural research and development more amenable to policy makers' analysis and guidance.

As a result of the foregoing points, FSR&D is beginning to appear at the national level in several ways:

- by programs that are part of a national research and development effort
- by FSR&D complementing ongoing research and development programs
- by the emphasis given to the approach by several of the International Agricultural Research Centers (IARCs) and other international organizations.

9.2. DEVELOPMENT POLICY AND THE ROLE OF FSR&D

The following examples illustrate the influence of FSR&D on national development policy:

In Indonesia, the productivity of small farms is being increased by using a farming systems approach to intensify cropping systems. In areas with irrigation and traditional double-cropping systems, triple-cropping systems are being tried. In partially irrigated and rain-fed areas, new double-cropping patterns are replacing single-cropping systems. As these multiple-cropping practices become widely accepted by small farmers, domestic food production will be increased thereby allowing the government to reduce its foreign exchange expenditures for food imports.

In Nicaragua, the small farmer traditionally engages in the single-cropping of beans during the growing season, although early dry periods frequently cause major crop losses. Using a farming systems approach, planners determined that more stable production could be achieved by intercropping sorghum with beans because sorghum is more drought-resistant than beans. Farmers accepted this innovation because of the stabilizing effect on seasonal production.

In an area in central Tanzania, frequent lack of rain in the middle of the growing period greatly reduces the reliability of maize yields. Traditionally, farmers have responded to this uncertainty by growing sorghum as a supplementary food source. Using a farming systems approach, research resulted in the introduction of a maize variety that matures earlier. While the new variety provided a lower yield because of its early maturation, the crop was less affected by the lack of rain, and therefore, offered a more reliable food source. More importantly, the farm production and income were increased with the introduction of this variety. With a more reliable harvest of maize, farmers could reduce their sorghum planting and grow more crops for market. Because of the earlier maize harvest, traditional follow-up crops could be planted sooner with generally increased yields.

Each of these examples illustrates the interaction between research for small farmers and national policy for increased or stabilized food production. In Indonesia, the emphasis was on increasing production; in Nicaragua, the emphasis was on stabilizing production; and in Tanzania, both objectives were emphasized.

In summary, an FSR&D approach shifts research to the farm where new technologies are tested and evaluated for their compatibility with national goals and the farming systems. Where new technologies cannot meet these criteria, they are laid aside in preference for those that do meet these criteria. This approach keeps research and development from becoming diffused, or diluted by concerns that do not relate to farmers' needs and national goals.

9.3. FARMERS' NEEDS AND THE ROLE OF FSR&D

Below are four more examples that illustrate FSR&D's response to small farmers' needs:

In Ethiopia, the International Livestock Centre for Africa (ILCA) has been experimenting with a mixed farming system. This effort includes (a) identifying more efficient means of silage making to conserve forage, thereby reducing the demands on the farmers' labor at a time when they normally weed their crops, (b) screening and testing leafy forage varieties to replace existing varieties that are too fibrous and low in protein, (c) testing the progeny oxen from improved dairy cows for their higher draught power and ability to pull more efficient implements and comparing their feed requirements with those of local oxen, and (d) exploring the feasibility of using low-production farm lands for rearing animals, thereby freeing up more productive areas for crops.

In one area of the Philippines, drought stress continuously reduces the yield of the dry-season rice crop. Using a farming systems approach, researchers determined that reducing the time between harvesting the wet-season rice crop and planting the dry-season rice crop would substantially increase the yields of the latter. On-farm research revealed that less intensive tillage, use of portable threshers, and planting older seedlings succeeded in greatly reducing the time between the two rice crops.

In one region of Zambia, little was known about farming practices and agricultural conditions. Thus, the researchers lacked sufficient knowledge to provide specific recommendations for increasing production. Field surveys revealed that farm production was limited by seasonal labor shortages that caused poor land preparation, late planting, and insufficient weeding. A closer analysis of the farming system indicated that changes could be introduced to overcome the seasonal labor shortage. The minor changes involved introducing early maturing varieties and improving weeding practices while the major changes required that farmers adopt new cropping regimes and use more efficient land preparation techniques.

In Indonesia, the government is establishing major resettlement schemes on the sparsely populated uplands of Sumatra. Many of these areas have not been cultivated, and those that have been cultivated are often farmed by traditional slash-and-burn methods. Major changes are needed in shifting existing agricultural systems to more intensive and sedentary systems. A farming systems approach is being used to design and test new intercropping sequences that will demand more intensive management in maintaining soil fertility and in planting and harvesting on time.

The above examples suggest various applications of FSR&D. The work in Ethiopia illustrates the linkages between crops, forages, and animals. In the Philippines, the FSR&D team learned that only minor technology changes were needed to increase crop production. In Zambia, the FSR&D approach revealed several opportunities for increasing production. In Indonesia, the FSR&D project is designing and testing the suitability of new farming systems within a varied and complex ecology.

The foregoing examples illustrate three ways in which FSR&D is responding to the needs of small farmers through (1) modifications to the existing farming system, (2) introductions of new farming systems, and (3) helping policy makers and researchers assess how much change to attempt. Each of these ways is discussed below.

9.3.1. MODIFICATIONS TO EXISTING SYSTEMS

In many countries the dominant farming systems may emphasize one or more major cereal crops or animal patterns, and the farmers and the government may need only to improve production within these systems. This is the case in the wheat and rice growing areas of Asia, in the bean and corn growing areas of Latin America, and in the mixed systems in parts of Africa. In addition, this approach works well in developing countries where farmers concentrate on either a subsistence or a cash crop. In these instances, FSR&D focuses less on changing the farmers' principal enterprises and more on enhancing productivity within the existing farming system. This requires careful examination of the farming system to learn how available technologies can serve farmers' needs.

9.3.2. INTRODUCTIONS OF NEW SYSTEMS

In some countries, the existing cropping or animal practices have reached production plateaus, and new production systems are needed. In other areas, sedentary patterns need to be designed for farmers who no longer practice shifting cultivation. Also, traditional farming practices may involve complex multiple-cropping and animal patterns that require considerable understanding and analysis before attempting to change them. In all these instances, researchers must have a high level of understanding to know when to integrate single and multiple technology components into the existing systems—an understanding that covers the social, economic, and technological variables of farming. This knowledge can then be used to determine where innovations can best improve the system's performance.

9.3.3. DECISIONS ON HOW MUCH CHANGE

FSR&D aids the policy maker and researcher in assessing the extent of change desirable for an existing farming system. In particular, many instances exist where policy makers do not know if production can be increased with minimal changes, or whether more substantial modifications are needed. Because of its focus on understanding how basic technology and management components interact within the farming system, FSR&D is being used to illuminate the kinds of changes required to achieve various production goals. When major changes in farming enterprises are being contemplated, FSR&D identifies available options and the risks and costs associated with such changes.

9.4. ORGANIZATIONAL CAPACITY AND THE ROLE OF FSR&D

FSR&D tries to integrate research and development through close relationships among the disciplines and among research, extension, and other organizations concerned with small farmer production. The agreement between ICTA and DIGESA (Sec. 8.6.1.) is an example of a formal approach toward integrating the efforts of research and extension.

A key organizational issue in agricultural research relates to how priorities are established in defining important research problems. In most countries, responsibilities for agricultural research are distributed among discipline and commodity programs. This arrangement, which allows considerable specialization, presents formidable management problems in producing a relevant and coherent research program. Where research is organized along disciplinary and commodity lines, each activity can become compartmentalized. Then, professionals from one discipline tend to become isolated from professionals in other disciplines. As a consequence, the professionals' activities are often defined in terms of their special interests, with less and less reference to the farmers' problems.

In response to these organizational issues, FSR&D has emerged as an approach that is not fragmented, insular, or divorced from farmers. Rather, it brings closer interaction among organizations and specialists in solving farmers' problems. FSR&D does this by focusing on the whole farm and thereby arriving at problems of the greatest concern to farmers. When problems are identified in this way, research management becomes easier because research activities are integrated and each research group knows how its work contributes to the overall effort.

9.5. FSR&D AND SUPPORTING ORGANIZATIONS

Many organizations can help support FSR&D activities. Because of the integration of extension with research, the extension service should, ideally, be formally associated with the FSR&D effort. Other organizations such as those dealing with credit and marketing can best be brought in as the needs of new technologies become apparent. If the characteristics of new technologies depend on modifications of the services currently available to farmers and if such changes are feasible, the appropriate services can be planned in anticipation of final research results. However, if adjustments are not feasible, then the research approach needs to be modified or abandoned. A hypothetical example suggested by Winkelmann (personal communication) illustrates the point:

Suppose local grain storage facilities are at capacity and announced price supports could not be maintained were grain production to be increased. Then, improved farming practices leading to greater grain production would cause the profit rates to drop. For the next several years, the current storage and unstable pricing conditions should become part of the assumed environmental conditions when planning the research program.

Without adequate coordination among research, extension, and supporting agencies, potential problems typical of this hypothetical example may not be realized until it is too late. Alternatively, when coordination is adequate and governments have the capability of supporting FSR&D's needs, they can help to relax or even remove the above type of restrictions. For example, a government (1) could encourage private industry to provide the needed storage facilities or provide them directly and (2) could take measures to improve its price support program.

9.6. ADOPTION OF AN FSR&D APPROACH

In adopting an FSR&D approach, policy makers need to consider personnel and other resource requirements and how these relate to existing programs. Responses vary according to the extent to which FSR&D is used as a minor or major factor in the overall research and development process. For example, Guatemala, Honduras, and Senegal adopted the farming systems approach as the principal means of reaching small farmers. This effort required restructuring agricultural research organizations and training programs to prepare staff for on-farm research. In Indonesia, Philippines, and Bangladesh, farming systems concepts and methods were introduced project by project and affect only one or two regions of each country. This gradual approach has left existing patterns of research and extension largely intact.

In brief, policy makers may tailor FSR&D to existing conditions in many ways. FSR&D need not mean major changes in organization and manpower; it can begin with existing organizations and by training only a few specialists. Such flexibility arises out of FSR&D's:

- emphasis on applied research
- adaptability to ongoing development programs
- the field teams' low demand for highly trained personnel

We will now consider each of these points in more detail.

9.6.1. EMPHASIS ON APPLIED RESEARCH

FSR&D strongly emphasizes using interdisciplinary teams to identify problems and opportunities for well-defined farming systems with specific environments. Critical to the effectiveness of the field teams is close and frequent association with senior scientists. These scientists aid the field team by (1) helping to set the direction of research, (2) helping to identify problems, (3) providing assistance as problems arise, (4) establishing procedures for the field team to follow, and (5) developing and implementing training programs in FSR&D.

The field teams' work ties back into more broadly applicable research through the requests the field teams make to experiment station staff and others concerning the problems they encounter. As a consequence, programs in other research areas benefit from the insights FSR&D offers. Generally and specifically applied research eventually come together by focusing on the needs of specific groups of farmers. In this sense, FSR&D strengthens other research and development programs.

9.6.2. ADAPTABILITY TO ONGOING DEVELOPMENT PROGRAMS

Governments have several options when considering where and how to locate FSR&D. These options can be divided into three categories. Those dealing with

- developing new technologies for small farmers
- adapting existing technologies to small farmers' conditions when government policies and support services cannot be changed easily
- adapting existing technologies to small farmers' conditions when government policies and support services can be changed (Norman, personal communication).

Where new technologies are needed in abundance, FSR&D's decision makers will probably want to arrange for FSR&D to be tied closely with ongoing experiment station activities. In contrast, where emphasis is on adapting existing technologies to farmers' conditions, the FSR&D teams will be more independent. This does not mean that FSR&D does not need other organizations. Inputs from research staff at the experiment stations, extension agents,

staff from production-oriented groups, and others can almost always be profitably used by FSR&D staff. Our statement simply means that FSR&D does not have to be so closely tied with these other organizations when technologies are available for adoption. And when the government is able to adjust its policies and influence organizations that provide supporting services, FSR&D's flexibilities and potential for accomplishments are even greater.

9.6.3. REQUIREMENTS FOR SKILLED PERSONNEL

Highly skilled and experienced personnel are not abundant in most developing countries. Therefore, policy makers need to make sure that the energies and skills of available scientists are put to good use. This concern applies equally well to FSR&D activities. Experiences of those countries having undertaken an FSR&D approach suggest that FSR&D is an effective way to use this scarce resource.

The approach is to develop effective teamwork between field teams and more senior staff. Younger, less highly trained field staff can become the focal point of the FSR&D approach when suitably complemented by senior staff. Since FSR&D's field staff members are often recent graduates with bachelor's degrees in agricultural or social sciences and are supported by assistants from technical schools, the bottleneck to expanding FSR&D programs has generally not been the lack of skilled personnel. Whatever bottlenecks may have arisen are more frequently due to the time needed to set up the FSR&D organization and to develop suitable training programs.

9.7. COST-EFFECTIVENESS OF FSR&D

FSR&D's cost-effectiveness may be of concern to a country's top decision makers. This is understandable, because they need to know if FSR&D will make good use of their country's resources. Unfortunately, the issue is not easily settled. To our knowledge, little has been reported that quantifies or otherwise substantiates alternative opinions. For the present, decision makers must rely on those who have had the closest contact with FSR&D and alternative methods of agricultural research and development.

The arguments that follow draw heavily on appraisals by Gilbert et al. (1980) and the CIMMYT workshop on methodological issues facing social scientists concerned with FSR&D (Harrington, 1980). Generally, these authors and those they cite conclude that FSR&D *is* cost-effective when compared with alternative approaches for reaching target groups of farmers. These authors' conclusions, which assume that a country already has an adequate research base in agriculture, center on three points:

(1) FSR&D's overall costs are less than conventional experiment station research.
(2) The rate of adoption of improved technologies by specific groups of farmers is higher.
(3) By following FSR&D procedures, large numbers of farmers can be reached.

9.7.1. COMPARISON OF EXPENDITURES

Occasionally, when deciding on whether to accept an FSR&D approach, policy makers will compare the relative costs of conventional experiment station research with the field orientation of FSR&D. In a sense such comparisons are irrelevant. FSR&D does not replace experiment station activities; rather, it complements them. Thus, the argument should really center on whether part of a country's available research budget should be allocated to FSR&D at the expense of experiment station activities.

To the extent that FSR&D reduces the rate of expansion of experiment station activity, future installations may not be built and the costs of equipment, operations, and staff will be less than otherwise. In its place will be more work on farmers' fields by generally less expensive staff that requires relatively larger expenditures for vehicle purchases and maintenance, field equipment, per diem, and incentives. We have the impression that the combined initial and recurring costs of FSR&D are less than the costs of comparable levels of activities conducted on experiment stations when full account is given to the initial costs of installation. In this sense, some displacement of experiment station activities, especially when new facilities must be built, seems justified.

9.7.2. COMPARISON OF RATES OF ADOPTION

This issue centers on the generation of new technologies acceptable to farmers. Proponents of FSR&D often cite the relatively high rates of adoption of new technologies in support of the effectiveness of the FSR&D approach. Hildebrand (1979) commenting on his involvement with ICTA says, "Considering the farmers from the beginning of the technology generating process has increased the speed and efficiency with which [ICTA] produces technology appropriate to [the farmers]. The probability of spending several years producing a new variety that has very limited geographical adaptability or that is rejected for not having characteristics important to the producers is greatly reduced under the methodology that has been developed."

9.7.3. COMPARISON OF NUMBERS OF FARMERS AFFECTED

A comparison of the numbers of farmers reached by the alternative methods is more difficult to resolve. By its very nature, much of the research traditionally undertaken at experiment stations has been intended for broad application. In contrast, FSR&D is more narrowly confined to specific groups of farmers selected during problem identification. Essentially, two different groups are the targets: one group comprises those who are able to adopt the more generally applicable results of the experiment station; the other group comprises those for whom the research has been specifically designed. The numbers of potential farmers tend to be larger for the first group than for the second group. But the missing parts of this comparison are the relative differences in the rates of acceptance and the significance of the changes. We simply have not seen studies or heard quantifiable arguments that would give us the type of information to carry such comparisons much further. We have, however, been given suggestions for making FSR&D more widely effective.

Gilbert et al. (1980) and Harrington (1980) suggest three ways for increasing FSR&D's breadth of applicability. One way relates to approaches that aim at reducing the time and expense of data gathering through informal, well-focused, and small sample surveys. A second way is to seek "better" not necessarily the "best" solutions for farmers. These two ways combine to make it possible to reach more farmers with acceptable results than spending additional time and money on somewhat larger benefits for smaller groups of farmers. The third way concentrates on studying the farmers' systems and the environments so that FSR&D results can be extrapolated more effectively to other areas. Complementary to this third point is the ability to eventually learn enough about the functional relationships between environmental gradients and farming systems that FSR&D could predict suitable technologies for study without going through the full FSR&D process; but this possibility lies in the future—especially for national progams.

9.7.4. CONCLUSIONS ON COST-EFFECTIVENESS

In summary, some may think FSR&D is more expensive than other forms of agricultural research because of (1) the apparently high cost of field work and the smallness of the target group of farmers and (2) the failure to appreciate the full costs of experiment station activities. Sometimes, the high cost of FSR&D is given simply as an excuse to avoid change. Overall, FSR&D practitioners feel its costs are not excessive, the acceptability of improved technology has been high, and ways can be found for reaching more farmers. In the final analysis, such cost comparisons may not be an issue, since FSR&D does not compete with conventional research, except for the last increments of the research budget.

9.8. A CONCLUDING COMMENT

Referring to the questions raised at the beginning of this chapter, FSR&D can be a positive factor in a country's research and development effort. If existing research and development activities are not producing results that are relevant to farmers' needs, FSR&D should help. By studying farming conditions, FSR&D aids in designing technologies for specific groups of farmers. Or, if existing research and development are not consistent with national goals, an FSR&D approach should help overcome this problem. One of the reasons is that FSR&D focuses on both the broad-scale and the narrowly-defined aspects of research and development for the small farmer.

In deciding whether to initiate FSR&D activities, the government will need to (1) decide on an approach to agricultural research for small farmers, (2) take steps to obtain cooperation from relevant organizations, (3) supply the needed funds and services, especially in support of field activities, and (4) be patient by giving the new program time to prove itself.

CITED REFERENCES

Gilbert, E.H., D.W. Norman, and F.E. Winch. 1980. Farming systems research: a critical appraisal. MSU Rural Dev. Paper No. 6. Dep. of Agric. Econ., Michigan State Univ., East Lansing, Mich.

Harrington, L. 1980. Methodological issues facing social scientists in on-farm/farming systems research. A paper presented at a CIMMYT workshop on Method. Iss. Facing Soc. Sci. in On-Farm/Farming Sys. Res. 1-3 April 1980. El Batan, Mexico.

Hildebrand, P.E. 1979. Incorporating the social sciences into agricultural research: the formation of a national farm systems research institute. ICTA, Guatemala, and The Rockefeller Foundation, New York.

Chapter 10

IMPLEMENTATION

If national leaders decide that an FSR&D approach is in the best interests of their country, their next step is to decide how to proceed. They and their staff need to consider such factors as an appropriate approach, organizational structure, staffing, off-site management, the roles of the field team, interdisciplinary teamwork, getting started, and evaluation of projects. This chapter covers these topics, and provides two examples and a summary.

10.1. DECIDING ON AN APPROACH

How FSR&D relates to other research and development activities depends on FSR&D's scale and the degree to which new institutions must be created. Furthermore, the scale of FSR&D will vary according to whether it is introduced as a project, a program, or a modification of existing research and development activities. The characteristics of projects and programs are discussed below because they influence the organization, staff, and management differently.

10.1.1. PROJECT APPROACH

A project approach to FSR&D involves initiating one or more projects that incorporate FSR&D procedures. Projects tend to have specific scopes of work to be completed by a certain time and, therefore, to have staff and organizations that are disbanded upon the projects' completion. Several FSR&D projects could be undertaken simultaneously or sequentially. These efforts, however, would not, necessarily, constitute a program. In many instances, projects are simply undertaken outside or independent of the implementing agency's regular activities. In brief, the project approach to starting an FSR&D activity requires little change in institutional philosophy, organizational structure, or programmatic agenda. International funding may allow FSR&D projects to be undertaken with only modest demands on the domestic budget.

10.1.2. PROGRAM APPROACH

In contrast to the above, a program approach to FSR&D comprises a more encompassing and continuous effort. Using a program orientation entails initiating FSR&D activities as a significant institutional effort with farming systems projects constituting a means of implementation. In this setting, FSR&D goals and activities officially become part of the country's efforts in agricultural research and development.

A government might choose the program approach to implement FSR&D procedures in a coordinated way throughout those organizations most concerned with small farmer production. Alternatively, the government might assign one organization the responsibility for implementing the FSR&D program and have that organization coordinate its efforts with the other organizations. The first approach would produce more uniform results throughout the government, but could lead to substantial delays in getting started.

In any case, changes in organizational structures will normally be required and FSR&D funding will become part of the annual budget and any development plans.

10.1.3. PROJECT VERSUS PROGRAM APPROACH

Policy makers encounter a range of opportunities and constraints when considering either a project or program approach. A project-by-project approach is particularly useful where the institutional environment is neutral or where important factions oppose FSR&D. A project approach, which makes fewer claims on existing financial and human resources, is less likely to be viewed as a major competitor for resources. Also, because a project approach normally operates within existing program philosophies of an existing organization, those with viable programs are less likely to oppose FSR&D.

In contrast, a new FSR&D program demands a significant and sustained response among organizations. The government must add new functions and structures to support the new activity. This addition can be accomplished by creating an entirely new, autonomous organization that assumes the task of implementing the FSR&D program, or the program can be undertaken within the existing organizational structure. Each alternative has its advantages and disadvantages.

Creating a new, autonomous organization for an FSR&D program presents some unique advantages. The activities are not constrained by the competing interests and the claims of other programs within the same organization. In this sense, FSR&D may become somewhat insulated from the usual intra-agency conflicts over staff and funds. The new organization can recruit additional staff, allocate resources, and define functions with greater freedom and discretion than when the program is embedded within a larger organization. This flexibility and the allowance for initiative favors decentralization—an approach important to FSR&D.

On the other hand, a new organizational structure can create administrative problems. A separate FSR&D organization outside and independent of the regular administrative structure can hamper close interagency cooperation. This is particularly true when an FSR&D program is launched independently of extension.

An alternative approach is to incorporate an FSR&D program within an existing organization. This approach may or may not entail undertaking significant restructuring of staff responsibilities and activities. In some cases, ad hoc administrative arrangements can be devised to support an FSR&D effort; whereas, in other instances, existing efforts may require redesigning and adding new organizational components.

Whether a country should use an FSR&D project or program approach depends on each situation. When senior decision makers agree that a farming systems approach should be a major activity, then an opportunity exists for designing and implementing a comprehensive and integrated program. When decision makers disagree or are indifferent, a project-by-project approach is more feasible. To attempt a program approach without sufficient top-level support could undermine the whole FSR&D effort.

10.1.4. GOVERNMENT SUPPORT

Whichever approach is taken, careful attention needs to be given to adequate governmental support for FSR&D activities. This attention is essential for several reasons. Specifically, FSR&D is comprehensive and must draw on the services of many disciplines and organizations that may not be under its jurisdiction. The government can help FSR&D's management by encouraging other organizations to provide the services of selected staff for limited periods and by giving access to their data, reports, and facilities. Such organizations are local, regional, and national, and include planning and administrative units, universities, research institutes, extension, production, and related service groups. Government incentives for such cooperation should include budget allotments and recognition for contributions to FSR&D activities.

Because of its nature, FSR&D calls for a strong commitment to interdisciplinary teamwork. This means learning other subjects, working with other disciplines, and leaving the security offered by an established discipline. Such commitment to interdisciplinary teamwork deserves sufficient governmental support to allow FSR&D time to demonstrate its effectiveness.

Finally, because the FSR&D staff spends so much of its time living and working in remote areas away from national and regional centers, the government should offer incentives such as higher salaries, per diem allowances, local and international training, and publicity for accomplishments, and should facilitate associations with FSR&D researchers in other countries. Equally important is to give field and regional headquarters teams freedom in implementing their programs as long as their work remains within general policy and program guidelines. Such autonomy is especially important in on-farm research, where timeliness of action is essential to successful experiments and relations with farmers.

10.2. ORGANIZATIONAL STRUCTURE

Whether FSR&D begins as a program or project, several questions must, eventually, be addressed concerning its integration into the government's ongoing activities. In many countries, agricultural research is divided into interrelated disciplinary and commodity groupings. Thus, a research organization could have several disciplinary specialties on the staff, and in turn, the staff could divide its time among several commodities.

The disciplines and commodities may be organized in several ways. In particular, differences occur in the extent to which research staff associate with one or more of the commodity programs, and the degree to which either the disciplinary specialties or the commodity programs control the research activities. These differences often reflect basic administrative realities concerning control of the research budgets and staff appointments and promotions.

The FSR&D approach draws on a variety of disciplinary and commodity programs and provides the technical coordination for bringing this about (Fig. 10-1). That is, FSR&D provides the framework for showing how the various disciplinary and commodity programs can be integrated in helping small farmers. In performing this function, the FSR&D team needs to identify the capabilities and bring together the mutual interests of specialists at the local, regional, and national levels. We provide further comments on this subject by project and program, as follows.

Figure 10-1. A schematic of the interactions between disciplinary specialists and commodity programs for FSR&D.

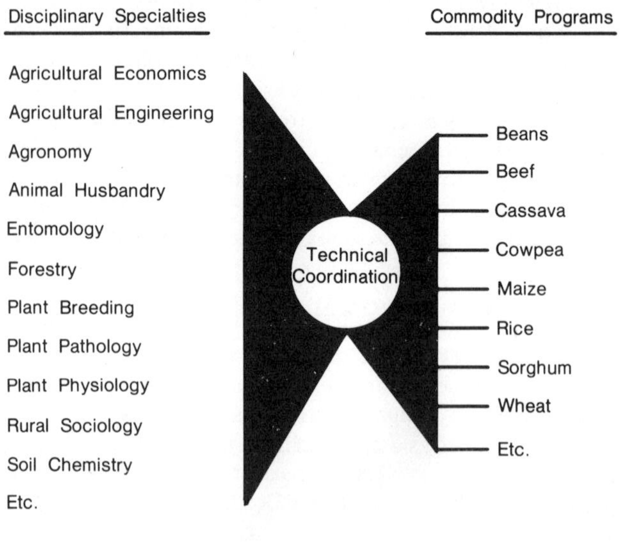

10.2.1. FOR THE PROJECT APPROACH

In the project approach, FSR&D activities are usually located within that disciplinary or commodity program offering the strongest support. If established within a discipline, agronomy, animal science, or even irrigation could be a logical choice. If within a commodity program, the farming systems effort is usually identified with the leading food crop. Thus, in Southeast Asia rice constitutes the major commodity for small farmers, whereas in Latin America, beans and maize occupy a central role for small farmers. These leading commodity programs usually have a secure role in the research community—a characteristic that usually allows greater flexibility in adopting and supporting new programs. When these commodity groups are staffed with experienced scientists, they can forge strong linkages with other commodity groups. This capability is particularly important in moving research programs from a single to a multiple cropping orientation.

The above choices allow an FSR&D activity to be integrated into an ongoing research operation in several ways. By using the project approach, the administrator has flexibility in working out informal arrangements to support FSR&D. In the short and medium term, the project leaders need not address the larger issue of structural change. The following illustrates a current cropping systems project in Bangladesh. Eventually, when responsibilities and procedures become more established and formalized, the project could evolve into a program. We now describe the Bangladesh project:

In Bangladesh seven disciplinary and commodity-based research institutions are associated with cropping systems research: the Bangladesh Rice Research Institute, the Bangladesh Agricultural Research Council, the Jute Research Institute, the Sugar Cane Research Institute, the Water Development Board, the Soil Survey Institute, and the Bangladesh Agricultural University. In many cases, organizations have experiment stations, but they jointly support a national cropping systems effort. The current cropping systems activity involves 12 research areas with each institute having primary responsibility for the work at one or more of them.

A cropping systems task force has been organized in each institute. Its members are recruited from the scientific staff of the disciplines represented in the specific institute. This task force reviews the research plans undertaken by its farming systems field teams and provides technical support for their activities. Inter-commodity collaboration is achieved by having each task force assist other task forces. Thus, the task force at the Soil Survey Institute conducts soils research for other research institutes.

To facilitate and strengthen interdisciplinary and inter-commodity research, a national Cropping Systems Working Group has been established. This group is headed by the chairman of the Bangladesh Agricultural Research Council. Its members are the leaders of the cropping systems task force at each of the participating research institutes. A senior member of the extension service is also represented in the group.

Periodically, the Cropping Systems Working Group meets and discusses the progress of the cropping systems effort and coordinates the various activities. To provide the group with continuous coordination, one member serves full time in that capacity. Administrative and scientific staff from the Bangladesh Agricultural Research Council assist the project coordinator.

The Bangladesh example illustrates several points concerning how an FSR&D effort can be introduced into a complex research establishment. Few structural changes were undertaken to accommodate the cropping systems effort. Rather, a coordinating committee enlists greater interdisciplinary and inter-commodity cooperation in FSR&D activities. Considerable management skill needs to be exercised to ensure that the participating institutions work together. In Bangladesh, cooperation is facilitated by offering incentives such as training abroad in farming systems methodology for the staff from the research institutions, and honorariums to researchers who spend additional time in the field. Such inducements are important when mobilizing interagency cooperation.

10.2.2. FOR THE PROGRAM APPROACH

Institutional change assumes immediate importance when launching an FSR&D program. The basic issue concerns how to develop an organizational capacity for an interdisciplinary, applied systems effort. An opportunity for change occurs when decision makers begin to question the rationale for continuing to organize research along disciplinary or commodity lines. In response, decision makers may redefine and merge conventional groupings to achieve an interdisciplinary approach. In other cases, traditional commodity and disciplinary operations may be left intact and newly defined FSR&D units added. Consequently, a range of organizational modes can be used to implement an FSR&D program, and the opportunity exists for imaginative variation in organizational structures. However, the manner of restructuring organizations can be reduced to two basic types: management centered modes and interdisciplinary-commodity modes.

Management-Centered Modes

In some instances, the responsibility for initiating and sustaining an FSR&D program is in the hands of senior managers, and the disciplinary specialists and commodity programs respond to directions from the organizational hierarchy. This approach generally arises when a separate organization is newly created for implementing an FSR&D program. In brief, the organization becomes a self-contained entity with the singular mandate for conducting FSR&D. Such a philosophy is infused throughout the organization so that a special unit to carry out FSR&D is not needed. Furthermore, traditional disciplinary and commodity-oriented divisions are not redefined because the entire program is based on a commitment to FSR&D. Management sees that all activities are reviewed and monitored from this perspective. In some instances, the focuses of the disciplinary and commodity groupings will be expanded and their roles will be subsumed within the larger interdisciplinary and multi-commodity perspective. In brief, a management-centered program fits those situations where FSR&D justifies the organization's existence.

The program of the Agricultural Science and Technology Institute (ICTA) is an example:

ICTA functions as the primary research institute for small-farm development in Guatemala, following the philosophy and goals of a farming systems approach. ICTA's management tries to ensure that this approach permeates the design and implementation of its research program. With FSR&D constituting such a central feature of ICTA's philosophy, its management feels that a separate farming systems group is not needed.

In ICTA, commodity and disciplinary groupings continue to function, but they work closely with and provide technical support to the regional field teams. Administrators and staff strongly emphasize the regionalization of research and development. Regional teams consist of commodity and disciplinary specialists and the field teams. The regional experiment station serves as the base of operation allowing the commodity specialists and the field teams to work closely together. ICTA's national and regional headquarters' teams coordinate the program.

Interdisciplinary-Commodity Modes

A second alternative for organizing an FSR&D program is to give the task of conducting FSR&D to a disciplinary or commodity group. Thus, a new discipline may be mandated to undertake FSR&D, and a counterpart commodity group may be established to support a multi-cropping or multi-animal orientation within the commodity programs. In other instances, a farming systems department may not be added, but a multiple cropping or animal program may be created for farming systems research.

Identifying a farming systems department and a counterpart commodity program provides the status and legitimacy needed to enable a farming systems program to be recognized as important in the organization's effort. Permanent staff are assigned to the program and its leadership can claim part of the budget. Under this arrangement an interdisciplinary, multiple commodity perspective is enhanced when those working in farming systems are trained in FSR&D and support its philosophy. An example from a proposal to change Senegal's agricultural research approach (ISRA, 1979) contains the following points:

- If the proposal is approved, agricultural research in Senegal will undergo major changes to emphasize FSR&D for small farmers. These changes will facilitate greater interaction among the disciplinary and commodity programs. Research activities of the Senegalese Institute for Agricultural Research (ISRA) would be organized around the following departments: (1) crop sciences, (2) animal sciences, (3) farming systems, (4) natural resources, (5) economics and sociology, and (6) research support services.
- Four regional teams organized under the Farming Systems Department would focus on the feasibility of new production technologies for small farmers. These teams would work out of the four regional experiment stations and be closely associated with staff from the other departments. Composition of the regional teams would include specialists in economics, sociology, agronomy, entomology, animal science, and extension.
- The regional teams would receive specialized help from commodity and disciplinary specialists. Included in the latter would be those with expertise in biochemistry, social sciences, agricultural engineering, pest management, post-harvest technology, and farm management. We present further details on the ISRA proposal in Appendix 10-A.

10.2.3. A GENERALIZED ORGANIZATIONAL DIAGRAM

Fig. 10-2 shows a generalized diagram for the organization of an FSR&D *program.* The diagram represents possibilities for a ministry or a semi-autonomous government corporation. This particular example draws heavily on ICTA's organizational structure, which is a government corporation (Ortiz, 1980). In this example, the technical director is in a position to bring training, the disciplines, and commodity programs into an integrated FSR&D program at the regional level. Extension is not shown in the diagram, but could be integrated into FSR&D activities in several ways. As a minimum, extension should (1) participate directly at the top level through representation by the minister, or the board of directors, (2) receive training in FSR&D as part of the activities of the national headquarters team, and (3) assign specialists to regional or to field teams.

A comparable diagram of an FSR&D *project* is not

Figure 10-2. A generalized organizational diagram for an FSR&D program.

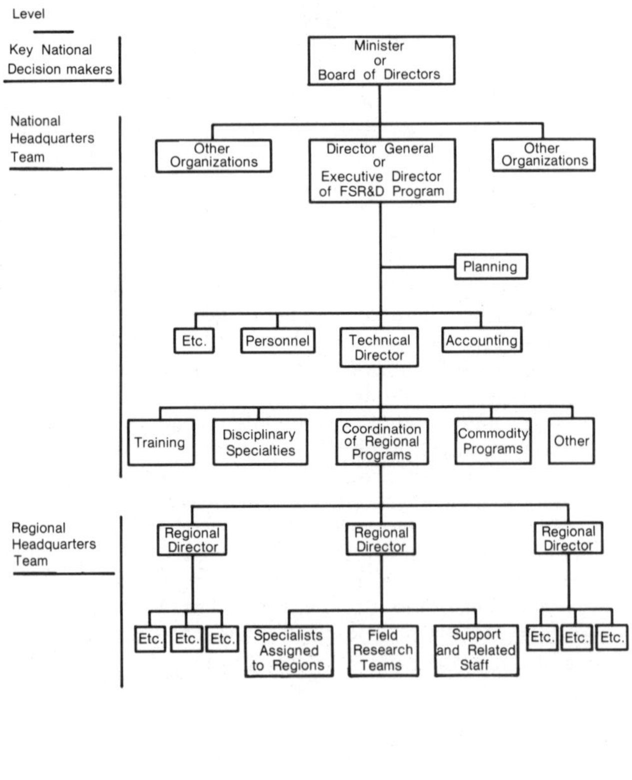

provided in this section, because the characteristics of the diagram depend so heavily on the type of project and the location of the project within the government. However, we provide additional examples of organizational diagrams for various situations in Appendix 10-B. We will now discuss more details of Fig. 10-2 according to the levels of activity.

Key National Decision Makers

The minister of agriculture or possibly natural resources or rural development could hold overall responsibility for the FSR&D program if located within a ministry. Alternatively, the top decision-making body could be a board of directors made up of representatives from key ministries and other organizations concerned with agricultural production and the small farmers' welfare. Possibilities for membership include agriculture (including extension), economics, finance, planning, rural development, universities, separate research and production organizations, organized farmer groups, and the like.

National Headquarters Team

The national headquarters team might be headed by a director general or an executive director who has broad responsibility for accomplishing the goals of the FSR&D program and establishing suitable relationships with other organizations. The heads of various activities and a planning group could assist this director. The planning group would help prepare the annual budgets and assist in planning the organization's development. Organizational activities include such areas as the technical program, accounting, personnel, publications, payroll, and other organizational needs. If the FSR&D program is part of a ministry, the ministry might conduct some or most of these functions.

The core of the FSR&D program will be the technical program, with critical responsibility resting with the technical director. This person would coordinate the regional programs, disciplinary specialties, commodity programs, training, and possibly other activities such as experiment stations, seed multiplication units, and

laboratories. The need for a coordinator of regional programs depends on the size of the program and the number of staff the technical director can effectively manage. Fig. 10-2 follows the management principle that the number of persons with major responsibility who report to a superior should be limited.

Regional Headquarters Team

The regional directors manage the FSR&D activities in their areas. They will have one or more field teams operating in the research areas, with disciplinary and commodity specialists either permanently or temporarily assigned to their regions, and support and related staff. If experiment stations are under the regional director's jurisdiction, the stations' staff would also be included. For such a possibility, commodity specialists would probably associate with or be part of the experiment station staff. Support staff include those needed to run a regional office.

Field Teams

The field teams would comprise those regularly assigned to one or more research areas within the region. As described in Sec. 3.3.1., the team's makeup is basically agronomy-livestock-social sciences plus technical assistants. This group is augmented, as needed, by specialists assigned to the region and by staff from national headquarters and other organizations. We will say more about the composition of the field team in Sec. 10.3.1.

10.3. STAFFING

This section provides information on staffing in addition to that provided in earlier parts of this book. While the levels of training and the numbers required vary depending on the types of FSR&D activities and availability of personnel, four types are required—field teams, disciplinary and commodity specialists, and ESFSs. In selecting staff, management should provide enough flexibility in the organization to allow adjustment for unforeseen needs as the FSR&D approach matures.

10.3.1. FIELD TEAMS

As noted previously, the field teams are frequently made up of young professionals and their technical assistants. These professionals are predominantly agronomists, livestock scientists, agricultural economists (preferably with training in farm management), and irrigation engineers when appropriate. These professionals should be university graduates, while the technical assistants should have degrees from schools with instruction in vocational agriculture. Where trained staff are scarce, those with less education can suffice if they have suitable agricultural backgrounds and are supervised adequately.

When FSR&D activities are just beginning, only one or two field teams may be needed. Later, after experience is gained and staff are trained, the number of teams can be expanded as FSR&D moves into other areas or becomes more concentrated within existing areas. At the rate of two visits per on-farm experiment per week and six visits per day, a two-member team acting individually could work with 30 farmers during a season. A five member team could, therefore, work with 75 farmers per season—i.e., 5 members × 6 visits per day × 5 days per week ÷ 2 visits per farmer per week = 75 farmers. These are only approximations; more precise estimates depend on such factors as the types of experiments, distances between farms, and ease of travel.

Field teams will normally be involved in the whole range of FSR&D activities—from the division of the target area into subareas until the multi-locational tests and pilot production programs. These field teams form the core of the FSR&D effort in that they are in closest contact with farmers and farmers' groups and are the ones most directly involved in the daily research activities. Therefore, members of these teams must be selected carefully to be sure that they can work effectively with farmers on farmers' fields. Their effectiveness will depend on how well they understand farmers' problems and opportunities, empathize with farmers' feelings, and treat them as equal partners in developing improved technologies. Moreover, the makeup of the team should be appropriate for the farmers with whom the team works, as with certain ethnic groups or with female farmers. In some cultures contact between women and men is easily made. In other cultures it is not, and female staff may have to be hired specifically for this reason (Townsend-Moller, personal communication).

10.3.2. RESEARCH SPECIALISTS

Because members of the field team are often young and inexperienced, they will need assistance from disciplinary and commodity specialists and from extension. The composition of these back-up specialists, including extension, depends on the types of activities carried out, the types of farming systems under study, and the particular problems encountered. These commodity and disciplinary specialists provide direction for the field teams, prepare procedures, set standards, carry out special studies, train the field team and others, and are available as special problems arise. The presence of such back-up support enables the field teams to operate with staff that is not highly trained in research methodology or in diagnosing farmers' problems. To the extent that the specialists assigned to the regional FSR&D team are part of the experiment station staff and the ESFSs are part of the regional or field teams, close ties with both the experiment station and extension should result. Where this is not the case, the FSR&D team should seek associations with the experiment station and extension organizations by some other means.

As the name implies, these specialists will normally be trained in a specialty. Ideally, FSR&D teams will cover the breadth of disciplines and commodities of concern to the area. Some of the disciplines can probably be satisfied by those holding bachelor's degrees, such as in agronomy, animal production, agricultural and irrigation engineering, sociology, and economics. But in other cases, those with

advanced degrees will be needed to adequately perform their functions, as with veterinary medicine, entomology, and plant and animal breeding. Where such capabilities cannot be found within the country, expatriates should be considered. Opportunities for part-time employment of staff from the universities and other organizations with highly trained scientists should not be overlooked.

As the FSR&D program matures, opportunities will arise for some of those with general backgrounds to receive specialized training in a discipline or commodity. For example, agronomists working with rice, beans, or maize might be sent to one of the IARCs for several months to learn specific research methods concerning pathology, physiology, economics, or other aspects of these commodities. Moreover, staff with particular promise can be sent abroad for advanced degrees; however, in doing this, care should be taken not to lose sight of the practical orientation of FSR&D nor to allow FSR&D to be used simply as a mechanism for those desiring advanced degrees. We will discuss training further in Chapter 11.

10.3.3. EXTENSION'S INPUT

Extension's inputs to the FSR&D process take two forms. One is through the extension service to the extent it operates within the target area. The other is through the ESFS as described in Chapter 8. When ESFSs are assigned to the FSR&D teams, they (1) become familiar with the research process, (2) aid in involving extension agents who can help in on-farm experiments, and (3) play one of the key roles during the multi-locational testing and any pilot production programs. ESFSs assigned to the FSR&D teams should have broad-based interests and capabilities in both extension and research, since they will act as the principal liaison between research and extension.

10.3.4. TEAM LEADERSHIP

Because FSR&D departs from several traditions and because of the entrenchment of many in long standing positions, the leadership of a new research approach such as FSR&D needs to be imaginative, dynamic, and dedicated. Leaders are needed who can relate the significance of FSR&D to the common interests of the others. These persons should have capabilities and inclinations toward applied, interdisciplinary research and be able to build a viable, enthusiastic research team. They will have responsibility for developing an effective organization, establishing the general approach to FSR&D, working out methodologies, developing and implementing a training program, and carrying out other FSR&D activities. If such leaders have not been exposed to FSR&D concepts, they should receive at least some form of accelerated training and orientation at one of the several institutions experienced in FSR&D methodology. In addition, expatriates can help develop the FSR&D organization and the leaders' capabilities—certainly during the initial years.

10.3.5. APPROACH WHEN TRAINED STAFF ARE SEVERELY LIMITED

Shortages of trained staff often constrain research and development in developing countries. Thus, considerable attention needs to be devoted to assuring that the energies and skills of existing scientific expertise are not wasted or that new programs do not place excessive burdens on existing financial and human resources.

Where demands are high and manpower limited, FSR&D requirements can be scaled down. For example, FSR&D activities can begin if an experienced agricultural scientist can be found who will accept the responsibility. This person should, preferably, have an agricultural specialization at the master's degree level and have the time and desire to guide the design and analysis of on-farm trials and tests. The field teams can be staffed by high school graduates, but these individuals should have practical skills in farming and knowledge of local conditions. These teams need the support of disciplinary and commodity specialists, perhaps from within established research organizations and from those brought in specifically to assist FSR&D activities.

Below are two instances in which a national government entered into FSR&D activities on a limited scale. While these two illustrate only partial applications of the whole FSR&D process, they lay the groundwork for future more ambitious programs:

> In one African country, field teams whose members have university training in social sciences are investigating three FSR&D research areas. Because agricultural researchers are scarce, the government decided to begin by identifying social and economic factors at the farm level. While waiting for agricultural scientists to become available, the teams will have, at least, initiated the FSR&D process and will be showing local farmers that something is being done for them.
>
> Another example of a starting strategy concerns a Middle Eastern country with a severe shortage of both agricultural and social scientists. In this case, not enough researchers were available to organize a field project. Therefore, an interim measure was adopted whereby local farmers participated in trials and tests undertaken at one of the research stations. Such an effort enabled farmers to participate in researcher-managed trials and farmer-managed tests, thereby bringing a more applied orientation to the activities of the research station. However, this was seen only as a temporary measure until more researchers became available.

These initial activities may look piecemeal and fragmentary. But with a proper base of long-term institutional support and commitment a comprehensive and integrated program can evolve. Accumulated experience suggests that the risk of failure under these conditions is much less than imagined and is outweighed by the benefits of having field teams that address farmers' problems.

In brief, we believe that a country can benefit from FSR&D even though funds are scarce and the staff is inexperienced. The lessons learned, even when the approach is limited, should provide valuable experiences that can accelerate accomplishments later.

10.4. OFF-SITE MANAGEMENT

When FSR&D field teams are located in different areas, overall supervision, administration, and monitoring become more important. For our analysis below, we divide these activities into general considerations and personnel management. We refer to them as being off-site because the central administration is usually located outside the research areas.

10.4.1. GENERAL CONSIDERATIONS

A number of management functions are needed to assist in initiating and sustaining on-farm research. When only a few field teams are operating, management can be ad hoc and informal. Once several field teams are dispersed throughout the country, formal institutional arrangements become necessary. In particular, the topics of institutional linkages, technical review, technical support, and logistics need to be addressed.

Institutional Linkages

Where a broad-scale FSR&D program is undertaken, national and regional advisory committees may need to be organized. The national-level committee usually consists of senior-level decision makers. Besides representatives from the ministries, mentioned earlier, other representatives could be from agencies such as irrigation, agricultural credit, cooperatives, extension, commodity purchasing and marketing, and fertilizer distribution. In some cases, the ministry of agriculture administers many of these functions so that the advisory committee would be largely from within the ministry. In other instances, these functions would be dispersed throughout the public and private sectors so that the committee would be composed of individuals from government and industry.

A high-level official of the ministry of agriculture, an agricultural research council, or some functional equivalent would likely chair the advisory committee. Similar groups could be at the regional level with committees consisting of high-level decision makers from the public and private sectors within each region. These regional advisory groups should have representation on the national committee, thereby helping to integrate national priorities and policies with regional priorities and policies.

The national advisory committee would (1) establish general priorities and policies for the FSR&D effort, and (2) help agricultural development agencies function consistently with program goals. Aside from these more programmatic functions, the national and regional committees help FSR&D in other ways. First, their existence provides visibility and recognition to FSR&D as a legitimate endeavor. This credibility is particularly important in the early stages when management is seeking to establish FSR&D as a viable and acceptable component of national and regional development. Second, with an interagency mandate, subordinate units can freely associate with the program. Local and regional agencies can provide crucial support for the FSR&D team's work in the research area. Conversely, where support is inadequate, the committees provide a forum for resolving these problems.

While these committees may not convene frequently, their importance should not be underestimated. FSR&D staff usually need data, specialized personnel, and administrative and program resources that can only be secured from a range of agencies. With these committees providing a mandate, the regional FSR&D directors and the field teams can anticipate support from these other agencies. The government might provide incentives for such support through the budgeting process. Also, the FSR&D teams can encourage cooperation by assisting these agencies occasionally in their technical and training needs. By building lines of communication and cooperation with other agencies, the FSR&D teams will be preparing for the transfer of their research results to implementing agencies.

Technical Review

As the number of FSR&D teams and activities increases, the field teams will not only be undertaking more on-farm experiments, they will also be requesting more off-farm research. Some mechanism is needed to review these requests and to assure that they relate to FSR&D's central purpose. In particular, the review should consider the following points:

- The field team's requests for off-farm research should be incorporated into the country's broad-based research program. The technical review helps to assure that national research programs and FSR&D are mutually reinforcing.
- Appropriate groups need to review the scientific soundness of both FSR&D's on-farm research and its requests for off-farm research. This will help in conserving funds and in developing a reputable research organization.
- On-farm and off-farm research should be reviewed periodically to assure that FSR&D's activities are successful in increasing small-farm production. Such a review (1) reduces the possibility that research becomes little more than a set of unrelated activities, and (2) decreases the possibility of duplication.
- Since budget and manpower are limited, not all FSR&D research initiatives can be accepted. The review process helps set priorities for allocating funds and personnel.

The technical review committee usually consists of the national research coordinators for the disciplinary and commodity programs and the leaders of FSR&D. Frequently, this committee will have counterpart committees within the regions. These committees also participate in the planning workshops to review research results and to plan for next season's activities.

Technical Support

Conducting on-farm research and development requires technical and scientific skills that are often not present among the regular field teams nor immediately available from the regional headquarters. Consequently, the national headquarters team can provide timely help from its technical staff and can secure experts from other sources.

Logistics

Because working in remote areas can be difficult, FSR&D management should emphasize field team support. Some possibilities are (1) providing satisfactory living conditions, (2) authorizing occasional trips to regional and national headquarters, (3) facilitating team members' contacts with others doing similar work, (4) publicizing the team's accomplishments, and (5) meeting the team's daily work needs. Some items especially important for the latter are

- reliable and timely transportation, including spare parts for vehicles
- adequate supplies—e.g., seed, fertilizers, and pesticides—for conducting experiments
- equipment for taking appropriate measurements—such as those related to climate, soil salinity, and water quality, and for making calculations—e.g., pocket calculators
- provision for travel advances so that team members are not forced to finance these expenses themselves.

When the field teams operate out of regional headquarters or an experiment station, many of these services will be provided. Nevertheless, the FSR&D management needs to ensure that the field teams' needs are adequately met.

10.4.2. PERSONNEL MANAGEMENT

Topics under this heading relate to recruitment, career development, and control of personnel assignments.

Recruitment of the Field Teams

Given the rigors of working in isolated areas, the general practice is to employ young men and women for these assignments. Many of them may be recent graduates from universities or vocational schools. However, some of them may be in their mid-careers with several years of field service. Those with previous experience in the area are especially valuable because of their knowledge of local agriculture, customs, language, organizations, and dignitaries. Upon recruitment, these members should receive training in FSR&D as discussed in Chapter 11.

Career Development

Besides the incentives and support mentioned above, consideration also needs to be given to the long-run careers of the field team members. Retaining competent and motivated staff will help the FSR&D effort materially. By the time an FSR&D member has been trained and has gained three or four years field experience, such a member becomes a valuable resource to the organization. A career path should be planned that allows members to use their skills and experiences in a range of management and scientific activities, both in the field and headquarters. These opportunities will be most easily provided by a healthy and growing FSR&D program.

Conventional incentives may have to be revised for FSR&D's university-trained staff, who often have strong if not sharply defined career objectives. The problem is that advancement among this group has tended to be based on an individual's record of publications in recognized journals. Moreover, journal reviewers have generally not favored articles describing applied or interdisciplinary topics. The most direct way to overcome this problem is for FSR&D management to promote its staff on the basis of individual and team accomplishments. Another possibility is for FSR&D leaders to establish a technical journal on FSR&D methods and results.

Control of Personnel Assignments

An important problem in developing countries concerns the competing demands placed on scarce scientific personnel. Under conditions of manpower scarcity, the efforts of one region to monopolize the services of disciplinary or commodity specialists could deprive others of their services. In brief, too much decentralization weakens a national research program.

A system needs to be devised that allows for an equitable, yet rational, allocation and interchange of scientific personnel in response to both regional and national priorities. Thus, when a field team needs help from a specialist and one is not within the region, these services should be provided by other regions or by a national center. Consequently, FSR&D's top management needs enough authority over its staff and the capability of acquiring the services of additional staff to meet the overall needs of the FSR&D effort.

10.5. THE ROLES OF THE FIELD TEAM

The field team has the principal function of developing improved technologies suitable to farmers' conditions. The team does this by working with farmers in their fields and by becoming thoroughly familiar with the farmers' systems and environment. The team also serves as the focal point for the interdisciplinary effort involving the farmers, the disciplinary and commodity specialists, and the extension service. The field team's role is further described in the following sections.

10.5.1. THE FIELD TEAM AND THE FARMERS

The first important relationship of the field team is with the farmers. The field team's success hinges on its ability to establish credibility with farmers through genuine interest and by accomplishments. The team needs to be aware that farmers may be expecting some improvements within one or two seasons. Should this not occur, the farmers could lose interest in further cooperation with the team. Besides improvements to the farmers' systems, these contacts offer the farmers a more systematic and institutionalized means for making their interests known to government agencies.

10.5.2. THE FIELD TEAM AND THE RESEARCH SPECIALIST

The second important relationship concerns the team and the disciplinary and commodity specialists. Because of its applied orientation, the field team sometimes constitutes a distinct group set apart from disciplinary and commodity programs. This usually results when FSR&D emerges somewhat independently of ongoing disciplinary and commodity research. In those instances where the field team is in need of technical assistance, it must enlist the services of whatever disciplinary and commodity specialists are available.

This condition contrasts with an FSR&D program in which the disciplinary and commodity specialists are part of the regional headquarters teams. In one of Guatemala's seven regions, the headquarters team consists of 28 persons—one half are university graduates and the remainder have had specialized training at or above the high school level.

FSR&D's mandate largely determines whether the field teams are an integral or separate part of the country's overall research effort. Where FSR&D emerges from a project-by-project approach, the field team will likely function more or less autonomously; where FSR&D has achieved full program status, an effort is usually made to integrate the field team activities with disciplinary and commodity programs. Both approaches can be effective.

10.5.3. THE FIELD TEAM AND THE EXTENSION SERVICE

The third important relationship involves the local extension service in the various FSR&D activities. The extension service contributes through its staff's knowledge of farmers, farmers' groups, and local conditions. This involvement is formalized when an ESFS is a regular member of the field team. At other times, ESFSs who are members of the regional headquarters teams may have temporary, but specific assignments to work with the field teams. At other times, extension staff may informally join the field teams as the need arises. This collaboration enables the extension service to become familiar with FSR&D and to prepare for the diffusion of the resulting technologies.

10.5.4. FUNCTIONAL ASSIGNMENTS FOR THE FIELD TEAM

Because of the nature of its work, the field team needs considerable autonomy over its daily activities. Within general policy guidelines set by the government and program direction set by national and regional headquarters, the field team should be given authority to define the approach to its work. Several reasons support a decentralized mode of decision making for the field team. In many instances, research areas are remote and communications difficult. A team should not be constrained from making decisions that sustain a project's momentum by having to wait for approval from central headquarters. Secondly, the team's level of effort and morale should be enhanced when it helps define its role and work style.

The field team should also be given sufficient autonomy to pursue alternative approaches for securing effective farmer participation. In this effort, the relationship between the field team and the farmers is crucial. Farmers need to know that the team values their insights and will use this information when designing on-farm experiments.

Should the field team become an appendage of an agency's mainstream activities with instructions being imposed from above, the FSR&D effort will be deflected from its original purpose. This does not mean that field teams should not be guided or supervised; it does mean that such guidance and supervision should be in harmony with the integrity of the team as a decision-making unit.

10.5.5. COMPOSITION AND ORGANIZATION OF THE FIELD TEAM

Since the field team is such a key element in FSR&D, further attention needs to be given to its staff and organization. This discussion focuses on team size and skills and internal management.

Team Size

In examining existing practices, the effective core groups generally range from two to five (sometimes up to seven) professionals occasionally with two field assistants for every professional. The size of the field staff is largely determined by these factors:

- If the technical demands of the project are relatively limited, the basic FSR&D tasks are easier to perform with a smaller field staff. When the FSR&D activities become complex, a larger staff is required.
- If enough trained personnel is not available, teams may have to remain small. Even though small, a team can be effective when the staff members are

sufficiently skilled and experienced.
- If the experiments are widely separated, a larger team may be needed with the members being divided into groups that serve different areas.

The recommendation on team size rests with the characteristics of the situation. An advantage of large teams is that greater interdisciplinarity occurs because of group dynamics. Group dynamics entails the mutually supporting effects of greater interactions among team members. A large field team also provides greater project visibility and legitimacy in the research area. On the other hand, some individuals work more effectively in small groups. Also, large groups require skilled leadership if efforts are to be effectively channeled.

Team Skills

The preferred mix of skills within a field team depends on several factors, including FSR&D's emphasis, the degree of change anticipated, and the level of interagency cooperation. More specifically:

- In those situations where many system components will be investigated, several specializations should be represented on the team. For example, changes involving interactions between cropping and animal systems will require both an agronomist and an animal scientist. Where the farm family is poorly understood, a social scientist should be included in the team.
- In many instances, certain technical problems will demand intensive and sustained attention. For example, in those situations where weed or insect problems loom large, a weed agronomist or entomologist should be on the team.
- Where major structural changes will be introduced into the system, a diversity of specializations will be needed.
- In many instances, an ESFS should be added to the field team to increase the direct involvement of the extension service.

The above factors suggest the variability of skills needed to conduct on-farm research. In some cases, where the team's tasks are defined in narrow production terms, the staff might include one to three agronomists and one or two agricultural economists. In other instances, where the team's mandate is to address all major components within the production system, the team's skills should be more diverse and specialized. For example, the suggested composition of a field team for agricultural research in Senegal was a general agronomist, an entomologist, an economist, a sociologist, a subject matter specialist—i.e., a specialist to act as a bridge between research and extension—and an animal scientist. In Appendix 10-A we provide additional details on the organization and staffing of agricultural research in Senegal.

In many cases individuals with specialized skills may not be available for full-time service. In these instances, the team may look to regional and national FSR&D headquarters, as well as to others such as local experiment stations, universities, research institutes, and development organizations. In some cases, the best source of help outside the FSR&D organization will be from regional administrators of agriculture or rural development when they have responsibilities for coordinating local activities and have strong linkages with national headquarters.

Besides their formal education, the core staff should have additional skills. First, they should have a basic understanding of farming. This capacity, which enables them to communicate with farmers in the farmers' terms, is essential in understanding the technical and management dynamics of small farming. Second, each team member should be proficient in learning about farmers' goals, needs, and perceptions. The value of such communication should not be underestimated, because constructive exchange between team members and farmers constitutes the basis for effective on-farm research.

Finally, where the numbers of scientists and technicians are severely limited, some initial scaling down of requirements will be necessary. Consequently, the recommendations in this section should be viewed as goals for attainment, rather than requirements.

Internal Management

Once a field team is in place, the tasks and workload are allocated among its members. In brief, the core team needs a management procedure for coordinating its activities. This entails appointing one member as team leader. This individual will perform three management tasks: (1) request assistance from various specialists and coordinate their work with the field team, (2) secure the cooperation of local government and other agencies, and (3) integrate the team's goals and activities with those of the FSR&D hierarchy. These activities can best be undertaken when the leader is mature and evokes the confidence of team members and local institutional counterparts. Likewise, the leader should understand how research fits into the larger process of agricultural and social change. (In Sec. 10.6.1., we provide additional comments on team leadership.)

When FSR&D activities increase in pace and scale or when on-farm research activities are dispersed, some management functions will need to be delegated to other members of the field team. When work is dispersed, some team members may wish to remain in their areas, thereby reducing travel time and increasing farmer contacts. The field team's leader can operate out of the team's central office and visit the other members as necessary.

10.6. INTERDISCIPLINARY TEAMWORK

We discussed interdisciplinarity as one of the distinguishing features of FSR&D in Sec. 2.4.4. and we have emphasized an interdisciplinary approach throughout this book. Indeed, FSR&D offers a rich opportunity for the application of interdisciplinary research techniques. Successful FSR&D requires many things, but an interdisciplinary approach is one of its fundamental requirements. In this section, we will first discuss a general model for successful interdisciplinary teamwork and then discuss its applicability to FSR&D.

FSR&D requires interdisciplinary teams—i.e., teams whose members represent different professions or disciplines. Although much is written about interdisciplinarity and its definitions, the key ingredient for true interdisciplinarity is *interaction*. This interaction invariably leads to synthesis and synergism. Synthesis of knowledge among interacting disciplines produces new ideas, concepts, and solutions. This productive interaction is called synergism. Synergism implies that the whole is greater than the sum of the parts. A distinction should be made between multidisciplinary, which simply means a combination of disciplines, and interdisciplinary which implies a combination of disciplines with frequent and significant interaction.

Generalizations pertaining to interdisciplinary management should be treated with the same caution as any management prescription. What works in one situation may not work in another. This is especially true in the international field where large cultural differences are encountered. Thus, the following "model" for interdisciplinarity is intended as a guide for integrating individual situations.

Figure 10-3. A model for interdisciplinarity.

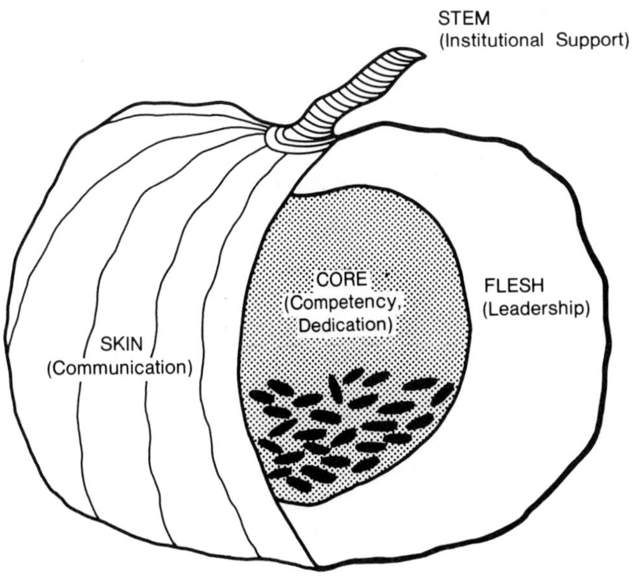

10.6.1. A MODEL FOR INTERDISCIPLINARITY

We shall use the squash drawn in Fig. 10-3 as our model for interdisciplinary teamwork. The essential components of this model are a core of competent and dedicated workers; fleshed out by adaptive and balanced leadership; held together and enhanced by frequent and open communication; and supported by an institutional framework that understands and rewards the extra time, effort, and costs associated with successful interdisciplinarity.

Competent and Dedicated Workers—The "Core"

Quality interdisciplinarity demands quality in the component disciplines, regardless of the disciplines of the team members. Below are several other reasons for seeking only those who are highly competent in their disciplines. Those who have a successful disciplinary record command the respect of others. Furthermore, a competent individual with a proven record is usually self-reliant and feels less threatened by interdisciplinarity, which often generates insecurity because it follows new courses of action or pioneers new techniques. Finally, the professional who is well-grounded in a discipline is in a good position to understand others' paradigms.[1] But whether the professional chooses to do so or not is partly a matter of personality. This respect for the knowledge that each team member brings to a joint effort is critical to the success of interdisciplinary teams. We shall discuss this point further under "leadership."

The second major ingredient of the core is dedication. This dedication must be focused on the goals of the interdisciplinary effort. Members of the team must have enough commitment to the project or program goals that they are willing to endure the difficulties of collaboration to achieve these goals. Thus, the team leader and those who share the responsibility of achieving the interdisciplinary goals should have a major role in selecting team members.

Although desirable, an agreeable personality is not a necessary attribute for an interdisciplinary team member. Unfortunately, high motivation and competency are not always combined with highly personable individuals. In contrast, assignment of individuals who are not fully commited to team goals or who lack competency in their own discipline almost ensures failure of an interdisciplinary team.

Leadership—The "Flesh"

Given staff who are competent in their disciplines and are strongly committeed to interdisciplinarity, the leader must develop a collaborative style that enables each member to fully contribute to the team's goals. Although many leadership characteristics are important, certain characteristics are critical to successful interdisciplinarity.

As with other team members, the leader must be competent in one of the disciplines relevant to the team's assignment and be committed to interdisciplinarity. The leader must be sensitive to differences between the members' disciplines. Each professional has, to some

[1]Paradigm is the set of concepts, methodologies and vocabulary associated with a particular group or discipline at a given time. For example, plant breeders work under a generally accepted set of assumptions, use common research approaches, and have a set of terms that have special and carefully defined meanings for them.

degree, a paradigmatic reference for viewing and solving problems. This paradigm is part of the professional's training and involves not only a mental framework for approaching problems but also the use of terms with specialized meanings. These two characteristics combine to form a set of mental "tools" common to that profession. An interdisciplinary leader should not only understand these professional differences, but should also seek to promote an appreciation for the differences among the team's members.

Kuhn (1970) presents a thorough discussion of paradigms, their relevance to science, and how they relate to changes in scientific thought. Among the important implications for interdisciplinary research are these:

- A paradigm is both determined by and determines those who share it.
- Because a paradigm tends to produce a strongly focused view by those who subscribe to it, communication across paradigms is difficult and may cause conflicts.
- Typically the more able scientist belongs to several paradigmatic groups—usually closely related—and is more open to the paradigms of others.
- Understanding a paradigm involves knowledge of the symbolic generalizations, shared commitment to certain beliefs, common values, and similar frames of reference for problem solving inherent in a specific paradigm.

In large measure the leader's success depends on the ability to bring together, in a synergistic manner, the unique perspective and competency of each member to common goals. Closely related to achievement of this task is the leader's ability to balance flexibility and authority in dealing with team members. The leader must be firm in bringing the team to a decision on the *what* and *when* of component tasks but should leave the *how* to the individual and collective ingenuity of the team. This requires a delicate balance of decisiveness and patience on the leader's part.

A good summary of the theories underlying and the characteristics associated with balancing task-oriented with people-oriented styles—i.e., concern over accomplishing tasks versus concern over people's feelings, motivations, etc.—can be found in Reddin (1970). Among the indicators Reddin gives for the "integrated" manager are (1) capability to harmonize institutional and individual goals, (2) reliance on common aims and ideals as the source of authority, (3) ability to use a variety of participatory techniques, (4) capacity to depersonalize authority, and (5) ability to develop a highly cooperative approach toward achievement of organizational goals.

Another important trait of a successful interdisciplinary leader is adaptiveness. The leader must use a variety of styles and techniques depending on the issues before the team and the stage of the team's effort. When a leader is not capable of such adaptiveness, then alternative strategies are needed, such as rotation of leadership among team members, or strong delegation of leadership tasks. These approaches, however, cost the team time and effort.

Team effectiveness can be increased through certain exercises. Examples of useful exercises are described in a handbook by Merry and Allerhand (1977). The basic steps they recommend in team development involve the following:

1) brief problem-sensing
2) examining effects of differences in perception
3) listening and clarifying
4) giving and receiving feedback
5) looking at process and content
6) developing interactive skills
7) personal contracting with team members
8) follow-up procedures.

Another excellent reference on team building that includes a description of 46 team building activities is the manual by Francis and Young (1979). To illustrate their approach, we provide two of these activities in Appendix 10-C. While useful, these exercises should be used judiciously since most are derived from Western experiences.

One of the most direct exercises in team building is to continuously refine objectives and tasks within the generally agreed upon goals for the team. For simplicity, we use a hierarchical arrangement of terms: "goals" are the general purposes of the team effort, "objectives" are the special targets to be achieved in reaching the goals, and "tasks" are the jobs undertaken to achieve the objectives.

Definition of goals can generally be achieved readily. Team members should be chosen who subscribe to these goals. Experience has shown that teams sharing common goals may still have difficulty defining and agreeing upon objectives. Thus, one of the major roles of the interdisciplinary leader is to bring about agreement on team objectives and to translate them into tasks. Delineating and assigning tasks and implementing work is also difficult. Consequently, a great deal of open discussion is needed during the initial stages. This process of objective and task refinement needs continual feedback throughout the team effort and requires skilled leadership in problem solving, decision making, and implementing the results. At the same time, this process provides an ideal opportunity for the leader to build and maintain a strong collaborative team.

An additional value of refining the objectives is the continual evaluation of individual and team progress. Careful identification of the measures of accomplishment for each objective offers bases for measuring individual and team performance. Because a team functions best when all members contribute, tasks and responsibilities should be clearly assigned, monitored, and corrected, when necessary.

Frequent and Open Communication—The "Skin"

Communication has been discussed somewhat under "leadership." Obviously much time must be spent in open give-and-take discussion in clarifying team objectives. Communication in interdisciplinary teams must be open and continual. The first and foremost responsibility

of the team leader is to ensure that this occurs, but each team member must make the extra effort to help. No clear choice as to the best way to achieve effective communication can be prescribed. However, one almost universal complaint from interdisciplinary teams is that the team members did not spend enough time during the initial stages of their work together. When available, a specialist with experience and graduate degree training in group dynamics and interpersonal communications should help the team develop these skills.

If sufficient effort has been invested initially, then the preferred model is to keep communications on an informal, need-to-know basis. Periodically, the leader should arrange a group review of progress toward objectives. Other actions by the leader to facilitate team interaction include common task assignments and having members work at the same location. The leader must provide the opportunity—i.e., time and space—for interaction.

Serious problems often develop when team members are at separate locations. When separate locations are necessary, the leader must exert extra effort and ingenuity to promote interactions among the team. Techniques include shared work plans, conference calls, site visits, delegation of leadership tasks, and social events.

Institutional Support—The "Stem"

Interdisciplinary efforts frequently cut across established institutional boundaries. In trying to foster interdisciplinarity, institutions should (1) assign a capable team leader, (2) delegate authority to the team leader in selecting team members, (3) allow sufficient time for teams to learn how to work productively before judging the results, (4) ensure adequate financial support, and (5) give a clearly defined reward for the team as well as for individual performance.

Selecting a good leader, from individuals with limited or no experience leading interdisciplinary teams, is not always possible. Those in authority must be prepared to change leadership whenever the leader does not perform satisfactorily. The team will provide plenty of indications when such action is warranted. If difficulties in leadership arise, an outside trouble-shooter may be needed to aid the team leader in resolving the difficulties. When the problem cannot be solved promptly, leadership may have to be changed.

A frequent mistake when establishing interdisciplinary teams is for higher authorities to force some of their staff to become members of the team. The team leader should resist such attempts and insist on having a major role in selecting team members. Because staff selection is so important, higher authorities should allow team leaders freedom in selecting their staff. Otherwise, the authorities cannot logically hold the leaders responsible for accomplishing their objectives.

Decision makers need to realize that while interdisciplinary approaches may take more time and resources than disciplinary approaches, the interdisciplinary approach may be the only effective way to produce satisfactory results. In this sense, the interdisciplinary teams can be truly cost-effective. Unless management believes in the approach and is willing to make a full commitment, interdisciplinary teams should not be formed. Although some of the complexities may be anticipated, too often the full range of complexities are not. Without an understanding of the time required to reach fully effective teamwork, interdisciplinary activities may be halted before they have had a fair chance of succeeding.

Finally, rewards should be provided for interdisciplinary team members. This is difficult because separating an individual's contribution from the team's is often complex. Several management practices can be used to offset these difficulties. Each team member should receive a reward based on the team's performance. Further, the team leader should rate each individual's contribution to the team's output. Where appropriate, team members may also be asked to rate fellow members' contributions. When interdisciplinary teams are temporary and members come from different organizations, the team leader and higher management should inform these other organizations of the team members' contributions to the team. Furthermore, provisions are needed to ensure that these other organizations do not penalize team members because they are not contributing directly to the parent organization's activities.

10.6.2. APPLICATION OF INTERDISCIPLINARITY TO FSR&D

Application of the foregoing concepts to FSR&D relates to a number of general and specific considerations. We present these considerations in the following two sections.

General Considerations

FSR&D is an interdisciplinary effort requiring at least one and usually more interdisciplinary teams. Therefore, FSR&D management needs to consider the implications at the different levels of operation and provide suitable conditions for interdisciplinary teamwork.

Leadership in FSR&D may be a critical factor in many developing countries because of the overload on existing staff. When qualified national leaders are not available, a country's decision makers will have to decide whether to entrust the leadership of a new FSR&D project or program to foreigners or to wait until the local staff can be properly trained. In view of the desirability of local leadership, such responsibilities should not be given to expatriates for more than the first one or two years of an FSR&D activity. By that time, the local staff should have received accelerated training in FSR&D and have learned FSR&D processes and methods under the expatriates.

Because FSR&D requires a long-term commitment by a variety of disciplines, the core teams in the research areas and in regional and national headquarters should be assigned to their positions for at least one and possibly several years. Rotations among these locations are useful in acquainting team members with a range of situations and activities.

When bringing new members into these groups, assignments should be on a trial basis. The judgment of the team leader along with team members should determine if the new member is to be given a regular position.

In Sec. 7.4.1., we described a helpful device when deciding on assignments whereby members of the FSR&D team are asked to write their job descriptions according to their interpretation of team goals and objectives. Team interaction leads to a workable set of descriptions and criteria for evaluating performance and, consequently, for setting responsibilities. Moreover, this device encourages better understanding of how each team member contributes to team goals and objectives. Any serious disagreement among the members should be resolved before finalizing the job descriptions. This same general process can also be used for (1) preparing lists of equipment, supplies, and related needs and (2) for adjusting team objectives and tasks in light of new or unexpected circumstances.

A work plan needs to be developed, which is the "glue" that holds the team together. This work plan should contain a summary description of the above activities, a schedule for their completion, and a set of times or conditions for staff meetings. Finally, the entire team needs to review and approve the work plan.

In some developing countries, problems among the disciplines are less acute than in the industrialized economies because of the staff's education. For instance, many of the professionals engaged in FSR&D in Guatemala have common agronomic training. Thus, having to learn the idiosyncrasies of other disciplines is not as frequent or urgent. Other approaches that help keep disciplines from dividing into self-interested groups is to focus on solving farmers' problems and to measure team success through farmers' acceptance of new technologies. With such a focus, individual disciplines are able to identify the nature and value of their contributions to the team effort.

On the other hand, problems can arise. One type of problem may occur when interdisciplinary teams are formed by assigning staff from more than one ministry. When this occurs, the team leader needs to be sure some members are not discriminated against because their parent ministry does not place as high a value on the team's activities as the other ministries. Another type of problem arises when (1) a discipline's members feel superior or inferior to the other members, or (2) members do not value small farmers or agriculture highly. These feelings can block effective team interactions and are especially critical when the discipline is key to the team's success. When these instances arise, FSR&D's leaders should take corrective action at the outset. Possibilities

include moving the project to another ministry or changing team membership.

Other problems can arise within interdisciplinary teams that are older and more established (Pelz and Andrews, 1966). These authors suggest two concerns relevant to FSR&D. One is that teams do not remain open and flexible to new ideas and approaches. This problem may be alleviated by adding new members from time to time and by presenting the team with new challenges. The nature of FSR&D is such that the latter will usually happen of its own accord. Another concern is that over time the team may become less action oriented when its leader becomes more facilitative and less directive. When this occurs, the leader's contribution as an orchestrator of group processes should probably be augmented by more active leadership from those with strong technical capabilities.

Specific Considerations

Each of the FSR&D activities offers a variety of opportunities for developing effective interdisciplinary teamwork. Following are some suggestions and examples from actual practice.

Selection of the target and research areas gives the FSR&D team an early opportunity to develop an interdisciplinary approach. Criteria for selection as well as consensus on final choices is an important team building opportunity.

During problem identification, members with different disciplines work together on a single task. Each draws on the experiences of the others in finding out what the farmers' most pressing problems really are.

In Guatemala, social scientists are paired with agronomists during the reconnaissance surveys (we provide more details in Appendix 5-Q).

Opportunities for interactions among the members are enhanced by having the pairings change each day. Researchers are encouraged to view the whole farm rather than the narrow perspective of their own discipline. Hildebrand (personal communication) has characterized the initial stages of these surveys as one in which entomologists look for insects and soil scientists gather soil samples; but by the time their work is finished, each has given up looking at just his or her own specialty and concentrates on identifying the key factors confronting the farmers. Thus, a shortage of labor during critical planting periods might be identified by any member of the team and not be left to an agricultural economist.

In the above example team members are encouraged to look beyond their disciplines by having them write about other disciplines. In writing these reports, they have access to specialists knowledgeable in the various subjects. The idea is to encourage members to view farmers and their environments from other vantage points and to begin to learn the terms and concepts of other disciplines. In this way, for example, agronomists will learn something about calculations of economic profitability and social scientists will learn something about the experimental designs for crop and animal studies.

Planning on-farm research begins with a workshop to ensure interaction of all relevant specialists. The team leader needs to see that some individuals do not exert undue influence on the team because of their status, experience in the region, or similar reasons. All members should look at the whole system and be free to offer suggestions on all components, not just on topics pertaining to their disciplines. Alternative hypotheses should be identified. One method for this is through brainstorming sessions in which ideas are quickly and freely offered without thought as to their eventual practicality. After an ample list of possibilities has been identified, further study will show which possibilities to consider more carefully. Such an approach broadens the range of possibilities and helps to avoid settling too quickly on an inferior plan of action.

On-farm experiments offer further opportunities for team building. Zandstra et al. (1981) suggests:

> "Development of strong [interdisciplinary] ties can also be assisted by the engagement of the whole team in field operations normally under the responsibilty of a single team member. For example, the entire team may participate in initial survey activities or selection of plots for pattern trials or design of specific component technology trials. Also, members should visit each other's trials and discuss the implications jointly. For example, the establishment of grain legumes after rice is an area where several disciplines overlap. Standing rice stubble helps suppress early season legume pests. It also changes water losses right after rice harvest and together with minimum tillage planting techniques can save residual soil moisture. Omission of tillage requires the development of special planting techniques and the evaluation of weed control requirements. Where planting techniques require substantial labor or specialized equipment, the opinion of economists about farmers' acceptance or limits to expenditures must be considered."

This example illustrates the need for interdisciplinary teamwork in FSR&D and some of the opportunities for its accomplishment.

Finally, other opportunities will naturally present themselves during this and the remaining activities in the FSR&D process. In some cases, daily interaction is needed between researchers dealing with rangelands, livestock, crops, soil conservation, farmer preferences, societal constraints, and economic matters. Such interactions are especially pertinent when designing experiments, drawing up terms of reference for studies, and evaluating the results. Frequent interaction and cooperation should help break down the too-often-encountered practice of individuals narrowly guarding their sources of data. These and similar suggestions should improve interdisciplinary teamwork, which is so essential to successful FSR&D activities. We provide a checklist for successful interdisciplinarity in Appendix 10-D.

10.7. GETTING STARTED

Because most decision makers know how to implement new projects and programs in their countries, this section will be short and will concentrate on technical and financial assistance.

Decision makers and their staff—possibly an ad hoc committee—will probably decide on an FSR&D approach after having met with various leaders who are concerned with research, extension, and agricultural production. The need for change and the relative advantages of FSR&D will, most likely, have been established through careful study. Should a country decide on a scaled-down approach or have sufficient funds of its own, the country can begin preparing for FSR&D immediately. More likely, the government will want to seek funds from external sources to assist in implementing the project or program.

A variety of sources are available to help in this process. These include

- the foreign aid programs of many, if not most, of the industrialized countries
- regional development banks and the World Bank
- regional agricultural research centers such as the Tropical Agricultural Research and Training Center (CATIE) in Costa Rica and the Group for Studies and Research in the Development of Tropical Agronomy (GERDAT) in France
- international agricultural organizations such as the Consultative Group on International Agricultural Research (CGIAR) and the Food and Agriculture Organization of the United Nations (FAO)
- private foundations such as Ford and Rockefeller.

A common practice is to have one or more consulting teams assist the country in preparing an evaluation of the technical, organizational, and economic feasibility of undertaking such activities. The United States Agency for International Development (USAID) customarily requires some form of preliminary evaluation that provides the basis for deciding whether or not to proceed. If this outcome is favorable, the next step is to prepare what USAID calls a Project Paper. Those preparing this document look into the above mentioned types of feasibility and related issues and come up with recommendations for action. If the FSR&D approach still appears favorable and the recipient government agrees with the findings, the Paper becomes the basis for drawing up a funding agreement between the government and USAID. This agreement may be a loan, grant, or both. Financial assistance of this sort is usually accompanied by technical assistance that aids the recipient in implementing the program and helps assure the lender or donor that the funds are used as intended.

Such financial assistance may provide for the costs of expatriate advisers, equipment and materials purchased abroad, sometimes domestic purchases, and training. Some of the training will be formal—often received abroad—but much of it will be informal. Funds for informal training can be used for (1) sending staff to places like CATIE and the International Agricultural Research Centers (IARCs) and to seminars and short courses at universities and (2) bringing in materials and specialists to give instructions within the recipient country. Training will be discussed more thoroughly in Chapter 11. Finally, a portion of the funds may be set aside for conducting one or more types of evaluation studies that report on progress and help improve performance.

10.8. EVALUATION OF PROJECTS[2]

Projects are evaluated to learn about their performance and how to improve them. More specifically, such evaluations (1) provide insight about project activities and effectiveness, (2) suggest ways to improve project designs and operations, and (3) provide a basis for reaching decisions on financial and technical support. Evaluation is aided when the bases for evaluation are established during the project design.

The discussion that follows relates to the evaluation of project accomplishments rather than to analyses of new technologies as discussed in Part 2 of Chapter 7. The distinction between analysis in Chapter 7 and evaluation in this section is a matter of scale. In Chapter 7, we concentrated on the acceptability of new technologies to individual farmers, whereas evaluation in this section refers to acceptability of project results to project managers, high-level decision makers in the government, and to funding organizations.

10.8.1. TYPES OF EVALUATION

Evaluations can be divided into three basic categories: built-in evaluations, special evaluations calling for intensive reviews, and impact evaluations. Which of these to use depends on available resources, the needs of the situation, the preferences of host country officials, and the requirements of donors or lenders. We discuss these and related issues below.

Built-in Evaluations

Built-in evaluations are intended to provide a periodic appraisal of project accomplishments and problems and to help improve project activities. These evaluations are designed to occur at prescribed intervals—usually after critical points of accomplishment and in time to make necessary changes. Evaluation plans should be prepared during project design and made an integral part of the overall project. Initially, the plans may be only rough drafts, with details developed later. The purpose of these early drafts is to identify the information needed so that data can be collected as activities evolve. Such data are then used in the evaluations.

These evaluations provide a series of links between those responsible for managing the projects and those responsible for their financial and institutional support. In this way, evaluation findings can improve current operations and can aid in future designs.

Special Evaluations

As the name implies, special evaluations are conducted for non-routine reasons, as when some part of the project needs intensive investigation. These occasions might arise when (1) management encounters problems that it cannot resolve by itself, (2) an opportunity arises that suggests possible changes in the scope or intensity of activities, or (3) something is sufficiently interesting to warrant special attention. Where unbiased appraisal of activities is needed, as when a project is in difficulty and the involved groups agree to this type of evaluation, a special review team may be contracted to do the study. In principle, this team of experts should have no vested interests in the outcome; however, finding such individuals may not be easy.

Impact Evaluations

These evaluations are usually conducted after the project is completed so as to comprehensively review the experiences and impacts of the project. Results can be used as a basis for policy formulation, design of future projects, and funding similar activities. This type of evaluation normally involves project members because they are familiar with the project's problems and accomplishments. Particularly difficult aspects of such evaluations are (1) lack of information about the conditions and aspirations for the project as originally designed, (2) what conditions would have been like had the project not been undertaken, and (3) the time lag for project actions to be felt. For these reasons, results of impact evaluations need to be judged cautiously and used primarily to gain impressions about the value of project activities. For a recent review of the World Bank's experiences in impact evaluations see Weiner (1981); and for examples of evaluations completed by USAID, see the studies of small farmer access roads in Colombia (1979), Kitale maize in Kenya (1980a), and the impact of rural roads in Liberia (1980b).

An effective approach to impact evaluation, which largely dispenses with base-line studies, is to use farm records to measure farmers' acceptance of change. Com-

[2]For easier reading we will refer, in this section, to projects and programs simply as projects.

bining rates of acceptance, the degree of change, and the potential number of farmers who might accept the change will give a good base for judging overall project impact. Hildebrand (1979) describes the process by ICTA in Guatemala as:

> "The evaluation of impact is being accomplished through the use of the farm records being kept in each one of the work areas. There are not enough resources in the Institute, nor especially in [the Rural Socioeconomic Unit], to conduct the census type survey that would be required periodically to monitor impact and use of technology on a more adequate basis. However, it is felt that the data accumulated over time from the farm records sufficiently demonstate trends in adoption of the technology being utilized and is an appropriate substitute for a benchmark study and follow-up studies for which the Institute has inadequate resources."

10.8.2. WHICH TYPES OF EVALUATIONS TO USE?

Perhaps the most frequent and useful approach is the built-in evaluation in which teams are brought in to appraise accomplishments and to offer suggestions for corrective action. Special evaluations should be undertaken only as the need arises. Impact evaluations have their usefulness in supporting future activities; however, future support is often forthcoming because of perceived needs rather than because of reviews of past activities. Possibly for this reason, widespread acceptance of impact evaluations has been slow.

10.8.3. DEVELOPING EVALUATION PROCEDURES

Evaluation procedures will vary according to (1) the type of evaluation, (2) the particular needs for conducting the evaluation, (3) data availability, (4) organizational requirements, and (5) the preferences of the evaluation team. USAID has developed an evaluation procedure called the Logical Framework that it has used since the early 1970's. The approach is a systematic means for linking project inputs with outputs based on assumptions of causality. The method forms one of the principal parts of project justification when preparing the Project Paper and is the basis for subsequent studies of project impacts.

In brief, the approach requires identification and narratives on project goals, purposes, outputs, and inputs. Goals refer to broad-gauged factors such as increases in the level of production or income of particular groups. Purposes refer to specific accomplishments that contribute to the attainment of the goals. Outputs are measures of performance in accomplishing the purposes. Finally, inputs are the factors needed to change the outputs.

As may be apparent, careful thought is needed to relate the contribution of inputs to outputs, outputs to purposes, and purposes to goals. In the process, the analyst is forced to identify indicators that can be used to verify accomplishments and to state the assumptions on which these relationships are based. While the process is complex when each aspect is considered in detail, in practice, USAID uses the approach for guiding the analyst through a logical thought process. The process produces better project designs and analyses than in the absence of such a method.

We provide further details on the Logical Framework in Appendix 10-E.

10.8.4. A CAVEAT ON EVALUATIONS

Project evaluations are important to the development process because they aid in identifying successes and failures and provide some of the reasons. Moreover, some form of evaluation will be made of a project whether or not procedures are established. Consequently, we favor at least some orderly and reasoned approach rather than none at all.

Nevertheless, evaluations are not easy, as implied earlier, and care must be taken not to confuse surrogate measures of accomplishments with the real thing. By that, we mean that project managers and their superiors may set up targets of accomplishment, such as (1) levels of expenditures against the budget, (2) number of full-time employees, (3) number of trials and tests conducted, and (4) number of contacts with farmers. But these are primarily measures of activities, not accomplishments. Where direct cause and effect relationships exist between activities and results—and sometimes they do—then these measures serve as meaningful surrogates for the true goals of a project. What is really wanted, however, are introductions of new technologies and other factors to improve farmers' welfare. The latter can be measured and evaluated, but the process is more complex and takes more time.

Another warning on evaluations is that those being evaluated may feel threatened by both the process and the results. Especially those evaluations directed to "finding out what went wrong" can create splits among the team members and cause other problems. Consequently, when considering project reviews, top management should be aware of these types of problems when it decides on (1) the need for review, (2) the type of review, (3) who should do the review, and (4) the use to be made of the results. One way to lessen potential problems from evaluations is to inform team members when they first come onto the project that evaluations will be made and about the purposes of the evaluations and the uses of the results.

10.9. PROJECT AND PROGRAM MANAGEMENT: A TWO COUNTRY PERSPECTIVE

We will now examine two cases to illustrate how some of the foregoing organizational and management aspects of FSR&D are being applied. In Honduras, agricultural research effort is being restructured to emphasize the FSR&D program and in the Philippines, FSR&D projects are being expanded. Each example provides some insight into different organizational modes.

10.9.1. AN EXAMPLE FROM HONDURAS

Recently, the government of Honduras reorganized agricultural research to focus on increasing small farmers' production. Because more agencies are providing services to small farmers than before, a farming systems approach was selected to help in securing adequate cooperation among the agencies.

A national level advisory committee for agricultural research has been proposed. Its members represent the extension service, credit agencies, farmers' organizations, the agricultural education institutes, and other relevant organizations. This committee serves the General Directorate by providing advice on basic policies and priorities relating to agricultural research. Similar interagency consultative committees are to be established for each of the country's seven regions. There, they would advise the regional directors for agricultural research.

The National Director of Research is responsible for the FSR&D program in general, but much of FSR&D's activities are under the control of the regional directors of agriculture. The National Director operates out of the capital city and is part of the General Directorate for Agricultural Operations, which in turn is part of the Ministry of Natural Resources. The National Director and his staff (1) facilitate linkages between agencies on major policy and administrative matters and (2) assume a major role in reviewing research proposals from the regions and in authorizing their budgets. A central research station is located at the country's main experiment station near the capital. Activities at the station include training and supervising, and supporting research undertaken in the seven regions.

Within each region, a farming systems team is being recruited and trained to undertake on-farm analyses and to test new technologies. These field teams interact closely with the regional experiment station, where disciplinary and commodity scientists provide technical support. On request of the field teams, these scientists undertake research relevant to FSR&D.

The administration of research is decentralized down to the regional level. Each regional director for agricultural research provides administration and support services for research in the region. Technical review committees are established, consisting of technicians from headquarters staff and the central experiment station. These committees review the research projects for each area in terms of technical soundness and overall impact. (We provide additional information on the Honduran program in Appendix 10-B).

10.9.2. AN EXAMPLE FROM THE PHILIPPINES

In the Philippines, cropping systems projects have been underway for several years. Administrators plan to expand the activities to eight new research areas. These areas will be located where the government has embarked on major development programs. Considerable attention is being devoted to assure that off-farm institutional linkages and management needs will support on-farm activities.

At the national level, a Multiple Cropping Committee has been formed with representatives from (1) the Philippine Council for Agricultural Research, (2) the University of the Philippines, (3) the Bureau of Plant Industry of the Ministry of Agriculture, and (4) the International Rice Research Institute. This Committee serves as a staff agency to the National Food and Agricultural Council. This Council is chaired by the Minister of Agriculture and includes the heads of all the important national agencies involved in agricultural development, such as credit and the national grain authority. The deputy director of the Council also serves on the Multiple Cropping Committee. In effect, the Council and the Committee generate interagency coordination and support for the FSR&D projects.

The function of the Multiple Cropping Committee is to plan and monitor the entire FSR&D effort. The Committee has a permanent secretariat of six who have bachelor's or master's degrees in the agricultural or social sciences. The Committee visits the field teams and monitors their work regularly. The field team members are permanent employees of the Bureau of Plant Industry, which has the responsibility for implementing research results. Since the Bureau is staffed primarily by agricultural scientists, it has a contract with agricultural economists to work with the field teams. Additional staffing needs are coordinated through the Multiple Cropping Committee that, in turn, draws on government and university staff.

In summary, the Bureau of Plant Industry implements the FSR&D activities, whereas the Multiple Cropping Committee provides high-level technical advice and back-up services to support the program. The Committee facilitates coordination at the national and regional levels and refers policy matters to the Council.

10.10. A SUMMARY PERSPECTIVE

This chapter concerns implementation of FSR&D once the government has decided to proceed. One of the first decisions concerns whether FSR&D will be undertaken as a project or as a program. The decision makes a difference in how other organizations in the government will relate to FSR&D. Next, we discussed the subjects of organizational structure. Staffing requirements referred to the field team, research specialists, extension's inputs, and team leadership, with a section on staffing when a country's resources are severely limited.

Next, we emphasized off-site management, giving attention to institutional linkages, technical reviews and support, logistics, and personnel management. We defined the field team's roles as they relate to farmers, research specialists, and extension workers, and commented on other aspects of the field team. Because FSR&D teams are made up of members from different disciplines, we provided a section on interdisciplinary teamwork. We covered "getting started" in a short section followed by a summary on evaluation of FSR&D activities. Finally, we closed the chapter with examples of FSR&D organization and management in Honduras and the Philippines.

CITED REFERENCES

Francis, D., and D. Young. 1979. Improving work groups: a practical manual for team building. University Associates, San Diego, Calif.

Hildebrand, P.E. 1979. Incorporating the social sciences into agricultural research: the formation of a national farm systems research institute. ICTA, Guatemala, and The Rockefeller Foundation, New York.

ISRA. 1979. Senegal agricultural research project. Government of Senegal, Dakar, Senegal.

Kuhn, T.S. 1970. The structure of scientific revolutions. The University of Chicago Press, Chicago.

Merry, U., and M.E. Allerhand. 1977. Developing teams and organizations. Addison-Wesley Publishing Company, Reading, Mass.

Ortiz D., R. 1980. [Draft]. Generation and promotion of technology for production systems on small farms: ICTA's approach and strategy in Guatemala. University of Florida Press, Gainesville, Fla.

Pelz, D.C., and F.M. Andrews. 1966. Scientists in organizations. John Wiley & Sons, Inc., New York.

Reddin, W.J. 1970. Managerial effectiveness. McGraw-Hill Book Co., Inc., New York.

USAID. 1979. Colombia: Small farmer market access. Proj. Impact Eval. Rep. No. 1. USAID, Washington, D.C.

______. 1980a. Kitale maize: the limits of success. Proj. Impact Eval. Rep. No. 2. USAID, Washington, D.C.

______. 1980b. Impact of rural roads in Liberia. Proj. Impact Eval. Rep. No. 6. USAID, Washington, D.C.

Weiner, M.L. 1981. Evaluating the Bank's development projects. *In* Finance and Development. Int. Monetary Fund and The World Bank, Washington, D.C. 1:1:38-40.

Zandstra, H.G., E.C. Price, J.A. Litsinger, and R.A. Morris. 1981. A methodology for on-farm cropping systems research. IRRI, Los Banos, Philippines.

OTHER REFERENCES

Bass, L.W. 1975. Management by task forces. Lomond Books, Mt. Airy, Md.

Cernea, M.M. 1979. Measuring project impact: monitoring and evaluation in the PIDER rural development project—Mexico. The World Bank, Washington, D.C.

OECD. 1975. AID evaluation: the experience of members of the development assistance committee and of international organizations. OECD, Paris.

______. 1972. Interdisciplinarity. OECD, Paris.

USAID. 1973. Selecting effective leaders of technical assistance teams. Bureau of Technical Assistance, USAID. Washington, D.C.

______. 1973. The logical framework: modifications based on experience. Bureau for Program and Policy Coordination, USAID. Washington, D.C.

Chapter 11
TRAINING

In most situations a country entering into FSR&D will need to train a cadre of researchers and extension staff in FSR&D's philosophy, concepts, and practices. Capabilities in FSR&D can be acquired in various ways. An approach of some organizations that we found appealing involves (1) training a small group in FSR&D; (2) having that group, with suitable assistance, develop in-country training programs for research, extension, and technical assistants; and (3) developing activities to complement these basic training programs. In planning the size of its training effort, the FSR&D team should consider staff replacements as well as additions to new staff. Replacements can be a significant factor because of the frequently high demand by other government organizations and by private companies for FSR&D staff members.

In this chapter we discuss the three points listed above, provide information on educational opportunities and materials, and close with a summary.

11.1. INITIAL EXPOSURE TO FSR&D CONCEPTS

Those who will be responsible for FSR&D in a country should receive concentrated training in FSR&D. Various possibilities exist. For instance, FSR&D activities can be started in the country with the help of an outside group experienced in FSR&D. Such an approach places the FSR&D staff alongside experts in the design and implementation of the FSR&D activities. Even with this approach, a country needs to train additional FSR&D staff who do not have the opportunity to work closely with the experts.

Another approach, successfully applied by some countries, is to send a core group of perhaps 5 to 10 to centers specializing in applied on-farm research methods. Centers like CATIE, CIAT, CIMMYT, and IRRI provide such training. Another possibility would be to send trainees to a country, such as Guatemala, Indonesia, or Senegal, in which FSR&D has been successfully implemented, provided suitable arrangements can be made with these countries. An advantage of working with national programs is that organizational and personnel arrangements, as well as financial realities, can be observed.

The members of this core group might be trained for 3 to 9 months and then return home to plan and implement FSR&D activities. Concurrently with this training abroad, the country could make arrangements to bring in FSR&D specialists who could help the core group develop the country's basic training programs in FSR&D.

11.2. DEVELOPMENT OF NATIONAL TRAINING PROGRAMS FOR FSR&D

Several national research organizations have integrated FSR&D training programs into their ongoing research activities. Two advantages of the national training programs are (1) they provide the trainees the opportunity to learn the practical concepts of FSR&D in the setting in which they will be working after they complete the course, and (2) the trainees will be contributing directly to the FSR&D effort while being trained. Waugh (personal communication) estimates that approximately 50 percent of the research trainees' time goes into activities supporting ICTA's program.

While those responsible for this training usually direct their attention initially to the training of researchers, they should also develop training programs for the extension staff who will work closely with the FSR&D teams. Below, we provide additional information on training programs for researchers, extension workers, and technicians. These comments draw heavily on Guatemalan and Honduran experiences.

11.2.1. PROGRAMS FOR RESEARCHERS

In this section we discuss (1) the purpose of the national training program for FSR&D researchers, (2) how to start the programs, (3) trainees, and (4) course content.

Purpose

The primary purpose of the training program is to instruct incoming staff in FSR&D's philosophy, concepts, and approach. Those being trained, especially the early ones, are expected to form the core of the FSR&D effort by (1) being able to develop concepts and approaches suitable to their conditions, (2) heading regional and field teams, and (3) training others.

How to Start

Those with knowledge and experience in FSR&D procedures are needed to plan the training program. Even though some of the country's more senior researchers may have attended one of the regional or international courses on FSR&D, they will usually need help from outside organizations. The primary reasons are (1) to provide the national researchers with additional help during this critical period of start-up, (2) to establish contact with one or more collaborating institutions, such as a regional or in-

ternational agricultural research center or a university, and (3) to gain access to training staff and materials. In ICTA's case, 11 Guatemalan researchers initially attended a training program at CIAT in Colombia. During this time, CIAT helped the Guatemalans plan their training program and then supplied CIAT staff for the implementation. In preparing the schedule of activities, the FSR&D leaders should allow team members enough time to adapt what they learned during training to their local conditions.

Trainees

Desirable qualifications of all candidates for training would be (1) to have a university degree in an agriculturally oriented curriculum, (2) to be willing to work with farmers in their fields and to live in a rural setting, and (3) to have an aptitude and desire for working as a member of an interdisciplinary team. In Guatemala, many of the candidates are relatively young and have either recently received their degrees or lack only the thesis. For the latter situation, ICTA helps the students complete their theses by collaborating with the university in (1) selecting the thesis topic, (2) providing counsel to the student, and (3) allowing the student time, materials, and facilities for conducting experiments.

ICTA customarily signs a 9 to 12 month contract with the trainees. This is enough time for the trainees to complete the national FSR&D training program. Then, ICTA can decide if a trainee should be asked to join that organization, be recommended to some other organization, or be dismissed.

Course Content

A general approach to training is to combine practical field experience with theoretical training in each of FSR&D's principal features. This training program is a full-time effort for the trainees in which part of their training contributes directly to the FSR&D program. The proportion of practical to theoretical training is roughly three or four to one. The programs take about 9 months in Guatemala and Honduras to adequately cover the planning and conducting of experiments and to analyze and evaluate the results.

The size of the groups to be trained at any one time depends on such factors as the capabilities of the trainers, available facilities, the urgency of developing trained staff, the needs of the ongoing programs, the ability of the trainees to contribute to ongoing programs, and the abundance and interest of qualified candidates. The training staff needs to balance the objectives of (1) providing trainees with adequate supervision and opportunities for field experience and (2) making the best use of qualified and scarce trainers. In this regard, the FSR&D leaders should use previously prepared training materials as long as these materials can be adapted to local conditions. FSR&D leaders can also seek help from expatriates to complement local trainers and to help prepare additional training materials.

To give the reader an idea about the size of groups, the first group scheduled to receive training in Honduras was seven; groups trained in Guatemala generally range from 10 to 15; and groups trained in production at CIMMYT and IRRI have ranged from 25 to 50. Some number between these extremes is probably best for most national training programs in FSR&D.

A well integrated training program should use a variety of instructional methods. Those used in Guatemala and Honduras include progressing through each of the FSR&D activities, such as characterizing and defining the research areas, identifying problems and opportunities for improvement, selecting collaborating farmers and experimental plots, obtaining survey data, designing and managing the experiments, designing farm records and supervising the data collection, monitoring climatic data, and analyzing and reporting on the results. The trainees conduct the experiments on farmers' fields and sometimes at the experiment station.

Waugh (personal communication) describes an instructional method in which the trainees call in both experienced specialists and farmers to obtain their advice in identifying and resolving problems. For example, trainees may first ask a group of specialists to study problems in the farmers' fields and to answer the trainees' questions. The group might meet in one or more fields for about an hour as part of the training process. If potential solutions to a problem are not identified, the trainees can undertake experiments to gain new information. Also, the trainees might call in a group of farmers to view the same problems and to obtain their responses. This approach: (1) helps establish good rapport between farmers and trainees, (2) gives trainees an indication of the farmers' understanding and ability to propose practical solutions to problems, (3) makes trainees more aware of expertise available to the region, and (4) provides trainees insight into the differences in the farmers' practical approach and the specialists' reliance on scientific knowledge.

A well conceived training program should include a variety of instructional methods; for example:

- Workshops and consulting periods can help the trainees with their experimental designs, statistical analyses, and economic calculations.
- Field days give the trainees the chance to exchange ideas with their instructors, specialists, and others.
- Seminars can provide the trainees with the opportunity to decribe their experimental procedures and results, problems encountered, and interesting observations.
- Integration of training into ongoing FSR&D activities allows the trainees close contact with their trainers under working conditions and offers the trainees the opportunity to accept increasing responsibility as their training progresses.
- Classroom instructions with assigned readings and problems can increase the trainee's knowledge in areas requiring strengthening, including instructions in oral and written communication.

The academic training portion can be held during slack periods between growing seasons. Trainees with good agronomic or animal science skills generally need help in the social sciences, while those from the social sciences need to learn more about agronomic and livestock practices.

The foregoing training activities provide the trainers with a variety of ways for evaluating the trainees' performances. Evaluations can be based on examinations covering theoretical materials, written technical reports, presentations and discussions in the seminars, and overall impressions. The trainees, in turn, evaluate the course by commenting on what they have learned and the methods of instruction.

By the time the training has been completed, the trainees should be well prepared for one of the positions on the field teams. They will have learned about the FSR&D approach, methods of applied research, and the sources of information and assistance. These sources include available literature, experiment station activities and results, and the various disciplinary and commodity specialists. In Appendix 11-A, we provide a more complete description of such a training program in Honduras.

11.2.2. PROGRAMS FOR EXTENSION

Somewhat different training programs are needed for those in extension who will work with FSR&D researchers. These programs should provide extension workers with a basic understanding of the FSR&D philosophy, concepts, and general procedures. Extension workers trained in FSR&D help most directly through their work with farmer-managed tests, multi-locational testing, and pilot production programs.

Except for the ESFSs, extension personnel will not be expected to have as much research expertise as other members of the FSR&D teams. In places where qualified personnel are scarce, some extension workers may have no more than the equivalent of a 10th grade education. Consequently, programs for extension workers will generally emphasize working with farmers, collecting data and monitoring activities, and diffusing information through demonstrations, field days, printed materials, radio, and possibly television.

The above types of training should help the extension workers for activities, such as

- understanding and applying criteria for identifying subareas
- selecting cooperating farmers
- conducting field surveys
- monitoring farm records and climatic data
- supervising farmer-managed tests and assisting with researcher-managed trials

- organizing and implementing multi-locational tests and pilot production programs.

An important link exists between the first and last points. To be effective in implementing the multi-locational tests and pilot production programs, the extension workers need to be able to distinguish subareas based on farmers' characteristics, farming systems, and environmental conditions.

ICTA has given three training programs for extension agents in Guatemala. These have been in-service training in which the trainees participate in the training program 1 or 2 days per week and spend the rest of their time with their regular extension work. Because most of the extension workers in Guatemala do not have bachelor's degrees, the technical content of the training material is at a different level from that of the researchers. Even so, ICTA has been able, at times, to use some training materials for both groups.

By being trained in FSR&D, the extension workers can materially help the FSR&D team in spreading its experiments over a broad area. This becomes possible by having the extension workers look after a major proportion of the farmer-managed tests. Not only can more tests be conducted, but the researchers' time is freed for researcher-managed trials and other activities calling for their expertise. But this is only one of the advantages of training extension workers. The extension service itself benefits. For example, the extension workers:

- become an integral part of the technology development process and are no longer simply receivers of new technologies
- can evaluate research results more objectively and can become more persuasive in their dealings with farmers, researchers, and others
- learn how to manage new technologies and thereby develop stronger ties with researchers
- can use the farmer-managed tests for holding small field days (Waugh, personal communication).

In Appendix 11-B, we provide additional details on the in-service training program that ICTA has prepared for DIGESA (extension).

11.2.3. PROGRAMS FOR TECHNICIANS

As FSR&D expands, additional technicians will need training. In many countries most of the technicians to be trained will have attended agricultural schools and many will have the equivalent of a high school education. However, some governments may have to use technicians with less education or exposure to agriculture. During and after training, these technicians can assist the field teams and in many cases will spend more time on the farms than other team members. In the Philippines, cadres of from 5 to 10 technicians support the efforts of the field team's researchers.

The technicians' training should concern collecting

data, keeping records, setting up field trials, sampling of yields, and related tasks. Instructions in FSR&D philosophy, concepts, and practices should augment the technician's training. Such training can be accomplished best through association with more highly trained and experienced staff and by sessions on special topics.

Experienced technicians should be consulted when identifying training needs and be asked to assist in the training. Such technicians are important in training other technicians due to their active role as members of FSR&D teams and to their generally good understanding of the conditions under which the trainees will operate. Moreover, the more capable and exerienced technicans enrich the training process and, by this involvement, may be able to advance more rapidly within the FSR&D organization.

11.3. COMPLEMENTING ACTIVITIES

In addition to the basic training programs directed to researchers, extension workers, and technicians the FSR&D leaders can also complement these programs in other ways. These include non-degree training, graduate degree training, and short-term activities. Each of these will now be discussed.

11.3.1. NON-DEGREE TRAINING

Two factors combine to highlight the importance of non-degree training. First, much of FSR&D is applied research that does not require large numbers of staff with advanced degrees. Second, most of the international agricultural research centers and at least one of the regional centers have very effective non-degree programs. Also, universities occasionally offer applied research and development courses at the non-degree level that may be suitable for FSR&D staff.

Instead of spending the time to obtain a master's or doctoral degree, the FSR&D programs are usually better served by taking advantage of the non-degree programs these institutions offer. For example, ICTA favors training that combines a discipline with a commodity. Thus a member of a field team might study bean diseases at CIAT. Even though specialized in this way, members returning to their field teams will once again have responsibilities that encompass the whole farm. Eventually, one so trained might become a specialist at a regional or national headquarters. Other training possibilities include a soils scientist studying soil and water conservation at the International Institute for Tropical Agriculture (IITA), an animal scientist studying livestock systems modeling at ILCA, or an ESFS studying communications and extension at IRRI. We will discuss these centers further in Sec. 11.4.

In this manner, the requirements of the various FSR&D teams can be met without serious disruption to program activities. In fact, when the courses occur during the "off-season," disruptions to the FSR&D activities may be insignificant.

11.3.2. GRADUATE DEGREE TRAINING

Even though FSR&D does not require large numbers of staff with advanced degrees, sound scientific research demands a core of well-trained researchers. When starting an FSR&D program, a country will have to use whatever staff is available and to perhaps borrow staff from other organizations. But once the program is established, FSR&D management will probably want to selectively send some of its staff to graduate school. Those so trained will, most likely, become the leaders and specialists at FSR&D's national and regional headquarters and can become the basis for developing a highly capable institution in applied agricultural research. One of the ways the Brazilian Agricultural Research Corporation (EMBRAPA) developed its research capabilities during the 1970s was to send a considerable number of its staff to graduate school.

However, FSR&D leaders need to be careful in the way they approach educating their staff at the graduate level, especially when such training occurs abroad. First, some of the graduate programs may not be suitable for FSR&D's needs. Second, time is needed to learn what type of graduate training to recommend and to observe which of the FSR&D staff are most suitable for the training. Third, sending too many of the FSR&D team's key staff can seriously obstruct progress, particularly during the early years when FSR&D is becoming established. Finally, some of those sent abroad are lost to the parent organization because of the changes in the individual's and organization's interests.

We provide further comments on graduate training possiblities in Sec. 11.5.

11.3.3. SHORT-TERM ACTIVITIES

The FSR&D leaders should consider ways to train their regular staff through short courses, workshops, seminars, and meetings. These activities help to maintain the staff's professionalism, interests, and capabilities in its work. Furthermore, such training can introduce new material or refresh attendees in key subject areas. The leaders may wish to conduct at least some of these sessions jointly with other organizations to ease the administrative and financial burden, develop cooperation with other organizations, and spread the FSR&D philosophy.

Short Courses

Considerable material can be offered and absorbed by attendees through short courses of from 1 to 5 days, provided the courses are carefully prepared, the topics are specific enough, and the attendees have suitable background in the subjects being presented. Topics might include instruction in experimental design, plant or animal diseases, weed control, statistical procedures, sampling techniques, and economic analyses.

FSR&D staff from national headquarters would normally be in a good position to present material in a way that is most useful to the trainees. Often, consultants can

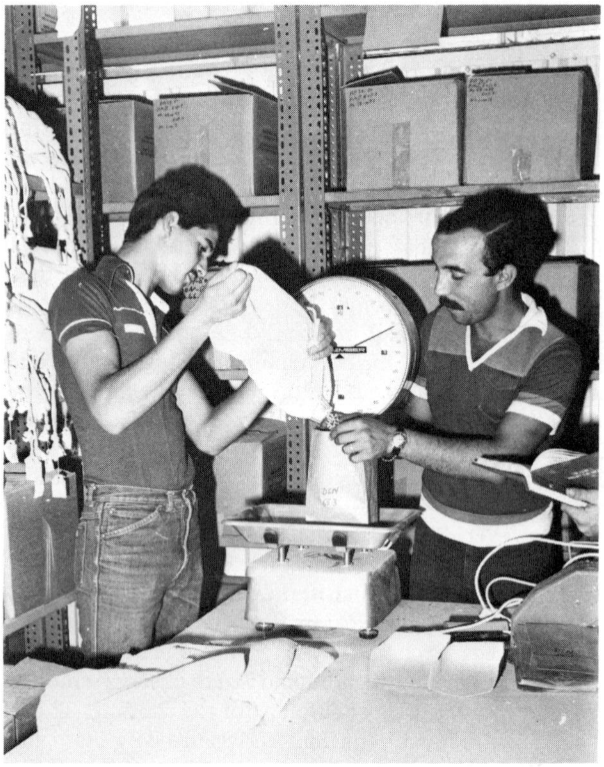

complement this effort, provided their assignments are carefully chosen and their presentations directed to the specific needs of the training program. Where possible, examples from the FSR&D program should be worked into the training material. Instructional help might also come from the faculty of universities within the country and abroad; from international and regional agricultural research centers; from international suppliers of materials and equipment, such as agricultural chemicals; and from others. Whether the instruction is provided in concentrated packages or spread out depends on staff availability and the urgency for learning the material.

Workshops

Workshops can provide the opportunity to learn how to apply various procedures to practical situations. For instance, FSR&D leaders might use workshops to help their staff learn the intricacies of a new statistical procedure, the application of a new agricultural chemical, or better ways to interview farmers. The setting could be in the field as well as in the classroom.

Seminars

Seminars can be used to report on FSR&D results. In Sec. 11.2.1., we spoke of trainees presenting the results of their experiments. But FSR&D leaders may also want to use seminars as a means for presenting research results and other topics by their regular staff, visiting scientists, and others. IRRI has a series of Saturday Seminars for presenting the research results of its staff. Finally, seminars can be used to acquaint others in the country with the FSR&D staff and activities.

Meetings

The Asian Cropping Systems Network uses working groups that meet in member countries about twice a year to arrange for collaborative action and to report on their research findings. These meetings, which last about one week, provide convenient opportunities for researchers in the host country to hear about FSR&D and to learn that researchers in other countries have similar interests. This last aspect has particular relevance when those promoting FSR&D concepts seek to interest others in their country in the legitimacy of the approach.

11.4. INTERNATIONAL AND REGIONAL CENTERS

Besides their regular research activities, the IARCs also stress training, as the following quote from CGIAR (1976) indicates

> "While research is at the heart of a center's activities, the training of research scientists and production specialists to serve in the less developed countries claims a large part of its time, its effort and its budget. The number of persons trained at the centers so far is well over 3,000, and the emphasis on this kind of activity continues to grow."

Most IARCs and at least one regional center (CATIE) offer a variety of training opportunities for FSR&D staff. Combined, the centers cover a wide range of commodities, subjects, and training arrangements:

- Examples of commodities are beans, cassava, cowpeas, ground cover, livestock, maize, rice, tropical pastures, and wheat.
- Subjects include communications, economics, intercropping, pest management, soil fertility and conservation, statistics, and technical writing.
- Training arrangements include postdoctoral fellows, visiting research associates, research scholars, postgraduate production interns, special trainees, and participants in short courses and professional meetings.

Opportunities are also present for training leading to graduate degrees, since most, if not all, of the centers have some form of cooperative agreement with universities. For example, Cornell and Kansas State Universities have such agreements with CIMMYT; IRRI cooperates with the University of the Philippines at Los Banos in granting master's and doctoral degrees; and CATIE offers a master's degree in cooperation with the University of Costa Rica.

The IRRI program cannot meet the demand for some of its courses, largely because of the expanding activities of the Asian Cropping Systems Network. For one of these courses, IRRI plans to reduce the training time from 6 to 3 months and thereby double the number of trainees accommodated in the 6-month period. During the longer training

period, the trainees did most of the work in growing and harvesting the crops. With the shorter period, the training must be more tightly organized and IRRI staff must help in growing the crops. Plantings have to be made earlier so that the trainees will have an opportunity to study the various stages of plant growth. The IRRI staff trains the group as a unit during the first two months. Then, the staff allows the trainees to specialize in an area such as agronomy or economics. One of IRRI's concepts for these programs is to train teams of four or five from the same country. This approach provides the team members with a common base of knowledge and experience when they return home.

A few other illustrations should suffice to indicate the nature of these programs. In addition to its training activities in Mexico, CIMMYT is setting up training centers in selected geographical regions of the world. CATIE's training and consulting work in farming systems research has benefited Central American countries for some time. Finally, postdoctorate fellowships at the IARCs aid in transferring FSR&D methodology to visiting scientists. An important aspect of this association is that these visitors work with senior researchers in planning and implementing research programs. This approach offers a few highly educated staff the opportunity to learn from those experienced in applied agricultural research.

We provide a fuller discussion of some of the training activities of CATIE, CIAT, CIMMYT, ILCA, and IRRI in Appendix 11-C, and a detailed listing of a 6-month training program in cropping sytems at IRRI in Appendix 11-D.

11.5. UNIVERSITY PROGRAMS IN THE UNITED STATES

We anticipate that the growing activity in FSR&D around the world will cause universities in the United States to begin adding FSR&D courses to their agricultural curricula. Cornell University, Kansas State University, Michigan State University, and the University of Florida have demonstrated such interest; and we expect other universities with strong agricultural programs to follow suit. Below, we propose tentative statements about the objectives and approaches to university programs in FSR&D in the United States. Our statements are tentative because these programs are still evolving.

11.5.1. OBJECTIVES FOR UNIVERSITY PROGRAMS

A suitable objective for a land-grant university in the United States concerning FSR&D might be the following: To provide a sound academic foundation for the growing interest in FSR&D that integrates the relevant disciplines (1) for understanding farmers, farming systems, and farmers' environment; (2) for identifying problems and opportunities relevant to farmers and society; and (3) for developing solutions in a way that allows the university to be more effective in its teaching, research, extension, and service activities.

11.5.2. APPROACH FOR UNIVERSITY PROGRAMS

Below are some thoughts concerning the approach for development of university programs in FSR&D. Some of these ideas come from Gilbert et al. (1980).

- Students should specialize in some agriculturally related discipline; as yet, FSR&D has not evolved into a separate discipline.
- An introductory course in FSR&D can be used to convey the philosophy, concepts, and methods of the approach.
- A graduate seminar would be useful in exploring FSR&D topics still in the formative stages, such as classification of areas to facilitate the transfer of research results, application of systems analysis tools, and the practicality of ecological concepts as an integrator of the physical, biological, and social sciences.
- A course is needed that stresses an interdisciplinary and applied approach to problem identification for mixed enterprise farms as typically encountered in the developing countries.
- Another course is needed that shows the diversity of farming systems throughout the world to aid students in identifying and understanding a farming system's most distinguishing features.
- FSR&D concepts should be integrated into traditional courses by teaching sections specifically related to FSR&D, such as multiple cropping in agronomy, crop-livestock interactions in animal science, survey techniques in rural sociology, and farm management in agricultural economics.
- Arrangements should be made for students to work in the field as members of interdisciplinary groups.
- Universities should explore the possibilities of having their graduate students conduct their field research in collaboration with one of the regional or international agricultural research centers.
- Because FSR&D concepts are new and evolving, initial curricula should be developed slowly and make effective use of faculty with demonstrated capabilities in FSR&D.
- Interested faculty should be provided with incentives to become more active in interdisciplinary research directed toward the problems of small farmers in developing countries. The United States Government is providing some help through the Title XII program to develop courses and interest faculty in this type of work.

Finally, an example of a group approach to the doctoral dissertation is that of six students who did their course work at Cornell University and their field work as a team with CIMMYT in Mexico. The students were required to produce both a research project in the areas of their specialty and a team research project. The degree of overlap was left to the discretion of the student and his adviser (Contreras et al., 1977).

11.6. TRAINING MATERIALS ON FSR&D

Besides those mentioned previously, training materials and references on FSR&D include *Farm Management Research for Small Farmer Development* (Dillon and Hardaker, 1980), various watershed management manuals from the Food and Agriculture Organization (FAO) of the United Nations, *Reaching Rural Families in East Africa: Handbook for Extension Workers* (FAO, 1973), *Region, Farm and Agroecosystem Characterization: The Preliminary Phase in a Farm System Research Strategy* (Hart, 1980), *Small Farm Development* (Harwood, 1979), and *Education of Development Technicians* (Trail, 1969). Mead (1955) proposed six principles for introducing technical change to small farmers which we present in Appendix 11-E. Also, two references in Spanish are *Agroecosistemas: Conceptos Básicos* (Hart, 1979) and *Agroecosistemas de México: Contribuciones a la Enseñanza, Investigación y Divulgación Agríola* (Hernandez X., 1977). Finally, additional references in French for Francophone, West Africa can be found in the works of ISRA (Institut Sénégalais de Recherches Agricoles) and the Groupement d'Etudes et de Recherches pour le Développement de l'Agronomie Tropicale (GERDAT) headquartered in Paris.

11.7. SUMMARY

Adequately trained FSR&D teams are the key to effective implementation of the FSR&D approach. While various means can be taken to provide this training, we favor an approach that involves (1) sending a small group to one of the IARCs or a regional center for concentrated training in applied on-farm research; (2) having the team return to develop national training programs in FSR&D with assistance of FSR&D specialists from the centers, the universities, or other organizations; and (3) providing additional training opportunities for the rest of the FSR&D teams.

The national training programs for the research staff should cover the range of FSR&D activities. Training programs for the extension staff should concentrate on extension's role in supporting research and in transferring technologies from research to extension. Training should also be considered for the technical assistants who work with the field teams. Those not being trained in one of these three programs can benefit from selected non-degree and graduate degree training and through short courses, workshops, seminars, and meetings.

The IARCs and regional centers can be a source of non-degree as well as degree training. In time, the universities are expected to provide a means for training students in FSR&D philosophy, concepts, and methods. Finally, trainers in FSR&D should take advantage of existing training materials of the type noted above.

CITED REFERENCES

CGIAR, 1976. Consultative group on international agricultural research. CGIAR, New York.

Contreras, M.R., D.L. Galt, S.C. Muchena, K.M. Nor, F.B. Peairs, and M.S. Rodriguez. 1977. An interdisciplinary approach to international agricultural training: the Cornell CIMMYT graduate student team report. Program in International Agriculture, Cornell University, Ithaca, New York.

Dillon, J.L., and J.B. Hardaker. 1980. Farm management research for small farmer development. FAO Agric. Ser. Bull. 41. FAO, Rome.

FAO. 1973. Reaching rural families in East Africa: handbook for extension workers. Man. ser. no. 1. FAO, Rome.

Gilbert, E.H., D.W. Norman, and F.E. Winch. 1980. Farming systems research: a critical appraisal. MSU Rural Dev. Paper No. 6. Dep. of Agric. Econ., Michigan State Univ., East Lansing, Mich.

Hart, R.D. 1980. Region, farm and agroecosystem characterization: the preliminary phase in a farm system research strategy. A paper presented at the 72nd Ann. Meeting of the A. Soc. of Agron. Detroit, Mich. 30 Nov.-5 Dec. 1980.

______. 1979. Agroecosistemas: conceptos básicos. CATIE, Costa Rica.

Harwood, R. 1979. Small farm development: understanding and improving farming systems in the humid tropics. Westview Press, Boulder, Colo.

Hernandez X., E. (ed.) 1977. Agroecosistemas de México: contribuciones a la enseñanza, investigación y divulgación agrícola. Colegio de Postgraduados, Chapingo, and CIMMYT, Mexico.

Mead, M. (ed.). 1955. Cultural patterns and technical change. Mentor Publishing Company, New York.

Trail, T. 1969. Education of development technicians. F. Praeger, New York.

ACRONYMS

ACSN	Asian Cropping Systems Network
AICRPDA	All India Coordinated Research Project for Dryland Agriculture India
AVRDC	Asian Vegetable Research and Development Center Taiwan
BAEX	Bureau of Agricultural Extension Philippines
BIFAD	Board for International Food and Agricultural Development Washington, D.C.
CATIE	Centro Agronómico Tropical de Investigación y Enseñanza (Tropical Agricultural Research and Training Center) Costa Rica
CENTA	Centro Nacional de Tecnología Agropecuaria (National Agricultural Technology Center) El Salvador
CGIAR	Consultative Group on International Agricultural Research Washington, D.C.
CIAT	Centro Internacional de Agricultura Tropical (International Center for Tropical Agriculture) Colombia
CID	Consortium for International Development Tucson, Arizona
CIMMYT	Centro Internacional de Mejoramiento de Maíz y Trigo (International Maize and Wheat Improvement Center) Mexico
CIP	Centro Internacional de la Papa (International Potato Center) Peru
CSU	Colorado State University Fort Collins, Colorado
CRIA	Central Research Institute for Agriculture Indonesia
DCI	Developed Country Institution
DIGESA	Dirección General de Servicios Agrícolas (General Directorate for Agricultural Services) Guatemala
EAP	Escuela Agrícola Panamericana (Pan American School of Agriculture) Honduras
EMBRAPA	Empresa Brasileira de Pesquisa Agropecuária (Brazilian Agricultural Research Corporation) Brazil
ESFS	Extension Specialist in Farming Systems
FAO	Food and Agriculture Organization of the United Nations Rome
FSR	Farming Systems Research
FSR&D	Farming Systems Research and Development
GERDAT	Groupement d'Etudes et de Recherches pour le Développement de l'Agronomie Tropicale (Group for Studies and Research in the Development of Tropical Agronomy) France
IADS	International Agricultural Development Service New York
IAR	Institute for Agricultural Research Ahmadu Bello University Nigeria
IARC	International Agricultural Research Center
ICA	Instituto Colombiano Agropecuario (Colombian Agricultural Institute) Colombia
ICARDA	International Center for Agricultural Research in the Dry Areas Lebanon
ICRISAT	International Crops Research Institute for the Semi-Arid Tropics India
ICTA	Instituto de Ciencia y Tecnología Agrícolas (Agricultural Science and Technology Institute) Guatemala
IER	Institut d'Economie Rurale (Institute of Rural Economics) Mali
IICA	Instituto Interamericano de Ciencias Agrícolas (Inter-American Institute of Agricultural Sciences) Washington, D.C.

IITA	International Institute for Tropical Agriculture Nigeria
ILCA	International Livestock Centre for Africa Ethiopia
ILRAD	International Laboratory for Research on Animal Diseases Kenya
INCAP	Instituto de Nutrición de Centro América y Panamá (Nutritional Institute for Central America and Panama) Guatemala
INIAP	Instituto Nacional de Investigaciones Agropecuarias (National Institute for Agricultural Research) Ecuador
IPPC	International Plant Protection Center Corvallis, Oregon
IRRI	International Rice Research Institute Philippines
ISRA	Institut Sénégalais de Recherches Agricoles (Senegalese Institute for Agricultural Research) Senegal
LDC	Less Developed Country
ORD	Organismes Régionaux de Développement (Regional Development Organizations) Upper Volta
PNIA	Programa Nacional de Investigación Agropecuaria (National Program for Agricultural Research) Honduras
ROCAP	Regional Office for Central American Program Guatemala
TAC	Technical Advisory Committee of the Consultative Group on International Agricultural Research Washington, D.C.
USAID	United States Agency for International Development Washington, D.C.
WARDA	West Africa Rice Development Association Liberia

GLOSSARY

Below are terms related to FSR&D as we have used them in this book. We selected for inclusion those (1) that are central to FSR&D, (2) that we have defined to meet our special needs and whose definitions may differ for other uses, and (3) that may not be familiar to the reader.

Agroclimatic environments: Areas with similar agroclimatic conditions but not necessarily contiguous, where a crop exhibits roughly the same biological expression so that we would obtain, for example, similar variety or fertility responses within a given environment, *everything else being equal.*

Baseline data: Data collected before a project begins—e.g., on crop yields, labor input, or market prices—against which a project's results can be evaluated.

Biological factors: Those factors, such as plant and animal characteristics and pest problems, that influence the health and vitality of plants and animals and the quality of harvested products.

Biological feasibility: An action or project that is biologically practical based on current knowledge of the attributes of the plants or animals.

Biological system: A term referring to cropping and livestock systems.

Climatic analysis: The analysis of data, over time, on such factors as precipitation, maximum and minimum temperatures, relative humidity, wind, and radiation.

Coefficient of variation (C.V.): A measure of the relative variability among experimental units of measure and/or plot sizes (Little and Hills, 1978). The coefficient of variation, denoted by C.V. (X), is the ratio of the standard deviation (S_x) to the mean ($\bar{X}$) expressed as a percent:

$$C.V.(X) = \frac{S_x}{\bar{X}} \cdot 100.$$

Collaborating farmers: Farmers chosen or who have volunteered to cooperate with an FSR&D project by (1) allowing researchers to conduct experiments on their farms, or (2) agreeing to test and evaluate new technologies themselves.

Commodity-oriented research: The focusing of research on one or more crops or animals by studying them in considerable detail. Commodities selected for emphasis should be the result of prior investigation demonstrating their importance to the farming system.

Commodity specialists: Researchers who have been trained to work with a specific crop or animal.

Component technology: The knowledge of individual technical factors involved in plant or animal production. The term includes studies of cropping patterns as well as management practices and comes largely from national and international experiment stations, where research is conducted with only a few variables in a well-defined environment.

Confidence interval: The population mean U_x is somewhere between $\bar{X} - zS_{\bar{x}}$ and $\bar{X} + zS_{\bar{x}}$. The interval is called a confidence interval for U_x. The $zS_{\bar{x}}$ is some multiple of the standard deviation of the sampling distribution $\bar{X}$. (Neter et al., 1973).

Cropping patterns: The crop species grown on a given field during a 12-month period. Cropping patterns include single, multiple, mixed, strip, and ratoon cropping.

Cropping systems: This term means subsystems within the farming system comprising one or more crops and all components required for production, including the interactions among crops, other household enterprises, and the physical, biological, and socioeconomic environments.

Cropping systems research: Such research concentrates on crops and cropping patterns, alternative management practices in different environments, and interactions between crops, between crops and other enterprises, and between the household and environmental factors beyond the household's control. The procedures are similar to farming systems research, but the breadth is generally less.

Cultural practices: Farmers' regular practices such as land preparation, seed selection, crop establishment, and fertilization.

Disciplinary research: The process of approaching the object of study—in this case the farming system or a portion of it—from the perspective of a particular discipline, e.g., economics.

Disciplinary specialists: Researchers who have been trained in a particular field of study, such as agronomy, animal husbandry, agricultural economics, and rural sociology.

Double cropping: Growing "two crops in sequence, seeding or transplanting one after the harvest of the other" (TAC, 1978). Such practices are also called sequential cropping.

Economic environmental factors: Factors such as (1) the availability of credit, (2) marketing potential and prices of farm products, (3) cost of hired labor, (4) costs of seeds, agricultural chemicals, and farm equipment, and (5) land ownership and tenant characteristics.

Economic feasibility: Activities and investments that produce benefits in excess of costs when the time value of money is considered. Other things being equal, investors and managers wish to maximize benefits over costs subject to various constraints, social pressures, and attitudes toward change and risk, leisure, and so on.

Enterprises: Activities undertaken to produce an output that contributes to total production or income of the farm family. Enterprises in FSR&D typically concern crops, livestock, processing or otherwise upgrading agricultural commodities produced on the farm, productive nonagricultural activities carried out on the farm—such as handicrafts—and productive off-farm activities by the household members.

Environmental factors: Those factors over which the farmer has little direct control. They include the physical, biological, and socioeconomic aspects of the farmer's setting.

Experimental variables: Those variables in an experiment that the researcher tests.

Extension specialist in farming systems (ESFS): An employee of the extension service who specializes in FSR&D. These specialists will be members of various FSR&D teams and act as an important liaison between research and extension.

Farmer-managed tests: On-farm experiments managed by farmers to learn how farmers respond to suggested improvements.

Farmer's environment: In FSR&D, those conditions under which the farmer operates that include physical, biological, economic, and sociocultural conditions.

Farming: An activity carried out by households on holdings that represent managerial units organized for the economic production of crops and livestock (Ruthenberg, 1971).

Farming systems: A unique and reasonably stable arrangement of farming enterprises that a household manages according to well-defined practices in response to the physical, biological, and socioeconomic environments and in accordance with the household's goals, preferences, and resources. These factors combine to influence output and production methods. More commonality will be found within the system than between systems. The farming system is part of larger systems and can be divided into subsystems.

Farming systems research and development (FSR&D): An approach to agricultural research and development that (1) views the whole farm as a system, and (2) focuses on the interdependencies among the components under the control of members of the farm household and how these components interact with the physical, biological, and socioeconomic factors not under the households' control. The approach involves selecting target areas and farmers, identifying problems and opportunities, designing and executing on-farm research, and evaluating and implementing the results. In the process, opportunities for improving public policies and support systems affecting the target farmers are also considered.

Farm records: For FSR&D, the type of accounts kept by members of the farm household on specified activities associated with individual crops or animal types. Input and output activities are often kept daily, with farmers sometimes receiving help from the field team's technical assistants.

Field teams: The field teams work with farmers in their fields. Such teams often consist of agronomists, economists, and supporting technicians. Where livestock is important, an animal scientist should be part of the team; where irrigation is practiced, an irrigation engineer can be a key member of the team; and where women are responsible for growing important crops or performing critical operations, field teams should include women.

Financial feasibility: A condition when cash resources are sufficient to meet cash requirements both in amounts and timing. Financial infeasibility means an insufficient amount of cash when needed.

Frequent interview survey: A type of survey involving the collection of data from a limited number of farms on a regular basis.

Homogeneous farmer groups: Those who farm under similar conditions and in similar ways so that recommendations for changes in technology will likely be accepted by the majority of them (see recommendation domain).

Household: The household comprises the farmer and other members of the farmily, is both a consuming and producing unit, and is a social organization. Households are often under the management of a single person, but sometimes operate collectively. Members normally live and sleep in the same place, share meals, and divide household duties.

Informal surveys: Surveys undertaken without formal sampling procedures, pretested questionnaires, and other means that permit statistical analysis of the data.

Infrastructure: The supportive features of an economy often provided by government, but sometimes provided by private industry, such as transportation, electricity, water, communications, and governmental organizations.

Iterative process: An approach that involves repeating activities and calculations to arrive at improved solutions through a series of successively better approximations.

Intercropping: Growing "two or more crops simultaneously in the same plot in different, but proximate stands. In this system, one crop system is part of the other crop's environment" (TAC, 1978).

Interdisciplinary: Involves frequent interactions among those from different disciplines who work on common tasks and come up with better results than had they worked independently.

Land type: A set of locations for which it is possible to develop common technologies.

Linear programming: A mathematical procedure that determines an optimal decision by maximizing or minimizing an objective function subject to a set of specified constraints. The objective and constraint functions are linear.

Livetock patterns: The animal species raised by the family over some period.

Livestock systems: Subsystems within the farming system made up of a set of one or more animals and comprising all components required for their production, including the interactions among the animals, other household enterprises, and the physical, biological, and socioeconomic environments.

Livestock systems research: A process similar to cropping systems research but with procedures that reflect the inherent differences between cropping and livestock systems—e.g., fewer numbers of animals than plants.

Management factors: Those factors the farmer can control through management decisions including such variables as cropping and livestock patterns, crop varieties, field cultural practices, fertilization, pest control, irrigation management, harvest data, sale of crop or animal products, use of labor, animal, or mechanical power, and post-harvest losses.

Mixed intercropping: Growing "two or more crops simultaneously intermingled in the same plot with no distinct row arrangement" (TAC, 1978).

Mixed systems: Cropping, livestock, and possibly other enterprises present within the farming system.

Mixed systems research: Follows a process similar to that for cropping systems research, except for the procedures that reflect the inherent differences between cropping and livestock systems—e.g., fewer numbers of animals than plants. Also, the researchers focus their attention directly on the interactions between crops and livestock.

Monoculture planting: Growing a single crop on the land at one time or the repetitive growing of the same crop on the same land.

Multidisciplinary: A combination of disciplines involved in an assignment not necessarily working in an integrated or coordinated manner.

Multi-locational testing: The process whereby new technologies developed in a research area are tested at other locations within the target area to learn what adjustments, if any, are needed before diffusing the technologies more broadly and intensively.

Multiple cropping system: "A system in which more than one crop is grown on the same plot of land in one year" (TAC, 1978).

Nonexperimental variables: Those variables in an experiment that the researcher is not testing. They are divided into those that the researchers cannot control, such as weather, and those that they can control, such as farming operations. For on-farm research, the FSR&D team generally tries to hold the latter at the farmers' level, but sometimes may use the level recommended by the extension service.

On-farm research: In FSR&D, the process of conducting researcher-managed and superimposed trials and farmer-managed tests on farmers' fields, and the process of conducting related studies of farm management and environmental conditions influencing the farmer.

Paradigm: A set of concepts, methodologies, and vocabulary associated with a particular group or discipline at a given time. For example, plant breeders work under a generally accepted set of assumptions, use common research approaches, and have a set of terms that have special and carefully defined meanings for them.

Partial budget analysis: A "form of marginal [incremental] analysis designed to show, not profit or loss for the farm as a whole, but the net farm income resulting from the proposed changes" (Brown, 1979).

Physical factors: The more important attributes of the climate, water, and land.

Pilot production program: A program designed to test how agricultural policies and support systems function when new technologies are introduced into an area on a large scale—e.g., 100 to 500 hectares.

Primary information. Data collected specifically for the current activity.

Program approach: An approach to FSR&D that involves institutionalizing FSR&D into the country's existing agricultural research and development programs through either coordinated efforts between all organizations that are most concerned with small farm production, or through one organization assuming primary responsibility for implementing the FSR&D program.

Project approach: An approach to FSR&D involving initiating one or more projects that incorporate FSR&D procedures. Projects tend to have specific scopes of work to be completed by a certain time by staff and organizations that are disbanded upon the project's completion.

Rain-fed farming: The growing of crops or animals under conditions of natural rainfall. Water may be stored in the crop field by bunding, as with lowland rain-fed rice, but no water is available from permanent water storage areas.

Random sample: A sample drawn so that every unit in the population or subpopulation has an equal probability of being selected.

Ratoon cropping: The development "of a new crop—without replanting—from buds on the root system, stubble, or stems of the preceeding crop. Some ratoon crops may be included in multiple cropping systems" (TAC, 1978).

Recommendation domain: "A group of roughly homogeneous farmers with similar circumstances for whom we can make more or less the same recommendation. Recommendation domains may be defined in terms of both natural factors (e.g., rainfall) and economic factors (e.g., farm size)" (Byerlee et al., 1980).

Reconnaissance survey: A field survey method of data collection that usually comes after secondary data collection. Also known as quick, informal or exploratory survey or in Spanish, *sondeo.*

Relay intercropping (relay cropping): "Growing two or more crops in sequence, seeding or transplanting the succeeding one some weeks before the harvest of the preceding crop" (TAC, 1978).

Replication: A treatment applied to more than one experimental unit (Little and Hills, 1978).

Research area: The location within the target area where the FSR&D team develops improved technologies. The location may represent the whole target area or only some of the subareas.

Researcher-managed trials: On-farm experiments managed by researchers to develop new technologies under farmers' conditions.

Resource feasibility: An action or project that is practical in terms of available resources—i.e., land, labor, capital, and management.

Row intercropping: The growing of "two or more crops simultaneously in the same plot in distinct rows" (TAC, 1978).

Secondary information: Published or unpublished data collected for purposes other than the current activity.

Sequential cropping: "One crop is planted after harvest of the first. (Sometimes called relay planting in West Africa.)" (Harwood, 1979).

Shifting cultivation: A method of cultivation in which "several crop years are followed by several fallow years with the land not under management during the fallow. The shifting cultivation may involve shifts around a permanent homestead or village site, or the entire living area may shift location as the fields for cultivation are moved" (Harwood, 1979).

Significance level: The greatest probability that a researcher is willing to accept that he or she will reject a hypothesis when it should be accepted. Biological researchers frequently set the significance level at 5 percent, but other values are also used.

Single crop system: A "system in which only one crop is grown on the same plot of land in one year" (TAC, 1978).

Single interview survey: A questionnaire or schedule administered only once to farmers selected by formal sampling procedures and conducted by trained interviewers.

Slash and burn: A type of shifting cultivation in high rainfall areas where bush or tree growth occurs during the fallow period. The fallow growth is cleared by cutting and burning; see shifting cultivation.

Small-scale farming: A situation in which farmers frequently have difficulty obtaining sufficient inputs to allow them to adequately use the available technology as would medium- and large-scale commercial farmers. Small does not, necessarily, refer to the area of land held.

Social scientists: Individuals who conduct research in such areas as economics, sociology, anthropology, political science, geography, and communication.

Sociocultural acceptability: In FSR&D, this term refers to the probability that a new technology will be acceptable to farmers within the context of their particular culture and community.

Sociocultural factors: The influences that community and culture exert on farmers.

Socioeconomic factors: Such factors include, for example, access to markets, available support services, norms and customs related to land use, and division of labor.

Sole cropping: The growing of only one crop at a time on a plot of land.

Sondeo: A Spanish word for a survey; see reconnaissance survey.

Standard deviation: A measure of the absolute variability among experimental units of measurement and/or plot sizes. The standard deviation, denoted by S_x, is defined as

$$S_x = \sqrt{\frac{\Sigma (x - \bar{x})^2}{n - 1}}$$

where $\bar{x}$ is the arithmetic average, n is the number of items, and x represents the values of individual observations (Neter et al., 1973).

Stratification: The process of dividing an area or population into relatively homogeneous subgroups to increase sampling efficiency.

Subareas: A subdivision of a target area or target groups of farmers, with common physical, biological, and socioeconomic factors.

Subsistence farmers: Farmers producing primarily for their needs, with the resulting low capacity to purchase inputs for production or consumption.

Superimposed trials: Relatively simple researcher-managed experiments applied across a range of farmer-managed conditions.

Support Services: Systems that will determine the appropriateness of a new technology—e.g., roads, transportation, markets, available credit, irrigation, and sources of supply for seed, herbicides and fertilizer.

System: Any "set of elements or components that are interrelated and interact among themselves. Specification of a system implies a *boundary* delimiting the system from its environment. Two systems may share a common component or environment, and one system may be a *subsystem* of another" (TAC, 1978).

Systems approach: An approach for "studying the system as an entity made up of all its components and their interrelationships, together with relationships between the system and its environment. Such study may be undertaken by perturbing the real system itself (e.g., via farmer-managed trials or by pre- versus post-adoption studies of new technology) but more generally is carried out via *models* (e.g., experiments, researcher and/or farmer managed on-farm trials, unit farms, linear programming and other mathematical simulations) which to varying degree simulate the real system" (TAC, 1978).

Target area: A geographical area selected for an FSR&D project based on the needs of the people living there or to take advantage of the area's agricultural potential. Criteria for selection are normally set by key national and regional decision makers.

Target populations: Populations of farmers with similar cropping and livestock patterns, methods of production, and potentials that FSR&D teams select for research and development; such populations are also called target farmers.

Technology: The combination of all the management practices used for producing and otherwise managing a given crop, crop mixture, livestock, or other farm activity.

Treatment: A "dosage of material or a method that is to be tested in the treatment" (Little and Hills, 1978). Introducing or testing a crop variety or an animal is a kind of treatment.

Variable: An element or factor subject to change or variation.

Variance: A term representing the square of the standard deviation ($S^2_{\bar{x}}$), which is the most common statistical measure of dispersion.

Whole farm analysis: A methodology designed to search for optimal solutions through incorporation of farmers' objectives, farming systems, and resources to arrive at improved cropping and livestock patterns and management practices for overall farming systems performance.

Whole farm approach: An essential characteristic of FSR&D in which FSR&D teams look at the whole farm setting to identify problems and opportunities, note interrelationships, design and conduct experiments, and evaluate results. This is not to be confused with whole farm analysis.

CITED REFERENCES

Brown, M.L. 1979. Farm budgets. World Bank Staff Occ. Paper No. 29. IBRD/World Bank, Washington, D.C.

Byerlee, D., M.P. Collinson, R.K. Perrin, D.L. Winkelmann, S. Biggs, E.R. Moscardi, J.C. Martinez, L. Harrington, and A. Benjamin. 1980. Planning technologies appropriate to farmers: concepts and procedures. CIMMYT, El Batan, Mexico.

Harwood, R.R. 1979. Small farm development: understanding and improving farming systems in the humid tropics. Westview Press, Boulder, Colo.

Little, T.M. and F.J. Hills. 1978. Agricultural experimentation. John Wiley and Sons, Inc., New York.

Neter, J., W. Wasserman, and G.A. Whitmore. 1973. Fundamental statistics for business and economics. Allyn and Bacon, Inc., Boston.

Ruthenberg, H. 1971. Farming systems in the tropics. Clarendon Press, Oxford, UK.

Technical Advisory Committee (TAC). Review Team of the Consultative Group on International Agricultural Research. 1978. Farming systems research at the international agricultural research centers. The World Bank, Washington, D.C.

APPENDIXES

APPENDIX TO THE PREFACE

APPENDIX P-A
PROJECT CONTRIBUTORS: FIELD CONTACTS, PARTICIPANTS IN THE WORKSHOPS AND PRETESTING, AND REVIEWERS

As noted in the Acknowledgments, we received substantial and timely assistance throughout the life of the project from those knowledgeable and interested in FSR&D. Below are the names of (1) the principal contacts we made during our trips overseas, (2) the participants in our August 1979 and June 1980 workshops, (3) the participants in the August 1980 pretesting of the book of guidelines, and (4) reviewers of the book's third draft.

CONTACTS DURING FIELD TRIPS

The following list contains only a single name for each of the organizations visited. However, the number of persons contacted during visits to these organizations was usually much greater. For all of those contacted, we offer our thanks for the time spent and information received. To help identify the organizations, we have broken the list into geographical regions. Some of these persons have changed positions since our visits, which mainly occurred from January 1979 to August 1980.

CENTRAL AMERICA

Hernan Ever Amaya, Head, Agricultural Economics Department, Centro Nacional de Tecnología Agropecuaria (CENTA), Ministerio de Agricultura y Ganadería, San Salvador, El Salvador.

Mario R. Contreras, Director of Research, Investigaciones Agropecuarias, Ministerio de Recursos Naturales, Tegucigalpa, Honduras.

Robert D. Hart, Crop Ecologist, Tropical Agricultural Research and Training Center (CATIE), Turrialba, Costa Rica.

Ramiro Ortiz D., Technical Director, Instituto de Ciencia y Tecnología Agrícolas (ICTA), Guatemala City, Guatemala.

NORTH AMERICA

Antonio Turrent F., Profesor de Suelos, Colegio de Postgraduados, ENA, Chapingo, Mexico.

Donald L. Winkelmann, Director, Economics Program, International Maize and Wheat Improvement Center (CIMMYT), El Batan, Mexico.

SOUTH AMERICA

Eliseu Roberto de Andrade Alves, President, Brazilian Agricultural Research Corporation (EMBRAPA), Brasilia, D.F., Brazil.

Douglas Horton, Head, Social Science Unit, International Potato Center (CIP), Lima, Peru.

Ruben Jaramillo, División de Estudios Socioeconómicos, Colombian Agricultural Institute (ICA), Bogota, Colombia.

John L. Nickel, Director General, International Center for Tropical Agriculture (CIAT), Cali, Colombia.

EAST AFRICA

Michael Collinson, Farm Management Specialist, International Maize and Wheat Improvement Center (CIMMYT), Regional Office in East Africa, Nairobi, Kenya.

John Liwenga, Chief Research Officer, Ministry of Agriculture, Dar es Salaam, Tanzania.

David Pratt, Director, International Livestock Centre for Africa (ILCA), Addis Ababa, Ethiopia.

WEST AFRICA

B.T. Kang, Acting Leader, Farming Systems Program, International Institute for Tropical Agriculture (IITA), Ibadan, Nigeria.

M. Louis Sauger, Director General, Institut Sénégalais de Recherches Agricoles (ISRA), Dakar, Senegal.

Dustin Spencer, Head, Development Department, West Africa Rice Development Association (WARDA), Monrovia, Liberia.

FAR EAST

Y. Hayami, Professor of Economics, Tokyo Metropolitan University, Tokyo, Japan.

J.C. Moomaw, Director, Asian Vegetable Research and Development Center (AVRDC), Shanhua, Taiwan, Republic of China.

Dong Wan Shin, Korea Rural Economics Institute, Office of Rural Development, College of Agriculture, Seoul National University, Seoul, Korea.

SOUTHERN ASIA

S.L. Chowdhury, Project Director, All India Coordinated Research Project for Dryland Agriculture (IDFP), Hyderabad, India.

Murray D. Dawson, Agricultural Research Program, Ministry of Agriculture, Joydapour, Bangladesh.

Wayne Freeman, Director, International Agricultural Development Service-Integrated Cereals Project (IADS-ICP), Department of Agriculture, Kathmandu, Nepal.

Zahidul Hoque, Bangladesh Agricultural Rice Institute (BARI), Bangladesh.

N.K. Jain, Director, Indian Agricultural Research Institute (IARI), New Delhi, India.

J. Kampen, Acting Head, Farming Systems Research Group, International Crops Research Institute for the Semi-Arid Tropics (ICRISAT), Hyderabad, India.

M. Rahman, Bangladesh Agricultural Research Council (BARC), Dacca, Bangladesh.

Dr. Seethanaman, Director, All India Coordinated Rice Improvement Program (AICRIP), Hyderabad, India.

M.S. Swaminathan, Director General, Indian Council of Agricultural Research (ICAR), New Delhi, India.

MIDDLE EAST

Harry S. Darling, Director General, International Center for Agricultural Research in the Dry Areas (ICARDA), Beirut, Lebanon.

SOUTHEAST ASIA

Kawi Chutikul, Dean of Agriculture, University of Khon Kaen, Khon Kaen, Thailand.

A. Gomez, Team Leader, Multiple Cropping Program, University of the Philippines at Los Banos (UPLB), Los Banos, Philippines.

Jerry L. McIntosh, Cropping Systems Agronomist, Cooperative Central Research Institute for Agriculture/International Rice Research Institute (CRIA/IRRI), Bogor, Indonesia.

Banpot Na Pompeth, Assistant Dean of Agriculture, Kasetsart University, Bangkok, Thailand.

Manu Seetisarn, Dean of Agriculture, University of Chiang Mai, Chiang Mai, Thailand.

Dr. Tamin, Director, The Malaysian Agricultural Research and Development Institute (MARDI), Serdang, Malaysia.

Hubert G. Zandstra, Head, Cropping Systems Research Program, International Rice Research Institute (IRRI), Los Banos, Philippines.

AUGUST 1979 WORKSHOP

The three-day workshop held in Fort Collins, Colorado, during August 1979, gave the project staff the opportunity to (1) discuss major FSR&D issues, (2) consider the book of guidelines' contents and method of presentation, and (3) obtain suggestions for conducting the project. A distinguished group of FSR&D practitioners, who represented an important part of the FSR&D effort around the world, attended. This meeting was helpful in setting our project on a proper course and in establishing working relationships with many of those actively engaged in FSR&D. Attendees were

FSR&D SPECIALISTS

Suryatna Effendi, Director, Bogor Research Institute for Food Crops, Bogor, Indonesia.

Robert D. Hart, Crop Ecologist, Tropical Agricultural Research and Training Center (CATIE), Turrialba, Costa Rica.

Richard R. Harwood, Director of Research, Rodale Press, Inc., Emmaus, Pennsylvania.

Peter E. Hildebrand, Agricultural Economist, Rockefeller Foundation Adviser to the Agricultural Science and Technology Institute (ICTA), Guatemala City, Guatemala.

Bert A. Krantz, Emeritus Soils Specialist, University of California at Davis, Davis, California.

Jerry L. McIntosh, Cropping Systems Agronomist, Cooperative Central Research Institute for Agriculture/International Rice Research Institute (CRIA/IRRI), Bogor, Indonesia.

David W. Norman, Professor of Economics, Economics Department, Kansas State University, Manhattan, Kansas.

Donald L. Plucknett, Deputy Executive Director, Board for International Food and Agricultural Development (BIFAD), United States Agency for International Development (USAID).

Donald L. Winkelmann, Director, Economics Program, International Maize and Wheat Improvement Center (CIMMYT), El Batan, Mexico.

Hubert G. Zandstra, Head, Cropping Systems Research Program, International Rice Research Institute (IRRI), Los Banos, Philippines.

OBSERVERS

Bruce H. Anderson, Executive Director, Consortium for International Development (CID), Logan, Utah.

Jerry B. Eckert, Team Leader, Lesotho Agricultural Sector Analysis Project, Colorado State University, Maseru, Lesotho.

Kutlu Somel, Deputy Director, Economic and Social Research Institute, Middle East Technical University, Ankara, Turkey.

A. Wayne Wymore, Professor of Systems and Industrial Engineering, University of Arizona, Tucson, Arizona.

USAID MONITOR

J. Kenneth McDermott, Associate Director, Office of Agriculture, Bureau for Development Support, United States Agency for International Development (USAID), Washington, D.C.

ADVISORY COMMITTEE

Gerald M. Burke, Assistant Academic Vice President, New Mexico State University, Las Cruces, New Mexico.

Frank S. Conklin, Professor, Department of Agricultural and Resource Economics, Oregon State University, Corvallis, Oregon.

William Furtick, Dean, College of Tropical Agriculture and Human Resources, University of Hawaii, Honolulu, Hawaii.

Jack Keller, Department Head, Agricultural and Irrigation Engineering, Utah State University, Logan, Utah.

James R. Meiman, Dean, Graduate School, Colorado State University, Fort Collins, Colorado.

Martin Waananen, Assistant Director of Resident Instruction, College of Agriculture, Washington State University, Pullman, Washington.

PROJECT STAFF

Jen-hu Chang, Professor of Geography, Department of Geography, University of Hawaii, Honolulu, Hawaii.

Ann Perry-Barnes, Research Assistant, Department of Agriculture and Resource Economics, University of Hawaii, Honolulu, Hawaii.

Perry F. Philipp, Professor of Agricultural Economics, Department of Agricultural and Resource Economics, University of Hawaii, Honolulu, Hawaii.

Michael D. Read, Graduate Research Assistant, Department of Agronomy, Colorado State University, Fort Collins, Colorado.

John S. Roecklein, Graduate Research Assistant, Department of Agricultural and Resource Economics, University of Hawaii, Honolulu, Hawaii.

Willard R. Schmehl, Professor of Agronomy, Department of Agronomy, Colorado State University, Fort Collins, Colorado.

Willis W. Shaner, Associate Professor, College of Engineering, Colorado State University, Fort Collins, Colorado.

Tom S. Sheng, Research Associate, Industrial Engineering Program, Colorado State University, Fort Collins, Colorado.

JUNE 1980 WORKSHOP

This two-day workshop held during June 1980 allowed a small group of FSR&D specialists to meet with project staff to review the first draft of the book of guidelines and offer suggestions for subsequent drafts. The invited specialists have impressive practical experience in developing FSR&D concepts and procedures. Consequently, they provided valuable inputs during the early writing stage. Those attending were

FSR&D SPECIALISTS

Peter E. Hildebrand, Visiting Professor, Food and Resource Economics Department, University of Florida, Gainesville, Florida.

David W. Norman, Professor of Economics, Economics Department, Kansas State University, Manhattan, Kansas.

Hubert G. Zandstra, Head, Cropping Systems Research Program, International Rice Research Institute (IRRI), Los Banos, Philippines.

CONSULTANT

George M. Beal, Research Associate, Communication Institute, East-West Center, Honolulu, Hawaii.

USAID MONITOR

J. Kenneth McDermott, Associate Director, Office of Agriculture, Bureau for Development Support, United States Agency for International Development (USAID), Washington, D.C.

PROJECT STAFF

Jen-hu Chang, Professor of Geography, Department of Geography, University of Hawaii, Honolulu, Hawaii.

Robert E. Dils, Professor Emeritus, College of Forestry and Natural Resources, Colorado State University, Fort Collins, Colorado.

Gary E. Hansen, Research Associate, Resource Systems Institute, East-West Center, Honolulu, Hawaii.

Helen Kreider Henderson, Lecturer in Anthropology, University of Arizona, Tucson, Arizona.

James R. Meiman, Dean, Graduate School, Colorado State University, Fort Collins, Colorado.

Perry F. Philipp, Professor Emeritus, Department of Agricultural and Resource Economics, University of Hawaii, Honolulu, Hawaii.

Michael D. Read, Graduate Research Assistant, Department of Agronomy, Colorado State University, Fort Collins, Colorado.

Willard R. Schmehl, Professor of Agronomy, Department of Agronomy, Colorado State University, Fort Collins, Colorado.

Willis W. Shaner, Associate Professor, College of Engineering, Colorado State University, Fort Collins, Colorado.

Tom S. Sheng, Research Associate, Industrial Engineering Program, Colorado State University, Fort Collins, Colorado.

Howard H. Stonaker, Livestock Consultant, Fort Collins, Colorado.

Derrick J. Thom, Associate Professor, Department of History and Geography, Utah State University, Logan, Utah.

Thomas F. Trail, Staff Development Specialist, Residential Instruction and Cooperative Extension Service, College of Agriculture, Washington State University, Pullman, Washington.

Donald E. Zimmerman, Assistant Professor, Department of Technical Journalism, Colorado State University, Fort Collins, Colorado.

AUGUST 1980 PRETESTING

We pretested the second draft of the book with a small group of potential users—both representatives of five developing countries and their expatriate advisers. Besides project staff and the USAID monitor, others interested in FSR&D also attended. The purpose of the review was to obtain reactions to and suggestions for the book from a group typical of potential users. We spent the first week at New Mexico State University reviewing the second draft, chapter-by-chapter. During the second week, the group visited the farming systems programs in Guatemala and Honduras. Trips to these two countries offered participants the opportunity to compare the book with actual situations. The reactions of these potential users helped us gain considerable insight into how to make the book more usable.

REVIEWERS

Robert O. Butler, Chief of Party, Farming Systems Project, Washington State University, Maseru, Lesotho.

George Bassili Hanna, Chairman, Agricultural Engineering Department, Cairo University, Cairo, Arab Republic of Egypt.

David W. James, Professor, Department of Soil Science and Biometeorology, Utah State University, Logan, Utah.

Deran Markarian, Chief of Party, Yemen Sorghum and Millet Project, University of Arizona, Sanaa, Yemen Arab Republic.

Bruno J. Ndunguru, Senior Lecturer and Head, Department of Crop Science, University of Dar es Salaam, Morogoro, Tanzania.

Winston P. Ntsekhe, Director, Research Division, Ministry of Agriculture, Maseru, Lesotho.

Mohamed H. Sharaf-Aldin, Director General, Department of Agriculture Services, Ministry of Agriculture, Sanaa, Yemen Arab Republic.

Donald Sungusia, Deputy Head, Extension and Technical Support, Ministry of Agriculture, Dar es Salaam, Tanzania.

Hassan Wahby-Aly, Director, Egypt Water Use Management Project (EWUP), Ministry of Irrigation, Cairo, Arab Republic of Egypt.

Edgar Zapata C., Chief of Research Unit, Bolivian Institute of Agricultural Technology, Ministry of Agriculture, La Paz, Bolivia.

HOSTS

Mario Contreras, Professor, Panamerican Agricultural School, Zamorano, Honduras.

S.B. Langham, Associate Director/Program Coordinator, ISAI/YEMEN Project, New Mexico State University, Las Cruces, New Mexico.

National Program for Agricultural Research Staff, Ministry of Natural Resources, Tegucigalpa, Honduras.

Region 1 Staff, Agricultural Science and Technology Institute (ICTA), Quezaltenango, Guatemala.

Jamie Solorzano, Director, Region 1, Agricultural Science and Technology Institute (ICTA), Quezaltenango, Guatemala.

Robert K. Waugh, Research Management and Policy Adviser (assigned by the Rockefeller Foundation), Ministry of Natural Resources, Tegucigalpa, Honduras.

OBSERVERS

John H. Foster, Project Manager, Agricultural Credit and Vegetable Cooperative Projects, USAID Mission to Egypt, Cairo, Arab Republic of Egypt.

Wilmer Harper, Assistant Professor, Department of Agricultural Economics and Agricultural Business, New Mexico State University, Las Cruces, New Mexico.

Larry Harrington, Economist, Economics Program, International Maize and Wheat Improvement Center (CIMMYT), El Batan, Mexico.

John D. Hyslop, Technical Assistance Officer, Office of International Cooperation and Development, United States Department of Agriculture (USDA), Washington, D.C.

Jerry L. McIntosh, Cropping Systems Agronomist, Cooperative Central Research Institute for Agriculture/International Rice Research Institute (CRIA/IRRI), Bogor, Indonesia.

CONSULTANTS

George M. Beal, Research Associate, Communication Institute, East-West Center, Honolulu, Hawaii.

Ramiro Ortiz D., former Technical Director, Agricultural Science and Technology Institute (ICTA), Guatemala City, Guatemala; presently Graduate Student, University of Florida, Gainesville, Florida.

USAID MONITOR

J. Kenneth McDermott, Associate Director, Office of Agriculture, Bureau for Development Support, United States Agency for International Development (USAID), Washington, D.C.

PROJECT STAFF

Robert E. Dils, Professor Emeritus, College of Forestry and Natural Resources, Colorado State University, Fort Collins, Colorado.

Gary E. Hansen, Research Associate, Resource Systems Institute, East-West Center, Honolulu, Hawaii.

Helen Kreider Henderson, Lecturer in Anthropology, University of Arizona, Tucson, Arizona.

James R. Meiman, Dean, Graduate School, Colorado State University, Fort Collins, Colorado.

Perry F. Philipp, Professor Emeritus, Department of Agricultural and Resource Economics, University of Hawaii, Honolulu, Hawaii.

Michael D. Read, Graduate Research Assistant, Department of Agronomy, Colorado State University, Fort Collins, Colorado.

Willard R. Schmehl, Professor of Agronomy, Department of Agronomy, Colorado State University, Fort Collins, Colorado.

Willis W. Shaner, Associate Professor, College of Engineering, Colorado State University, Fort Collins, Colorado.

Tom S. Sheng, Research Associate, Industrial Engineering Program, Colorado State University, Fort Collins, Colorado.

Howard H. Stonaker, Livestock Consultant, Fort Collins, Colorado.

Thomas F. Trail, Staff Development Specialist, Residential Instruction and Cooperative Extension Service, College of Agriculture, Washington State University, Pullman, Washington.

Donald E. Zimmerman, Assistant Professor, Department of Technical Journalism, Colorado State University, Fort Collins, Colorado.

REVIEWERS OF THE THIRD DRAFT

Those who reviewed all or major portions of the third draft and submitted detailed and very helpful comments were

Frank S. Conklin, Professor, Department of Agricultural and Resource Economics, Oregon State University, Corvallis, Oregon.

Cornelius de Haan, Director of Research, International Livestock Centre for Africa (ILCA), Addis Ababa, Ethiopia.

T. Scarlett Epstein, Director, Action-Oriented Study of the Role of Women in Rural Development, School of African and Asian Studies, University of Sussex, Brighton, England.

Richard R. Harwood, Director of Research, Rodale Press, Inc., Emmaus, Pennsylvania.

Helen Kreider Henderson, Lecturer in Anthropology, University of Arizona, Tucson, Arizona.

Peter E. Hildebrand, Visiting Professor, Food and Resource Economics Department, University of Florida, Gainesville, Florida.

John D. Hyslop, Technical Assistance Officer, Office of International Cooperation and Development, United States Department of Agriculture (USDA), Washington, D.C.

David W. Norman, Professor of Economics, Economics Department, Kansas State University, Manhattan, Kansas.

David Nygaard, Leader, Farming Systems Research Program, International Center for Agricultural Research in the Dry Areas (ICARDA), Beirut, Lebanon.

David D. Rohrbach, Economic Policy and Planning Division, Office of Agriculture, Bureau for Development Support, United States Agency for International Development (USAID), Washington, D.C.

Derrick J. Thom, Associate Professor, Department of History and Geography, Utah State University, Logan, Utah.

Thomas F. Trail, Staff Development Specialist, Residential Instruction and Cooperative Extension Service, College of Agriculture, Washington State University, Pullman, Washington.

Warren H. Vincent, TP Consultant, Integrated Agricultural Production and Marketing Project, Kansas State University Office, Diliman, Quezon City, Philippines.

Robert K. Waugh, Research Management and Policy Adviser (assigned by the Rockefeller Foundation), Ministry of Natural Resources, Tegucigalpa, Honduras.

Donald E. Zimmerman, Assistant Professor, Department of Technical Journalism, Colorado State University, Fort Collins, Colorado.

Other reviewers who also contributed important suggestions were

Chris D.S. Bartlett, c/o British Council Division, Calcutta, India.

George M. Beal, Research Associate, Communication Institute, East-West Center, Honolulu, Hawaii.

Gerald M. Burke, Assistant Academic Vice President, New Mexico State University, Las Cruces, New Mexico.

Kathleen Cloud, Project Director, Women and Food Information Network, Office of International Agriculture, University of Arizona, Tucson, Arizona.

Michael Collinson, Farm Management Specialist, International Maize and Wheat Improvement Center (CIMMYT), Regional Office in East Africa, Nairobi, Kenya.

Robert E. Dils, Professor Emeritus, College of Forestry and Natural Resources, Colorado State University, Fort Collins, Colorado.

Gary E. Hansen, Research Associate, Resource Systems Institute, East-West Center, Honolulu, Hawaii.

Larry Harrington, Economist, Economics Program, International Maize and Wheat Improvement Center (CIMMYT), El Batan, Mexico.

Bert A. Krantz, Emeritus Soils Specialist, University of California at Davis, Davis, California.

Max K. Lowdermilk, Associate Professor of Sociology/Social Work, Department of Sociology, Colorado State Univesity, Fort Collins, Colorado.

Gerald Matlock, Director, Office of International Agricultural Programs, University of Arizona, Tucson, Arizona.

Jerry L. McIntosh, Cropping Systems Agronomist, Cooperative Central Research Institute for Agriculture/International Rice Research Institute (CRIA/IRRI), Bogor, Indonesia.

Donald L. Plucknett, Scientific Adviser, Consultative Group for International Agricultural Research (CGIAR), Washington, D.C.

John S. Roecklein, Graduate Research Assistant, Department of Agricultural and Resource Economics, University of Hawaii, Honolulu, Hawaii.

Kenneth H. Shapiro, Michigan State University, East Lansing, Michigan.

Kathleen A. Staudt, Assistant Professor, Department of Political Science, The University of Texas at El Paso, El Paso, Texas.

Richard L. Tinsley, Associate Professor, Department of Agronomy, Colorado State University, Fort Collins, Colorado.

Darlene A. Townsend-Moller, Assistant to the Director of Cooperative Extension and Associate Director of the Lesotho Farming Systems Project, Washington State University, Pullman, Washington.

C. Peairs Wilson, Kansas State University Team Leader, Integrated Agricultural Production and Marketing Project, Kansas State University Office, Diliman, Quezon City, Philippines.

APPENDIX TO CHAPTER 2

APPENDIX 2-A
MATHEMATICAL MODELING

Small-scale farming, especially subsistence farming in developing countries, is complex because (1) farm households typically engage in many enterprises, (2) interactions among enterprises are closely linked, and (3) the households have varied objectives. When a system is complex, researchers and others have many opportunities for suggesting improvements to the system. However, identifying optimal solutions, even when data are adequate, is beyond the capabilities of simple, straightforward approaches. Under such conditions, mathematical modeling aids researchers by helping them to (1) understand how the system functions, (2) identify key factors in the system, (3) consider alternatives for improvement, and (4) search for optimal or substantially improved solutions.

As noted in Sec. 2.7.2. of Chapter 2, this book contains little on mathematical modeling. We omitted modeling because we found little to indicate that mathematical modeling in FSR&D at the national level will become widespread in the near future. Whatever optimization work we encountered was being conducted mainly by other organizations for other purposes. Closely associated with mathematical modeling is the researcher's use of schematic or conceptual models. These do have applicability for this book, which we discuss in Sec. 5.4.3. and in Appendix 5-J.

Although mathematical modeling does not have much application to national level FS&RD efforts presently, we believe that it will as FSR&D becomes better established and as FSR&D teams become more sophisticated in their approaches. With this possibility in mind, we will report briefly on some modeling work at IRRI, ILCA, and CATIE and then list some of the studies reported in the literature.

MODELING AT IRRI

Our review of IRRI's modeling efforts revealed work in two basic areas. First, IRRI uses simulation models of rice production to relate the influence of weather on (1) rice growth and yields and (2) pest and disease damage. IRRI also uses these models to estimate water available to plants based on rainfall and evaporation. These models are of the simple water balance, toposequence[1] water balance, single crop, and toposequence crop types. Even though IRRI is continually testing and refining these models, its staff reports that much work is still needed in making them more useful for rice research.

Second, IRRI is working with linear and goal programming as applied to whole-farm analyses of experiments with rice farms in some Southeast Asian countries. When these whole farm procedures are verified and implemented, they are expected to help central research institutes identify new technologies and adjust existing technologies to farmers' systems.

MODELING AT ILCA

ILCA has made modeling of livestock systems one of the cornerstones of its activities. It uses modeling in four basic ways. First, herd-level models are being used to study animal growth, fertility, and death rates in complex environments. Results from these models provide ILCA with information for making preliminary assessments of research and development options.

Second, ILCA is applying planning models to farming systems in which livestock is integrated with crop production. These linear programming models provide improved resource allocation alternatives at the farm and regional levels, including new technology possibilities in different regions of East Africa.

Third, ILCA plans to use national policy models for (1) regional evaluation of land use; (2) study of economic trends, such as future supply and demand for livestock and livestock products; and (3) evaluation of the terms of trade between the pastoral and cropping sectors.

Fourth, as part of its initial conceptualization of the systems with which it works, ILCA is using schematic models to trace out the development paths of alternative research strategies.

MODELING AT CATIE

Of CATIE's modeling activities, we are most familiar with Hart's work. He has taken the ecosystems concepts of Odum (1971) and applied them to farming

[1]Relates to the lateral flow of water through a series of adjacent fields in a way that facilitates growing crops in sequence (Angus and Zandstra, 1979).

Figure 2-A-1. A generalized qualitative model of a farm system with socioeconomic and agroecosystem subsystems and inputs, outputs, and between-subsystem flows of money, materials, energy and information (Hart, 1980a).

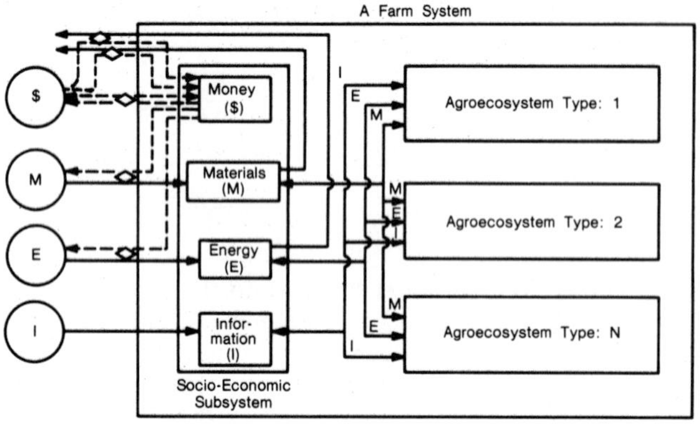

systems. Hart's approach involves breaking systems into subsystems and tracing through the flows of money, materials, energy, and information, as shown in Fig. 2-A-1. The boxes within the farm system are the household (labeled socioeconomic subsystem) and the cropping and livestock systems (labeled agroecosystem types 1, 2, . . . n).

One of Hart's (1980a) activities at CATIE was to use such a model as a basis for gathering weekly data for an entire year on the activities of a farm family in Honduras. The results of this study were used to identify the farming system's main components and the physical and money flows of the system. Hart has also used the same approach as part of CATIE's training program (Hart, 1979 and 1980b).

To our knowledge, Hart has not applied simulation or optimization procedures to these data. Nevertheless, the data have allowed Hart and others to gain considerable knowledge about the structure and money and material flows of the farming systems under study. More will be said about Hart's work in Appendix 5-J, which is on conceptual models.

REFERENCES ON FSR&D MODELING

We close this appendix with a listing of some of the modeling activities that may interest researchers in FSR&D. The list is for readers who wish to gain an idea about the types of studies and information currently available. We have broken the list into major topics and have included a few general references for the reader who is not familiar with mathematical modeling's terms or concepts.

AERIAL RECONNAISSANCE

- *Landsat Computers and Development Projects* (Adrien and Bumgardner, 1977)
- *LACIE: A Proof-of-Concept Experiment in Global Crop Monitoring* (MacDonald and Hall, 1977)

CLIMATE

- *Climatic Factors and the Modeling of Rice Growth and Yield* (Angus and Zandstra, 1979)
- *A Model for Predicting Soybean Yields from Climatic Data* (Hill et al., 1978)
- *Simulation of Influence of Climatic Factors on Rice Production* (Kenlen, 1976)

ECONOMICS

- *Measuring the Economic Benefits of New Technologies to Small Rice Farmers* (Barlow et al., 1979)
- *New Cropping Patterns for Iloilo and Pangasinan Farmers: A Whole Farm Analysis* (Jayasuriya, 1979)
- *Optimal Multiple Cropping Systems for the Chiang Mai Valley* (Thodey and Sektheera, 1974)
- *Economics and Design of Small-Farmer Technology* (Valdes et al., 1979)

EVAPOTRANSPIRATION

- *Model for Predicting Evaporation from a New Crop with Incomplete Cover* (Ritchie, 1972)
- *Modeling of Potential Evapotranspiration and Solar Radiation for Different Regions of the Philippines* (Tamisin et al., 1979)

LINEAR PROGRAMMING

- *Linear Programming Applications to Agriculture* (Beneke and Winterboer, 1973)
- *Linear Programming Methods* (Heady and Candler, 1958)
- *A Farm Level Linear Programming Analysis of Dry-Land and Wet-Land Food Crop Production in Indonesia* (McCarl, 1978)
- *Linear Programming for Smallholder Agriculture* (Nelson, 1974)

LIVESTOCK

- *Initial Application of Modeling Techniques in Livestock Production Systems Under Semi-Arid Conditions in Africa* (Anderson and Trail, 1978)
- *A General Cattle Production Systems Model. I: Structure of the Model and II: Procedures Used for Stimulating Animal Performance* (Sanders and Cartwright, 1979a and 1979b)

MISCELLANEOUS

- *Incorporating Multiple Objectives into Plans for Low-Resource Farmers* (Flinn and Jayasuriya, 1979)
- *One Farm System in Honduras: A Case Study* (Hart, 1980a)

- *Operations Research* (Hillier and Lieberman, 1974)
- *Rice and Risk: Decision-Making Among Low-Income Farmers* (Roumasset, 1976)
- *Rational Farm Plans for Land Settlement in Indonesia: A Study Using Programming Techniques* (Wardhani, 1976)

SIMULATION

- *A Primer on Simulation and Gaming* (Barton, 1970)
- *Small Farm Improvement Strategies: The Implications of a Simulation Study of Indigenous Farming in South East Ghana* (Low, 1975)

APPENDIX TO CHAPTER 3

APPENDIX 3-A
REVIEW OF FSR&D ACTIVITIES

In this appendix, which augments our discussion in Chapter 3, we summarize some of the more pertinent facets of current FSR&D activities under the headings of national and international FSR&D programs. The national programs are the most relevant for this book, but those international programs that are actively cooperating with national governments in developing programs and activities appropriate to their needs are relevant too.

Additional details on these programs may be found throughout this book. Also, recent reviews of FSR&D programs throughout the world are contained in the Technical Advisory Committee (TAC) report (1978) and Gilbert et al. (1980). References for some of the national programs are: for Brazil, EMBRAPA (1978); for Guatemala, ICTA (1976); for Honduras, PNIA (1981); for Indonesia, CRIA (1979); for Nepal, His Majesty's Government Department of Agriculture (1979); and for Senegal, ISRA/GERDAT (1977). Additional references for some of the international centers are CATIE (1979), CGIAR (1976), CIAT (1980), CIMMYT (1978), CIP (1977), ICARDA (1978), ICRISAT (1979), IITA (1978), ILCA (1980), IRRI (1977a and 1977b), and WARDA (1978).

NATIONAL FSR&D PROGRAMS

Several national programs have been initiated during the past decade. Two of these are the programs of the Agricultural Science and Technology Institute (ICTA) of Guatemala and the Senegalese Institute for Agricultural Research (ISRA). Other important FSR&D efforts are by countries that are members of the International Rice Research Institute's (IRRI) Asian Cropping Systems Network (ACSN) headquartered in the Philippines. Recently, additional FSR&D programs have been established while other programs are being planned and still others have elements of FSR&D in their approaches.

Agricultural research in Guatemala was reorganized in 1973 to bring improved technologies to the small-scale farmers throughout the country. The program is coordinated out of Guatemala City with offices in seven of the eight regions of the country. Programs are developed within the regions under the direction of regional directors and the general guidance of the technical director and program coordinators from the central office. Using rapid surveys, ICTA has developed an effective means for stratifying groups of farmers and identifying their problems. The emphasis is on on-farm experimentation and evaluation that relies heavily on understanding farmers' conditions and applying technologies previously developed on experiment stations.

Among the oldest of the national FSR&D programs is that involving the Experimental Units (*Unités Expérimentales* in French), which are part of the ISRA program in Senegal. Twenty villages were selected at two sites for studying farm systems and identifying constraints to the acceptance of new innovations, technologies, and practices being developed at nearby experiment stations. Interdisciplinary teams of researchers work with farmer cooperatives and extension representatives to evaluate farmers' problems and responses to change. The program's key is its intensification of agriculture through land improvement that embodies land consolidation, improved cultivation through deep plowing using animal traction, phosphate replacement, fertilizers, improved seed, diversification of crops, four-year rotations, incorporation of organic matter into the soil, and agroforestry. This approach takes advantage of the limited rainfall, maintains soil fertility, and raises overall productivity.

Member countries of the ACSN, organized in 1974, have several programs with an FSR&D emphasis. The Network stresses systematic and integrated research on specific problems of small farmers in various target areas. One of the older and more active of these is the National Multiple Cropping program of the Central Research Institute for Agriculture (CRIA) in Indonesia. The program started with locations in West Java and Central Lampung as the target areas. The objective of CRIA's program is to identify and remove constraints to more intensive cropping patterns using combinations of improved component technologies and new or modified cropping patterns. The results will help the government to resettle farming familes in less crowded areas.

A more recent FSR&D program is that of the Ministry of Natural Resources in Honduras. This program centers around an interdisciplinary group of graduates from Cornell University, who completed coordinated doctoral dissertations on tropical maize production.

In 1977 Nepal began its integrated cereals project that follows a farming systems approach.

EMBRAPA (Brazilian Agricultural Research Corporation) is using interdisciplinary teams at Petrolina, in the

Northeast, to identify appropriate cropping systems for semiarid areas.

Only recently, several other FSR&D projects have been integrated or are being planned. In Lesotho, team leaders are emphasizing the extension component. Mali is receiving funds from several donors for an FSR&D project that will be initiated in the southern part of the country. The United Nations Food and Agriculture Organization (FAO) is assisting the Egyptian government in the development of an FSR&D program in that country. Upper Volta has a farming systems unit that is coordinating its efforts with the regional activities of the SAFGRAD (Semi-arid Food Grain Research and Development) project. Kenya and Zambia have initiated training programs in response to the International Maize and Wheat Improvement Center's (CIMMYT) efforts to integrate farming systems concepts into that region; and Tanzania is considering a large program in FSR&D. Swaziland is designing an FSR&D project. And the government of Gambia has requested an FSR&D project. Since our review is not comprehensive, other FSR&D activities are, undoubtedly, underway or are being planned.

INTERNATIONAL FSR&D PROGRAMS

Several of the International Agricultural Research Centers (IARCs) and at least one regional center have been active in FSR&D. IARCs were given a material boost by the findings, recommendations, and CGIAR (Consultative Group on International Agricultural Research) acceptance of the TAC report. This report strongly supported continued efforts of the IARCs in FSR&D. IRRI, ICRISAT (International Crops Research Institute for the Semi-arid Tropics), IITA (International Institute for Tropical Agriculture), ILCA (International Livestock Centre for Africa), and ICARDA (International Center for Agricultural Research in the Dry Areas) have farming systems in their mandates; CIMMYT is following a commodity approach to farming systems; CIP (International Potato Center) is carrying out some aspects of FSR&D; and CIAT (International Center for Tropical Agriculture) formerly had a farming systems program, but has absorbed these activities into its commodity programs. Many of these centers provide valuable training for those from the national programs. This training ranges from cropping systems to commodity and disciplinary research—both on-farm and on-station.

The program with the most far-reaching influence on national programs is the cropping systems research of IRRI. Ten countries have been organized into the Asian Cropping Systems Network to promote national programs capable of absorbing improvements generated both by IRRI and by the interactions of the member countries. In addition to IRRI's manuals (e.g., Zandstra et al., 1981), publication of the proceedings of its Saturday Seminars and the Network's workshops are also very useful in understanding the intricacies of cropping systems research.

The programs of ICRISAT and IITA come closer to the conceptualization of "farming systems in the large" (see Sec. 3.4.1.) in that they cover the range of small-farmers' activities, including preservation of their environments. ICRISAT has focused on means of capturing and preserving rainfall so as to obtain higher yields on the main crop and open up the possibilities for a second crop. These practices include, among other things, improved tillage, use of broadly graded beds and furrows, and surface water storage. Such efforts could lead to a substantial alteration to the system, because farmers in the semiarid tropics customarily have enough moisture for only one crop a year.

IITA has concentrated on developing a replacement for the present system of shifting cultivation in the humid and subhumid tropics. The focus of its efforts along these lines includes minimum and no tillage, ground covers, alternative means of clearing, improved implements, multiple cropping, and maintenance of soil fertility. The traditional systems are being threatened by population increases and the pressure to reduce the periods of fallow, with the accompanying problems of erosion and fertility loss.

Both the approaches of ICRISAT and IITA attempt to increase production substantially more than if farmers follow their present patterns and trends. As a consequence, the programs are comprehensive and long-run in scope. Most of the work has been confined to on-station research because, as they put it, they are still attempting to learn the technical relationships among the system's components. Neither program, however, has gone far in studying livestock or mixed-farming systems.

CIMMYT, through the lead of its Economics Program, is using a cropping systems approach in finding ways to improve the efficiencies of wheat-based and maize-based systems. As long as these two commodities are an essential part of the small-scale farmers' cropping system, CIMMYT will assist in improving their overall cropping patterns and farm management practices. The approach of their regional offices, such as those in Ecuador and East Africa, takes the socioeconomic and physical environments about as they are. That is, the researchers operate within the constraints the farmers customarily face and try to improve factors under the farmer's control. On the national level, two CIMMYT manuals (Perrin et al., 1976 and Byerlee et al., 1980) are directed toward conditions specific to developing countries and are therefore particularly useful. In East Africa, CIMMYT has recently begun regularly publishing a farming systems newsletter.

The other international centers have been less active in FSR&D because of their late entry into the field, the complexity of the tasks, or the emphases of their mandates. ILCA's program centers on research applied to the basic livestock systems of Africa. Its program is in the initial stages and the ILCA staff is still concerned with methodological procedures. One procedure involves the monitoring of livestock projects to determine some of the institutional and social reasons for acceptance and rejection of alternative livestock technologies. ILCA's program includes the study of small ruminants in Nigeria, investigations of the cropping-pastoral situations in Mali, a study of trypanotolerant livestock across the center of Africa, modeling in Botswana, and experiments in Ethiopia.

ICARDA, which was established in 1976, has had a delayed start due to the unsettled conditions in the Middle East, where its principle activities are centered. Its overall efforts are directed toward working with the drier regions of Western Asia and Northern Africa. Of its several programs, FSR&D is the largest. ICARDA has decided to approach farming systems with few constraints on the analysis. That is, most everything is subject to investigation should the initial analysis so indicate. In pursuing this approach, ICARDA has undertaken a lengthy baseline study of conditions in its area of responsibility.

CIAT used to have a farming systems program as such, but has now placed its FSR&D efforts in the commodity programs for beans, beef, cassava, rice, and swine.

After significantly improving potato varieties and management practices, CIP has begun on-farm experiments on how to persuade farmers to accept these improvements. CIP began research in 1977 that resembles CIMMYT's approach in concentrating on a commodity within the farmer's cropping system.

Finally, for the group financed by CGIAR, the West Africa Rice Development Association (WARDA) is coordinating and supporting technically some on-farm studies. This work is still formative. The extent to which significant farming systems programs emerge—except for Senegal and Upper Volta—remains to be seen.

The regional program of the Tropical Agricultural Research and Training Center (CATIE), headquartered in Costa Rica, is conducting both fundamental and applied research in FSR&D. CATIE staff is exploring fundamental concepts in FSR&D, such as ecological systems, case studies of farm families, soil texture gradients, climatological analogs, multiple cropping, and plant efficiency in use of energy. CATIE also conducts non-degree and degree training at the master's level and directly assists its Central American members in implementing FSR&D programs.

APPENDIXES TO CHAPTER 4

APPENDIX 4-A
ILLUSTRATION OF THE USE OF GENERAL FARM DATA TO HELP ESTABLISH RECOMMENDATION DOMAINS

Byerlee et al. (1980) define a recommendation domain as a group of farmers with about the same characteristics and environmental conditions for whom similar recommendations can be made (see Sec. 4.2.1. in Chapter 4). Below, is an illustration of the use of general farm data to help in distinguishing among different recommendation domains. Table 4-A-1 (see p. 244) and the description are taken from Byerlee et al. (1980) without alteration:

> "The table shows an example of assembling descriptive statistics by recommendation domains in a tropical maize producing area. Initially three recommendation domains were distinguished a) large-scale farmers, b) farmers of the land reform program growing maize on flat land and c) farmers of the land reform program growing maize on sloped land. The research program decided to focus on the latter two groups where potential pay-offs in terms of production and income equality were greater. The basic difference in practices of farmers is seen to be in land preparation where farmers with flat land generally use tractors while farmers on steep land use hoes or simply slash with a cutlass. Other practices are essentially the same for both recommendation domains. The base practices for on-farm experiments would then consist of tractor ploughing and harrowing on flat land and hoeing on steep land, the local variety planted at a density of 35,000 plants/ha, weeded by hand with the first weeding about four weeks after planting and use of insecticide but no fertilizer."

APPENDIX 4-B
GROUPING FARMERS INTO HOMOGENEOUS POPULATIONS

In Sec. 4.2.1. in Chapter 4, we discussed methods for identifying those within similar environments who farm in similar ways. In this book, we refer to similar environmental conditions as subareas. Others refer to farmers who follow similar practices and are alike in other ways as recommendation domains. Appendix 4-A shows how Byerlee et al. (1980) used a description of various farmers' practices to identify two recommendation domains. In this appendix, we provide further information on grouping farmers into relatively homogeneous populations as the basis for designing and recommending improved technologies.

This appendix reports on a situation in the Central Province of Zambia that involved approximately 350,000 people living within an area of about 120,000 sq km. The following sections are quoted from Collinson (1979):

> "The smallest administrative division in Zambia is the Ward. There were 72 Rural Wards (1975 boundaries) in Central Province averaging about 1000 households each. (Numbers varied widely, mainly within the 200 – 3000 range, about 50% were within ± 500 of the average.) Information collected at the ward level is aggregated into relatively few farmer groupings and the aggregation process smoothes out small inaccuracies arising in the ward level information. For programme implementation purposes the resulting groupings can be specified in terms of Districts and Wards, units already in use in agricultural planning and administration.
>
> "A questionnaire was developed to collect descriptive information about farming in the wards. The questionnaire is shown as [Table 4-B-1]. It sought to tap the experience of agricultural staff locally involved in day-to-day agricultural administration in the areas to be covered. Foreknowledge of their likely biases was used to try to ensure balanced information. Data collection was organised through the four District Agricultural offices in the Province. The questionnaire was administered by the research economists to Station Officers, that is agricultural extension staff, each in charge of 5 - 10 wards with several Camp Officers subordinate to them.
>
> "Before the survey proper the questionnaire was tested in two locations to evaluate the relevance of the questions and to improve the phrasing in putting the question to the respondent. During the survey proper the economist first discussed the wards with the District Agricultural Officer seeking information on:
>
> (1) The proportion of Traditional, Emergent and Large Scale farmers in the ward, a hierarchial

Table 4-A-1. Tabulation of farmer practices by recommendation domain — tropical maize (Byerlee et al., 1980).

General Farm Data	*Recommendation Domain* *Flat Land*	*Steep Land*
Average Farm Size (ha)	11.1	10.2
Area in Maize in August (ha)	4.6	2.6
Area in Tree Crops (ha)	3.5	2.7
Annual Cropping Pattern in Selected Field		
Per cent Maize-Maize	31	37
Per cent Maize-Maize-Beans	33	33
Per cent Maize-Squash-Maize	12	10
Per cent Other Systems	24	20
Land Preparation		
Per cent Plough-Harrow (with tractor)	38	0
Per cent Harrow Only (with tractor)	24	0
Per cent Hand Hoe	34	68
Per cent Chop Only or Chop and Burn	3	27
Planting		
Per cent Plant "Improved" Variety	18	3
Distance Between Rows (cm)	103	102
Distance Between Hills (cm)	92	94
Average Seeds per Hill	3.7	3.9
Per cent Replant	26	35
Weeding		
Per cent Weed with Horse or Tractor	15	3
Per cent Weed with Hoe	85	97
Per cent Weed Twice	83	80
Average Time of First Weeding (weeks after planting)	4.5	3.8
Other Inputs		
Per cent Apply Insecticides	86	82
Per cent Use Fertilizer	2	0
Production		
Average Yield (ton/ha)	1.2	1.1
Per cent Maize Sold	63	56

division already recognised by the Ministry of Agriculture.

(2) For each of these farmer categories:
  (a) The main power source used.
  (b) The approximate, typical area cultivated.
  (c) The main crops grown for food and for cash.

"This initial information was used as a check on Station Officers responding to the detailed questionnaire for each ward. Discrepancies were identified and taken up with the Station Officer concerned. If he stuck to his response the discrepancy was taken up with the District Agricultural Officer.

"Several approaches to data tabulation were tried to facilitate interpretation. A straightforward tabulation is essential. Its value for interpretation is enhanced if, as far as possible, wards which are contiguous on the ground are also continuous in the table. Some compromise is inevitable. The ordering of the ward data in the table is helped by drawing a grid on the ward maps. The wards are numbered as they are touched by the grid moving either North to South or East to West. Two other approaches to tabulations followed the identification of the major sources of variation between wards in different parts of the Province. First information on these major sources was written onto the map in each ward to help crystallise the boundaries between zones. The second approach was to build 'trees' on the main sources of variation. For example wards were first grouped on the basis of power source. These groups were then split on the basis of the main starch staple food. These sub groups were split on the basis of major cash sources and so on. The process is easiest if wards are first grouped on variables which have fewest categories. The process gets too complex after three or four variables have been considered but helps to improve the understanding of sources of variation which are related.

"The key step in interpretation is deciding the sources of variation which are critical in dictating resource allocation in farming systems of the area. Identifying these key variables reduces the collected information to manageable proportions. However, the key variables will vary from area to area. Due to the homogeneity in climate and altitude in Central

Table 4-B-1. Farm system zoning questionnaire, Central Province, Zambia (Collinson, 1979).

DISTRICT__________ WARD NO.__________ FARMER GROUP__________

A. ANIMALS KEPT BY *MOST* FARMERS	1. THREE MAIN TYPES OF ANIMALS KEPT	1	________
		2	________
		3	________
	2. IF CATTLE, MAIN PURPOSES FOR KEEPING	1	________
		2	________
		3	________
B. FOODS *GROWN (G)* OR *BOUGHT (B)* BY *MOST* FARMERS	1. STARCH STAPLES	1	________
		2	________
		3	________
	2. RELISH CROPS TO FLAVOUR STAPLES	1	________
		2	________
		3	________
	3. ANIMAL PRODUCTS FOR FOOD	1	________
		2	________
C. MAIN CASH SOURCES FOR *MOST* FARMERS (OVERALL RANK)	1. NEW CASH CROPS AND % GROWING	1	________
		2	________
	2. CROP SALES AS A CASH SOURCE	1	________
		2	________
	3. LIVESTOCK AS A CASH SOURCE	1	________
		2	________
	4. OFF FARM CASH SOURCE	1	________
		2	________
D. LAND USE AND METHODS AND TIME OF MOST FARMERS	1. YEARS CULTIVATED	1	________
	2. TYPICAL AREA (HA)	1	________
	3. MAIN METHODS OF LAND PREPARATION	1	________
		2	________
	4. MAIN MONTHS OF LAND PREPARATION	1	________
		2	________
E. HIRE AND PURCHASE OF RESOURCES BY MOST FARMERS	1. TYPES OF HIRED LABOUR & PAYMENT	1	________
		2	________
	2. WORK DONE BY HIRED LABOUR	1	________
		2	________
	3. MAIN INPUTS PURCHASED AND CROPS USING	1	________
		2	________
		3	________

Province differences in timing over the crop calendar were unimportant. In mountainous regions in particular, varying crop calendars and cropping opportunities would compound the number of identified groups.

"The cost of identifying homogeneous farmer groups for Central Province, using this methodology, are low. In terms of professional mandays three stages can be distinguished.

(a) Preparation: 6 - 8 mandays.
   (i) Developing and testing the questionnaire
   (ii) Arranging the programme of District visits
   (iii) Preparing background material and maps.
(b) Data collection: 6 - 10 mandays.
   Administering a questionnaire for some 100 Ward/farmer category combinations to some dozen station officers called in to their District Offices.
(c) Domain identification: 8 - 12 mandays.
   (i) Tabulation of the collected data (can be done by clerks)
   (ii) Interpretation of the data
   (iii) Deriving, describing and mapping Recommendation Domains.

"Some 20 - 30 professional mandays are required for the whole exercise. Allowing for the need to arrange Station Officers meetings, to travel to District Offices, and for delays in mapping, a turn around time of two months for some 100 enumeration units is feasible."

APPENDIX 4-C
SELECTION OF FSR&D AREAS

We continue in this appendix with the topic of area selection as presented in Sec. 4.2. of Chapter 4 and in Appendixes 4-A and 4-B. This appendix, quoted from CRIA (1979), outlines procedures for selecting target areas, subareas (sub-districts), and a research area (villages) for cropping systems research in Indonesia.

The team follows three steps in selecting a rural community as the research area:

> "First, a target area is identified which is a relatively homogenous agro-climatic area including several districts . . . and several thousand hectares. Then the Cropping Systems Research Coordinator must decide which edaphological conditions to study such as rainfed, . . . irrigated . . . (full, 7-9 months or 5 months irrigation), . . . and other conditions. Second, one or several sub-districts . . . are selected from among these districts that include a large area in the desired research environment. Next, one or more villages . . . characteristic of each desired environment are selected."

The following paragraphs describe target areas, sub-district selection, and village selection:

"TARGET AREAS

"The selection of target areas for cropping systems field research is based on four criteria. First, target areas are usually regions identified by the Government as priority agricultural development zones. Second, the area must be representative of a large agro-climatic zone so that the research results will have widespread applicability. Third, the environment must be of a type in which the research staff believes there exists agricultural technology that, with slight modifications, it will be possible to increase yields and cropping intensity. Finally, the target area must have some marketing and infrastructural development or is in the process of being developed.

"SUB-DISTRICT SELECTION

"In selecting the sub-districts . . . , the primary consideration is to identify an area which has a large number of hectares of the desired land use type. The research staff visits each [district] extension office and collects secondary data for each [sub-district] about the number of hectares in rainfed, technical irrigation, semi-technical irrigation, simple irrigation, annual crop upland, and perennial crop upland. Based on these data, the [sub-district] with the largest area of the desired land use type is selected.

"VILLAGE SELECTION

"The selection of the [villages] involves several considerations. The research staff visits each of the chosen [sub-districts] and collects from the extension office the secondary data listed in Table [4-C-1].

"Once the secondary data is collected, a matrix is prepared for each [sub-district], with the [village] forming the rows and the data forming the columns, as shown in Table [4-C-2].

"After transforming the [village] secondary data to the 'Data matrix' [Table 4-C-2], the mean value for each characteristic is calculated. These mean values taken together may be interpreted as a description of the 'typical or representative [village].' To identify the [village] which is most representative of the population of [villages], first the mean value for each characteristic is subtracted from the respective values associated with each [village

Table 4-C-1. Data required for systematic selection of rural communities as research areas in Indonesia (Adapted from CRIA, 1979).

Data	*Purpose*
Distance from main road (km)	To guarantee that the rural community is easily accessible
Area in each land use class (ha)	To permit the selection of rural communities with the largest area in the desired land use class
Relative area in each slope class (percent)	To avoid rural communities with atypical topography
Relative area in each soil texture (percent)	To avoid rural communities with atypical soils
Area planted to each crop, by month (percent)	To identify current production level
Population, by economic activity (number)	To determine importance of agricultural employment
Rainfall by month for past 10 years (mm)	To determine number of months 100 mm or more of rain and probability of less than 100 mm at beginning and end of cropping season
Participants (number) in the rice production program of the government	To determine the availability of credit and level of technology in the rural community
Months during which irrigation water is available (percent of area with less than 5, 6-7, 8-9, and 10-12 months of irrigation per year).	To identify areas according to irrigation regimes
Draft animal population (number)	To determine the availability of draft power
Tractor population (number)	To determine the availability of mechanical power

Table 4-C-2. Characteristics of sample rural communities in sub-district (Adapted from CRIA, 1979).

Village	Distance	Land Use (ha)			Soil (%)			Cropping (%)*			Yield (t/ha)*			Farmer	Bimas†(%)	Power
	(km)	Irrigated	Rainfed	Upland	Clay	Silt	Sand	LLR	C	CV	LLR	C	CV	popula-tion (%)	members	(ha/animal)
	(1)	(2)	(3)	(4)	(5)	(6)	(7)	(8)	(9)	(10)	(11)	(12)	(13)	(14)	(15)	(16)
1. Maritengae	6	600	5,000	700	55	30	15	60	30	10	3.0	0.7	6.7	75	45	10
2. Panca Rijang	10	4,000	1,000	600	50	20	30	70	20	15	2.8	0.5	5.4	63	33	15
3. Branti	15	8,000	2,000	1,000	90	5	5	80	15	5	4.1	1.3	10.6	81	68	6
4. Watang Pulu	7	3,000	100	2,000	75	13	12	68	25	7	3.4	0.8	8.4	68	60	21
5. Dua Putue	4	600	900	6,000	85	5	10	75	5	20	3.5	1.0	9.0	74	50	9
Mean	8	3,200	1,800	2,100	71	15	14	71	19	11	3.4	0.9	8.0	72	51	12

*LLR = lowland rice, C = maize, and CV = cassava.
†Bimas is the acronym for a production organization.

Table 4-C-3. Absolute deviation from the mean of each characteristic (Adapted from CRIA, 1979).

Village	Distance	Land Use (ha)			Soil (%)			Cropping (%)*			Yield (t/ha)*			Farmer	Bimas†(%)	Power
	(km)	Irrigated	Rainfed	Upland	Clay	Silt	Sand	LLR	C	CV	LLR	C	CV	popula-tion (%)	members	(ha/animal)
	(1)	(2)	(3)	(4)	(5)	(6)	(7)	(8)	(9)	(10)	(11)	(12)	(13)	(14)	(15)	(16)
1	2	2,600	3,200	1,400	16	15	1	11	11	1	0.4	0.2	1.3	3	6	2
2	2	800	800	1,500	21	5	16	1	1	4	0.6	0.4	2.6	9	18	3
3	7	4,800	200	1,100	19	10	9	9	4	6	0.7	0.4	2.6	9	17	6
4	1	200	1,700	100	4	2	2	3	6	4	0.0	0.1	0.4	4	9	9
5	4	2,600	900	3,900	14	10	4	4	14	9	0.1	0.1	1.0	2	1	3

*LLR = lowland rice, C = maize, and CV = cassava.
†Bimas is the acronym for a production organization.

Table 4-C-4. Rank-order of rural community characteristics for all rural communities in the sub-district sample (Adapted from CRIA, 1979).

Village No.	*Characteristic** (1)	(2)	(3)	(4)	(5)	(6)	(7)	(8)	(9)	(10)	(11)	(12)	(13)	(14)	(15)	(16)	*Total*
1	2	3	5	3	3	4	1	5	4	1	3	2	3	2	2	1	44
2	2	2	2	4	5	2	5	1	1	2	4	3	4	4	5	2	48
3	4	4	1	2	4	3	4	4	2	3	5	3	4	4	4	3	54
4	1	1	4	1	1	1	2	2	3	2	1	1	1	3	3	4	31
5	3	3	3	5	2	3	3	3	5	4	2	1	2	1	1	2	43

*See Tables 4-C-2 and 4-C-3.

[Table 4-C-3)]. This difference is the deviation from the mean for each characteristic. Next for each characteristic the [village] with the smallest deviation from the mean is assigned the value of 1, the [village] with the second smallest deviation is assigned the value 2, etc., until all [villages] have been ranked in terms of deviation from the mean [Table 4-C-4]. Finally, after ordering all [villages] for all characteristics each row (representing one [village]) is summed. This gives a single index value for each [village]. The [village] with the smallest index value will be most representative of the population of [villages]."

From Table 4-C-4, village No. 4 is the most representative one. Unless this village has some characteristic that precludes its selection, the team chooses it as the research area.

APPENDIX 4-D
CLIMATIC ZONES IN SOUTHEAST ASIA

IRRI (1974) distinguishes four major rainfall zones and several subzones in Southeast Asia; part of these areas are shown in Fig. 4-4 in Chapter 4 of this book. The IRRI researchers selected these zones and subzones according to their importance for growing rice. They explained their selection process as follows:

"By grouping available information on rainfall profiles together, several of the most prominent rainfall patterns were selected. Some countries like Indonesia and the Philippines have done this type of study (e.g., based on more than 4,000 sites in Indonesia, around 150 dominant rainfall profiles were selected). The first criterion used for selecting major climatic zones was monthly rainfall. An arbitrary boundary was set at 200 mm. This amount is based on two assumptions: (1) Losses due to evapotranspiration, although variable over the year, generally amount to around 100 mm per month; and (2) Losses due to percolation and seepage, although variable depending on soil characteristics, are generally set at around 100 mm per month.

"The second criterion was the number of months with 200 mm or more rainfall. An arbitrary boundary was set at five to nine consecutive wet months. If there are less than five consecutive wet months the possibilities of growing two crops are limited. If there are more than nine consecutive wet months the Southeast Asian farmer is most likely to grow two crops of puddled rice.

"Based on these criteria, the following zones were delineated.

"1) Zone I. Areas with more than nine consecutive wet months with more than 200 mm rainfall per month [Fig. 4-4 in Chapter 4]. This zone includes the major part of Kalimantan and East Malaysia, central Sumatra and along the northwest coast of Sumatra, and isolated spots in Java, primarily the southwestern part. In addition, isolated spots exist near the mountains. Eastern Mindanao, the Visayas, and Luzon in the Philippines come into this group.

"2) Zone II. Areas with five to nine consecutive wet months with more than 200 mm rainfall. This zone covers the major part of Sumatra, the western and central Java, the major part of West Malaysia, the southern and eastern part of Thailand, and the southern part of Burma. Eastern and central Luzon, the Visayas, and Mindanao in the Philippines come into this group. A major part of Vietnam and Laos are also classified as Zone II.

"Since this zone is of major interest for multiple cropping, it is divided into four subdivisions:

a) Zone II.1. Areas with five to nine consecutive wet months and with 100 to 200 mm rainfall per month during the remaining part of the year. . . . South Sumatra, Kalimantan, west Java, north Sulawesi, northeast Malaysia, southwest Malaysia, and Mindanao have this type of climate. Year-round cropping is possible with puddled rice.

b) Zone II.2. Areas with five to nine consecutive wet months and with 100 to 200 mm rainfall per month during the remaining part of the year and with another minor rainfall peak. . . . These areas are found only north of the equator but below 10° N (e.g., north Sumatra, north Borneo, northwest

Malaysia, and the southern tip of the Thailand peninsula). This zone is equally suitable for multiple cropping although farmers are likely to grow two crops of puddled rice . . . .

c) Zone II.3. Areas with five to nine consecutive wet months and with at least 2 months of less than 100 mm rainfall . . . . This area covers large parts of central and east Java, southern Thailand, eastern and southeast Thailand, southern Burma, and major parts of the Philippines.

d) Zone II.4. Areas in the southeast and east of Thailand are characterized by a sharp end to the rainy season; the dry season in these areas and in parts of Burma is very pronounced with virtually no rain during 2 to 3 months . . . . The other extreme is found on the west coast of Burma where, during the rainy season, peaks of over 1000 mm of rain per month occur.

"3) Zone III. Areas with two to five consecutive wet months. Although the wet season may often be too short to grow two crops, areas that receive 100 mm or more of rainfall per month during the dry season have been separated from those that receive less than 100 mm per month during the dry season:

a) Zone III.1. Areas with two to five consecutive wet months but with at least 100 mm rainfall per month during the remainder of the year . . . . These areas are located in major parts of Sulawesi, Malaysia, and northern Sumatra.

b) Zone III.2. Areas with two to five consecutive wet months and a pronounced dry season with at least 2 months less than 100 mm rainfall per month. . . . East Java and the major part of central and north Thailand are covered by this pattern. The island of Palawan in the Philippines, the central part of Vietnam and a large part of central Cambodia and southern part of Laos.

"4) Zone IV. Areas with less than two consecutive wet months. These areas are located in northeast Thailand, north Burma, east Kalimantan and some spots in Sulawesi, central Cambodia and central Visayas in the Philippines. They are not suitable for any type of agriculture unless additional water is available. . . ."

APPENDIX 4-E PHYSIOGRAPHIC REGIONS IN SOUTHEAST ASIA

In Southeast Asia, IRRI (1974) uses the following four physiographic classifications with which different forms of rice cultivation are closely associated; see Fig. 4-5 in Chapter 4 for part of this area. The classifications are as follows:

"1. COASTAL PLAINS OF MARINE, DELTAIC, AND FLUVIAL ORIGIN

a. Regions which flood deeply during the wet season, where cropping, other than with rice, requires extensive drainage. Examples are part of the Bangkok Plain and the Mekong Delta.

b. Regions which do not flood deeply or where sufficient drainage has been installed to control deep flooding. With their better water control, these land regions have potential for intensified cropping. Examples are the Bangkok Plain, Mekong Delta, and northwest Malaysia.

"The major soils are Inceptisols, Entisols, Vertisols, and Histosols.

"2. INLAND TERRACES AND PLAINS OF ANCIENT ORIGIN, USUALLY WELL DISSECTED

These are not divided on the basis of flooding because most of them probably do not flood deeply. An example is Northeast Thailand. The major soils are Ultisols and Alfisols.

"3. INLAND TERRACES, FANS, AND VALLEYS OF RECENT ORIGIN

a. Regions which flood deeply during the wet season and where cropping other than with rice would require extensive water control in the form of dams in the catchments, and river embankments. Such areas are intersected by major rivers and have limited potential for multiple cropping except in areas where dry season irrigation is possible. Large areas of Bangladesh fall into this category.

b. Regions which do not usually flood deeply. These are normally the terraces, valleys, and fans of smaller rivers with limited catchments. Such areas are common within all of the region's rice growing areas.

"The major soils are Inceptisols, Entisols, and Vertisols.

"4. SLOPING LANDS

a. Terraced for paddy rice production. Depending on soil stability, terraces may be made on slopes up to 100 percent. The stability of terraces depends on soil character, Oxisols generally giving stable terraces. Major areas are Java and adjacent islands. The major soils are Oxisols and Entisols.

b. Unterraced lands. Upland rice is grown on these lands. Sometimes as part of a stable agricultural system, but more commonly as part of a shifting cultivation system. These unterraced lands are widely scattered in all of the region's rice growing areas. The major soils are Ultisols and Alfisols."

APPENDIX 4-F RESEARCH AREA SELECTION IN ICRISAT'S VILLAGE LEVEL STUDIES

When a target area or subarea is large, the FSR&D team may want to break the subarea into subdivisions using administrative units (see Sec. 4.3.6. in Chapter 4). Such a process was followed for a village study in India (Jodha et al., 1977). In selecting research areas representative of the target area—an agroclimatic zone in this case—ICRISAT (International Crops Research Institute for the Semi-Arid Tropics) went through three steps:

1) selection of a district
2) selection of a subdivision of a district
3) selection of villages (research area).

We quote from this study, with the permission of Jodha et al. (1977):

"SELECTION OF DISTRICTS

"Since the purpose of the [Village Level Studies] was to understand the factors affecting the traditional system of farming in different agroclimatic zones, it was decided to purposefully select three districts representing major agroclimatic zones within the semi-arid tropics of India. Basic factors considered for selection of districts were soil types, pattern of rainfall, and relative importance of crops like sorghum, pearl millet, pulses and groundnuts—crops in which ICRISAT is primarily interested. District selection was also influenced by the availability of a nearby agricultural university/research station from where planning and logistical assistance could be obtained. Other things being equal, distance from Hyderabad was a factor considered in district selection, as it was felt important that the senior staff of the Economics Program be able to regularly visit and stay in the villages."

SELECTION OF A SUB-DIVISION

"For selection of villages for study, the first step was selection of a [sub-division] in each of the selected districts [administrative units within Indian states]. There is a real possibility of fairly large inter-[sub-division] differences within a district; these differences often become obscured when district averages are presented. To guard against the consequences of such a possibility and also to take full advantage of availability of [sub-division] data, the following procedure was adopted for [sub-division] selection.

"Relevant [sub-division]-level details of all the districts adjoining the selected district and/or showing broad similarities to the selected district were compiled and compared. The [sub-division] within the selected district reflecting the situation of the majority of the [sub-divisions] in the region (comprising the districts considered) was finally selected. Thus the selected [sub-divisions] represent the situation of a broad homogeneous region rather than the administrative district in which it happens to be located.

For sub-divisions, "details of about 40 different characteristics were collected from district census hand books, district statistical reports, and unpublished records of the Indian Bureau of Economics and Statistics. The variables selected broadly represented the natural and man-made resource base and their utilization patterns in the [sub-divisions] and included density of population, extent of literacy, density (number per hundred hectares of net sown area) of cultivators, agricultural laborers, cattle, buffaloes, sheep and goats, iron and wooden plows, electric pumps plus oil engines, number of tractors per 1000 hectares of net sown area, percentage of forest area, barren and uncultivable lands and net sown area to total geographical area, cropping intensity, extent of net sown area irrigated, contribution of different irrigation sources, average rainfall and rainfall intensity, percentage shares of wheat, rice, sorghum, pearl millet, chickpea, pigeonpea, cotton, groundnut, total cereals and total pulses to the gross cropped area, and proportions of villages of different size.[1]

"By looking at the numerical values of each of the characteristics, various class-interval ranges were set up. On the basis of these classifications, frequency distributions of the forty variables across [sub-divisions] were derived. The frequency-distribution tables were used to determine the modal class ranges wherein the majority of the [sub-divisions] in the region fell. [Sub-divisions] of selected districts which fell within the modal range class of values of a particular variable were considered as representative of the region as far as the variable under question was concerned. [Sub-divisions] were thus examined for their representativeness with respect to each of the considered variables (or characteristics qualifying the majority of the [sub-divisions] in the region). Two or three [sub-divisions] which scored the highest (i.e. had a maximum of features characterizing the majority of the [sub-divisions] in the region) were chosen; one of these was then selected as the final choice. Before final selection, it was ensured that the selected [sub-division] qualified on the basis of crucial variables like net sown area, average rainfall, extent of irrigation, and proportions of important crops. When more than one [sub-division] was found to qualify for selection, their position was more closely examined before final selection."

---

[1]Details about average size of land holding were proposed for consideration but the required [sub-division] data were not readily available for many [sub-divisions] and could not be done except for Mahbubnagar and its adjoining districts. Similarly, comparison of [sub-districts] in terms of communication and educational facilities, extent of electrification, etc., could not be done for want of comparable data for all the [sub-divisions].

SELECTION OF VILLAGES

"For selection of *villages* within the selected [sub-division/sub-divisions], the predominant characteristics of the latter were kept in mind. A number of villages were chosen to represent typical characteristics of the [sub-division] (in terms of cropping pattern, land-use, irrigation, etc.). In order to select the village truly representing the traditional situation, those having special programs or more than normal support or resource transfers from outside, or those located nearer towns and highways, were not considered. Some 12 to 20 villages were visited for each of the villages finally selected."

Members of the local research station and the district agricultural departments cooperated informally with the ICRISAT team in the final village selection.

APPENDIXES TO CHAPTER 5

APPENDIX 5-A
PHYSICAL RESOURCES OF THE RESEARCH AREA AFFECTING BIOLOGICAL PRODUCTION

In Sec. 5.4.1., we discussed the farmers' physical setting. In this appendix, we outline data forms for the FSR&D team's use in summarizing secondary information of physical resources in the research area. Tables 5-A-1 to 5-A-3 are typical of those that might be used for climate, land types, and soil types, respectively.

For detailed information on climate, see Appendix 6-A; and for land types and land evaluation, see Appendix 5-B. For summary sheets when observing physical and biological conditions for crop and livestock conditions during reconnaissance surveys, see both the Tables 5-1 and 5-2 in Chapter 5 and Appendix 5-P. Finally, for data forms to identify, characterize, and monitor physical and biological conditions during on-farm cropping and livestock experiments, see Appendixes 7-A and 7-B, respectively.

Table 5-A-1. Information from climatic records.

Rainfall
- Annual averages and variations ____________
- Monthly distribution ____________
- Critical flooding or drought periods ____________
- Variability from average during cropping season ____________

Temperature
- Monthly averages and variations ____________
- Maximum-minimum during critical periods of the cropping season ____________

Wind
- Prevailing direction ____________
- Average velocity and variations ____________
- Critical periods and conditions ____________

Solar radiation
- Monthly distribution ____________

Table 5-A-2. Information on land types from published sources.

*Descriptive term**	*Percentage of research area*	*Major soil type*	*Irrigated or nonirrigated*	*Significant features for crop production†*
________	________	________	________	________
________	________	________	________	________
.	.	.	.	.
.	.	.	.	.
.	.	.	.	.

*For example, hill or ridge tops, hillsides, valley bottoms.

†For example, very erodible, floodplain, high water table.

Table 5-A-3. Information on soil types from surveys.

Name	________	________	. . . .
Texture of surface soil	________	________	. . . .
Soil depth	________	________	. . . .
Water table depth	________	________	. . . .
Slope	________	________	. . . .
Drainage characteristics	________	________	. . . .
Distinctive production limitations*	________	________	. . . .

*For example, strongly acidic, salt affected, low fertility, thin soil, poor aeration, compact subsoil, and slow infiltration rates.

APPENDIX 5-B
LAND TYPES AND LAND EVALUATION

An important part of characterizing the research area is to identify land types and to evaluate their potential for agricultural production. We describe how such characterization fits into problem identification under the

Physical Setting in Sec. 5.4.1. In this appendix, we provide additional details on these two subjects. The first part of this appendix describes "land types." A land type is a set of locations for which it is possible to develop common technologies. To divide a given area into such land types, the FSR&D team must first have a good knowledge of that area. Such information may be available from secondary sources. If not, a land evaluation survey must be made. The second part of this appendix lists six principles FAO suggests for guiding land evaluation surveys.

LAND TYPES

Because environments differ with areas, an area should be analyzed in terms of the major land types represented. According to Zandstra et al. (1981) "A set of sites that have similar cropping pattern determinants is defined as an environmental complex or cropping systems land type." They explained:

> "Land types must be sufficiently different to merit the development of a different technology for each of the land types. Ultimately extension services will introduce any new cropping systems technology developed for a region to the farmers. Recommendations cannot be tailored to individual fields, but have to be generalized to a considerable extent. This unavoidably implies a loss of adjustment of the recommendation. The best division of an area into land types is then that division that provides the greatest fit of the recommendations used for the area with the [fewest number of] land types.
>
> "Recommendations (and therefore land types) may be stratified according to differences in farm types (large or small, with or without bullocks), water supply, soil characteristics, cropping history, infrastructural features or others. It is useful to consider possible extension strategies with respect to the contemplated land type divisions, as the capacity of the extension services can influence the impact of cropping system research results."

Zandstra et al. (1981) reported that the factors for distinguishing land types in an area are those that "most strongly influence performance of cropping patterns in the area. Careful observation and study of existing cropping systems in the area generally give important indications of what these factors may be."

Zandstra et al. (1981) offered the following general guide for classifying land types for crop-oriented research which, they pointed out:

> "will need to be modified to suit specific conditions of a site and the information available.
>
> "1. Separate land types into *dryland* and *wetland.*
>
> "2. Differentiate between irrigated and rainfed land. Rainfall will normally not vary sufficiently from location to location within a site to consider stratifying the area on the basis of rainfall. Irrigation, however, can vary greatly. . . . With respect to irrigation the *source* (tank, river diversion, shallow well, deep well) and the duration of irrigation (total duration and the weeks before the onset of the rainy season and after the end of the rainy season during which irrigation is available) can be important.
>
> "3. The next most important land qualities for identifying different land types are landscape and geomorphology. Although they do not intrinsically influence crop production, they are associated with many determinants, such as depths of water table, water enrichment potential, slope, soil texture, and fertility.
>
> "4. In wetland areas, the lowest and highest position of the *water table* can be of great relevance to the type of cropping pattern suited for that land type. An area with a shallow water table (< 1m) during the dry season may have a vastly different production potential than one with a deep water table (> 2m). In areas subject to flooding the water table will be above ground level for part of the year and *duration of flooding* will become an important determinant.
>
> "5. Because of its effect on soil water relationships, *soil texture* is probably the next most important determinant of cropping systems. Substantial differences in clay content may justify the recognition of a different land type and the development of a different technology for it.
>
> "6. Soil fertility and soil chemical conditions can often be corrected through management inputs. Where differences in such factors are great, or difficult to correct, an additional stratification associated with these factors may be used. This may be particularly of interest to areas subject to soil salinity, extreme acidity or toxicities.
>
> "7. Identify major socioeconomic differences that occur within the site. These may be substantially different farm types or market conditions. Such differences can often be expressed in different recommendations through analytical means and may not require stratification of experimental activities. They should, however, be reflected as different land types for extension purposes."

Zandstra et al. (1981) considered a sketch illustrating the typical spatial relationships of the land types found at the site to be valuable (Fig. 5-B-1). Zandstra et al. (1981) also stated that a "brief table showing rough estimates of land that occur in each type and the dominant current use and possible potential uses of each, will help convey the land type divisions to others [Table 5-B-1]. It also provides a preliminary guide to the land types on which research will have the greatest payoff. Only land types which occupy a major portion of the target area of the site and which present good prospects for improved cropping patterns should be considered in the research program. . . ."

Figure 5-B-1. Schematic sketch of typical relationships of the land types found in the research areas (Zandstra et al., 1981).

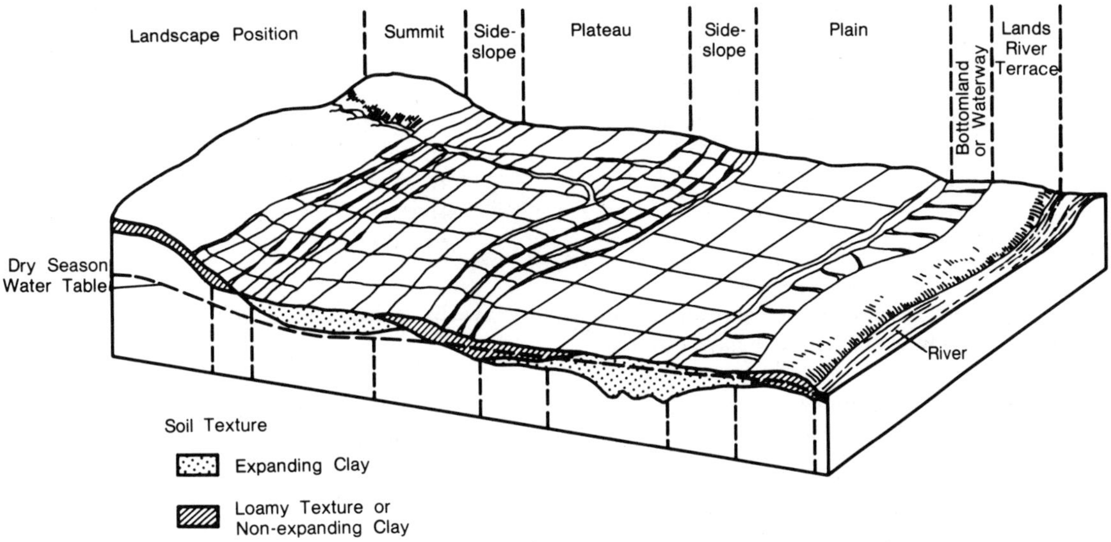

Table 5-B-1. Land types, their major characteristics and present and potential uses (Adapted from Zandstra et al., 1981).

Land type No.	Land type Name	% area	Major soil types	Rainy season water table depth (m)	Hydrology*	Flooding hazard	Major present use	Potential use
1	Summit	15	Balo loam	>5	Pluvic	Absent	Tree crops	As is
2	Sideslope	10	Balo loam	>2	Fluxic	Absent	Corn-rice	Corn-rice/mung
3	Plateau (wetland)	25	Ablo Sicl.	2-3	Fluxic	Absent	Rice-fallow	Rice-(soybean, mung)
4	Plateau (dryland)	5	Balo loam	2-3	Cumulic	Absent	Pasture	As is
5	Plain	27	Albo Sicl.	1-2	Cumulic fluxi-cumulic	2 in 10	Rice-fallow	Rice-rice Mung-rice-sorghum
6	Bottomland	7	Albo Sicl. Loba clay	<0.5	Cumulo-delugic	4 in 10	Rice-rice	Rice-rice-rice Rice-rice
7	River terrace	5	Olab sandy loam	>3	Fluxic	1 in 10	Corn + rice, pulses	Cassava + corn + rice-pulses Corn + pulses
8	Home garden	5	Albo Sicl.	>3	Fluxic	1 in 10	Fruits, herbs, vegetables	As is
9	Bed and furrow fields	1	Loba clay	<0.5	Cumulo-delugic	1 in 10	Rice + vegetables	Expand area

*These terms follow the hydrological classification as defined in the Glossary to Zandstra et al. (1981). They are

"CUMULIC - derived from accumulation. Descriptive of a wetland type where 100 mm of accumulated water will stay for more than 7 days, when the soil has been puddled, even without any water addition from rain or irrigation.

"DELUGIC - derived from deluge. Descriptive of a land type where the water level stays for more than 2 weeks duration at a depth greater than 30 cm, which is above the normal height of bunds or dikes, during high rainfall months.

"FLUXIC - derived from the flux or passing through. Descriptive of a wetland type where free water remains in the field when the soil has been puddled, but the depletion rate of free water, including evapotranspiration losses, is more than 10 mm per day.

"PLUVIC - derived from pluvia or rain. Descriptive of a land type where water contributed by rain or irrigation does not stay for more than 3 hours on the soil even if the soil has been worked wet . . . ."

LAND EVALUATION

Techniques have been developed for evaluating the suitability of land according to its characteristics, potential uses, and the situation. One of the more widely used approaches is that developed by FAO (1976),[1] which the reader may wish to consult. Following are six basic principles from that publication:

"i. LAND SUITABILITY IS ASSESSED AND CLASSIFIED WITH RESPECT TO SPECIFIED KINDS OF USE

This principle embodies recognition of the fact that different kinds of land use have different requirements. As an example, an alluvial floodplain with impeded drainage might be highly suitable for rice cultivation but not suitable for many forms of agriculture or for forestry.

The concept of land suitability is only meaningful in terms of specific kinds of land use, each with its own requirements, e.g. for soil moisture, rooting depth, etc. The qualities of each type of land, such as moisture availability or liability to flooding, are compared with the requirements of each use. Thus the land itself and the land use are equally fundamental to land suitability evaluation.

"ii. EVALUATION REQUIRES A COMPARISON OF THE BENEFITS OBTAINED AND THE INPUTS NEEDED ON DIFFERENT TYPES OF LAND

Land in itself, without inputs, rarely if ever possesses productive potential; even the collection of wild fruits requires labour, whilst the use of natural wilderness for nature conservation requires measures for its protection. Suitability for each use is assessed by comparing the required inputs, such as labour, fertilizers or road construction, with the goods produced or other benefits obtained.

"iii. A MULTIDISCIPLINARY APPROACH IS REQUIRED

The evaluation process requires contributions from the fields of natural science, the technology of land use, economics and sociology. In particular, suitability evaluation always incorporates economic considerations to a greater or lesser extent. In qualitative evaluation, economics may be employed in general terms only, without calculation of costs and returns. In quantitative evaluation the comparison of benefits and inputs in economic terms plays a major part in the determination of suitability.

It follows that a team carrying out an evaluation requires a range of specialists. These will usually include natural scientists (e.g. geomorphologists, soil surveyors, ecologists), specialists in the technology of the forms of land use under consideration (e.g. agronomists, foresters, irrigation engineers, experts in livestock management), economists and sociologists. There may need to be some combining of these functions for practical reasons, but the principle of multidisciplinary activity, encompassing studies of land, land use, social aspects and economics, remains.

"iv. EVALUATION IS MADE IN TERMS RELEVANT TO THE PHYSICAL, ECONOMIC AND SOCIAL CONTEXT OF THE AREA CONCERNED

Such factors as the regional climate, levels of living of the population, availability and cost of labour, need for employment, the local or export markets, systems of land tenure which are socially and politically acceptable, and availability of capital, form the context within which evaluation takes place. It would, for example, be unrealistic to say that land was suitable for non-mechanized rice cultivation, requiring large amounts of low-cost labour, in a country with high labour costs. The assumptions underlying evaluation will differ from one country to another and, to some extent, between different areas of the same country. Many of these factors are often implicitly assumed; to avoid misunderstanding and to assist in comparisons between different areas, such assumptions should be explicitly stated.

"v. SUITABILITY REFERS TO USE ON A SUSTAINED BASIS

The aspect of environmental degradation is taken into account when assessing suitability. There might, for example, be forms of land use which appeared to be highly profitable in the short run but were likely to lead to soil erosion, progressive pasture degradation, or adverse changes in river regimes downstream. Such consequences would outweigh the short-term profitability and cause the land to be classed as not suitable for such purposes.

This principle by no means requires that the environment should be preserved in a completely unaltered state. Agriculture normally involves clearance of any natural vegetation present, and normally soil fertility under arable cropping is higher or lower, depending on management, but rarely at the same level as under the original vegetation. What is required is that for any proposed form of land use, the probable consequences for

[1]Reprinted from *A Framework for Land Evaluation:* Soils Bull. 32 by FAO. 1976. By permission of the Food and Agriculture Organization of the United Nations, Rome.

the environment should be assessed as accurately as possible and such assessments taken into consideration in determining suitability.

"vi. EVALUATION INVOLVES COMPARISON OF MORE THAN A SINGLE KIND OF USE

This comparison could be, for example, between agriculture and forestry, between two or more different farming systems, or between individual crops. Often it will include comparing the existing uses with possible changes, either to new kinds of use or modifications to the existing uses. Occasionally a proposed form of use will be compared with non-use, i.e. leaving the land in its unaltered state, but the principle of comparison remains. Evaluation is only reliable if benefits and inputs from any given kind of use can be compared with at least one, and usually several different, alternatives. If only one use is considered there is the danger that, whilst the land may indeed be suitable for that use, some other and more beneficial use may be ignored."

APPENDIX 5-C MARKETING FACTORS AFFECTING SMALL FARMERS

In Sec. 5.4.1, we outlined some of the economic factors that make up the farmers' environment. Among the more important economic factors are those dealing with farmers' marketing of their outputs and their purchases of productive inputs. In this appendix, we present a checklist for such inputs and outputs, which the FSR&D team can use as background material for developing and analyzing alternative technologies. The team should use the checklist selectively by concentrating on those portions relevant to its situation. In time the team will probably want to develop its own checklist. We caution the team not to try to answer each of the questions in the list before initiating on-farm research. Rather, the list should be used as a reminder of factors that may be important when considering how alternative technologies are influenced by the farmers' economic environment.

For additional information on marketing, we refer the reader to the sources we used in preparing this appendix. They include *Planning Technologies Appropriate to Farmers: Concepts and Procedures* (Byerlee et al., 1980), *The Design of Rural Development* (Lele, 1975), *The Economics of Agricultural Development* (Mellor, 1966), *Getting Agriculture Moving* (Mosher, 1966), and *Assessment of the Capacity of National Institutions to Introduce and Service New Technology* (Smith, 1977).

The checklist that follows is divided into two parts: one dealing with markets for farm products and the other dealing with inputs to farm production.

MARKETS FOR FARM PRODUCTS

This section deals with market factors important when farmers sell their products. These factors relate to general market characteristics, pricing, transportation, storage, processing, information, farmer organizations, regulation, and other government involvement.

GENERAL MARKET CHARACTERISTICS

For each of the major markets in which the farmers normally trade, the team should consider checking

- commodities traded in the market—e.g., the major grains and livestock and other important commodities
- periods when the market operates—e.g., days of the month or week, hours of the day
- daily volume traded by commodity
- available physical facilities—e.g., market stalls, roofed storage areas, cold storage, livestock pens
- available services—e.g., grading, inspection, weighing, feeding, market information, packaging, transportation
- market fees
- farmer access to the market—e.g., can farmers sell their products themselves or must they go through intermediaries
- types of traders operating in the market—e.g., private traders, cooperatives, government agencies
- number and size of traders—e.g., only small traders, small and large traders, only large traders
- degree of monopoly power, collusion, farmer exploitation
- reasons for controlling traders—e.g., as related to control of storage, processing, transportation, credit, relations with central marketing groups
- market channels—e.g., farmer-consumer, farmer-peddler-consumer, farmer-itinerant trader-consumer, farmer-retailer-consumer, farmer-village trader-central market receiver, farmer-processor
- degree of specialization in the market by product and function
- kind and importance of contract arrangements between buyer and farmer—e.g., farmers contract to deliver a specified quantity and quality of product at a given time and the buyer guarantees the price of the product and supplies the farmers with inputs.

PRICING

Market prices for farmers' products can be critical in determining farmers' potential responses to change. For the farmers' major markets, the team should consider the following:

- price stability by commodity—e.g., seasonal, yearly
- frequency and extent of price fluctuations
- reasons for instability of seasonal prices—e.g., insufficient storage facilities, lack of buffer stocks
- price spreads between markets and reasons—e.g., poor transportation, inadequate market reporting
- steps taken to increase price stability

- government's role in helping private traders—e.g., financing, giving technical assistance, providing forecasts of future supply and demand
- price stabilization schemes by the government—e.g., price guarantees, price floors and ceilings, marketing boards
- methods used by government to influence prices—e.g., using buffer stocks, setting quotas on production or marketing, subsidizing local demand, promoting exports.

TRANSPORTATION

Transportation facilities and costs influence which products are produced in an area. Transportation affects marketing margins, arrival time, quality of commodities, and consequently the prices farmers receive for their output. Thus, the team should consider

- suitability of access roads for motorized transport—e.g., seasonal and year-round
- availability, quality, dependability, and cost of transportation services—e.g., bus, truck, boat, train
- availability of refrigeration facilities, which are often essential for perishable products—e.g., fish, milk, fresh fruits and vegetables.

STORAGE

Storage comes into play because most farm products are harvested seasonally whereas demand is often year-round. Farmers usually attempt to store most of their nonperishable crop production intended for family use. Also, where feasible, both governments and farmers can frequently benefit by increasing storage to provide buffer stocks. These stocks help dampen seasonal and annual price fluctuations. In addition, perishable products require cold storage to extend the period of their availability. In this regard the team should check

- type, capacity, quality, and cost of nonrefrigerated on-farm and off-farm storage
- type, capacity, quality, and cost of refrigerated storage and supporting ice and refrigeration plants.

PROCESSING

The commercial production of many agricultural products such as palm oil, tea, coffee, rubber, sugar, and dairy products requires nearby processing plants. The raising of some perishable products may become profitable only if seasonal surpluses or low grades can be processed. However, processing is not limited to perishable crop and livestock products. Even if sold in their original form, agricultural products may need to be cleaned, dried, artificially ripened, graded, and packaged. They may be shelled, ground, mixed, baked, pasteurized, or otherwise transformed.

Factors for the team to check are

- capacity, level of use, and modernization of processing plants
- specialization of processing plant equipment
- possibilities for using processing plants for other crops and during other seasons than the plants' normal use
- quality and cost of plant outputs.

INFORMATION

When local merchants are the farmers' only source of market information, farmers are at a disadvantage when selling their products. On the other hand, governments can help provide farmers with information. Items to consider in this regard are

- existence of market information by the government
- reliability and currency of the information
- method of conveying information—e.g., radio, mail, posters, bulletins
- types and grades of products for which information is provided
- types of information given—e.g., today's prices, supply and demand conditions in local and central markets; today's information compared with information from the previous weeks, months, years; expected shipments; weather, current crop and livestock conditions; and the outlook
- farmers' understanding of the terms and quality grades used in the market reports
- information farmers obtain from government sources compared with that from merchants.

FARMER ORGANIZATIONS

Farmer organizations can also be important to small farmers. Thus, the team might look into

- the kinds of farmer organizations operating in the market—e.g., cooperatives providing a single service, such as selling, buying, bargaining, and credit; and multipurpose cooperatives providing combinations of these services
- the efficiency and fairness of each organization's operations
- each organization's share of the market
- farmers' understanding of the concept and functions of these organizations and the farmers' role in them.

REGULATION

Governments frequently regulate markets and certain marketing activities. Thus, the team might look into the following aspects of these regulations:

- trading rules
- standard weights and measures
- quality and grade standards for selected commodities
- contracts and agreements.

OTHER GOVERNMENT INVOLVEMENT

So far we have mentioned government activities in pricing, information, and market regulation. In addition, the government may perform other functions such as buying, selling, transporting, storing, and processing.

The team should consider checking

- marketing activities in which government agencies are directly involved
- ways in which government agencies compete with private traders and farmer organizations
- use of nongovernmental channels by government agencies—e.g., the Taiwanese Government's use of farmers' associations for buying rice
- government monopolies in marketing.

INPUTS TO FARM PRODUCTION

In this section, we discuss economic aspects of farm inputs such as purchased supplies, tools and equipment, labor and traction, and credit. Below are items an FSR&D team may want to check:

PURCHASED SUPPLIES

- adequacy of inputs at time when farmers need them
- packaging in quantities suitable for farmers' needs
- adequacy of financing
- degree of traders' monopoly power, collusion, and farmer exploitation
- role of government in market surveillance, provision of input information, and subsidies.

TOOLS AND EQUIPMENT

- availability of reliable suppliers of tools and equipment
- adequacy of stocks, spare parts, and servicing facilities
- availability of local mechanics for making repairs
- availability of custom services such as plowing, harvesting, and spraying.

LABOR AND TRACTION

- on-farm and off-farm employment opportunities
- rates of pay for such employment
- supply of labor for hire
- opportunities, rates, and conditions for renting tractive power such as oxen and tractors.

CREDIT

- how farmers finance their purchases of productive inputs
- farmers' short-term needs—e.g., cash, hired labor, and storage facilities
- farmers' intermediate needs—e.g., buying animals, equipment, and financing consumption following serious crop failures
- farmers' long term needs—e.g., land improvement and wells
- share of total farmers' needs in each of the above categories financed from their own resources and with the help of relatives and friends
- amounts and purposes of farmers' borrowings in each of the above categories from different types of lenders—e.g., money lenders, dealers, processors, cooperatives, banks, and government agencies
- interest rates, repayment conditions, and security requirements
- rates of defaulted loans by type of loan and lender.

APPENDIX 5-D
THE SOCIOCULTURAL ENVIRONMENT

As indicated in Sec. 5.4.1. in Chapter 5, FSR&D teams need to be aware of the sociocultural environment of the research and target areas. Farmers who operate within this sociocultural setting are restricted by it in some matters, and make good use of it in others. Those trained in the social sciences are needed to interpret the subtleties of this environment; however, we feel this short appendix can alert FSR&D teams to some of the concepts. Below, we have included sections on religion, norms and customs, and social institutions.

RELIGION

Religion plays an important role in agriculture because it contains a culture's ideas about the relationship between people and nature. For example, religious ceremonies may accompany first planting and harvest. Some religious practices are related to the ecosystem. Harris (1966) found that cattle are more important to Hindus in India as a source of milk, plow traction, and dung for fuel and fertilizer than as a source of meat protein. Rappaport (1968) found religion instrumental in regulating the pig population and redistributing pig surpluses in the protein hungry region of New Guinea.

Some religious practices do not appear to have any beneficial influence on farming, but are, nonetheless, important to farmers. For example, the Mayans put seven corn kernels in each hole at planting (seven is a holy number for the Mayans), measure seven *quarto* between corn hills and rows, and show reluctance to thin young corn plants because corn is regarded as a form of traditional deity among these Central American Indians.

The FSR&D team should be aware of how religious values relate to agricultural systems. Although religious beliefs are difficult to change by applying external pressure, most religious systems are general enough to accommodate substantial changes in farming methods.

NORMS AND CUSTOMS

Norms are the acceptable standards of behavior of a society, while customs are the ways people actually behave. FSR&D teams need to be aware of the positive and negative influences that both can have on their activities. For example, most communities have norms that govern how quickly change should be accepted. In some communities, rapid acceptance of a new technology gives farmers status, while in other communities the same behavior is not respected. In either case, the teams should not underestimate the persistence with which local communities retain their norms and customs.

SOCIAL INSTITUTIONS

Every society has established institutions for governing interpersonal relationships. Of particular interest to an FSR&D team are the traditional social institutions for controlling land, stratifying the populations, and exchanging labor. These include corporate kin groups, social classes, agricultural associations, leadership, and factions.

CORPORATE KIN GROUPS

Corporate kin groups—i.e., people with a common ancestry—own and control land in many traditional societies. These groups allocate land to members depending upon the availability of land and a variety of other criteria.

If land is abundant, every heir may receive a portion of the corporate kin group's land. But, if land is scarce, the land may pass undivided to a single heir. Communal land holding systems, such as these, may create problems for FSR&D teams because individual farmers may be unwilling to make permanent improvements in land they do not control.

SOCIAL CLASSES

Wealth, power, and education may stratify communities into social classes. FSR&D teams should be aware of the social classes in the research area because the social classes may determine who works for whom and who controls agricultural inputs.

AGRICULTURAL ASSOCIATIONS

Most traditional societies contain associations with agricultural roots—e.g., work exchange and water management groups. These associations frequently provide services that individual farmers cannot provide on their own.

LEADERSHIP

Established leaders may be concerned about the influence an FSR&D team may have on their community. Therefore, the team should identify formal and informal political and social leaders to learn the source and limits of their authority. These leaders can be approached for sanction and advice and should be kept informed of the team's progress. The team needs to exercise considerable judgment in its dealings with local leaders. For example, if a team becomes closely identified with a particular leader and that leader loses his or her authority, the team's credibility could be lost.

FACTIONS

The team may find itself in a community divided by factions. In such a situation, the team should try to remain neutral. However, the team may need to move to another community if the factions begin to nullify its work.

APPENDIX 5-E
AN EXAMPLE OF SOCIOLOGICAL RESEARCH
Adoption of Agricultural Technology by the Queche Indians of Guatemala

Sometimes an FSR&D team may want to learn more about how farm households make decisions, as we discussed under the Household as an Integrating Unit in Sec. 5.4.2. in Chapter 5. This appendix discusses the application of scientific methods in sociology to learn about farmer acceptance of new technologies. Below, we provide a brief description of a study by Beal and Sibley (1967) to illustrate the approach. This illustration, which makes use of statistical procedures to test hypotheses, is more formal than many studies about farmers' acceptance of new technologies. This research effort occurred in Guatemala in one of the areas of ICTA's concentration; however, the study preceded the ICTA program.

We present this synopsis of the Beal and Sibley (1967) study to illustrate

- sociological research methodology
- how this type of research can contribute to an understanding of why farmers act as they do and how this understanding can aid in gaining farmer acceptance of improved practices.

The sections that follow are Beal's account of their work.

RESEARCH METHODOLOGY

For this research process:

1) The social scientist starts with a research problem. In this case, the objective was to understand and to predict why and how farmers accept or do not accept tested and recommended farming practices and technology.
2) The social scientist translates the problem into generally researchable terms. In this case, the dependent variable—that which we are trying to predict—was the adoption of recommended farming practices and technology. Next, we selected those independent variables that we thought should have been related to, accounted for, or predicted farmer adoption. Social scientists usually derive independent variables in one or more ways: (a) from theories, which in this case are theories about human behavior; (b) from a review of past research in the area; or (c) from observation and explanation of farmer behavior.

Drawing from these three sources, we illustrate our conceptual framework in Fig. 5-E-1. In a

Figure 5-E-1. Conceptual framework for analysis of individual behavior (Beal and Sibley, 1967).

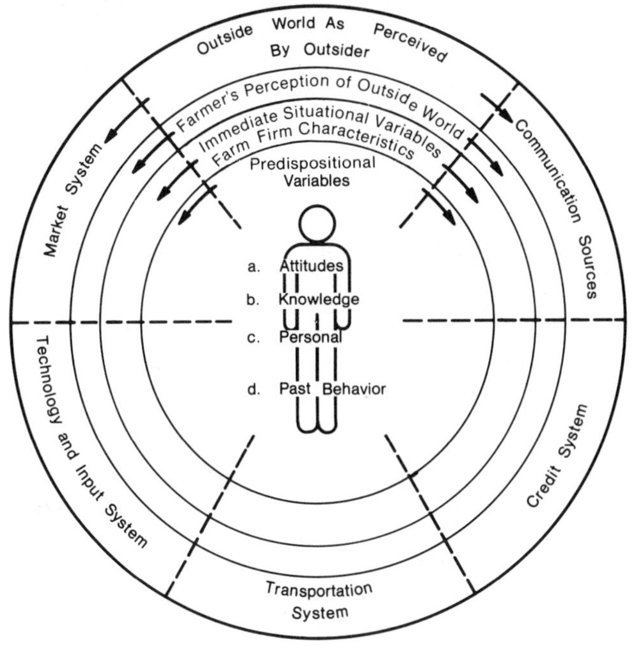

greatly oversimplified form, we theorized that the individual has certain predispositions to act a certain way—the inner circle of Fig. 5-E-1. If we understand those predispositions, we assume we should be able to determine relationships, and understand and predict behavior.

First, we assumed four categories of predispositions: attitudes, knowledge, past behavior, and certain personal characteristics that directly or indirectly predispose a person to act in a certain way. Second, we theorized the farmers' behavior concerning farm practices that would be articulated through the farm firm—the second circle. Third, we theorized a farmer's behavior would be conditioned or influenced by the farmer's perceptions, beliefs, and feelings about the relevant environment in which he or she acts—e.g., market, technology, and input systems such as transportation, credit, communications, and information.

In addition, as researchers, we objectively collected information on what the situation was in the relevant environment—the outside circle. Thus, we assumed that the difference between farmers' perceptions and researchers' observations can be determined and used for educational purposes and planning strategies for change.

3) The social scientist delineates subconcepts within the general level concepts. For example, within the general level concept of attitudes, we chose attitudes toward (a) scientific methods in farming versus traditional methods, (b) taking risks, (c) being able to control nature, (d) profit accumulation, and (e) the government.

Under knowledge subconcepts, we focused on markets, transportation, credit, and farm inputs such as fertilizers. Under past behavior, we chose such things as nonfarm jobs, visiting outside the community, use of information sources, and marketing behavior. Under personal characteristics, we chose such subconcepts as education, literacy, and age. Finally, we used the same process to specify subconcepts under firm characteristics, as well as environmental variables such as markets, input structure, and credit.

4) The social scientist hypothesizes about the relations among these concepts based on theory, past research, and observations. For example, we would expect to find a positive correlation between education and adoption and between perceived fair treatment from a credit source and obtaining credit.

5) The social scientist develops measures for each concept and subconcept. For example, we measured the dependent variable—adoption—by several scores composed of the use and degree of use of 19 farming practices recommended by the Guatemala Extension Services and the Ministry of Agriculture. The practices related to such factors as fertilizer and chemicals, planting rates, animal vaccination, and improved seeds. In terms of the independent variables, our measurements covered a variety of conditions, such as the very simple, e.g., age; the relatively straightforward, e.g., the credit-orientation score; and the relatively complicated, e.g., attitude scales based on pre-survey sample data meeting acceptable scientific criteria for scale acceptance.

6) Finally, the social scientist (a) develops a questionnaire, including all the necessary items and measures, then pretests and revises it, (b) delineates the population of farmers for generalization and draws a sample to be interviewed, (c) secures and trains interviewers, then legitimizes and manages the collection of field data, (d) codes data and places the data on analysis cards, (e) analyzes data using frequency tables, central tendency and zero order correlation methods, regression analyses, analysis of variance, and other statistical tests, (f) interprets the data and statistical tests, and (g) writes a report.

SOME FINDINGS

In all, we used 51 independent variables in this study. Taken in total, these 51 variables explained about 78 percent ($R^2 = 0.78$) of the variation in farmers' adoption of farming practices. We will now discuss a few examples that illustrate some of our findings. We list these findings according to subconcepts, with the general level concept shown in parentheses.

CONTROL OVER NATURE (ATTITUDES)

We found a wide range in attitudes among the farmers interviewed regarding their control over nature. Many farmers felt a person's life was predetermined, God controls the person's life and farming, God gives special powers to certain individuals so they can farm well and become rich, and they cannot do much to improve their lot by using fertilizer, weed killers, and new methods—a fatalistic attitude set. Other farmers believed they could control their environment by using new technology. As hypothesized, those who felt they could control their destiny—control over nature—were more prone to adopt new technology—a zero order correlation of r = 0.41 with adoption. Thus, researchers and extension specialists working with farmers should recognize these feelings, note the range among farmers, and then decide either to try changing these values through education, demonstration, field tests, and the like, or to try working around them.

Other attitudes were significantly and positively related to adoption in the hypothesized direction.

KNOWLEDGE OF CREDIT (KNOWLEDGE)

Farmers who thought of credit as a source of funds for investment, rather than for consumption, could (1) correctly specify the source of credit, (2) describe what purchases could be made with the credit, and (3) tended to adopt new farming practices to a much higher degree than did those who had a low knowledge of credit (r = 0.20). Thus, an FSR&D project might experiment with a program to educate farmers on (a) the use of credit as a production factor, (b) legitimate sources of credit, (c) types of credit available, and (d) strategies for obtaining credit. Then, the researchers could assess the extent of increased use of new technologies and increased production.

COSMOPOLITAN (PAST BEHAVIOR)

Those farmers that *more often* traveled from their village and visited Guatemala City, rather than staying in the local village, were much more prone to adopt new technology—zero order correlation of r = 0.45. Thus, an FSR&D program might try to give farmers a more outward-looking perspective to see if this changes their behavior at home.

INFORMATION COMPETENCE (PAST BEHAVIOR)

Those farmers who used knowledgeable sources of information—e.g., Ministry of Agriculture and Extension—regarding farming practices rather than neighbors, relatives, and friends had much higher adoption scores—zero order correlation of r = 0.43. Reaching farmers with highly credible sources of information appears to be important to acceptance.

AGE (PERSONAL CHARACTERISTICS)

Younger farmers will adopt more practices (r = 0.24). It will be more difficult to reach older farmers and get them to adopt new practices. A recognition of this generality could lead researchers and extension agents to concentrate on younger farmers when diffusing research results.

TILLABLE ACRES OWNED (FIRM CHARACTERISTICS)

As might be expected, those farmers who owned the largest number of tillable acres had higher adoption scores (r = 0.30). This finding suggests another basis for categorizing farmers into separate, more homogeneous groups.

MARKET ORIENTATION SCORE (PERCEPTIONS)

We found a wide range of farmer perceptions about the market for corn and wheat if farmers doubled their yields; the difficulty in selling the increase; whether the farmer would get a good price; and how the individual is treated when buying inputs and selling crops. Those who possess positive perceptions toward these phenomena have a higher adoption score (r = 0.24). For example, farmers who saw the marketing structure, price, and treatment as incentives rather than constraints were more likely to have adopted the technologies. Thus, an FSR&D project might propose ways (a) for improving the marketing structure and the treatment of the farmer, and (b) for helping the farmer understand the marketing structure as an opportunity rather than a constraint. Also, if farmers have reason to think they will be treated "very fairly" when they go for credit, they are more likely to adopt a new technology.

CONCLUDING COMMENTS

These are examples of the findings and potential use of more formal sociological studies. For an excellent summary of findings and a large number of generalizations from a pool of over 6,000 adoption-diffusion studies for many parts of the world, see Rogers and Shoemaker (1971). For books on methodology in the social sciences, including statistical techniques, see Kerlinger (1973) and Lin (1976). Finally, two publications that provide insights into the communication aspects of development and diffusion are a selected bibliography by Rahim (1976) and *Communication and Development* by Rogers (1976).

APPENDIX 5-F
DECISION MAKING BY SMALL FARM FAMILIES

Many factors influence the decisions farm families make about adopting new technologies. Those hoping to influence farmers' decisions should understand and accept the many-faceted nature of the farmers' decision making process. In Sec. 5.4.2. in Chapter 5, we presented six factors that help to explain this process. These concerned farmers' characteristics, knowledge, beliefs, attitudes, behavior, and goals. In this appendix, we provide additional factors influencing farmers' decisions. These are the family structure, labor supply, income, savings, and attitudes toward risk.

FAMILY STRUCTURE

Two basic types of family structure are the nuclear family and the extended family, both of which are strongly influenced by the family's requirements for labor. The nuclear family consists of the husband, wife, and their children. This form commonly occurs if productive land is both limited and located close to villages. The extended family is a grouping of two or more nuclear families. This form includes (1) members of several generations, (2) members of one generation as when married brothers and sisters live in the same household, and (3) a man or a woman with more than one spouse. The extended family provides a larger labor force and usually occurs when land holdings are scattered and fertile land is abundant.

Sahlins (1957) contends that patterns of land use determine the familial structure in Moala, Fiji, and that the extended family structure found there is specifically adapted for exploiting the family's scattered resources. Netting (1965) discussed the Kofyar of Northern Nigeria who, when changing from intensive to extensive cultivation methods, enlarged their household work groups by changing from a nuclear to an extended family structure. While this example may suggest flexibility in family structure, new technologies are more readily adopted when such structural changes are not required.

FAMILY LABOR SUPPLY

In developing countries, farming typically involves the productive labor of most household members. When looking at the workers in the farm household, the team must be aware of the roles of women and children. Children gather wood and water, perform various agricultural tasks, tend animals, and care for babies, thus freeing mothers for agricultural work. In this way, children contribute substantially to the household's labor supply. In Java, White (quoted in Biggs, 1978) found that women 15 years old and older worked more than 25 percent more hours than did men.

Family members generally have their individual responsibilities for working in the fields, processing agricultural products for home and the market, and caring for livestock. And, at times, when labor demands are heavy, as with planting, weeding, and harvesting, the whole family may work together.

FAMILY INCOME

Fluctuating incomes resulting from periodically poor harvests or forages are a way of life for most small farmers. Families traditionally have been able to smooth out their incomes and labor requirements by exchanging labor whereby farm families help each other with planting, weeding, and harvesting, and by sharing in the output.

Recently, however, more cash transactions are replacing reciprocal labor arrangements. Consequently, farm families are having to devise means for obtaining cash for these needs. In some cases, cash has become necessary to pay for even some of the family's subsistence needs.

Opportunities for obtaining cash include (1) labor for hire, both in agriculture and in off-farm employment, (2) sale of agricultural produce, including animals, (3) sale of handicrafts, (4) renting out draft animals, and (5) renting or selling land.

FAMILY SAVINGS

Although considerable variation occurs among small farmers, even those with good managerial skills may be unable to save money because of their low income levels and fluctuations in environmental conditions. Often, small farmers will put what small savings they do generate into livestock—as a means of savings, gaining prestige, and for other purposes.

ATTITUDES TOWARD RISK

The decisions of small farmers concerning the adoption of new technology are influenced by their perceptions of risk. Decisions are frequently based on ensuring survival when conditions are unfavorable. Thus, small farmers generally prefer known technologies over new technologies, even though the new technologies may, in fact, be better in terms of both average results and variability of results. An advantage of the on-farm experiments in FSR&D is that farmers have an opportunity to observe the results of new technologies firsthand. In this way, the spread between farmers' perceptions about the risks of new technologies can be brought closer to the new technologys' actual risks.

APPENDIX 5-G
DECISION TREES: A METHOD FOR LEARNING ABOUT FARMERS' DECISIONS

FSR&D teams need to continually seek a better understanding of how farmers make decisions about new technologies. In Sec. 5.4.2., we discussed the farming system and some factors to consider about the way farmers make decisions. In this appendix, we present a summary of an analytical method Gladwin (1979)[1] used involving decision trees. An approach using decision trees subdivides farmers' decisions into their basic elements. Gladwin's work is "part of a larger attempt to view [a project] through the eyes of the proposed adopters of the new technology—the farmers." Below, is a summary on decision trees, Gladwin's methodology, and a section on discovering decision criteria:

[1]Reprinted from *Economic Development and Cultural Change*, 28:1:155-173, by C.H. Gladwin by permission of The University of Chicago Press. Copyright 1979 by The University of Chicago. 0013-0079/80/2801-0009$01.58.

Figure 5-G-1. A hypothetical example of a decision tree (Gladwin, 1979).

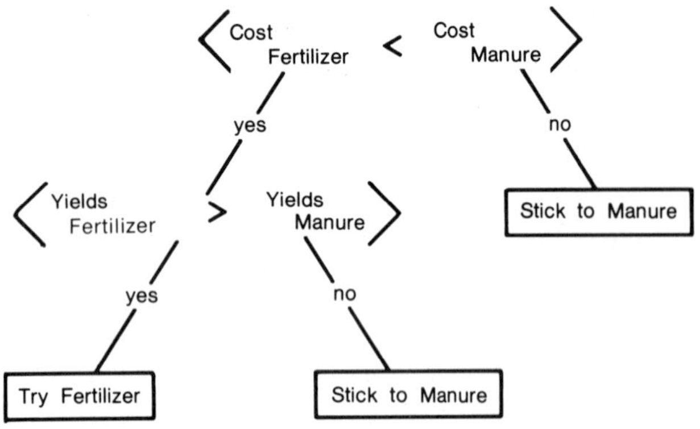

"DECISION TREES

". . . to empirically study farmers' adoption decisions, it was necessary to posit both a theory about how people make choices and a methodology for studying a particular choice. The theory of choice used in this study assumes that people, in choosing between two alternatives, do not make complex calculations of the overall worth or utility of each alternative. Rather, people tend to use procedures which *simplify* their decision-making calculations [Cyert and March 1963; Simon, 1969; Tversky, 1972; Quinn, 1976]. Hierarchical models or trees, with decision criteria at the nodes or branching points of the tree, can represent such procedures. The decision criteria can either be simple orderings of alternatives on some aspect or dimension or factor of the alternatives (e.g., Is $\text{cost}_{\text{fertilizer}} < \text{cost}_{\text{manure}}$?) or they can be constraints that must be passed or satisfied (e.g., is $\text{cost}_{\text{fertilizer}} \leq \$70$?). In either case, the criteria or constraints are *discrete* rather than continuous; that is, the alternative "fertilizer" either passes the criterion or constraint or it does not. A decision tree is thus a sequence or series of discrete decision criteria, all of which have to be passed along a path to a particular outcome or choice. For example, in the simple, hypothetical, decision-tree model of the choice between chemical fertilizer and manure in [Fig. 5-G-1], fertilizer must pass both cost and productivity criteria for the farmer to choose the outcome, 'try fertilizer.' If fertilizer fails either constraint, the model predicts that the farmer will keep on using manure.

"METHODOLOGY

"Given a form of decision model, a plan to 'discover' the particular criteria decision makers use in a specific decision context is necessary if one is interested in more than the psychological processes underlying choice. The major methodological concern is thus to find the specific aspects and constraints decision makers are using. One must determine what information is actually used in making decisions, as opposed to information unused in decision making which may nevertheless be given in response to interview questions.

"Producing a decision-process model can be considered to have five steps:

"1. The researcher observes the decisions made by a small sample of 20-30 decision makers. Given their decision behavior, he or she observes and elicits the site-specific decision criteria that decision makers use in their decision rules. Discovering decision criteria is the most difficult part of the model-building process and will be discussed in more depth in the next section.

"2. Using the language and categories that the decision makers themselves use, the researcher puts the decision criteria into a flowchart.

"3. To be sure that the flowchart is descriptively adequate, the researcher tests its predictions against the decisions made by the sample of decision makers used to built it.

"4. The researcher models the flowchart, that is, gives a more general, less site-specific interpretation to the language and categories decision makers use. From the flowchart, a purely descriptive statement of the decision process, the researcher should find more general, cross-cultural decision criteria underlying the specific categories used by the decision makers. In other words, the researcher should find the 'deep-structure' rules underlying the 'surface' rules. An example of a deep-structured rule in this paper is: I will adopt if, with the recommendation, I will maximize profit subject to risk, knowledge, and capital or credit constraints. An example of a surface rule is: I will not adopt the recommendation if I plant *en seco* in *arenal* soils and do the first cultivation before 'the rains come.' Clearly the 'model' of the decision is the deep-structure rule.

"5. The researcher tests the model by using it to predict the decisions made by a new, different, representative sample of decision makers. . . .

"Since the aim of a decision model is to predict the farmer's decision, the proper test of the model is percent predictability—the ratio of correct predictions of farmers' choices divided by the total number of farmers who made the choice.

"DISCOVERING DECISION CRITERIA

"Before the general cross-cultural decision criteria can be assembled into a decision model, the site-specific categories and surface rules used by decision makers in a given locality must be elicited and or observed. The method used in step 1 to discover the site-specific decision criteria has five steps: (a) The interviewer holds other decision criteria constant and isolates the criterion of interest. (b) With the sample of decision makers whose behavior varies only on the criterion of interest, the interviewer looks for contrasts over decision makers, space or time. (c) Once a contrast in decision behavior is found, the interviewer elicits the decision criterion. (d) Once the decision criterion is elicited, the interviewer tests it on another decision maker. The criterion is correctly specified if it 'cuts' the sample of decision makers into two subsets: those who pass or satisfy the criterion and decide to do X and those who fail the criterion and decide not to do X. (e) If the criterion is not correctly specified (i.e., if it does not cut) the interviewer must acquire more information. In fact, he must acquire the cognitive strategies used by farmers to grow the crop in question. With more knowledge and understanding of the farming system, the interviewer then revises the criterion and tests the new version on another farmer. This process is repeated until the criterion cuts or predicts decision behavior."

APPENDIX 5-H DESCRIBING EXISTING CROPPING SYSTEMS

This appendix, quoted directly from Zandstra et al. (1981), gives step-by-step instructions in how to prepare tables of data describing the most important features of existing cropping systems in a new research area. The purpose is to aid the reader in knowing how to collect data for problem identification, as we described in Part 1 of Chapter 5. Zandstra et al. (1981) wrote:

"1. First record major crops and varieties for each land type recognized at the site, and time periods when they are grown [Table 5-H-1]. If more than one crop schedule is followed, specify each and number them. If the same varieties and same crops are grown at the same time but on different land types, crops should be listed separately and identified by different numbers.

"2. For each land type record major cropping patterns, and include idle land, tree crops, pasture, etc. as patterns if land where they are grown is cultivable [Table 5-H-2]. Denote each pattern by capital letters and show the approximate percentage area of cultivable land planted to each. Use the same crop definitions as those used in [Table 5-H-1]. Area in each crop and cropping intensity in a site will be computed from this data. To denote planting arrangements in time and space, use a hyphen (-) if crops are sequenced; use a plus (+) if crops are planted simultaneously (more than 2/3 of growing season overlaps); use a slash (/) if crops are planted in relay (less than 1/3 growing season overlap). For example, a cropping pattern of dry-seeded rice followed by a sorghum mungbean intercrop, in which melons are interplanted into the sorghum after mungbean is harvested would be presented as follows:

[Dry-seeded] rice - sorghum + (mungbean - melons)

A crop of sorghum in which mungbeans are relayed would be presented by:

Sorghum / mungbean

When appropriate, the multiple cropping index (MCI) or land use intensity can be calculated for each land type. It is often useful to present cropping patterns in diagramatic form, indicating planting and harvesting times of each crop (Example [1]).

Example 1. "Preparation of a cropping pattern diagram

"Each crop in the pattern should be presented in this diagram. Begin by indicating the first month of the growing season below the diagram. Indicate the planting dates of each crop with a single line and the

Table 5-H-1. Crops produced in each land type, their growing period and yield (Zandstra et al., 1981).

Crop	Varieties	Land type	Time period	Estimated yield (t/ha)
Rice 1	RD3, Bahagia, IR8	Irrigated I	15/4-15/9	3.5, 2.7, 3.0
Rice 3	Fastvar, IR30	Irrigated II	1/4-15/7	3.2, 4.5
Rice 4	Fastvar, IR34	Irrigated II	1/8-15/11	3.2, 4.3
Rice 2	Fastvar, IR30	Rainfed wet land	15/4-15/9	2.7, 3.0
Cassava 2	Local early maturing variety	Rainfed wet land	15/9-15/4	9
Cassava 1	Local variety	Dryland	15/4-30/12	13
Corn 1/ Cassava 3	Local varieties	Dryland	15/4-1/8-30/12	1.8, 10.0
Corn 2	DMR 2, Local var.	Dryland	15/4-11/8	1.9, 1.0
Corn 3	DMR 2, Local var.	Dryland	15/8-1/12	2.4, 1.4
Corn 4	DMR 2, Local var.	Dryland	15/1-11/4	1.6, 0.9

Table 5-H-2. Cropping patterns, their land use duration, land type association and frequency (Zandstra et al., 1981).

Pattern	*No. of months*	*Landtype*	*% Cultivable land*	*MCI* computation*	*Land use intensity*
A. Rice 1	5	Irrigated I	20	100 × .20 = 20	5/12 × .20 = .08
B. Rice 2-Cassava 2	12	Rainfed wetland	20	200 × .20 = 40	12/12 × .20 = .20
C. Cassava 1-Corn 4	12	Dryland	15	200 × .15 = 30	12/12 × .15 = .15
D. Corn 1/Cassava 3-Corn 4	12	Dryland	10	200 × .10 = 20	12/12 × .10 = .10
E. Corn 2-Corn 3-Corn 4	12	Dryland	10	300 × .10 = 30	12/12 × .10 = .10
F. Rice 3-Rice 4	7.5	Irrigated II	5	200 × .05 = 10	7.5/12 × .05 = .03
G. Tree crops	12	Dryland	5	100 × .05 = 5	12/12 × .05 = .05
H. Idle land	12	Rainfed wetland	5	0 × .05 = 0	0/12 × .05 = 0
I. Other	6	Dryland	10	100 × .10 = 10	6/12 × .10 = .05
Aggregate for the site †				165	.76

*Multiple Cropping Index
†Weighted on the basis of % cultivable land

harvesting dates with a double line. The acceptable range of planting dates for each crop should be indicated by a diagonal line covering the range of planting dates. A double line indicating the expected range of harvesting dates (not necessarily the same as range of planting dates) defines the period over which this crop is expected to occupy this plot. Write the name of the crop between the two lines. Then proceed with the next crop. In the case of a cropping sequence, use the same line, indicating again the range of planting dates and harvesting dates expected for this second crop. Again indicate the type of crop between the two lines. Continue this on the same line if a third crop is planted in sequence. In case any or more than one of these crops is combined with a crop planted in sequence or in relay, use the remaining rows in the diagram. Again indicate the range of planting and harvesting dates for each crop. The first example [below] shows a 'trans-

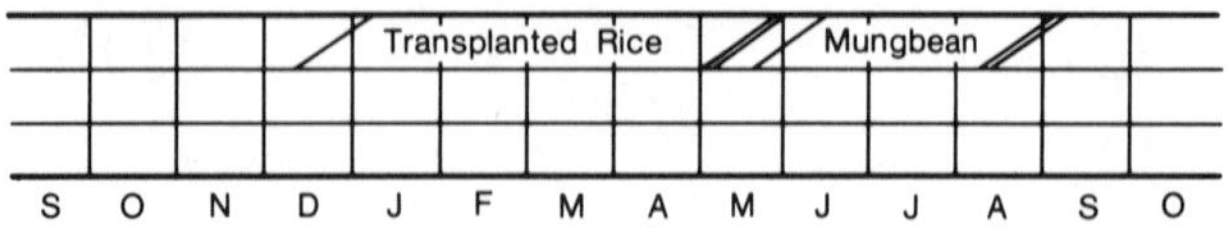

planted rice followed by mungbean' pattern in a region where the growing season starts towards the end of October. In this case the period of transplanting (not seeding) is indicated, as . . . when the cropping pattern will start to occupy this plot. The second example provided [second column] shows a cropping pattern of dry-seeded rice followed by a sorghum mungbean intercrop, in which melons are interplanted into the sorghum after

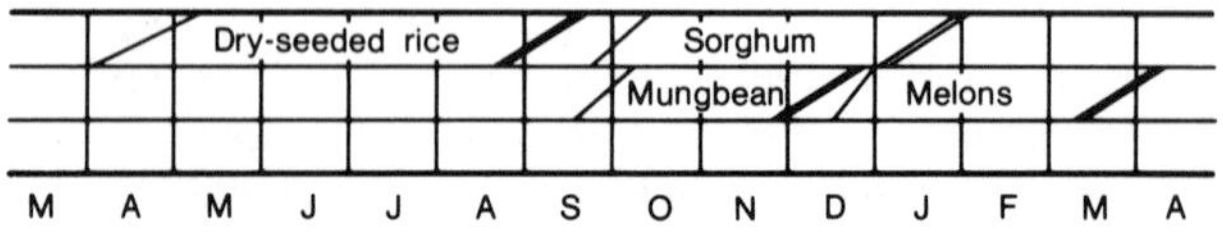

mungbeans are harvested. The growing season in this example starts in April.

"3. Enumerate the principal cropping system (combinations of cropping patterns on a farm) and percentage of all farms at the site that follow each system. Number each system by Roman numerals and check to insure that all patterns included are identified in [Table 5-H-2]. For ease of reference, the system may be named according to an important feature, as shown in [Table 5-H-3]."

Table 5-H-3. Description of cropping systems at a site (Adapted from Zandstra et al., 1981).

Cropping system	*System's name**	*Cropping pattern*	*% of farms on which observed*
I	Irrigated rice	A, F	20
II	Wetland rice-cassava	A, B	15
III	Mixed wetland-upland	B, C, D	25
IV	Mixed upland	C, D, E, G, I	30
Other			10

*This is any name convenient to label the system.

Table 5-I-1. Yearly recurring variations in wage rates and cash availability for a typical farm household (Zandstra et al., 1981).

		Month											
		Jan	*Feb*	*Mar*	*Apr*	*May*	*Jun*	*Jul*	*Aug*	*Sep*	*Oct*	*Nov*	*Dec*
What daily wage rates are usually paid during each month in the area?	(Highest												
	(Normal												
	(Lowest												
In which months is there additional labor (fill in A) or Loss of Labor (L) in the area?*													
When are farm labor requirements highest (H) and lowest (L)?													
How many adult family members are available for work?													
In what months is it most difficult to meet expenses?													

*From migrating labor, school holidays, religious or cultural reasons.

APPENDIX 5-I SAMPLE FORMS FOR DESCRIBING ON-FARM RESOURCES USED IN CROP PRODUCTION

"On-farm resources are factors available for crop production that can be modified and allocated by farmers and can be identified and measured within farm boundaries" (Zandstra et al., 1981). Tables 5-I-1 through 5-I-4 from Zandstra et al., 1981, illustrate formats for presenting data about on-farm resources that are needed in problem identification—see Part 1, Chapter 5. We have included tables for recording (1) yearly recurring variations in wage rates and cash availability for a typical farm household, (2) crop production capital of a typical farm, (3) sources of credit for agricultural production in the area, and (4) the technical experience and practices of a typical farmer.

Table 5-I-2. Crop production capital of a typical farm (Zandstra et al., 1981).

	% of farmers		Rental cost*
	Who own	*Who rent*	*(M̸)*
Water buffalo			
Oxen			
Sprayer			
Tractor			
Tiller			
Irrigation pump			
Thresher			
Rice blower			
Drier			
Other			

*Per ha or per day, as appropriate; M̸ = Monetary unit.

Table 5-I-3. Sources of credit for agricultural production in the area (Zandstra et al., 1981).

Source	*Size of credit**	*% of farm households†*	*% cost/ annum‡*
Official bank			
Family			
'Friends'			
Commercial lenders			
Others:			

*Average amount per loan transaction.

†Identifies how many farmers use each source (Can add to more or less than 100)

‡Calculated as:

$$\frac{\text{Interest + other costs } (\not{M})}{\text{Value of credit } (\not{M})} \times \frac{12}{\text{Duration (months)}} \times 100$$

Table 5-I-4. Technical experiences and practices of a typical [farmer] (Zandstra et al., 1981).

	Most have heard of	*Some have tried*	*Usual practice*
Rice planting methods			
Dry broadcast			
Dry furrows			
Wet broadcast			
Transplant: Uses seeder or transplanter			
Upland planting			
Row planting			
Intercropping			
Relay cropping, zero tillage			
Weeding after planting			
Harrowing			
Interrow cultivation			
Handweeding: Uses animals			
Handweeding: Uses herbicide			
Fertilization			
Basal			
Broadcast			
Band or hill application			
Split application			
Special equipment			
Insect control			
Manual			
Commercial insecticides			
Locally produced insecticides			
Use of sprayer			
Chemical disease control			

APPENDIX 5-J CONCEPTUAL MODELS

We spoke of conceptual models in Sec. 5.4.3. and briefly in Appendix 2-A. In this Appendix, we will go into more detail about alternative types of conceptual models. We begin with two examples that are little more than layouts of the farm and homestead and then provide two functional diagrams—one a mixed system in Nepal and the other one on livestock in Africa. Finally, we present two examples of rather detailed drawings of farming activities.

LAYOUTS OF FARMS AND HOMESTEADS

In this section, we present Figures 5-J-1 and 5-J-2 that represent increasing detail about the farm layout and the farming system. Fig. 5-J-1 is representative of a small farm in Guatemala (McDowell and Hildebrand, 1980). There, we can observe how the authors show the types of crops, location, and relative size of land devoted to each purpose. As a minimum, FSR&D teams should prepare such drawings when describing the more relevant types of farming systems in the research area.

Fig. 5-J-2 is another drawing from McDowell and Hildebrand (1980) that shows animal-keeping facilities, and trees. Such drawings can be sketched rapidly and are useful in studying the types and efficiencies of household activities.

FUNCTIONAL DIAGRAMS

Functional diagrams such as those in Figures 5-J-3 and 5-J-4 aid the researcher in understanding the processes affecting crops and animals. Fig. 5-J-3 is an example that is a bit more abstract than Figures 5-J-1 and 5-J-2. This diagram shows the variety of farming enterprises, the interactions of the farm household (Farm Resources) and the local community ('Community' Owned Production Resources), and the farmer's market transactions (Off-farm Interactions). Preparation of a diagram such as this requires understanding the farmers' activities, how the farm household fits into the local community, and the extent the farmers trade in the market. By the time the FSR&D team understands the underlying factors in this type of diagram, it will have a good start on being able to describe the farming system. This will aid the team considerably in identifying farmers' problems and opportunities and subsequently in planning on-farm research.

Fig. 5-J-4, which concerns water and energy flows through a plant-eating animal in Africa, aids the researcher in understanding the processes affecting crops and animals. Through presentation of flows of key elements—in this case water and energy—researchers are in a good position to understand some of the more complicated aspects of the systems they are studying. When quantitative values are placed on these flows and interactions, the researchers have a basis for setting up more complex models and for designing research activities. This diagram

Figure 5-J-1. Land use of a small farm typical of the western highlands of Guatemala (McDowell and Hildebrand, 1980).

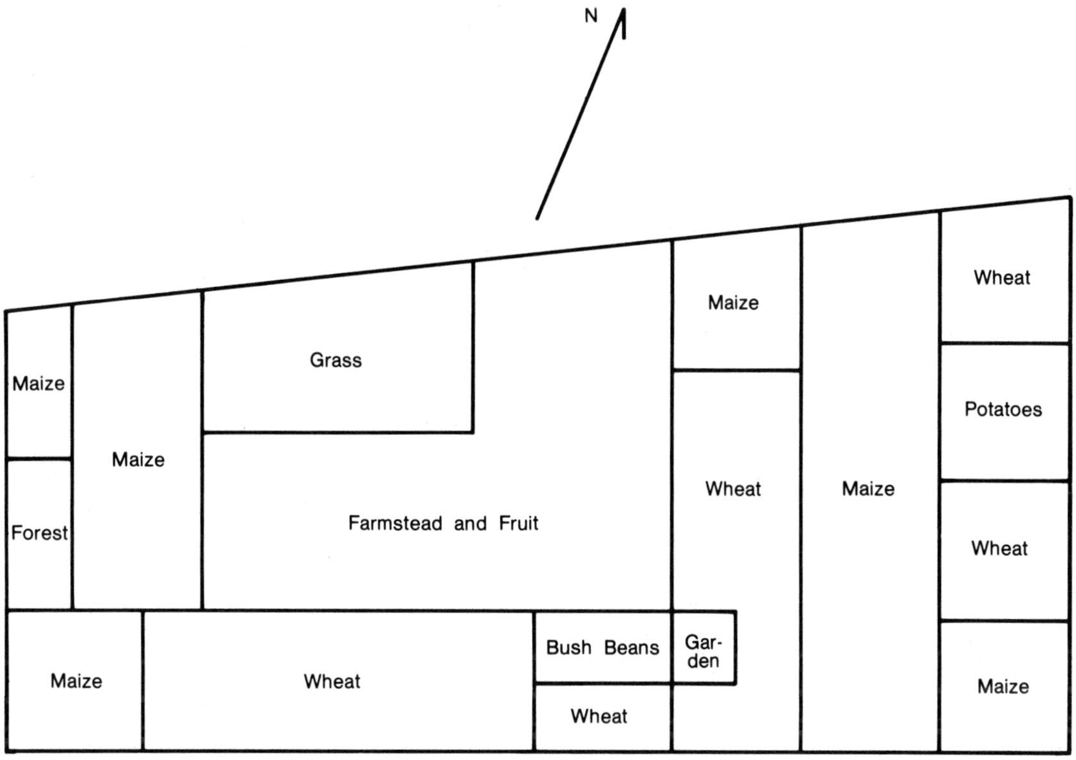

Figure 5-J-2. Family living quarters and animal-keeping facilities on a farm in the western highlands of Guatemala. [This figure is an enlargement of the farmstead and fruit area found in Figure 5-J-1] (McDowell and Hildebrand, 1980).

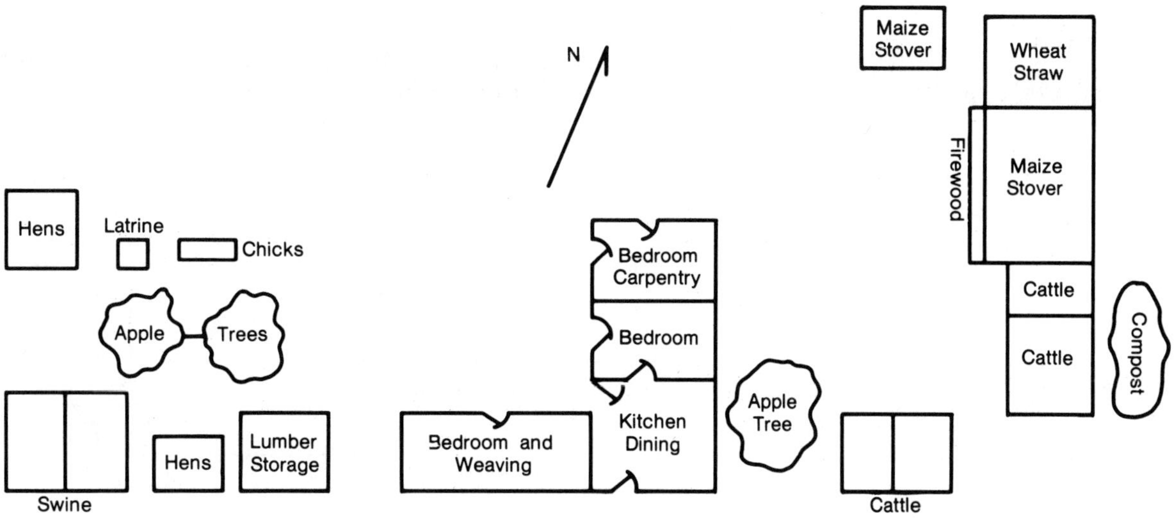

Figure 5-J-3. Conceptual model of the production system of a Nepalese hill farm (Harwood, 1979).

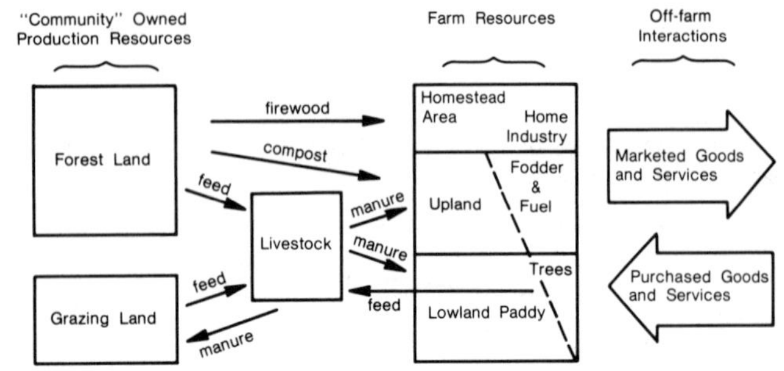

illustrates concern over flows of key resources, which is typical of other efforts, such as the measure of water balances—i.e., water inflows and outflows—for irrigation systems.

DETAILED DRAWINGS

Figures 5-J-5 and 5-J-6 are the result of considerable effort to understand the nature of the systems under study. Fig. 5-J-5 by McDowell and Hildebrand (1980) elaborates on the same type of figure presented in Chapter 5—i.e., Fig. 5-2—by providing data for each of the household activities. The values shown in the boxes represent the percentages associated with these activities. For example, the box labeled *crops* shows that the farm produced maize and the other boxes show how the maize was distributed: namely, 70 percent as *feed* for the animals, 19 percent as *food* for the family, 10 percent going to the *market*, and 1 percent as *seed*.

Fig. 5-J-6 by Hart conforms to the ecological means of presentation, as introduced in Appendix 2-A. More detailed work by Hart (1980a) allows these interactions to be quantified also. But in this case, the quantification is in terms of physical quantities such as kilograms of maize fed to the chickens and consumed by the family and the kilograms of potatoes sold in the market, and the corresponding cash payments. An interesting finding from a similar study of a Honduran family (Hart, 1980b) is that the family stored maize much as one would money, selling small amounts to cover household requirements and major amounts to cover production requirements.

Figure 5-J-4. Water and energy flow through a herbivore in tropical Africa (ILCA, 1980).

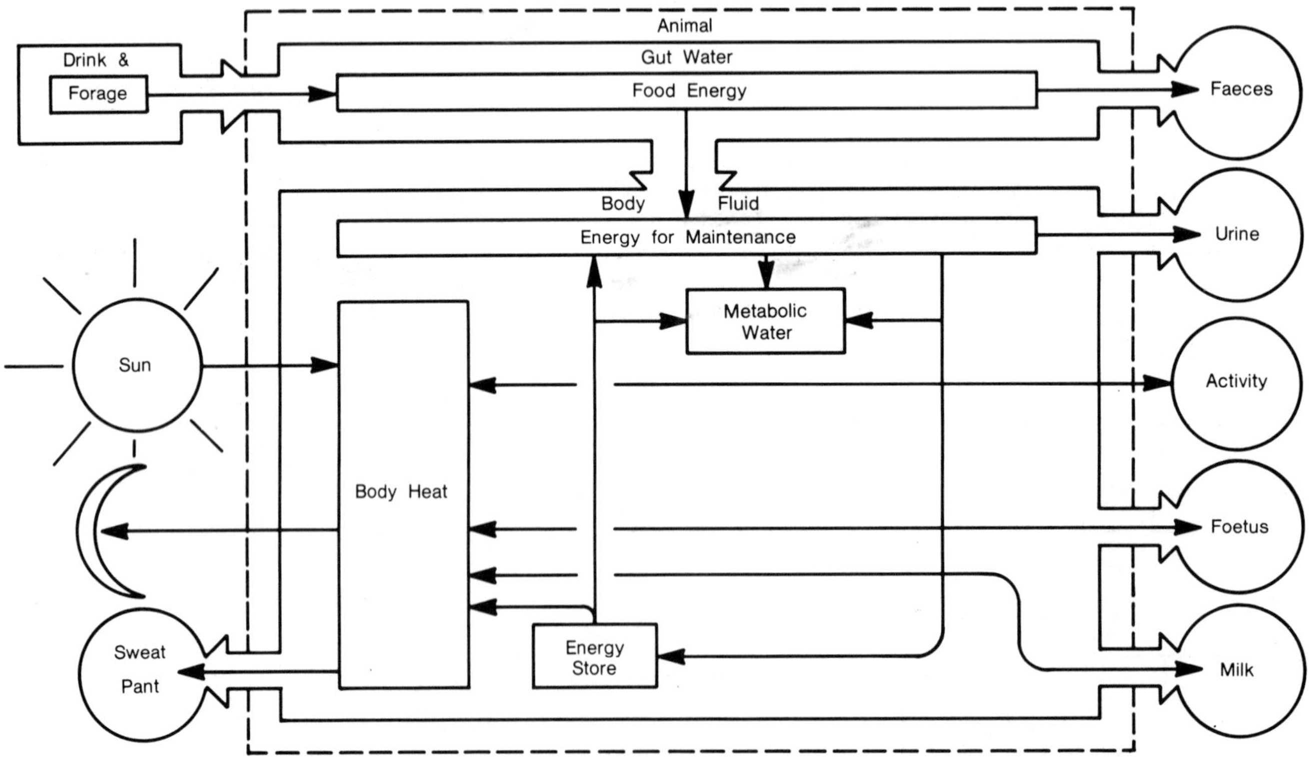

Figure 5-J-5. Distribution of labor, income (sales of products or off-farm labor), and purchases of exogenous sources for small crop/livestock farm in the western highlands of Guatemala (McDowell and Hildebrand, 1980).

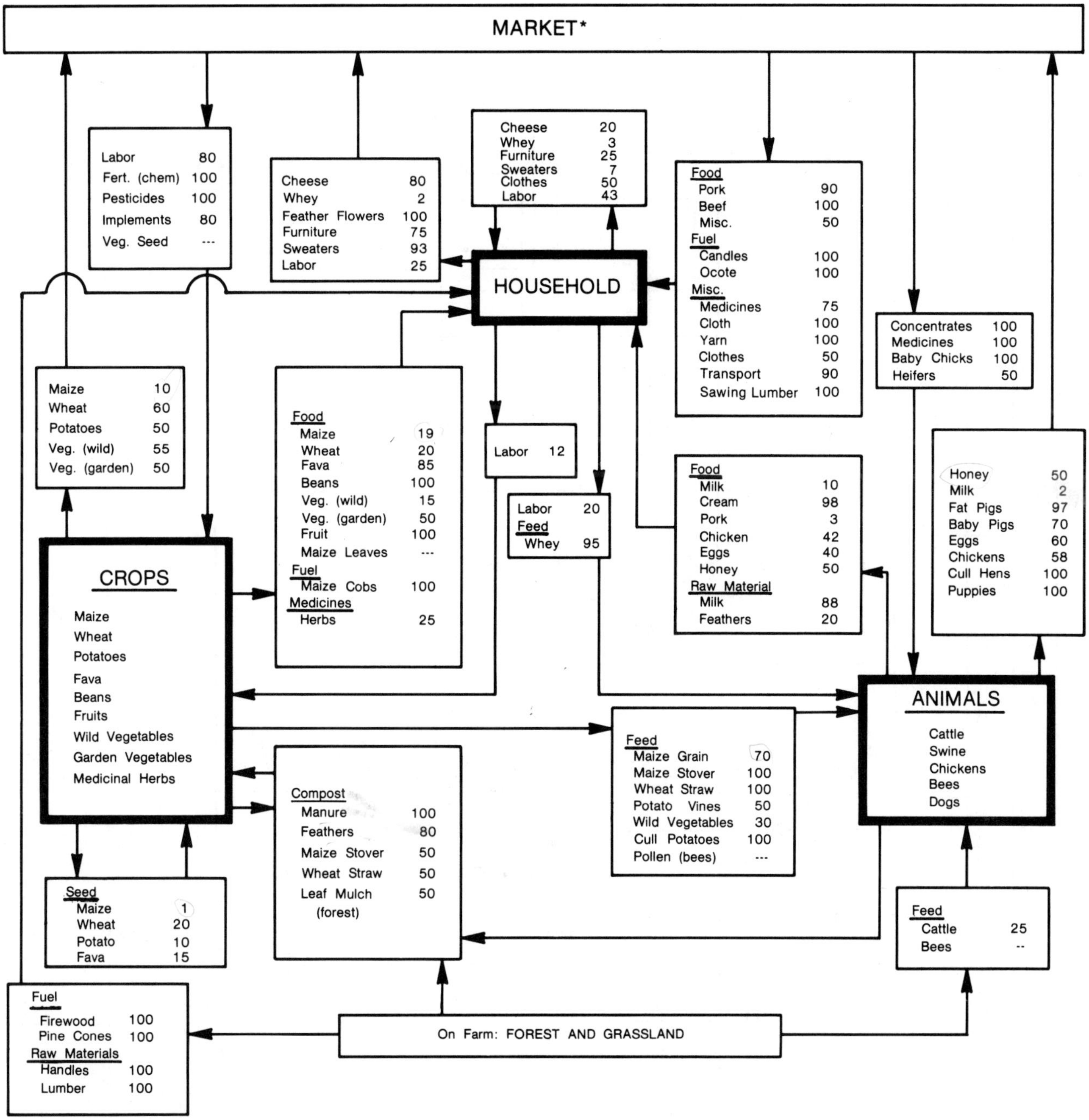

*Values are percentages

Figure 5-J-6. A typical small commercial farm system in La Esperanza, Honduras (Hart, 1980a).

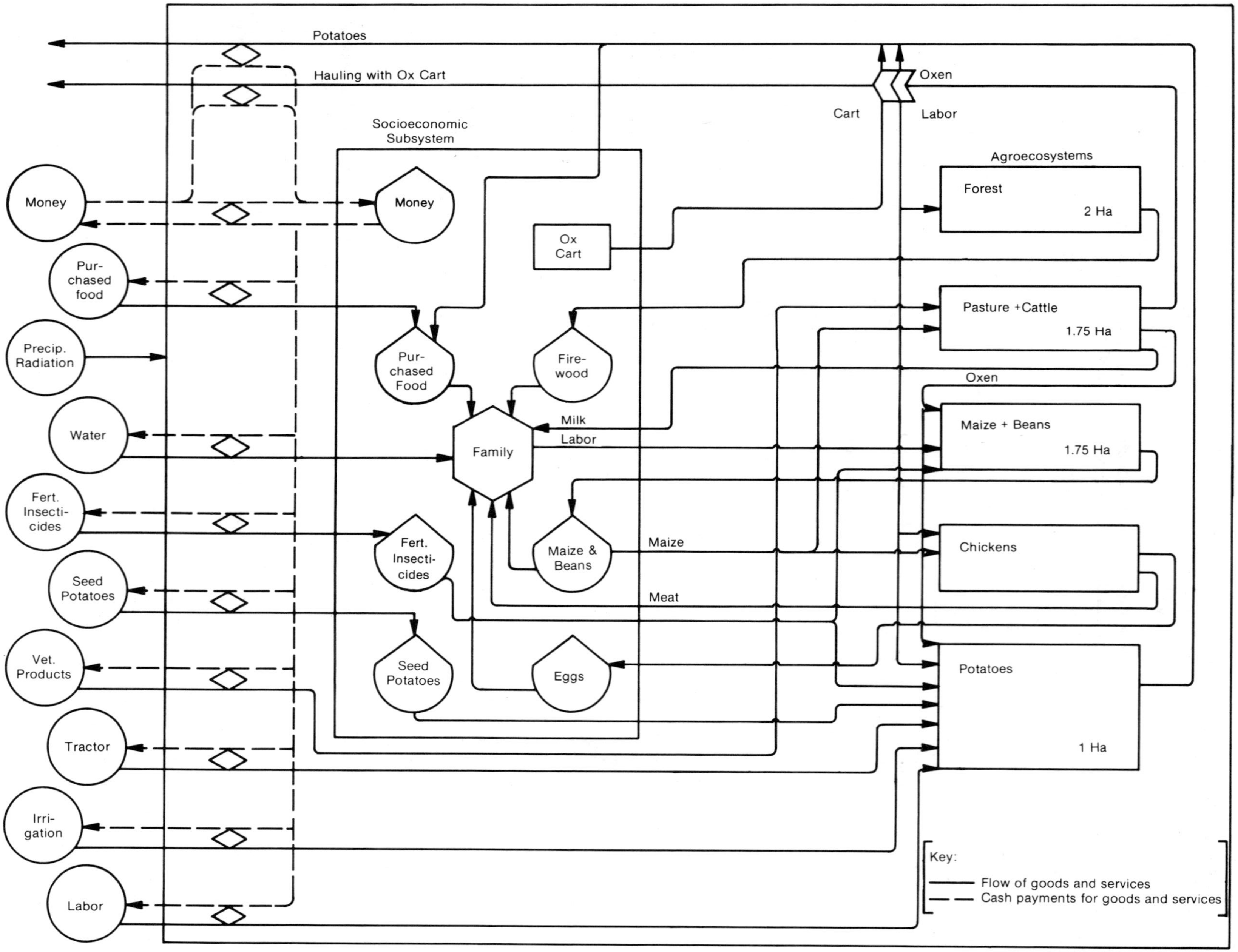

APPENDIX 5-K
MIXED CROPPING AND LIVESTOCK SYSTEMS

In Sec. 5.4.3. in Chapter 5, we discussed how conceptual models could be used in analyzing farming systems. Fig. 5-2 in that section depicted a cropping and livestock system typical for humid upland conditions in Asia (McDowell and Hildebrand, 1980). That figure showed the linkages between crops and livestock, the household, and the market. To provide more background on both the value of conceptual modeling and the model's linkages between components, we provide a discussion of this Asian system in Part 1 of this appendix; and in Part 2 of this appendix, we provide a more general discussion of the linkages between cropping and livestock systems typical of semi-arid Africa.

PART 1: DESCRIPTION OF A HUMID UPLAND FARMING SYSTEM IN ASIA

In this part, we quote directly from McDowell and Hildebrand (1980) concerning their description of the mixed farming system for humid upland Asia:

> "The upland system [Fig. 5-2] is widespread over the humid tropics of Asia. There are well-developed farmsteads with permanent, cleared fields but with no bunding and no irrigation. The major crops are rice, maize, cassava, wheat, kenaf, sorghum, and beans. Most households have small numbers of several species of animals, with swine and poultry prevalent. Following these in popularity are cattle and buffalo. Sheep and goat numbers are normally low. Where tall-growing crops (maize and sorghum) are cultivated, cattle are kept to utilize crop residues. In rice areas buffalo predominate. Frequently, one or two buffalo or cattle are kept for use in land preparation and to provide transport for crops, crop residues, and to some extent members of the family. Swine are tethered or penned, and cattle or buffalo are tethered at night in order that manures may be collected and to avoid theft. The manures are frequently composted with crop residues. Poultry are usually free-roving.
>
> "Fuel is not yet a severe problem in much of the humid-upland systems but is becoming increasingly so as more and more forests are cleared.
>
> "The farm infrastructure is variable, developed for some areas but extremely limited for others. Land tenure and social services are also variable. Many upland areas are distant from markets.
>
> "The land ranges from rolling hills to steep slopes. The soils have moderate fertility, and in general drainage is good. Erosion hazards are classed as moderate. The rainfall is seasonal and erratic within the rainy season, thus periods of moisture stress are frequent.
>
> "Among the assets of this system are some possibility for multiple cropping, excellent potential for crop and animal integration, good potential for small-holder dairying with crop rotation, and feasibility of cooperative production and marketing. Rice is milled at the village level; therefore, rice bran and other by-products are available for supplementary feeding of animals. Some of the current limitations to increased output are inadequate or absent credit and animal health services, insufficient power for tillage [Duff (1978)], and limited access to markets. In addition, farms are often so geographically fragmented that much potential for grazing is lost. Considering the assets and liabilities, the potential appears good for change through increased cropping intensity, especially of fodder crops for animal feeding; increased animal holdings in order that farmers could have scheduled outputs for marketing; expanded farm infrastructure; extended use of draft power; and larger milk supplies.
>
> "With time, the upland areas of Asia promise to meet a rising demand for milk and meat through greater crop/animal integration [De Boer and Weisblat (1978)].
>
> "Integration on small farms will minimize the need for feed concentrates in animal production, and there is some potential for on-farm self-sufficiency in power (gasohol, biogas, and animal draft) based on conversion of sweet potatoes and cassava."

PART 2: LINKAGES BETWEEN CROPPING AND LIVESTOCK SYSTEMS IN SEMI-ARID AFRICA

To illustrate the importance of understanding the linkages between cropping and livestock, we have excerpted from McCown et al. (1979),[1] which contains examples for semiarid Africa. In these excerpts, the authors have isolated various interactions between cultivation and livestock production. Following are sections on (1) the basic differences between the two production systems, (2) interactions when agricultural and pastoral production take place in different management units, and (3) interactions when agricultural and pastoral production take place within the same management unit. McCown et al. (1979) reported:

> "Although the majority of people in semi-arid Africa sustain themselves primarily by growing crops, this means of production is not practiced by all, nor is it the sole means practiced by the majority. Because of inadequate rainfall and high evaporation rates, average crop yields are low, and the risk of crop failure is high. Traditionally the inhabitants of these regions have relied on domestic grazing

[1]This material is reproduced by permission of the authors and the publisher. Copyright (c) Springer-Verlag, Berlin, Heidelberg, 1979.

animals to supplement their food supply. The ultimate degree of this dependence is embodied in pure pastoralism; however, semi-arid Africa has a wide range of variation, both in the nature and the degree of economic dependency on livestock.

* * *

"BASIC DIFFERENCES BETWEEN THE TWO PRODUCTION SYSTEMS

". . . crop production and animal production tend to take place in different management units, typically belonging to different ethnic groups. This dichotomy is, however, far from absolute: most pastoralists grow some crops, and most farmers keep some animals. Variation in the balance struck between the two activities and the nature and degree of interdependencies between the activities must be viewed as different forms of adaptation. Understanding the patterns of variation in these forms in space and time requires, in addition to information on environmental conditions, an appreciation of the major differences between the two production systems."

The following are some comparisons:

"1. Both systems require the same kind of production factors: land (fields, pastures), capital (seeds, herds), and labor.

"2. In neither system does the right to use of land usually require capital expenditure. (In this geographical region, grazing rights are almost universally free.) The situation with respect to cultivation rights is more variable, but purchase or rent of land is exceptional.

"3. The amount of capital required for subsistence in pastoralism is high relative to that in agriculture. This is related to the differences in the annual rates of returns on seed (10-50x) and on animals (0.1-1x, in terms of reproduction).

"4. In agriculture, labor is invested in the land, and the requirements are highly seasonal; in pastoralism, labor is invested in the herd, and the requirements are relatively constant.

"5. Agriculture is labor-intensive, with increase in labor during the peak requirements periods providing the greatest marginal returns; pastoralism is capital-intensive, with increase in herd size providing greatest marginal returns.

* * *

"In the light of these differences between agricultural and pastoral production systems, it is to be expected that successful farmers in semi-arid regions might invest surplus in livestock, and that unsuccessful pastoralists might resort to farming. In the discussion of various types of interdependencies which follows the importance of both forms of adaptation can be seen.

"INTERACTIONS WHEN AGRICULTURAL AND PASTORAL PRODUCTION TAKE PLACE IN DIFFERENT MANAGEMENT UNITS

"Historically, the interactions between pastoralists and agriculturalists involving agricultural economic consequences can be classified as nonexistent, positive, or negative. Examples are given here of each class, as well as some indication of recent trends in the nature and importance of each interaction, or linkage.

"No Supporting Linkage

"In this case, pastoralists and agriculturalists are self-sufficient units, i.e., the pastoralists rely on animal products and the agriculturalists have a consumption profile based on agricultural products. Traditional relations between Masai and Kikuyu might illustrate this situation, but because of pacification and expanded consumption profiles resulting from improved communication and education, it is unlikely that distinct examples could still be found today.

"Ecological Linkage

"In this case, the practice of one activity influences the other through its effects on the ecosystem. For example, during the dry season natural forage is in short supply, and the quality is normally very low. The residues of most crops, which are of little or no value to the cultivator, provide a superior diet for the pastoralist's herds. At the same time, manure deposited on the fields as the cattle graze is beneficial for the subsequent crop.

* * *

"Exchange Linkage

"This linkage is mediated by the transactions between agriculturalists and pastoralists involving goods, or both goods and services. Exchange of goods is the typical linkage when agriculturalists and pastoralists have similar consumption profiles, and the livelihood of one group is partially dependent on the products of the other. Examples are numerous. Traditionally, nomadic pastoralists such as Moors, Tuareg, Fulani (Peulh), and Baggara camp for at least part of the year in close proximity to agricultural areas, during which time they exchange products, e.g., milk, ghee (butterfat), meat, and hides, for millet and sorghum.

* * *

"Competition Linkage

"This occurs when the same resource (land) is ecologically suitable for both agricultural and pastoral production. Where this is the case, relative political power is likely to determine the land use pattern. During the centuries just prior to colonization, much control was exercised by belligerent pastoralist groups. With pacification and increased central authority, the balance of power has shifted decisively to the cultivators.

"At present, two basic trends are profoundly altering the land use relationships between pastoralists and agriculturalists. To begin with, land is becoming increasingly scarce. The rapid growth of the rural population is expanding cultivation at the expense of the best grazing land. Increased cultivation of industrial crops has hastened this trend, as has government reservation of public lands (van Raaij, 1974).

"The second trend is that of increased individualization of land tenure. Although this rarely means that land is individually owned, the traditional 'free range' philosophy whereby livestock has free access to water and fodder reserves of rangelands, fallow lands, and harvested fields is increasingly challenged by cultivators who want to control access to their holdings. Cultivators in northern Nigeria will admit that there long existed a system of tacit consent whereby Fulani pastoralists could graze their cattle on fields as soon as the harvest was completed, but there is a general feeling among cultivators that times have changed. Particularly in densely populated and commercialized cropping areas tensions and conflicts arise during the post-harvest period, and court records reveal an annual spate of cases against pastoralists for damage to late-maturing crops (van Raaij, 1974:36). Competition is minimal in the rainy season when supplies of feed and water are generally ample. Pastoralists normally go to drier areas unsuited for cultivation, in part because the feed value of range species in these areas is superior. It is in the dry season that the competition is acute, with permanent water sources as the focal points.

* * *

"INTERACTIONS WHEN AGRICULTURAL AND PASTORAL PRODUCTION TAKE PLACE WITHIN THE SAME MANAGEMENT UNIT

"Livestock ownership benefits a cultivator either directly, or indirectly via benefits to crop production. This section discusses first investment and food linkages, which are directly beneficial, and then manure, draft, and fodder linkages.

"Investment Linkage

"If . . . livestock are present in the economy, this implies an opportunity for investment and consequently an incentive to increased agricultural production. One would thus expect that the cultivators would be stimulated to achieve a higher level of production in order to convert value from the agricultural to the pastoral sector. This strategy implies growth in the enterprise; it gives security both in terms of food (animals can be eaten or transacted), and in terms of mutual support (animal distribution can be used as a means for obtaining assistance from other people). The fact that livestock, especially cattle, are so highly valued as an investment all over Africa makes this the most typical form of integration of livestock into agricultural enterprises. A great many, perhaps most, of the cattle owned by cultivators serve little other purpose.

"It is common practice among the Wolof (Ouolof) of Senegal, the Mossi of Upper Volta, the Bambara and other tribes of southern Mali, the Hausa of northern Nigeria, and the Fur of the Sudan to entrust their cattle to the care of neighboring pastoralists. Since the benefits of keeping the cattle close at hand are appreciable (regular milk supply, lower risk to the investor, and, in some cases, manure production), one would expect to find contractual herding only where local resources such as forage, labor for herding, and animal husbandry skills are traditionally inadequate. Nevertheless, this practice is found also in communities that possess enough grazing resources to maintain modest herds but which have invested in more cattle than these local resources can support.

* * *

"Food Linkage

"Usually the investment linkage is supported by a food linkage, as in the case where the agricultural producers supplement their diet with meat and milk products from their herds and flocks. Some groups rely heavily on milk from large animals. The Songhai of Mali, for example, rely on cattle, and the Somalis rely on cattle and camels. Even in communities that lack a major cattle sector, the more successful farmers often keep cattle (Hill, 1972:217). More commonly, however, cultivators keep small stock, such as poultry, sheep, and goats, to supply themselves with animal products.

* * *

"Manure Linkage

"In traditional agriculture the manure linkage is associated with intensive cultivation. Intensive cultivation systems are usually associated with high population densities. This can be seen in refugee hill areas of Cameroon, Nigeria, Mali (Morgan and Pugh, 1969:104), and Wakara Island, Lake Victoria (Ruthenberg, 1971:118). Marginal returns of labor are frequently less under these intensive systems than under many extensive systems of cultivation.

"With increasing political security, population pressures have been relieved through migration from the hill areas to the plains. Such migration is typically accompanied by a change in the cultivation system. Given the preferred balance between labor and consumption, easier access to land implies that a point will be reached sooner or later where optimization of returns on land and labor give way to extensive shifting cultivation with less need for manure. With increasing population pressure and better agricultural technology, this trend is being reversed again, and there is increasing reliance on manure to increase soil fertility.

* * *

"Draft Linkage

"This linkage was rarely found in traditional agriculture in semi-arid regions. There were, however, some interesting exceptions. Along the Nile, oxen provided the draft power both for plowing and for the Persian wheels used in irrigation. In the Kheiran depressions in Kordofan (Sudan), an unusual farming system has evolved. The system involves a combination of intensive agriculture based on irrigation from groundwater with ox-drawn water wheels in the dry season, and extensive shifting hoe cultivation in higher-lying areas in the rainy season. With increasing involvement in the market economy, the draft linkage has grown in importance, especially in West Africa.

* * *

"Fodder Linkage

"For livestock as well as for humans, the dry season is a nutritional bottleneck. Except in the permanently wet lowlands, no pasturage grows until the onset of the rainy season; meanwhile, herds consume the stores of standing hay. To make matters worse, the nutritive value of this forage steadily decreases as its tissues cure and weather. There is, therefore, a great need to produce forage crops superior in dry-season quality to range forage. Moreover, a sown leguminous fodder crop or ley would be more effective in restoring soil fertility than a bush or grass fallow. This practice, more than any other, is the key to mixed farming along the lines developed in Europe and North America. The main question is whether the value of livestock products and service and enhanced soil fertility outweigh the value realized from alternative uses of labor and land. In most of the dry tropics of Africa the pressure on land is not yet sufficient, nor is the price structure favorable, for a fodder linkage based on sown leys.

"An important fodder linkage of a different sort does occur widely where a crop that provides either human or industrial products also provides dry-season feed as a by-product. Normally, of course, the primary products guarantee sufficient returns on land and labor to justify cultivation."

APPENDIX 5-L
DATA COLLECTION IN A RURAL SETTING

Collecting data in rural settings can present problems that, if not considered at the outset, will hamper FSR&D's effectiveness. Under Sec. 5.8 in Chapter 5, we provided general comments about data collection. In this appendix, we provide additional background that is appropriate for both informal and formal methods of data collection.

Following are sections on securing local leaders' support, obtaining cooperation from farmers, incentives for interviewees, conducting interviews, additional comments on interviews, and problems in interviewing. Much of this appendix is based on the work of Bernsten (1979), Byerlee et al. (1980), and Kearl (1976).

SECURING LOCAL LEADERS' SUPPORT

Most farmers will be suspicious and yet curious about researchers who enter their village. FSR&D researchers need to recognize this fact and understand that small villages are complicated social systems.

Thus, the FSR&D team needs to determine how best to approach farmers within their setting. This often means identifying local leaders and seeking their support. These leaders may be officials of the affected ministries, local authorities, traditional leaders, merchants, and others. Also, social scientists who are familiar with the village structures within the research and target areas can help the team in this regard.

In some cases the team will find that village leaders are highly esteemed and in other cases the leaders are not respected. For cases where the leaders are highly esteemed, the team needs to seek the leaders' support. Gaining their support

- reassures the farmers of the team's intentions, and thereby reduces farmers' suspicions and encourages them to cooperate with the team
- helps elicit meaningful answers from the survey's respondents
- allows the team to check the accuracy of the interview data with the leaders.

When talking with leaders, the researchers should

- explain the project
- emphasize that the farmers' answers will be kept confidential
- stress that no names or individual responses will be given to governmental agencies
- explain how the farmers will be selected or sampled
- keep the leaders informed on the project's progress.

Some researchers will sample—i.e. select—the names of farmers with the leaders present. This approach assures the leaders of the techniques being used. If the names of the leaders are not pulled in the samples, some researchers interview the leaders anyway and then exclude those interviews from the data analyses (Ogunfowora, 1976a).

Generally, gaining the support of the leaders takes time and researchers should be patient. Some researchers have found staying in the village for some time before beginning the survey, attending social gatherings, eating with the farmers and leaders, becoming known, and otherwise associating with leaders helps the survey go more smoothly (El Hadari, 1976a).

Failure to make the necessary contacts with leaders in advance of surveys has produced situations of uncooperativeness by village farmers because some leaders have reacted negatively to the surveys. El Hadari (1976b) reported an incident in which the first day's interviewing went smoothly. On the second day, the villagers were uncooperative. The researchers finally learned that the key merchant had threatened the villagers. In this case, the survey was designed to collect information on buying, selling, and lending. The merchent was trying to protect his interests and keep outsiders from learning of his transactions.

OBTAINING COOPERATION FROM FARMERS

Once the researchers gain the leader's support, the researcher should then seek the support of the farmers before beginning the interviews. Some researchers increase the awareness of the project through radio. Other researchers begin with a village meeting in which they explain the project.

From here, the interviewers must then contact specific farmers who were sampled and are to be interviewed. In Appendix 5-W, we discuss the importance of carefully selecting and training interviewers. Researchers need to be aware that the farmers' perceptions of the interviewers may influence the farmers' cooperativeness. Many of the interviewers' characteristics, such as the sex, age, dress, tone of voice, politeness, respectfulness, and social and ethnic group, influence the farmers' answers (Bernsten, 1979).

In the initial contact with farmers, the interviewers try to gain the farmers' cooperation and to schedule the interview. The interviewers need to explain how the farmer was selected, what information will be needed, how the information will be used, the confidentiality of the information, and the potential benefits to the farmer. The interviewers should encourage questions from the farmers and answer their questions.

In scheduling the interview, the interviewers need to arrange a time and place convenient for the farmer. Generally, researchers should not schedule a survey when the farmer is busy. For this, the interviewers should know the farmers' daily and weekly work cycles (Bernsten, 1979).

The place of the interview should be where the farmer can be relaxed and not distracted. Some interviews work best in the home, others under a shade tree, and others in the farmers' fields. Interviewing farmers in their fields provides the interviewers with a chance to observe the crops. In some cases, the interviewers might offer to help the farmer for awhile before beginning the interview (Byerlee et al., 1980).

INCENTIVES FOR INTERVIEWEES

Generally, experienced researchers avoid giving incentives to individual farmers for being interviewed. Such incentives can create budget problems and dissatisfaction in the village among those who were not interviewed.

However, some researchers have found it necessary to reward individuals who spend excessively long times with the researchers, and tradition, in some cultures, dictates a token gift for the chiefs, priests, or elders. Such gifts are indirectly meant for the land or the earth god (Nabila, 1976a). Norman (1976a) justified the giving of something to the whole community rather than to individuals. For example, the team might ensure the community receives fertilizers or credit. To approach the issue of incentives realistically, the team needs to know the culture and farmers' customs.

CONDUCTING INTERVIEWS

The interviewer should first greet the farmers according to local custom and treat the farmers with respect. If the local language requires it, the interviewers should use the "polite" form of address. The interviewer should not interview the farmer from a vehicle, and the interviewer should try to avoid positioning himself above the farmer. If possible, the interviewer should try to talk to the farmer alone, with no other farmers, friends, or family present.

The conversation should open with locally accepted "polite talk" and then the interviewer must explain the purpose of the research. In some cultures taking notes in the respondent's presence may be resented. In the case of lengthy, detailed questionnaires, this cannot be avoided. For short questionnaires or informal interviews, the interviewer can memorize the questions and later record the results.

In beginning, the interviewer should make clear to the farmer that there are no "correct" or "incorrect" answers—that the interviewer is interested in what the farmer does and thinks. The interviewer should be careful of his or her responses or reactions to the farmer's answers and avoid unwittingly guiding the farmer by indicating approval or disapproval of an answer. If a question causes silence, or the farmer seems unwilling or unable to answer, the question might be rephrased.

The interviewer must be sensitive to the farmer's available time. In general, 1½ hours is the maximum for a single interview, while interviews of 35 to 40 minutes are best. Of course, if the farmer is in a talkative mood, the interviewer may want to continue longer.

After all relevant topics have been covered, or the farmer's attention is needed elsewhere, the interview should end. As at the beginning of the interview, the interviewer must be courteous and respectful. The farmer should be thanked for his or her time, and left with the proper local farewell. And before leaving, the interviewer may arrange to meet the farmer again, or to explain the next steps of the investigation.

ADDITIONAL COMMENTS ON INTERVIEWS

The team members must be willing to accept local customs, living conditions, and culture if they want to interview farmers and obtain useful information. If team members complain about living conditions, the local people may consider the interviewers or researchers as being hypocrites. Furthermore, Nabila (1976b) suggested that the interviewers and researchers be willing to conduct the interviews in a relaxed manner. This includes being ready to sit on the bare ground and pay respects to the leaders if the tradition demands it.

Other researchers have found spending time at the beginning of the interview listening to farmers' problems puts the farmer in a more cooperative mood (Gafsi, 1976a).

Sometimes researchers or interviewers will encounter individuals who want to be interviewed but whose names were not sampled. Refusing to interview them may antagonize them and create problems when interviewers seek information from those whose names were sampled. To handle such situations, Nabila (1976c) reported training the interviewers to go through the motions of interviewing those who insisted on being interviewed. Thus, questions were asked and notes taken, but the data were not added to the sample.

PROBLEMS IN INTERVIEWING

Several problems might develop during the interviews, which we will now discuss as being interviewer-based, farmer-based, and miscellaneous.

INTERVIEWER-BASED PROBLEMS

Some interviewers begin paraphrasing the questions, others change the order of the questions and still others rush through the interview (Bernsten, 1979).

Each of the changes an interviewer makes in the questionnaire can introduce bias into the data collected. In the training sessions, and in the interviewers' instruction manual, the researchers must instruct the interviewers how to read each question, how to rephrase questions, how to probe when necessary, and how to pace an interview. We discuss the training of interviewers and the interviewers' instruction manual in Appendix 5-W.

FARMER-BASED PROBLEMS

Some farmers may refuse to cooperate, others will be reluctant to cooperate, and still others will be missed (Bernsten, 1979).

Even after working with local officials and leaders, learning the local customs and going through all other suggested steps, some farmers or individuals will refuse or be reluctant to cooperate.

Often farmers are uncooperative for reasons, such as (1) they do not see any relationship between the research and their well-being, (2) they believe the interviewers are seeking information for tax purposes, (3) they detest the invasion of their privacy concerning income and indebtedness, or (4) they want some tangible reward for being interviewed (Ogunfowora, 1976b).

In some cases adequately explaining the purpose of the project, how farmers were selected, and the criteria for the interviews overcomes the problem. An alternative approach is to assign a different interviewer to the farmer who is reluctant to cooperate. Furthermore, having a leader force farmers to be interviewed usually produces poor results.

Sometimes interviewers cannot locate the farmers they are to interview. Again, the procedures to handle such situations should be included in the interviewers' instruction manual and in the training sessions. Bernsten (1979) recommended first trying to find out when the farmer would return and then returning at that time to reschedule the interview. The researchers need to establish a standard rule for all interviewers to follow for such problems. A commonly used guideline is to make three attempts at different times of the day on different days of the week. If the farmer cannot be reached, then an alternative individual is interviewed. This is where sampling with replacements comes in, as we discuss in Appendix 5-V. Such changes usually do not create serious problems unless the number of missing individuals exceeds 15 percent of the individuals randomly selected (Bernsten, 1979).

MISCELLANEOUS PROBLEMS

Several problems might delay or hamper the speed of the survey. First, inappropriate individuals may be selected despite the random sampling designs. If, after beginning the interview, the interviewer learns an inappropriate individual has been selected, the interviewer should terminate the interview (Bernsten, 1979). Second, interviewers and researchers need to be prepared for poor weather conditions that prevent or delay reaching the individuals to be interviewed. In such cases, the researchers need to decide whether to delay or to eliminate the interviews from the sample. However, eliminating names may introduce biases into the data. Third, on occasion, the interviewers have transporation that breaks down, in which case the interviewer needs to arrange for alternative transportation or wait for repairs.

The aforementioned situations and circumstances are common to many surveys. Thus, the researchers need to be prepared to solve these problems as they arise. The best approach is for the researchers to (1) try to anticipate such problems, (2) train the interviewers to handle as many of the problems as possible, and (3) come to the interviewers' aid for those problems they cannot handle.

APPENDIX 5-M GATHERING DATA ABOUT WOMEN

In many cultures, women play a major role in small-scale agriculture. They produce crops and tend livestock, as well as care for the family. Consequently, FSR&D teams must understand how women participate in agricultural activities if the new technologies are to be fully relevant.

This appendix, which ties in with the discussion in Sec. 5.8. on data collection, begins with background information on women and then provides selections on collecting data about rural women from *Learning About Rural Women*, a recent publication edited by Zeidenstein (1979).

MEASURING WOMEN'S PARTICIPATION IN AGRICULTURE

The necessary information on women varies from situation to situation. But as a minimum, researchers should learn the following:

- women's role in making decisions about agricultural production and marketing
- the amount and type of agricultural work women perform
- women's level of knowledge of agricultural technologies and their production tasks
- women's control over and use of agricultural resources
- women's access to services such as extension, credit, and training.

To obtain the most accurate information, researchers should formulate specific questions and, whenever possible, interview women directly.

Because cultural and religious practices sometimes keep male researchers from talking directly with women, FSR&D teams may need to have women interview women within the household. Additional information about rural women can be obtained from rural health centers, family planning clinics, and rural development groups.

Another problem linked to gathering accurate data on women is that researchers as well as women tend to be influenced by their ideas of what women ought to do. This will influence how they perceive women's roles. Consequently, FSR&D teams should obtain answers to the following questions:

- Who participates in decision making regarding specific agricultural tasks?
- Who is generally in charge of carrying out the task?
- What is the average intensity of participation?
- How much time is spent on the task?

These and related issues are discussed in the remainder of this appendix.

EXCERPTS FROM *LEARNING ABOUT RURAL WOMEN*

As part of the Population Council's Studies in Family Planning, Zeidenstein (1979) prepared a special issue on rural women's roles and status in different societies. We have selected portions of five articles from this issue that deal with methods of gathering information about farm women.

FRAMEWORK FOR ANALYZING TIME ALLOCATIONS

McSweeney (1979) reported on an approach for gathering data on women's activities. This was adapted from the work of the African Training and Research Center for Women of the United Nations' Economic Commission for Africa. This approach compares time spent by women and men on agricultural and household activities. Results of such a study for the village of Zimtenga in the north-central zone of Upper Volta are shown in Table 5-M-1.

Table 5-M-1. Comparisons of time allocations to rural activities between women and men for the first 14 waking hours in Zimtenga, Upper Volta. Reprinted with the permission of the Population Council, from Brenda Gael McSweeney, "Collection and analysis of data on rural women's time use," *Studies in Family Planning* 10, no. 11/12 (November/December 1979), p. 381.

Activity	*Average time allocated (minutes)** *Women*	*Men*
A. Production, supply, distribution	367	202
1. Food and cash crop production	178	186
Sowing	69	4
Weeding, tilling	35	108
Harvesting	39	6
Travel between fields	30	19
Gathering wild crops	4	2
Other crop-production activities	1	47
2. Domestic food storage	4	1
3. Food processing	132	10
Grinding, pounding grain	108	0
Winnowing	8	0
Threshing	4	0
Other processing activities	12	10
4. Animal husbandry	4	3
5. Marketing	4	0
6. Brewing	1	0
7. Water supply	38	0
8. Fuel supply	6	2
B. Crafts and other professions	45	156
1. Straw work	0	111
2. Spinning cotton	2	0
3. Tailoring	2	10
4. Midwifery	41	0

Table 5-M-1 (cont.)

Activity	*Average time allocated (minutes)**	
	Women	*Men*
5. Other crafts/professions (e.g., metal work, pottery, weaving cloth, beekeeping, etc.)	0	35
C. Community	27	91
1. Community projects	27	0
2. Other community obligations	0	91
D. Household	148	4
1. Rearing, initial care of children	18	0
2. Cooking, cleaning, washing	130	1
3. House building	0	0
4. House repair	0	3
E. Personal needs	158	269
1. Rest, relaxing	117	233
2. Meals	21	29
3. Personal hygiene and other personal needs	20	7
F. Free time	77	118
1. Religion	2	6
2. Educational activities (learning to read, attending a UNESCO meeting or class)	17	4
3. Media (radio, reading a book)	0	14
4. Conversation	14	69
5. Going visiting (including such social obligations as funerals)	43	19
6. Errands (including going to purchase personal consumption goods, such as kola, next door)	1	6
G. Not specified†	18	0
Total work (A, B, C, D)	587	453
Total personal needs and free time (E, F)	235	387

*Based on time budgets prepared by direct observation.
†When observation did not last the full 14 hours.

MEASURING RURAL WOMEN'S ECONOMIC ROLES AND CONTRIBUTIONS IN KENYA

Smock (1979) described special survey forms used in Kenya's Integrated Rural Survey to gather information on the contributions of each household member to agricultural production and household tasks. The questionnaire used to augment the survey had three sections: (1) background data on the respondent and the household, (2) participation of family members in agricultural production, and (3) participation of family members in other household work. We now present a summary of the headings and categories of responses for the three forms:[1]

Form 1: Background Data.

Question	*Possible Responses*
1. Identity of respondent	Female head, only wife of male head, senior wife of male head, junior wife of male head, only wife of son of male head, senior wife of son of male head, junior wife of son of male head, other.
2. Age of respondent	------------
3. Respondent's highest level of formal education	No formal schooling, 1-2 years primary, 3-4 years primary, 5-8 years primary, 1-2 years secondary, 3-4 years secondary, 4+ years secondary.
4. Respondent's marital status	Never married, formerly married, currently married.
5. Number of years married	Less than 1 year, 1-3 years, 4-7 years, 8-15 years, 16-25 years, more than 25 years.
6. Age of respondent's husband	------------
7. Respondent's husband's highest level of formal education	No formal schooling, 1-2 years primary, 3-4 years primary, 5-8 years primary, 1-2 years secondary, 3-4 years secondary, 4+ years secondary.
8. Number of generations in household	One, two, three, four.
9. Number of generations that regularly work on holding	One, two, three, four.
10. Number of residents in household	Females 15 and over, males 15 and over, females 6-14 not at school, males 6-14 not at school, females 6-14 at school, males 6-14 at school.
11. Number whose main occupation is working on holding	(Same categories as above.)
12. Who takes the major share of the responsibility for preparing the land for farming?	Adult females, adult males, adult females and children, adult males and children, entire family, family and hired labor, hired labor, tractor service.

Form 2: Participation of Household Members in Agricultural Production.

This form is concerned with the participation in agricultural activities of the following groups: females 15

[1]Reprinted with the permission of the Population Council from Audrey Chapman Smock, "Measuring rural women's economic roles and contributions in Kenya," *Studies in Family Planning* 10. no. 11/12 (November/December 1979), pp. 386-389.

and over, males 15 and over, females 6-14 not at school, males 6-14 not at school, females 6-14 at school, and males 6-14 at school. Participation is broken down into planting, weeding, harvesting, and marketing. Participation is measured as one who does not work, works regularly, and works sometimes. Information is gathered for 10 major crops.

Form 3: Involvement of Household Members in Other Household Work.

This form seeks information on the same six household groups as Form 2. For these groups, participation in the following tasks is recorded: fetching firewood, fetching water, buying food, caring for children, cleaning house, preparing and cooking food, herding sheep and goats, or caring for cattle and poultry.

CIRCUMVENTING PROBLEMS OF ACCESSIBILITY TO RURAL MUSLIM WOMEN

Safai (1979) described a study undertaken by the Government of Iran to determine the type and amount of agricultural labor performed by rural women. The team used Persian women for interviewers. In addition, a man originally from the area who was familiar with the customs and dialect was on the team. The team preceded its appearance in each village by a visit to the village headman or some other village official, who helped the team select sample families.

The team administered a computer coded questionnaire to selected women. The questionnaire concerned a question regarding every task relevant to rice and tea—the two principal crops. Each woman was asked her degree of participation in each task. Researchers, after checking the answers in several ways, used the results to determine how much women contributed to various stages of the agricultural cycle.

WHAT RURAL WOMEN KNOW: EXPERIENCES IN BANGLADESH

Martius-von Harder (1979) focused on the activities in which women in rural areas have an important role (the specific activities in which they have a key function), the demands agricultural production places on them, and the approaches for integrating women into the process of rural development.

Martius-von Harder (1979) found

- women who cannot move about freely could not answer correctly questions about the household's property, soil quality, irrigation, or marketing
- women could not answer questions related to the minutes or hours to complete a task, since they are not accustomed to thinking in these terms
- when people were asked for information outside their areas of knowledge, they gave immediate, but generally incorrect answers
- both men and women could give correct answers when the quantities were two or three, but not when they were 10 or more.

The findings of Martius-von Harder (1979) supported the wisdom of pretesting questionnaires and interpreting results carefully.

WOMEN IN RICE CULTIVATION: SOME RESEARCH TOOLS

Mencher et al. (1979) used diaries and personal interviews to collect data on women's participation in a rice crop cycle in India. Literate women kept diaries of all their agricultural activities and a few women who did not cultivate rice also kept diaries for comparison purposes. Illiterate women kept charts, similar to those designed by Hatch (Appendix 5-X). Also, literate village women visited selected households to collect additional data on women's agricultural activities.

APPENDIX 5-N ASSESSMENT OF SECONDARY DATA

In Sec. 5.9., Chapter 5, we stated that secondary data could be a good source of information. For this to be true, the FSR&D team needs to ascertain the accuracy of the data. This appendix provides guidelines for checking on possible errors in data because of improper collection, analysis, and interpretation.

Below are two lists of questions an FSR&D team should ask about the secondary data. Depending on the answers, the team can then decide whether to use the data and, if so, what limitations the team should place on the data's use. When the team is unable to answer some of the questions, it will have to rely on its judgment about the data's accuracy.

- When were the data generated?
- What was the source for the data and what is the general reputation of the source?
- Do the data have any stated limitations about their use?
- What were the stated assumptions concerning conclusions drawn from the data?
- Are the methods used for collecting, tabulating, and analyzing the data stated? If not, can they be ascertained?
- Are the methods still reliable or have subsequent practices indicated certain weaknesses?
- Are those who took part in generating the data still available for consultation?
- What methods were used to collect the data? Did the researchers use observations, physical measurements, analyses of other secondary data, or interviews by researchers or enumerators?
- Did the basis for collecting data change during the period for which the data are reported and, if so, how does this affect estimates of trends?
- Do the data contain gaps that could cause statistical errors in their use?

When data are gathered by interview some additional questions need to be asked of those who might be familiar with the way the interviews were conducted. Below are some questions.

- What was the physical setting for the interviews?
- Were other members of the interviewee's family present or were any friends present?
- How were the individuals selected for interview?
- What percent of the target population of farmers was interviewed?
- Were the questions open-ended or close-ended?
- In what sequence were the questions asked?
- Was a sampling error calculated and is it known?

By asking the above types of questions, the researchers should be in a better position to evaluate the usefulness of the secondary data. If the answers to the above types of questions lead the researchers to have doubts about the data, they should seek advice from those experienced in evaluating the validity of secondary data.

APPENDIX 5-O ANALYSIS OF THE CONTENT OF INFORMAL INTERVIEWS

One of the disadvantages normally associated with informal surveys, as discussed in Sec. 5.10.1., Chapter 5, is the lack of quantified data. One way to turn part of the information obtained through informal surveys into quantified data is through content analysis. This appendix provides a brief overview of this process—a technique that can be applied to the analysis and quantification of responses to open-ended questions. Open-ended questions are those in which respondents are free to elaborate on a subject as they wish.

If the FSR&D teams write brief summaries of farmers' responses to informal surveys, these summaries provide the raw material for content analysis. An advantage of converting what appears to be qualitative data into quantitative data is that the teams gain further insight into the farmers' responses and can thereby draw firmer conclusions from the data.

Let us assume the researchers are looking for the principal categories of farmer responses to a question. First, the researchers write definitions for each category in such a way that all anticipated responses are covered.

Once the definitions are prepared, coders analyze the responses. Using the definitions, the coders separate the responses by category. The researchers must check the coders' work to be sure that the coders agree on how to categorize the responses. This is accomplished by having the coders categorize a number of the same responses. For this check, called intercoder reliability, researchers generally seek a correlation of at least 80 percent.

If a correlation of 80 percent is not achieved, the researchers rewrite the definitions and test the new definitions until the desired level of correlation is reached. Then, the coders analyze all responses and tabulate the frequency of each category.

Here, we have suggested the unit of analysis was the category of response. Other bases for analysis can be selected depending upon the problem and topic being undertaken. For further details on content analysis, see *Content Analysis in Communications Research* (Berelson, 1971) and *Advances in Content Analysis* (Rosengren, 1980).

AN EXAMPLE

We will illustrate the method using one of the questions that an FSR&D team might ask of a farmer during an informal interview. For example, the team might have asked farmers about their maize yields and the farmers might have said they wanted to increase their yields. Then the team members might ask farmers, "What is limiting your maize production?" After each interview, the team summarizes the response and estimates the percentage of time the farmer devotes to each reason. At the end of the day, the team decides what the major reasons are based on and the amount of time devoted to each reason.

Thus, the team might find that farmers mentioned insects, lack of fertilizer, or inadequate water as the major production constraints. The team could then frame questions for subsequent interviews more definitively. For example, a formal questionnaire might include the following question:

What keeps you from obtaining more maize from your fields?

Reason	*Importance*
Insects	________
Lack of fertilizer	________
Inadequate water	________
Other	________

The importance of each reason to the farmers could now be estimated by asking farmers to rank the four possibilities from most to least important.

By applying this approach to the more relevant questions of the informal survey, the team will have established a firmer basis for quantifying farmers' activities and reactions.

APPENDIX 5-P ILLUSTRATIVE TABLES FOR COLLECTING DATA DURING RECONNAISSANCE SURVEYS

In our discussion of reconnaissance surveys in Sec. 5.10.1., Chapter 5, we discussed various means for gathering data about farmers, their means of production, and their environment. The reader may recall that reconnaissance surveys, as we described them, are relatively short explorations that help FSR&D teams (1) understand the farming systems in the research area and (2) identify the more relevant problems and opportunities for improve-

Table 5-P-1. General information on typical farms in the research area.

A. *Dates of data collection* ________ *Collector* ________

B. *Farm access*
- Distance to nearest road usable by: e.g., motorcycle, 4-wheel drive vehicle, truck ________
- Distance to nearest all-weather road ________
- Distance to other transportation facilities: e.g., river, canal, airfield, railroad ________

C. *Farm land status*	*(hectares)*	D. *Farm land use*	*(hectares)*
Privately owned	________	Crops	________
Rented	________	Other tillable purposes	________
Other: e.g., village or tribal land	________	Pasture	________
Total	________	Forest	________
		Other: e.g., fish ponds	________
Note: Totals for C and D should be equal.		Wasteland	________
		Total	________

E. *Other information about land*
- Number and sizes of land parcels in farm ________
- Distances from farmhouse to fields ________
- Percentages of total farmland suitable for:
  - motorized equipment ________
  - irrigation ________
- Access to other land and resources: e.g., community pasture and forest, roadside pasture ________

F. *Farm enterprises*

	Major enterprise	*Minor enterprise*
Crops: e.g., species, varieties, principal uses		
Cereals	________	________
Root crops	________	________
Vegetables	________	________
Fruit	________	________
Other tree crops	________	________
Other: e.g., sugarcane, sisal	________	________
Animals: e.g., breeds, sex, number, principal uses		
Cattle	________	________
Buffalo	________	________
Sheep	________	________
Goats	________	________
Swine	________	________
Poultry	________	________
Nonagricultural enterprises: e.g., spinning, weaving, pottery-making	________	________

G. *Farm labor*: ________

H. *Power, equipment, and tools*
- Power ________
- Equipment ________
- Tools ________

I. *Farm buildings and facilities*
- Storage facilities ________
- Processing facilities ________
- Livestock housing and yards ________
- Irrigation facilities ________
- Other ________

J. *Marketing of output*: ________

K. *Acquisition of inputs*: ________

L. *Estimates of income, expenditures, and savings*
- Farm income ________
- Farm expenditures ________
- Off-farm income ________
- Savings ________

Table 5-P-1 (cont.)

M. *Type of household*: e.g., nuclear or extended ________
- Ethnic background ________
- Numbers in household ________
- Rights and obligations of members, by age and sex ________
- Characteristics of members:
  - Literacy and education ________
  - Health ________
  - Knowledge: e.g., farming and off-farm experiences ________
  - Beliefs: e.g., what the person thinks is true ________
  - Attitudes: e.g., feelings, emotions, sentiments ________
  - Behavior: e.g., past actions ________
  - Goals ________
  - Other ________

N. *Miscellaneous*: e.g., help from others, obligations ________

---

ment. Researchers participating in the surveys gather information through observations and informal discussions with those in the area.

Tables 5-P-1 to 5-P-5 in this appendix are included to illustrate the type of information practitioners have found useful during this phase of FSR&D. The tables cover information about the household, the farm, farming enterprises, crops and cropping patterns, farm management practices, and livestock. We urge the readers to consider these tables as checklists that can be used in recording relevant information as it becomes available. These tables *are not* designed as formal questionnaires in which information is sought for each item and the data are subjected to statistical analyses. Such an approach would divert the team's attention from the true nature of the reconnaissance survey, which is to learn about farmers' conditions in a quick and informal way. More detailed and statistically valid information is generally obtained later from formal surveys, as we discuss in Sec. 5.10.2. Even as a checklist, these tables contain considerably more information than is gathered by those who follow the *sondeo* approach (see Appendix 5-Q).

Finally, because these tables are examples of the types of information the teams may wish to collect, FSR&D teams will undoubtedly need to modify them to meet their specific needs. This modification includes altering table format to provide sufficient space to record the data.

Table 5-P-2. Principal sole cropping patterns in the research area.

Dates of data collection __________ Collector __________
Description of each cropping pattern ____________________
Estimated percentage of the research area in the cropping pattern ______________________________

Species	_____	_____	_____	. . . .
Variety	_____	_____	_____	. . . .
Dates of seedbed preparation	_____	_____	_____	. . . .
Range of planting dates	_____	_____	_____	. . . .
Range of harvesting dates	_____	_____	_____	. . . .
Crop yields	_____	_____	_____	. . . .
Turnaround times (days)	_____	_____	_____	. . . .

Table 5-P-3. Intercropping pattern.

Dates of data collection __________ Collector __________
Description of the intercropping pattern ____________________
Estimated percentage of the research area in the intercropping pattern ______________________________

Crops in the pattern	*First species*	*Second species*	*Third species*	
Variety	_____	_____	_____	. . . .
Range of planting dates	_____	_____	_____	. . . .
Range of harvesting dates	_____	_____	_____	. . . .
Crop yields	_____	_____	_____	. . . .

Table 5-P-4. Farm management practices and costs for principal crops in the research area.

Cropping pattern ______________________________ Crops ______________________________
Dates of data collection ______________________________ Collector ______________________________

Activity	*Description*
Land preparation	Power sources: hand _______ animal _______ tractor _______ custom _______ Land preparation methods ______________________________ Dates: begun ____________________ ended ____________________ Requirements: labor_____materials_____tools & equipment_____other_____ Costs: labor_______materials_______tools & equipment_______other_______
Seeding and emergence	Power sources: hand _______ animal _______ tractor _______ custom _______ Seeding or planting methods ______________________________ Seeding or planting rates ______________________________ Dates of seeding or planting ______________________________ Soil moisture at germination ______________________________ Plant stands ______________________________ Requirements: labor_____materials_____tools & equipment_____other_____ Costs: labor_______materials_______tools & equipment_______other_______
Chemical fertilizer	Types of fertilizer ______________________________ Power sources: hand _______ animal _______ tractor _______ custom _______ Methods of application ______________________________ Dates of application ______________________________ Rates of application ______________________________ Requirements: labor_____materials_____tools & equipment_____other_____ Costs: labor_______materials_______tools & equipment_______other_______
Pest Control	Principal weeds, insects, diseases, other pests ____________________ Power sources: hand _______ animal _______ tractor _______ custom _______ Methods of controls: hand__________chemical__________other__________ Dates of control ______________________________ Control procedures ______________________________ Kinds of chemicals ______________________________ Requirements: labor_____materials_____tools & equipment_____other_____ Costs: labor_______materials_______tools & equipment_______other_______
Irrigation	Sources of water: canal _______ pump _______ well _______ other _______ Power sources: hand ___________ animal ___________ motor ___________ Water availability: daily ___________ weekly ___________ monthly ___________ Methods of irrigation ______________________________ Requirements: labor_____materials_____tools & equipment_____other_____ Costs: labor_______materials_______tools & equipment_______other_______

Table 5-P-4 (cont.)

Activity	*Description*
Harvest	Power sources: hand ______ animal ______ tractor ______ custom ______ Methods ______ Time period from planting to harvest (days) ______ Harvest losses ______ Requirements: labor ______ materials ______ tools & equipment ______ other ______ Costs: labor ______ materials ______ tools & equipment ______ other ______
Threshing, cleaning, & drying	Power sources: hand ______ animal ______ tractor ______ custom ______ Methods ______ Requirements: labor ______ animals ______ tools & equipment ______ other ______ Costs: labor ______ materials ______ tools & equipment ______ other ______
Crop	Total yields ______ Percent in each quality grade ______
Crop disposal	Method of disposal & percent of total: sale ______ barter ______ home use ______ farm use ______ other ______ Place of sale: standing in field ______ at farm gate ______ to wholesaler ______ to processor ______ to consumer ______ Sales prices for each quality grade ______ Marketing operations: cleaning ______ sorting ______ grading ______ packaging ______ transporting ______ other ______ Requirements: labor ______ materials ______ tools & equipment ______ other ______ Costs: labor ______ materials ______ tools & equipment ______ other ______
Farm storage	Type of products stored ______ Storage facilities: type ______ quality ______ capacity ______ Amounts and periods of storage ______ Storage losses: quantity ______ quality ______ timing ______ Losses due to: insects ______ diseases ______ rodents ______ water ______ other ______ Requirements: labor ______ materials ______ tools & equipment ______ other ______ Costs: labor ______ materials ______ tools & equipment ______ other ______
Crop residue disposal	*Composts for field application* Types of residue ______ Methods of preparation ______ Methods of application ______ Rates of application ______ Dates of application ______ Requirements: labor ______ materials ______ tools & equipment ______ other ______ Costs: labor ______ materials ______ tools & equipment ______ other ______ *Home uses of organic residues* Types of residue ______ Used as heating fuel: how managed ______ Used as building material: how managed ______ Other ______
Application of animal manures	Type of manure ______ Power sources: hand ______ other ______ Methods of application ______ Rates of application ______ Dates of application ______ Requirements: labor ______ materials ______ tools & equipment ______ other ______ Costs: labor ______ materials ______ tools & equipment ______ other ______

Table 5-P-5. Management practices for livestock in the research area.

Kind of animal ________________ Location ________________
Dates of data collection ________________ Collector ________________
Inventory: ________________

Reasons for keeping animal

Products: e.g., quantity, quality, value
Milk ________________
Meat ________________
Breeding ________________
Manure ________________
Hides ________________
Other ________________
Power: e.g., plowing, water pumping ________________
Transportation ________________
Other: e.g., keeping down weeds, threshing, savings ________________

Methods and costs of marketing and utilization: ________________

Breeding and offspring

Numbers in breeding herd: females ________________ males ________________
Natural or artificial insemination ________________
Breeding seasons ________________
Breeding efficiency ________________
Number born per year: litters ________________ offspring ________________
Number of offspring weaned per year ________________
Number sold, consumed, given away per year ________________
Other ________________

Diseases, parasites, and metabolic problems

Infectious diseases by organisms
Bacteria: e.g., brucellosis, leptrospirosis ________________
Virus: e.g., BVD (bovine virus diarrhea), Newcastle disease, hog cholera ________________
Fungi: e.g., ringworm and other systemic fungus infections ________________
Noninfectious diseases: e.g., tetanus, black leg ________________
Parasites
Internal: e.g., flukes, tapeworms ________________
External: e.g., flies, lice, ticks ________________
Metabolic problems: e.g., faulty nutrition ________________
Degree of occurrence and methods of control ________________
Controls
Effectiveness ________________
Requirements and costs: labor ________________ materials ________________
tools & equipment ________________ other ________________

Feeding

Method of feeding: e.g., range herding, roadside herding, grazing in fenced pasture ________________
Kinds of feed
Pasture and forage ________________
Green feed ________________
Hay, silage ________________
Root crops: e.g., cassava, sweet potatoes, taro ________________
Grains ________________
Grain by-products ________________
Protein concentrates: e.g., soybean meal, fish meal, bone meal ________________
Commercial feed mix ________________
Additives: e.g., vitamins, minerals, salt ________________
Surplus fruits or vegetables: e.g., breadfruit, bananas ________________
Household garbage ________________
Other ________________

Table 5-P-5 (cont.)

Indicate the importance of each method of feeding and the importance and quality of each kind of feed by
Season ______
Sex of animal ______
Age of animal ______
Condition of animal and animal use: e.g., breeding, lactating, working, idle ______
Feeding problems ______
Requirements and costs: labor ______ materials ______
tools & equipment ______ other ______

Keeping and housing

Keeping animals on open range ______
Keeping animals in enclosure at night, open pasture during day ______
Keeping animals in fenced pasture
Types, qualities, and conditions of fencing ______
Amount of separation by sex and age ______
Amount of herd rotation ______
Amount and timing of closing pasture for haying ______
Watering facilities ______
Pasture upkeep ______
Requirements and costs: labor ______ materials ______
tools & equipment ______ other ______
Keeping animals in stable
Kind of materials used in building stable: e.g., wood, cement ______
Condition of stable and quality of maintenance ______
Kind of stable and construction ______
Availability of open-air pens ______
Facilities for young, lactating, and sick animals ______
Other ______
Requirements and costs: labor ______ materials ______
tools & equipment ______ other ______

Sanitation

Animal sanitation ______
Housing sanitation: e.g., methods of washing, rotating, disinfecting ______
Water sanitation ______
Manure and urine sanitation ______
Product storage and processing sanitation ______
Requirements and costs: labor ______ materials ______
tools & equipment ______ other ______

Other: ______

APPENDIX 5-Q
SUMMARY OF THE *SONDEO* METHODOLOGY USED BY ICTA[1]

Under Sec. 5.10.1., Informal Methods, we discussed the reconnaissance survey and concluded with a note on the *sondeo*, the Spanish term for a reconnaissance survey. Hildebrand (1979a) summarized ICTA's approach to the *sondeo* methodology in a report, which we quote:

"INTRODUCTION

"The *sondeo* is a modified survey technique that has been developed by ICTA as a response to budget restrictions, time requirements and the other methodology utilized to augment agricultural information in a region where technology generation and promotion is being initiated. The purpose of the *sondeo* is to provide the information required to orient the work of the technology generating team. The cropping systems are described, the agro-socioeconomic situation of the farmers is determined and the restrictions they face are defined so that any proposed modifications of their present technology are appropriate to their conditions. In order to understand the methodology, it is first necessary to understand how ICTA is organized.

"Without entering into the organization at the national level, the regional organization will be discussed. Each of the regions in which the Institute

[1]Reprinted from *Agricultural Administration*, Vol. 8, 1981, by permission of Applied Science Publishers Ltd.

functions has a Regional Director who is the representative of the Director General of the Institute and of the Technical Director. Within the region, each area in which work is being carried out is [supervised by] a 'Sub-regional delegate', a technician who has a minimum amount of administrative responsibilities. All the technicians, from whatever discipline or program, who work in the area are responsible to him. This multi-disciplinary team is usually comprised of some or all of the following: plant breeders, pathologists, a technician from socio-economics and approximately 5 general agronomists who are the Technology Testing Team. This group, backed up by the national Coordinators of Programs (corn, beans, etc.) and Support Disciplines (socioeconomics, soil management) are responsible for orienting and conducting the generation and promotion of technology in the area. The work includes basic plant breeding and/or selection on the (usually small) experiment station in the area, farm trials, tests by farmers of promising technology, evaluation of acceptability of the technology tested by farmers, and economic production or farm records maintained by farmers with the help of the technicians. In order to provide the original orientation to the team, the *sondeo*, or reconnaissance survey, is conducted by members of the Technology Testing Team who are going to work in the area, sometimes personnel from the appropriate Program, and a team from socioeconomics comprised of one or more of the following: anthropologists, sociologists, economists, agricultural economists and/or engineers. Usually, there are 5 people from Socioeconomics and 5 from the Technology Testing Team who form a 10 man *sondeo* team for an area.

"If ICTA is to work in an area not previously defined, such as by the bounds of a land settlement or irrigation project, one of the objectives of the *sondeo* is to delimit the area. This is done by first selecting the predominant cropping system used by potential target farmers in the area and later determining the area in which this system is important. The reason that a homogeneous traditional or present cropping system is used is that it is this cropping system that ICTA will be modifying with new or improved technology. Hence, having a well-defined, homogeneous system with which to work simplifies the procedure of generating and promoting technology. The premise on which selection of a homogeneous cropping system is based is that all the farmers who presently use it have made similar adjustments to a set of restrictions which they all face, and since they all make the same adjustments, they must all be facing the same set of agrosocioeconomic conditions.

"Besides delimiting the area of this homogeneous system, the tasks of the *sondeo* team are to discover what agro-socioeconomic conditions all the farmers who use the system have in common, then determine which of them are the most important in determining the present system, and therefore, would be the most important to consider in any modifications to be made by the team in the future. Finally, the end product of the *sondeo* is to orient the first year's work in farm trials and plant selection and it also serves to locate future collaborators for the farm trials and for the farm record project.

"Because the farm trials are conducted under farm conditions, during the first year they provide an additional learning process into the conditions that affect the farmers and are invaluable in acquainting the technicians with the realities of farming in the area. The farm records which are also initiated the first year, provide quantifiable technical and cost information on the technology being used by the farmers. At the end of the first year's work, then, the technicians have not only been farming under the conditions of the farmers in the area, but they also have the information from the farm record project. For this reason, it is not necessary to obtain quantifiable information in the *sondeo*. Additionally, the *sondeo* is not a benchmark study. More reliable information for evaluation of the impact in the area is available from the farm records, which gain in value each year.

"THE *SONDEO*

"The primary purpose of the *sondeo*, then, is to acquaint the technicians with the area in which they are going to work. Because quantifiable information is not needed, it can be conducted rapidly and no lengthy analyses of data are required following the survey to interpret the findings. No questionnaires are used so farmers are interviewed in an informal manner which does not alienate them. At the same time, the use of a multidisciplinary team serves to provide information from many different points of view simultaneously. Depending on the size, complexity and accessibility of the area, the *sondeo* should be completed in from 6 to 10 days at a minimum of cost. Areas of from 40 to 150 km^2 have been studied in this period of time. Following is a description of the methodology for a 6 day operation.

"DAY 1

"The first day is a general reconnaissance of the area by the whole team as a unit. The team must make a preliminary determination of the most important cropping system that will serve as the key system, get acquainted in general terms with the area and begin to search out the limits of the homogeneous system. Following each discussion with a farmer, the group meets out of sight of the farmer to discuss what each one's interpretation of the interview was. In this way, the team members begin to get acquainted with how each other thinks.

Interviews with farmers (or other people in the area) should be very general and wide-ranging because the team is exploring and *searching for an unkown number of elements.* (This does not imply, of course, that the interviews lack orientation.) The contribution or point of view of each discipline is critical throughout the *sondeo* because the team does not know beforehand what type of restrictions may be encountered. The more disciplines that are brought to bear on the situation, the greater is the probability of encountering the factors which are, in fact, the most critical to the farmers of the area. It has been established that these restrictions can be agro-climatic, economic or socio-cultural. Hence, all disciplines make equal contributions to the *sondeo.*

"DAY 2

"The interviewing and general reconnaissance of the first day serve to guide the work of the second day. Teams are made up of pairs: one agronomist from the technology testing team and one person from Socioeconomics who work together in the interviews. The 5 teams scatter throughout the area and meet again either after the first half-day (for small areas or areas with good access roads) or day (for larger areas or where access is difficult and requires more time to travel). Each member of each team discusses what was learned during the interviews and tentative hypotheses are formed to help explain the situation in the area. Any information concerning the limits of the area are also discussed to help in the delimitation. The tentative hypotheses or doubts raised during the discussion serve as guides to the following interview sessions. During the team discussions, each of the members learns how interpretations from other points of view can be important in understanding the problems of the farmers of the region.

"Following the discussion, the team pairs are changed to maximize interdisciplinary interaction and minimize interviewer bias and they return to the field guided by the previous discussion. Once again following the half-day or day's interviews the group meets to discuss the findings.

"The importance of these discussions following a series of interviews cannot be over-stressed. Together the group begins to understand the relationships encountered in the region, delimit the zone and start to define the type of research that is going to be necessary to help improve the technology of the farmers. Other problems such as marketing are also discussed and if solutions are required, appropriate entities can be notified.

"However, it is important to understand the effect that these other limitations will have, if not corrected, on the type of technology to be developed so that they can be taken into account in the generation process.

"DAY 3

"This is a repeat of day 2 and always includes a change in the makeup of the teams after each discussion. At least a minimum of 4 interview discussion cycles is necessary to complete this part of the *sondeo.* If the area is not too complex, the cycles should be adequate. Of course, if the area is large enough that a full day is required for interviewing between each discussion session, then four full days are required for this part of the *sondeo.*

"DAY 4

"Before the teams return to the field for more interviews on the fourth day, each member is assigned a portion or section of the report that is to be written. Then, knowing for the first time what topic each will be responsible for, the teams, regrouped in the fifth combination, return to the field for more interviewing. For smaller areas, this also is a half day. In the other half day, and following another discussion session, the group begins to write the report of the *sondeo.* All members should be working at the same location so that they can circulate freely and discuss points with each other. For example, an agronomist who was assigned the section on corn technology may have been discussing a key point with an anthropologist and may need to refresh his memory about what a particular farmer said in a brief discussion with him. In this manner, the interaction among the disciplines continues.

"DAY 5

"As the technicians are writing the report, they invariably encounter points for which they have no answer, nor does anyone else in the group. The only remedy is to return to the field on the morning of the fifth day to fill in the gaps that were felt the day before. A half day can be devoted to this activity and finishing the writing of the main body of the report.

"In the afternoon of this day, each team member reads his written report to the group for discussion, editing and approval. The report should be read from the beginning just as it will be when finished. As a group, the team should approve and/or modify what is presented.

"DAY 6

"The report is read once again, and following the reading of each section, conclusions are drawn and recorded. When this is finished, the conclusions are read once again for approval and specific recommendations are then made and recorded both for the ICTA team who will be working in the area and for any other agencies that should be involved in the general development process of the zone.

"The product of the sixth day is a single report generated and authored by the entire multi-disciplinary team and should be supported by all of the members. Furthermore, after participating for all six days with each other, each member should be able to defend all the points of view discussed, the conclusions drawn and the recommendations made.

"THE REPORT

"To a certain extent, the report of the *sondeo* is of secondary value because it has been written by the same team that will be working in the area. But just the fact that they have written it, is where most of the value lies. By being forced into a situation where many different points of view had to be taken into consideration and coalesced, the horizons of all will have been greatly amplified. On the other hand, the report can serve as orientation for non-participants such as the Regional Director or Technical Director in discussing merits of various courses of action. However, it is also obvious that the report will appear to be one written by 10 different persons in a hurry, which is just exactly what it is! It is not a benchmark study with quantifiable data that can be used in the future for project evaluation; rather it is a working document to orient the research program and that served one basic function in just being written.

"The exact format and content of a report of a *sondeo* will vary according to the area being studied and the nature of the crops or livestock enterprises included. Following is a brief description of an outline of a report recently completed in one area of Guatemala where small grains and vegetables were of primary interest.

"PURPOSE

Describes the reason the *sondeo* was undertaken and the dates.

"HOMOGENEOUS TECHNOLOGY

Describes the principal characteristics of the technology regarding the crops of interest found within the limits of the area and the important differences outside the area that changed the nature of the cropping system and defined the limits of the area.

"DESCRIPTION OF THE DELIMITED AREA

Geographical limits, altitude, soils, other important features and includes a map drawn with the boundaries as precise as possible.

"LAND

Land tenure and farm size were important restrictions in the cropping system and were described.

"LABOR

General labor availability and periods of scarcity and the special tasks performed by women in the homogeneous system were described.

"CAPITAL

The capital flow in the traditional system which provides the funds for investing in both the basic grains and the vegetables was described and the poor functioning of the small farm credit system was noted.

"CORN

The most important components of the corn production system were described.

"BEANS

The role beans play in the system and their lack of general importance were described.

"VEGETABLES

The production system and the marketing of vegetables were described.

"LIVESTOCK ACTIVITY

The special importance of livestock and the livestock-crop interaction were discussed.

"CONCLUSIONS

Conclusions for each one of the above sections were drawn with special emphasis on their meaning to the future work of ICTA.

"RECOMMENDATIONS

Those relevant to ICTA and to other entities in the Public Agricultural Sector as well as the private sector.

"COORDINATING THE *SONDEO*

"The disciplinary speciality of each member of the *sondeo* team is not critical so long as there are several disciplines represented, and, if the *sondeo* is in agriculture, a significant number of them are agriculturalists. At least some of these should also be from among those who will be working in the area in the future. The discipline of the Coordinator of the *sondeo* is probably not critical either, if he is a person with a broad capability, an understanding of agriculture (if it is an agricultural *sondeo*) and experience in surveying and survey technique. However, the coordinator must have a high degree of *multi-disciplinary tolerance* and be able to interact with all the other disciplines represented on the team.

"The coordinator, in a sense, is an orchestra director who must assure that everyone contributes to the tune, and that in the final product, all are in harmony. He must control the group and maintain discipline. He arbitrates differences, creates enthusiasm, extracts hypotheses and thoughts from each participant, and ultimately will be the one who coalesces the product into the final form. It is, perhaps not indispensable that he has had prior experience in a *sondeo*, but it would certainly improve his efficiency if he had."

APPENDIX 5-R GUIDELINES FOR PRE-SURVEY SEQUENCE

Reconnaissance surveys, as discussed in Sec. 5.10.1. and the *sondeo* as detailed in Appendix 5-Q, provide researchers with a way of informally collecting information about farmers. We now present a series of guidelines from Collinson (1979) that researchers should find helpful in focusing their attention on specific aspects of the farming system:

"DETAILED GUIDELINES FOR PRE-SURVEY SEQUENCE: BY DISCUSSION WITH FARMERS

"1. DESCRIPTION OF THE LOCAL FARMING SYSTEM

"(1) Enterprise pattern and end uses

(a) List the crops grown and livestock kept by local farmers. Note for each one whether it is grown by the majority or few local farmers. If a few only, what is special about those few, e.g. large with plenty of land and capital, close to specialized markets or processing facilities, old and traditional, etc.

(b) For each major crop list the varieties grown, give the local name and, where possible, relate to known variety names. Assess whether each variety is important to most farmers, to a few, or to all on particular occasions. Detail why it is important.

(c) For all major crops, varieties and animals, list the *end uses* to which they are put. This should include the fruit in the case of crops and any other part of the plant used as a product. Animal products and by-products are equally important. Where different varieties of the same crop may be grown by the same farmer it is particularly important that differences in end uses are described.

(d) For each identified product, including varieties with different end uses, detail the sequence which is followed when it is taken from the plant in the field or from the animal. Include *when* it is taken in the life cycle of the plant, how it is prepared or processed or used, and if sold what exact use is it then put to. In describing these end uses it is important to be detailed right through the sequence.

(e) Note particularly any crops, crop varieties or animals
  (i) that used to be widespread among farmers of the area but are now disappearing. Assess why such crops, varieties or animals are disappearing
  (ii) that have recently become popular with farmers of the area and appear to be spreading. Assess the reason for their popularity.

"(2) Food supply and preferences

(a) Detail the main dishes eaten by farm families in the area. The constituents of the dishes and the preferred *state* of each constituent. What alternative constituents are used when preferred ones are not available?

(b) List the preferred starch staples and relishes, and the substitute staples and relishes used when preferred ones are scarce. Indicate on a chart:
  (i) the months when each is readily available from farm production.
  (ii) months when supplies may be uncertain.
  (iii) months when supplies are definitely not available from farm production.

(c) Assess whether any new foods are becoming popular and replacing traditional ones.

(d) List foods commonly purchased by farm families
  (i) all the year round
  (ii) at certain periods of the year which should be specified.

(e) If major foods are bought at certain periods assess whether
  most farm families buy some major foods at particular periods in some years.

(f) If families have to resort to buying only

in some years assess how frequently this is and the reasons for this problem arising in those years.

(g) See whether prices of the major foods vary over the year. Give an indication of price levels at seasons when food is:
   (i) plentiful
   (ii) scarce.

(h) Which is the most difficult period of the year for feeding livestock and why?

"(3) Cropping calendar

For each crop, and where different varieties are grown by the farming community, for each variety, indicate

(a) the usual planting time for the crop
(b) the range in possible planting times, including the latest time that farmers will consider it worthwhile to plant that crop or variety
(c) the length of time the crop spends in the ground
(d) the usual harvest time for that crop.

Also assess the major reasons for farmers planting each crop variety at the time they do.

"(4) Cash sources and uses

(a) List the major crops and livestock products sold by farmers in the area and the main channels through which each is sold.

(b) Assess whether prices earned through the major outlets are subject to large variations
   (i) between seasons
   (ii) within seasons.

   Seek to identify reasons for large variations, examples of the extent of variations and, for within season variations, the periods of high and low prices.

(c) Assess the 'usual' level of cash incomes from the major products sold on local farms.

(d) For products that are foods as well as sources of cash evaluate the different circumstances in which farmers will decide to sell rather than store for food. Assess which circumstances are most common in sales decisions.

(e) Assess how common is off farm employment among farmers and farmers' families, the main types of off farm employment and the usual level of cash income earned from these sources. Distinguish temporary and permanent off farm work. For temporary work identify the periods of the year when it is undertaken. Evaluate whether this is because opportunities arise then, or farmers' need cash at these times.

(f) Assess what are farmers' main cash expenditures during the year. When do these arise?

(g) List the purchased inputs recommended to farmers in the area, assess how far farmers know of them and what proportion use them. When are the major inputs purchased during the year? Assess whether the farmer has cash at this time.

(h) How much does the typical local farmer spend on purchased inputs in a year?

"(5) Husbandry

Detail the husbandry practices which *most farmers* follow for their maize crop. It is important that the description is as detailed as possible.

(a) How does the farmer decide *where* he will plant his next maize crop? What factors does he consider in the decision?

(b) Land preparation
   (i) What is the method of land preparation?
   (ii) When, in relation to the start of the rains and to planting time, does the preparation start?
   (iii) What sequence of work is involved if there is more than one operation?
   (iv) How does the farmer work; does he prepare a whole field before planting, or prepare and plant a bit the same day, or what?
   (v) What is the final form of seedbed?
   (vi) Are there alternative methods of land preparation?

(c) Planting
   (i) What is the arrangement of plants in the field: maize and any mixtures?
   (ii) Where other crops are mixed in, it will be important to describe in what sequence all the crops are put in the ground.
   (iii) How do farmers plant in relation to rainfall: dry planting before rain, the same day as rain falls, within a limited period after rains?
   (iv) Do farmers just make one planting of maize each season or are there usually several?
   (v) Do farmers commonly have to replant or fill in fields?
   (vi) What is the method of putting the seed in the ground, and how

many seeds are put per hole?

(d) Weeding and thinning
  (i) What implement or implements are used for weeding and what pattern of work is followed between the plants in the ground?
  (ii) How soon after planting is the first weeding done? Does the timing vary very much with conditions, if so how much and which conditions?
  (iii) How many weedings will normally be done? Will this vary with the date of planting, the weather or the soil in the field selected?
  (iv) Do they thin the maize plants either in the row or from each planting hole? If so at what age? Do they use the thinnings for cattle feed?

(e) Pest control
  (i) major pests for which control is sought
  (ii) timing and method of control.

Assessment of proportion of local farmers using pest control.

(f) Use of fertilizer on maize (if any)
  (i) type of fertilizer, source
  (ii) usual rate, method and time of application.

Assessment of proportion of local farmers using fertilizer.

(g) Use of leaves, tops and stalks for cattle feeding
  (i) proportion of local farmers using
  (ii) method of feeding to animals
  (iii) for leaves, number of pickings made, number of leaves taken and the timing in relation to plant growth
  (iv) for tops, stage of plant growth that the top is taken. Is this a critical time for cattle feed?

(h) Method and timing of harvesting and storing
  (i) At what stage does harvesting begin?
  (ii) What method is followed in picking cobs, dehusking, shelling and disposing of stover?
  (iii) How is the crop stored? Is any preservative used?

(i) Seed selection and preservation
  (i) Do the farmers usually select seed in the field or from their stored harvest? If from store, when is it selected?
  (ii) What criteria do local farmers use when they choose next year's seed from their own crop?
  (iii) Do they process and preserve the chosen seed in a special way?

Is the crop treated in any other way, either while in the field or in the household?

"2. IDENTIFICATION OF RESOURCE CONSTRAINTS

"(1) Land

(a) Are farms in the area registered or held under traditional custom?
(b) What proportion of the area of land held by the typical local farmer is cultivated in any one season and what proportion is under grass or fallow?
(c) Is the arable area changed periodically and allowed to fallow?
(d) Are crops rotated, if so what crop sequences are followed?
(e) Can farmers get new land: by clearing, by renting, by purchase? If so how far away would new land for clearing be? How much money would be needed to rent or purchase an acre? Would this vary by the type of soil and location of the piece of land?
(f) Soil types and maize management
  (i) Do farmers prefer a specific type of soil for growing maize, if so which and why?
  (ii) Do farmers prefer a special location for their maize crop?
  (iii) Do farmers vary the soil type and/or location where they grow their maize depending on the sort of season they expect? If so what influences their decision?

"(2) Labour

(a) What is the busiest month of the year for local farmers? During this month what work are they doing mainly and with which crops?
(b) Is this the busiest month every year, or does it vary from year to year?
(c) Which is the second busiest time of the year for local farmers and what work are they doing then and on which crops?
(d) Do many local farmers hire any labour?
  (i) permanently throughout the year
  (ii) temporarily for a particular job or particular period?
  (iii) When farmers hire casual labour what month or months is it mainly hired and for what type of work?

(e) Do many farmers hire machinery? If so is it tractor or ox driven, which operations is it mainly hired for, at which time of the year and for which crops?

(f) How much money will a typical farmer spend on hired labour and machinery in a year—if any?

"3. FARMERS ASSESSMENT OF HAZARDS

"*(1) Yield variability*

(a) What variation do farmers expect in maize production from season to season:
   (i) What sort of production would they expect in a 'bad' year?
   (ii) How frequently do such bad years occur in the area?
   (iii) What are the main factors that make a year bad for maize?
   (iv) What sort of production would they expect in a 'good' year?

Repeat for the three or four major crops in the system.

(b) As far as the farmers are concerned low yields in which crop are the most serious for them and their families?

(c) What measures do they take to combat the effects of low yields when they occur—how do they manage when production of this vital crop fails?

"*(2) Rainfall problems*

(a) Which crops sometimes give poor results because of rainfall?

(b) With reference to maize, which type of rainfall problem is most serious?
   (i) late start to the rains
   (ii) too little rain during the growing season
   (iii) early finish to the rains
   (iv) too much rain.

(c) When did this type of rainfall problem occur on a widespread basis in the area and give a poor maize crop?

(d) Discuss with farmers how they react to this type of failure; i.e. they know their next maize harvest will be poor:
   (i) in preserving food supplies in the household
   (ii) in managing their farms to offset the effect on their food supplies?

It may be important to go through this sequence with reference to another major starch staple—sorghum where grown—and a major relish crop.

"*(3) Pests and diseases*

(a) What do local farmers consider as their major pest and disease problems? Specify
   (i) crops and pests
   (ii) frequency with which the problems occur.

(b) Do local farmers believe they have any means managing their farms to *prevent* these pests and diseases occurring? Discuss them one by one.

(c) Do local farms have any way to treat the crops or the land once they see these pests and diseases appearing? Discuss them one by one.

"4. FARMERS OPINIONS ON COMPONENTS OF CURRENTLY RECOMMENDED MAIZE TECHNOLOGY

"From the Ministry of Agriculture write out, in full, the current recommendations for growing maize in the area. Taking one component of the improved management at a time discuss it with local farmers. Attempt to assess the problems which each component presents to them in their situation."

APPENDIX 5-S
SUGGESTIONS FOR DEALING WITH FARMERS' RECOLLECTION OF INFORMATION

Having to rely on farmers' recollection of information has its disadvantages, especially for frequently repeated activities, such as those occurring daily or weekly. Below are suggestions by Norman on factors that influence recall and by Gucelioglu on two methods of inquiry. Both are from Kearl (1976). The topic of farmers' recall is taken up in Sec. 5.10.2. in Chapter 5. Norman (1976b) reported:

"FACTORS THAT INFLUENCE RECALL

"In many developing areas, and northern Nigeria is no exception, nearly all farmers are illiterate and consequently no records on farming transactions are kept. Therefore memory recall is critical in collecting data. Lipton and Moore [1972] have drawn a useful distinction between *single point* and *continuous* data and between *registered* and *nonregistered data.* The continuum ranging from single point to continuous data refers to the length of time taken to complete an activity. The continuum ranging from registered to nonregistered refers to the extent to which circumstances influence the respondent's ability to remember the quantities of an activity. Securing reliable data in the *continuous nonregistered* class requires frequent interviewing if measurement errors are to be kept at a reasonable level, since memory recall will not be good.

"Unfortunately, while measurement errors can be reduced by more frequent interviewing of relatively small samples, sampling errors are reduced through using large samples. The research worker with fixed resources must invariably face an unpleasant choice between either trying to minimize measurement or sampling errors.

"When we have tried to reduce measurement errors to a reasonable level, frequent interviewing has been supplemented with direct observations on certain critical variables. Without frequent interviewing we find it almost impossible to obtain reasonable estimates of labor utilization, particularly of family labor, which is a major input in traditional agriculture.

"What 'frequent interviewing' should mean in the northern Nigeria context was subjectively determined. Ideally one should decide the acceptable degree of measurement error and then determine the minimum frequency of interview and necessary research resources to meet the requirement. This becomes impossible when little is known about the environment and data are being collected on many different variables, as in a farm management study. We collected data at two levels of frequency:

Class 1—Data collected twice weekly (e.g., labor, seed and fertilizer inputs by field, etc.).

Class 2—Data collected infrequently (e.g., farm inventory, retail prices, crop rotation and land tenure patterns, conversion ratios, etc.)."

Gucelioglu (1976) reported:

TWO METHODS OF INQUIRY

"In studies concerning household expenditures there are two main methods of inquiry:

1. Asking families to enter daily or weekly income and expenditures in a special notebook.

2. Collecting information about income and expenditures by periodic interviews.

"Both methods have advantages and disadvantages. The first depends primarily on the family's understanding of budgeting and its cooperation in maintaining a daily or weekly accounting. The main responsibility nevertheless falls upon the interviewer. Because questions on expenditure always cover a period of past time, the problem of faulty recall always exists.

"In choosing the collection method to be used in the rural areas in Turkey, we tried to combine the two systems and notebooks were given to the literate member of the household who was asked to write down income and expenditures. In addition, interviewers visited the households weekly to obtain information on income and expenditures. What is the optimum frequency of visits? We visited each household once a week, on the day following the weekly village open-air market. Thus the interviewers were able to visit each household four or five times a month."

APPENDIX 5-T
VALIDITY FROM THE SOCIAL SCIENCE PERSPECTIVE

When the FSR&D team obtains information by means of interviews, it needs to be aware that some of the information may not be valid. We discussed some of these issues as part of Frequent Interview Surveys in Sec. 5.10.2. In this appendix, we introduce the concept of validity from the social science perspective, identify some of the threats to validity, illustrate the concept with examples, and provide selected references.

CONCEPT OF VALIDITY

Why is this appendix needed? FSR&D's interdisciplinary nature suggests that using insights and approaches from many fields can advance the art and science of FSR&D work.

The question of validity arises whenever researchers try to make causal inferences about the results of the introduction of a new technology. At this point, we stress that validity must be considered not only in the biological context, but in the sociological context as well.

Suppose a farmer increases crop production after receiving advice on the application of fertilizers to the crop. How much of this yield can be attributed to the fertilizer? The prudent researcher will ask, "What factors besides the fertilizer might have contributed to the increased yield? Could the farmer or the environmental setting have contributed to the increase?"

To begin to answer these questions, we turn to sociology, psychology, communication, and evaluation research, and consider how these disciplines make cause and effect inferences. One of the major issues in making causal inferences about a social program is the validity of the program's results. Validity can be explained by asking, "Are we measuring what we think we are measuring? Can the differences observed be attributed to the treatment? Can some of the differences be attributed to factors other than the treatment?"

The social sciences discipline offers explanations based on statistical, construct, internal, and external factors. These factors, which we now explain, often have relevance for FSR&D:

- Statistical validity deals with conclusions about the association of a presumed cause and a presumed effect (Cook et al., 1977).
- Validity constructs are usually abstract concepts that social scientists try to measure, for example, a farmer's attitude about some factor in an experiment. Of concern is the accuracy of the inferences

that researchers attach to experimental results (Cook et al., 1977).

- Internal validity deals with whether the experimental treatment caused the observed results.
- External validity deals with the ability to extend the treatment's results to other settings.

In FSR&D, internal validity must be considered during the on-farm research and analysis, while external validity must be considered when extending the results of the project to other farmers.

If researchers do not consider the validity issue, they may conclude a treatment or technology works when it does not, or conclude that a treatment did not work when it did. Social scientists have developed categories for different threats to external and internal validity. Specifically, Campbell (1975) identified nine threats to internal validity and six threats to external validity. The internal threats are history, maturation, instability, testing, instrumentation, regression, artifacts, selection, experimental mortality, and selection-maturation interaction. External threats are interaction effects of testing, interaction of selection and experimental treatment, reactive effects of experimental arrangements, multiple-treatment inference, irrelevant responsiveness of measures, and irrelevant replicability of treatments.

EXAMPLES OF THREATS TO VALIDITY

To minimize this appendix's length, we have selected six threats to validity to illustrate the points. Readers seeking additional information can consult this appendix's references.

HISTORICAL THREAT

FSR&D researchers might find historical threats to validity a problem. Events, situations, or factors that occur during a study could cause a result that might mistakenly be attributed to a treatment. For example, prices might rise sharply during the study. Because of these increased prices, farmers might weed, cultivate, and harvest their crops more carefully than during the previous season. Such factors might account for some or all of a crop's increased yield rather than the technological treatment being tested.

MATURATION THREAT

Maturation, another threat to internal validity, is a change in the farmers during the study. For example, farmers might learn a new cultivation technique and apply this technique without the FSR&D researchers' knowledge of the change. Maturation is simply changes in the farmers over time that might account for some of the differences in yields.

TESTING THREAT

Another internal threat to validity is that of testing. By knowing an FSR&D project is going on in their area or on their farm, the farmers might behave differently. For example, knowing they are part of a project or study may influence farmers to give more attention and care to their crops and animals than the farmers would under normal conditions.

SELECTION THREAT

The selection threat to validity could be an issue, if, for the previous year, the FSR&D researchers measured yields from the poorest of the poor farmers. By the laws of chance, yields from the group will be higher the following season—even without any treatment.

INTERACTION THREAT

In any study, the question arises as to how far the researchers can generalize the results. One of the threats to external validity of a study's results is that of the interaction effects of testing. Simply stated, farmers might react to being involved in a study and this might influence the treatment. If farmers know an FSR&D project is going on, they may become more sensitive to the ways they farm and change their farming practices. These changes might interact with the treatment and account for some increases in productivity. Thus, when the technology is extended to other areas, the results might be significantly less than the study's results.

EXPERIMENTAL TREATMENT THREAT

Another validity issue is the reactive effects of the experimental treatment. For example, the researchers might provide inputs—fertilizers, additional oxen, tractors, or labor—to work the fields and then apply an experimental treatment. Under normal conditions, the farmers may not have the inputs and thus the study's results could not be generalized to farmers without the same inputs.

DEALING WITH THREATS TO VALIDITY

How then does the researcher deal with external and internal threats to validity? The answer is by using experimental and quasi-experimental research designs. In FSR&D, researchers apply experimental designs when they randomly assign farmers, their fields, or other units to particular treatments or controls. The researchers apply quasi-experimental designs when they nonrandomly assign farmers, fields, or other units to particular treatments or controls.

Researchers will seldom be able to use a design that rules out all threats to validity. Instead, the researchers can concentrate on plausible threats to validity and seek the best possible design under the given conditions.

A HYPOTHETICAL EXAMPLE

At this point, we present a hypothetical example to

Table 5-T-1. A hypothetical design illustrating different potential gains of rice under treatment and no treatment.

Group	*Time One Yield kg/hectare*	*Treatment kg Zn per hectare*	*Time Two Yield kg/hectare*	*Apparent Gains (kg)*
A	1000 (A_1)	5	1600 (A_2)	600
B	1000 (B_1)	0	1100 (B_2)	100
C		5	1450 (C_2)	450
D		0	1050 (D_2)	50

illustrate how the design can be used to identify some of the threats to validity.

Consider a group of farmers with relatively homogeneous characteristics who farm on the same soil types, grow similar crops, and follow similar management practices. In the initial surveys, the FSR&D team found that yields averaged about 1000 kg of rice per hectare for the last growing season. As a result of further analysis, the researchers hypothesized that an application of zinc sulfate at 5 kg Zn per hectare would increase yields by 30 percent or more.

To test this hypothesis, the researchers randomly selected four groups of farmers within the project's area. For groups A and B, the researchers asked the farmers to report the last season's yield, which was 1000 kg per hectare.

For groups A and C, the researchers provided zinc sulfate for the farmers' crops. Later, researchers asked all farmers to measure and report their rice yields for the season (see Table 5-T-1).

The question arises as to how much of the 600 kg gain for group A and the 450 kg gain from C should be attributed to the zinc sulfate and how much can be explained by other factors?

By comparisons of the before and after yields among the groups, the more plausible yields can be determined. This experimental design provides procedures for testing the various threats to validity—e.g., history, maturation, testing, instrumentation, regression, selection, mortality, the interaction of selection and maturity, and the interaction of testing and treatment (Campbell and Stanley, 1966).

In this appendix, we identify each group by letter and subscript indicating the time of the measurement of the yield. For example, A_1 = group A measured at time one and B_2 = group B measured at time two.

Elaborating on the data and threats to validity and comparing D_2 with A_1 or B_1 suggests that the yields at time one were lower. Several plausible explanations can be offered for the 50 kg gain. First, the yields at time one may be lower than normal and time two yields will naturally increase (regression). Second, some factor may account for the difference—possibly more favorable weather conditions or farmers had more time or assistance to work their fields (history). Third, the farmers may have changed their measuring techniques and the yields appear to increase (instrumentation).

Thus, the basis for determining the yields would be to subtract D_2 from C_2 (1450-1050), which gives a yield increase of 400 kg per hectare attributable to the zinc sulfate. Then how can the 150 kg difference (600-450) per hectare be explained?

By subtracting the 50 kg gain attributed to regression, maturation, history, and instrumentation from the 100 kg gain (B_2 - B_1), we can speculate that 50 kg of the 150 kg difference may be the effect of testing. By knowing an experiment was going on, the farmers may have changed some farming methods that boosted yields slightly.

At this point we have explained 50 kg of the 150 kg difference. How do we explain the remaining 100 kg difference? The testing-treatment interaction may have accounted for the differences. By knowing they were part of the treatment group, the farmers who received zinc sulfate might have thought the chemical must improve the crops. Then, they began giving the crop additional attention, or when harvesting the crop or measuring the yield, they may have been more careful than usual.

What should the FSR&D team expect when it begins extending zinc sulfate to other farmers who are not part of the program? Most likely, the increased yields will be less than 400 kg per hectare.

This discussion has not explained all threats to validity, nor discussed the many experimental and quasi-experimental designs that might be used. More can be learned about this subject from Campbell and Stanley (1966), Cook and Campbell (1976), Haskins (1968), and House (1980), as well as the two references mentioned at the outset of this appendix.

APPENDIX 5-U QUESTIONNAIRE DESIGN

In Sec. 5.10.2., Chapter 5, on formal methods, we suggested several points to consider in designing questionnaires. This appendix, based primarily on Kinnear and Taylor (1979), Bernsten (1979), and Byerlee et al. (1980), provides additional guidelines on the topic. Below are sections on (1) preliminary considerations; (2) question content, format, wording, numbering, and sequence; (3) physical layout and length; and (4) pretesting and revisions.

PRELIMINARY CONSIDERATIONS

To begin, the FSR&D team needs to determine what the questionnaire should produce. Depending upon the situation, a variety of information might be needed. Here are some examples:

- information about the farmers' practices on a particular crop—from land preparation to post-harvest operations
- information about the harvest including what proportion the farmers market, what proportion the

farm family consumes, and how the farmers use the crop residues
- information about farmers' knowledge of plants, soils, weather, animals, hazards, and crop damages
- information about those factors of the total farming system that bear on a particular crop—e.g., labor, bottlenecks, crop sequences, rotations, the family's food preferences, seasonal consumption patterns, and cash flows.

One way of determining the data needed is to formulate tentative hypotheses to focus the team's attention on specific topics. Some of the hypotheses will be specific while others will be general. For example, specific hypotheses might concern factors used to determine yields and input variability, and general hypotheses might be related to management strategies if little is known about the farmers' management practices.

Next, the FSR&D team needs to consider at least two problems that might arise in designing the questionnaire. First, some team members may feel that information related to their field is more important than that of other team members. Here is where the team leader must resolve the issue by encouraging an interdisciplinary approach.

The second problem arises as to whether all farmers are asked the same questions or whether some farmers are asked more detailed questions. For example, if the sample of farmers is stratified by farm size, the team may want to ask farmers with the largest farms detailed questions related to mechanization possibilities, practices, and problems. Small farmers, on the other hand, might be asked questions that focus on draft animals and hand labor.

QUESTION CONTENT

Next, the team must develop questions for each of the variables in the hypotheses and determine the farmer's ability and likelihood of answering the questions. Here, the objective is to sensitize the team and to identify problems that could develop related to the questions. A question could be too sensitive to be answered truthfully, if at all—for example, questions related to cash flow analysis and profitability. The social scientists on the team might be able to point out potential conflicts and suggest methods of dealing with such problems.

Three basic issues must be acknowledged

- Farmers may not have thought about the issue raised by the question. If used, such questions will result in the collection of pseudo—i.e., misleading—data.
- Farmers may not know the answer to the questions—e.g., total number of hours spent in land preparation—or they may not recall the exact information required.
- Farmers may know the answer to the question, but give incorrect information because the question was poorly phrased—e.g., the farmer may not include the days of labor provided by helpers if this question were asked, "How long did it take you to plant this field of wheat?"

QUESTION FORMAT

Questions can usually be broken down into four categories: open-ended, multiple choice, dichotomous, or tabular. An open-ended question requires the interviewer to write out the response, while the multiple choice question allows him to check the appropriate response category. A dichotomous question, an extreme case of the multiple choice question, allows only two alternative responses—e.g., yes-no, daytime-nighttime, buy-grow, etc. Tabular questions start with a question and proceed to a table where the interviewer fills out row after row of information.

We suggest the advantages and disadvantages of these question formats in Table 5-U-1.

QUESTION WORDING AND NUMBERING

The three basic issues mentioned earlier can be overcome by carefully selecting the variables for which the questions will be written, by carefully wording the questions, and by carefully sequencing the questions. To illustrate these points, we quote examples from Bernsten (1979).

1. Every question should focus on one point and have only one answer:

Poor wording: "During the 1977-78 wet season, did you operate a farm and engage in nonfarm work?"

Clearer wording: "During the 1977-78 wet season, did you operate a farm?"
"During the 1977-78 wet season, did you engage in nonfarm work?"

2. Questions should not contain vague words such as many, often, and frequently.

3. Every question should use terms the farmers commonly use rather than the technical terms of the team's researchers.

4. Every question should be neutrally phrased to avoid biasing the respondent. This includes words or phrases that are emotionally charged or those that suggest approval or disapproval.

Poor question: "Do you believe [the President's] land reform program has helped the farmers in this village achieve ownership of land?"

Clearer question: "Do you believe land reform has helped farmers in this village achieve ownership of land?"

5. Every question should be phrased so that the respondent cannot feel which answer is preferred. Leading the respondents can be a serious problem created by the wording of questions and the tone of the interviewer's voice.

Poor question: "During the 1978 wet season, did you follow the superstitious practice of giving food

Table 5-U-1. Advantages and disadvantages of four question-response formats (Based, in part, on Marketing Research by T.C. Kinnear and J.R. Taylor; copyright © 1979 McGraw-Hill Book Company; used with the permission of McGraw-Hill Book Company).

Type	*Open-Ended*	*Multiple Choice*	*Dichotomous (two responses)*	*Tabular (Data recorded in columns or tabular form)*
Advantages	1. Good first question; allows general attitude to be expressed	1. Faster and less costly to administer than open-ended questions	1. Easy to administer	1. Much information can be gathered with few questions
	2. Establishes rapport (gains cooperation)	2. Less chance of interviewer bias	2. Fast and easy to code	
	3. Provides researchers with good insights into questions			
Disadvantages	1. Potential for interviewer bias	1. Design of questions takes longer—answers must be exhaustive and mutually exclusive	1. Can get substantial measurement error	1. Interviewers must be highly skilled
	2. High costs and much time involved in coding	2. Potential of introducing bias by choice of alternative presentations or by the order of the questions	2. Many questions are not suitable for "yes/no" responses	
	3. Extra weight given inadvertently to more articulate respondent	3. Restricts answers to only choices provided by the researchers		

to the rats so they would not damage your rice crop?"

Better question: "During the 1978 wet season, did you give food to the rats so they would not damage your rice crop?" "If you did give food to the rats, what was your purpose?"

6. Every question should specify the relevant time period for consideration.

Poor question: "Did you plant high yielding varieties on the rice parcel you operated?"

Clearer question: "During the 1977-78 wet season, did you plant only high yielding varieties on the rice parcel you operated?"

7. Questions that concern management practices often can be clarified by increasing the similarity in the way the questions are asked.

Poor format: "Did you buy any fertilizer during the 1977-78 wet season?"

"During this wet season of 1977-78 did you purchase any insecticide?"

"Was any rat poison purchased for use on your farm during the 1978 wet season?"

Clearer format: "During the 1978 wet season, did you purchase any fertilizer?"

"During the 1978 wet season, did you purchase any insecticide?"

"During the 1978 wet season, did you purchase any rat poison?"

8. Frequently, a pre-qualifying question will be necessary to verify that the question of interest applies to the respondent. Such pre-qualifying questions can reduce the chances of collecting misleading—pseudo—information.

Poor question: "During the 1977-78 wet season, was fertilizer available at the local supply store when you wanted to fertilize your rice land?"

Clearer question: "During the 1977-78 wet season, did you grow rice?"

"During the 1977-78 wet season, did you apply fertilizer on your rice land?"

(Omit the following question if the answer to either of the above two questions was negative.)

"During the 1977-78 wet season, was fertilizer available at the local supply store when you wanted to fertilize your rice crop?"

9. Many questions relate to farming practices in general. Small farmers often have numerous separated

parcels of land on which different practices are followed. Questions on management practices should be asked with respect to individual parcels. A tabular question might work well in such a case.

10. Questions that require accurate answers for aggregate values will be more reliable if the amounts are first collected on an individual basis—e.g., on a weekly, monthly, or per parcel basis—and then summed by the researchers.

Poor question: "From your 1977-78 wet season harvest, how many kg of paddy did you sell?"

Clearer question: "From your 1977-78 wet season harvest, how many kg of paddy did you sell during the following months?"

March	______
April	______
May	______
etc.	______
Total	______

11. Unrealistic questions should not be asked. These refer to questions about information the farmers are not likely to know, recall, or wish to divulge.

12. Each question should be numbered. This is done to ease the interviewer's task and to aid in the processing of data. Main sections are given a number, then questions within that section receive the section number as well as a number to indicate its position within the section. An example follows:

1.0. Farmer's Socioeconomic Characteristics
- 1.1. "How old are you?"
- 1.2. "How many children live in your household?"
  - 1.2.1. "How many are males under 15 years?"
  - 1.2.2. "How many are females under 15 years?"

2.0. Employment Activities of Farmer
- 2.1. "During the 1977-78 wet season, did you work for pay on anyone else's farm?"
- 2.2. "During the 1977-78 wet season, did you work for pay in nonfarm activites?"

QUESTION SEQUENCE

Sequence the questions in an orderly flow that will seem logical to the farmer. Usually the team can divide the questions into separate areas of interest. Byerlee et al. (1980) suggested the following progression for topic sections on the farmer's activities:

1) screening questions to determine if the farmer fits the requirements of the sample

2) facts about management practices used on a particular crop—i.e., from land preparation to post-harvest operations, including the farmer's use of inputs

3) facts about disposition of crop—e.g., yields, marketing, storage, and crop residues

4) opinions about specific management practices and the severity of hazards, problems, and constraints for the crop

5) important facts and opinions about the total farming system that bear on a particular crop—e.g., labor bottlenecks, crop sequences and rotations, livestock, manure for crops, food preferences, seasonal consumption patterns, and cash flows.

Such an arrangement should avoid frequent changes of topics or flipping the pages of the questionnaire backward or forward.

PHYSICAL LAYOUT AND LENGTH

The first page of the questionnaire shows the official name and address of the project, which often assures farmers of the questionnaire's authenticity. A short, written statement explains the survey to the farmer. It should include

1) the objectives of the program and of this survey
2) explanations of how and why the farmer was chosen
3) how the information will be used
4) an explanation of the confidential nature of the survey.

The interviewer should begin with a section on the farmer's name, village, location, and any other information that might be required should the farmer need to be contacted again. Each topic section should have a brief written introduction preceding it. Sensitive questions or groups of questions may require a preliminary explanation as to their necessity. Once again, stress the confidentiality of the information.

As a general rule leave a sufficient amount of blank space for open-ended questions. A blank space for answers to multiple choice questions should be placed near the right-hand margin for each question. This facilitates coding and editing. All instructions should be in bold face or capital letters so as to attract the interviewer's attention.

The length of the questionnaire is important. The longer the questionnaire, the more difficulty the interviewer will have in holding the farmer's attention. Ideally, administering a questionnaire should take no longer than one hour. If a questionnaire takes longer, it should be divided and administered on separate days.

PRETESTING AND REVISIONS

Pretesting involves five activities:

1) Select the personnel. Only a few experienced interviewers and the research team should do the pretesting.

2) Translate the questionnaire, if required. If the researchers on the team do not speak the language of the farmers, the questionnaire must be translated. The following process helps ensure that the translation is correct. Each question is explained in detail to the translator. This

translator then writes a version of each question in the farmer's language. A second translator converts the translated version back into the original language. Finally, the original question is compared with the twice translated version and any inconsistencies are worked out by the translators and the research staff.

3) Develop an operations manual. With a large number of questions and a number of codes for each question, an operations manual must be assembled that will guide the interviewer in the field. The instructions are a supplement to the definitions or information found in the questionnaire itself and explain how to ask and fill out the responses for each question.

4) Move to location. The pretest team must move to the site and locate farmers that have the same socioeconomic characteristics as do the farmers in the sample. The team should exclude any farmers who are included in the sample.

5) Check for problems in the questionnaire. Problems may occur in the questions, the sequencing of the questions, the format, or the coded responses. Problems with a question usually exist when:

- the same answer is given by each farmer
- the farmer cannot answer the question
- answers indicate that the question is misunderstood
- when the respondent takes a long time to answer.

These problems may have resulted from an illogical ordering of the questions, poor translation, unfamiliar terms, or complicated grammatical constructions.

The format should be simple enough so that the average interviewer has no difficulty administering the questions. When problems do arise, experienced interviewers can usually provide appropriate solutions.

The pretest should be viewed as an exercise that, among other things, seeks to elicit all the possible responses that will be given for each question. Comprehensive codes can then be developed for the formal survey, thus speeding the data processing phase.

APPENDIX 5-V SAMPLING

In Sec. 5.10.2., Chapter 5, on formal methods of data collection, we concentrated on surveys and said additional information would be provided in an appendix on sampling. Thus, this appendix elaborates on why researchers should sample, the sampling process, sampling unit, lists, sampling methods, deciding on sampling methods, and determining the sample size. For this, we draw on the works of Slonim (1960), Babbie (1973), Kearl (1976), and Bernsten (1979).

WHY SAMPLE?

Several points emerge from survey research methodologies that explain why researchers should sample the items (in statistics, called the population) they are studying. First, consider the differences between a census and a sample. A census is a complete enumeration of all units within the population while a sample is generally a small number of the population's units. For both a census and a sample, the researchers then look at predetermined characteristics—in the case of FSR&D, the researchers want to know about characteristics of the farming system.

If done properly, sampling can accurately reveal the characteristics of the population under study and can require much less time, money, and personnel than does a census. From a sample, the researchers can then generalize about the whole population.

Like the other scientific concepts and techniques upon which FSR&D is based, statistical sampling requires an understanding of the procedures if they are to be applied properly. Thus, this appendix introduces some fundamental concepts of sampling. For further information, the FSR&D teams may want to review the references noted above, or other references suitable for the developing countries. Also, the teams may want to consult with empirical sociologists, communications researchers, or others with similar expertise.

THE SAMPLING PROCESS

The sampling process requires the researchers to

1) specify the sampling unit
2) obtain an adequate list of these units
3) determine the acceptable degree of accuracy of the sample
4) determine the sample size
5) select the sampling method
6) select the sample.

Then, the researchers complete the remaining steps of the survey.

SAMPLING UNIT

Whenever FSR&D researchers want to know more about the farming systems being considered, they usually focus on units such as farmers, households, or villages. In survey methodology, these units are called the sampling unit or sometimes the unit of data collection. The researchers must specify and define the sampling unit they want to study. This step is critical because it influences the sampling methods a researcher uses and facilitates identifying lists from which researchers sample (Bernsten, 1979; Beirut Seminar Working Group, 1976a).

LISTS

Most researchers sample from population lists—commonly called sampling frames. To be assured of the quality of the final data, the researchers must evaluate the quality of the lists before they begin sampling.

The units on the list should include all units of the population to be studied. If not, the purpose of sampling is defeated—the researcher cannot generalize to the individuals beyond the list from which the sample was selected. Too often, lists will be inaccurate because they:

- are old and do not include individuals who have moved into or left the area, or who have died
- are incomplete
- contain individuals atypical of the populations to be studied
- do not contain the characteristics important for the study.

Following are several problems in using lists at the village level, as identified by Bernsten (1979):

"1) *Extension agent lists* may include only farmers visited by the agent. Use of this frame will probably produce data that suggest an unrealistically progressive, well informed, and successful farm population.

"2) *Irrigation parcel lists* usually include an entry for each parcel in the system. Thus, farmers with large, fragmented holdings would have a greater chance of being selected than individuals farming only one parcel.

"3) *Land tax lists* often include only land owners. Consequently, leaseholder and share tenant farmers would be excluded.

"4) *Head tax lists* may exclude women, nonresident farmers and may be incomplete.

"5) *Census lists* may exclude individuals who have arrived since the last census and nonresident farmers.

"6) *Voter lists* are often out-of-date and may not include nonresident farmers, farmers who don't own land (when property ownership is a condition for voting), and women in areas where only men are permitted to vote.

"7) *Farmer lists* are often out-of-date and may exclude nonresident farmers.

"8) *Government production program participant lists* will exclude farmers who aren't program participants.

"9) *Government pump and tractor owner lists* may exclude both farmers who obtained the unit outside the official loan program and equipment renters; and will include farmers who no longer own or use the equipment. . . ."

And Bernsten (1979) indicates maps or aerial photographs can be useful, but some sampling schemes may overrepresent the farmers with large land holdings. In closing his discussion of sampling frames, Bernsten (1979) indicates a census might be used where no sampling frame exists. However, censuses are time-consuming, expensive, require many enumerators, and often miss some of the population.

In some cases, researchers may want to develop their own lists. At this extreme, developing a complete list means doing a census, while, at other times, it requires careful compilation using existing lists. The researchers' decision will hinge on the need for the lists and the costs of generating them.

SAMPLING METHODS

In this section, we discuss three types of nonrandom sampling—purposive, accidental, and quota; and five types of random sampling—simple random, systematic ordered, stratified random, cluster, and multi-stage.

NONRANDOM SAMPLING

Although nonrandom sampling cannot produce data from which researchers can safely generalize to larger populations, some researchers use nonrandom sampling methods. If nonrandom sampling is used, researchers have no way of knowing whether the individuals interviewed represent the population being studied. However, interviewing such individuals does provide researchers with a "feeling" about the population, which, at times, is sufficient justificiation for the method.

Purposive Sampling

The researchers establish a series of characteristics of the sampling units they want to study. And, without using a population list or specifying selection procedures, the researchers select and interview the sampling units that supposedly conform to the desired characteristics of the population (Bernsten, 1979).

A hypothetical example of a purposive sample would be to categorize farmers according to those who raised maize and beans, had more than four children, two cows, and 15 chickens, came from a common village, and lived on moderately sloping lands. The researchers would then seek, in no predetermined way, farmers meeting these characteristics. When such a farmer would be found the researchers would interview the farmer and then continue seeking similar farmers until the desired number had been interviewed.

Accidental Sampling

The researchers, in a haphazard way, select and interview anyone they meet.

A hypothetical example of an accidential sample would be to interview the first 10 individuals the researchers met when walking along a path to a village.

Quota Sampling

Bernsten (1979) suggests using quota sampling where researchers want to stratify their samples and have lists for the entire population, but not according to strata.

Using the overall population list, the researchers draw a random sample. Then interviewers are given the names and sent out to conduct the interviews. To begin, the interviewers ask a series of screening questions to place the individual into a particular stratum. If the screening question places the individual into a stratum and the quota has not been filled, the interviewer completes the interview. The process is repeated until the quota for the stratum is filled. If the screening questions reveal an individual that would be placed in a stratum whose quota has been filled, the interviewer terminates the interview (Bernsten, 1979).

Although individuals from the master population were selected at random, the researchers cannot generalize about the strata's characteristics because the number of individuals to be interviewed in each stratum was specified by the research design and is not, necessarily, representative of the whole population (Bernsten, 1979). We should point out that some writers such as Babbie (1973) and the Beirut Seminar Working Group (1976a) call this nonprobability sampling.

With nonrandom sampling, unbeknown to the researchers, biases may enter into the selection of the individuals to be interviewed. Thus, researchers must be cautious because they have no way of identifying and eliminating biases. Simply, a selection bias creates a serious threat to the validity of the data. In Appendix 5-T, we discuss the validity issue further.

In some cases, the expense of probability sampling may be prohibitive and thus nonprobability sampling might be used if the researchers are careful in their approach and interpretation of the data (Babbie, 1973; Kearl, 1976).

RANDOM SAMPLING

With random sampling, the selection bias and threat to validity are minimized; and researchers can estimate the sampling error and can generalize about the whole population.

With random sampling (1) each unit in the population has a chance of being selected and (2) researchers will know the probability of selecting a particular unit and can control for sampling error (Slonim, 1960; Bernsten, 1979).

The sampling error is the researcher's measure of how close the statistics represent the larger population. The sampling error includes two elements: the confidence interval and the confidence level. For example, researchers might report that they are 95 percent confident that between 40 and 50 percent of the farmers in the area use insecticides. This means that 95 percent of the time the intervals so computed will actually contain the true value for the whole population. In the above example, 95 percent is the confidence level and 40 to 50 percent of the farmers using insecticides is the confidence interval – i.e., 45% ± 5%.

We will now elaborate on some of the methods of random sampling.

Simple Random Sampling

By using this method, the researchers assure themselves that each unit in the population has the same chance of being selected. Although simple random sampling is easy to implement, obtaining an adequate population list can be difficult (Bernsten, 1979).

Starting with a population list, the researchers determine the number of units in the population and the sample size – the number of units to be selected for study. Researchers then number the units in the population from 1 to N. Using procedures outlined in most general statistics textbooks, the researchers generate a random number and select the corresponding numbered unit from the population. This process continues until the whole sample has been selected (Bernsten, 1979).

Systematic Ordered Sampling

For a systematic ordered sample, researchers start with a random number on the list and then take every k^{th} unit on the list. If the researchers want to draw a sample of 20 farmers from a population list of 500 farmers, the researchers would divide the population size by the desired sample size to obtain the sampling interval, k. In this case, $k = 500 \div 20 = 25$, which means that every 25th unit on the population list is to be sampled. Using a table of random numbers, the researchers first select a number between 1 and 25. For this example, the randomly selected number is 9. The researchers then select farmer number 9 and then every 25th farmer thereafter.

Generally researchers prefer systematic random sampling over simple random sampling because systematic random sampling is quicker and easier to use (Bernsten, 1979). However, researchers must be sure the units are listed randomly. Ordered lists – for example, a list ranking farmers from those producing the most to the least rice per hectare – would introduce bias into the sample and should not be used.

Stratified Random Sampling

Here, the researchers separate the units into strata having the desired characteristics and randomly select samples of units from each stratum (Bernsten, 1979). Tollens (1976) suggested various strata possibilities for dividing populations. These include crop production systems, climate, soil types, farm size, sex, and tribal differences.

The strata the researchers select should be relevant for the research objectives and the population, be easy to use, and contain enough units that statistically reliable comparisons can be made (Bernsten, 1979). Furthermore, the population should be homogeneous within the stratum and heterogeneous among strata.

The primary disadvantage of a stratified random sample is that the researcher must know enough about the population being studied to place the units into the proper strata (Bernsten, 1979).

Cluster Sampling

Cluster sampling entails first obtaining a list of grouped units – for example, villages – and sampling these units. Then, the researchers obtain lists of subunits – for example, farmers – from each of the selected clusters and samples from the subunit lists for each cluster. Bernsten (1979) points out that cluster sampling reduces the costs of interviewing farmers because they tend to be close together; however, sampling bias may be introduced because geographically close units may be more similar to each other than are dispersed units.

Multi-stage Sampling

The previously discussed sampling schemes are relatively straightforward and do not, on their surface, appear complex. Multi-stage sampling is more complex in that researchers combine several sampling techniques to draw their sample.

Bernsten (1979) explained this procedure as follows:

> "After the population is initially sampled, the resulting subpopulation is again sampled. This procedure can be repeated as many times as desired to build up a survey sample with the required characteristics. For example, suppose we want to draw a sample of farmers in Central Luzon, Philippines, but one-third must be rainfed, canal irrigated, and pump irrigated. In order to reduce travel costs, we list all of the provinces and randomly select three provinces. In each of these three areas we list the population of villages and then randomly select two. In each village, we obtain a list of farmers in each irrigation class. Then, we randomly select a given number of cases from each of the three sub-populations. If we desired to have 20 cases in each of the three irrigation classes in each of the two villages in each of the three provinces, a total sample size of 360 cases would be required."

The advantages of multi-stage sampling are that researchers usually find obtaining the list of the large units relatively easy; the focus on subunits reduces the costs of having to have population lists for which only a limited number of subunits will be drawn; and the cases, when selected, are usually close together (Bernsten, 1979).

The main disadvantages of multi-stage sampling include the complex sampling procedure and the difficulty in generalizing to the clusters from the samples (Bernsten, 1979).

DECIDING ON SAMPLING METHODS

The researcher must consider a number of factors before selecting the sampling scheme for the particular study. First, researchers need to consider the data needed and the generalizations they would like to make about the target population. This, in turn, influences the research design for the study.

With the desired research design in mind, the researchers must then consider the availability of sample lists. The most decisive factors are available funding, the time required for the research design, and the expertise of key researcher staff.

After weighing these factors, researchers will need to select that sampling method that best fits their needs. We have not tried to provide more detailed guidelines to the selection of sampling methods because conditions and needs vary so widely.

SAMPLE SIZE

Bernsten (1979) pointed out that determining the sample size using the various formulas provided in statistical textbooks is academic and that the usefulness of such formulas for projects similar to those found in FSR&D may be quite limited. He stressed that researchers are often interested in several different parameters and each requires a different sample size to obtain a particular level of precision. Thus, he recommended that researchers sample enough units from the population to have a minimum of 20 cases for each major sampling category.

Using this approach, one way to arrive at the total sample required is to prepare dummy tables for comparing those variables the researchers want to consider. While each cell within the table should have a minimum of about 20 units, more would be better. The comparisons among variables that need to be made will then determine the total sample size.

Researchers generally increase the sample size to reduce the standard error—a measure of sampling error. But, as the sample size increases, so do sampling costs, and often a high level of precision is not worth the additional cost. For example, Collinson (1976) reported that using a sample of 100 farmers would produce a standard error of 10 percent. To reduce the standard error by 25 percent (to 7.5 percent) would (1) require increasing the sample size by 80 percent (to 180 farmers), (2) increase the costs for a single interview by 45 percent, and (3) increase the costs for multiple interviews by 60 percent. Thus, for many projects budget restrictions have more influence on sample sizes than does the researchers' desire for precision.

Finally, researchers should draw additional units when preparing the sample so that they will have enough should some of the units not be available or suitable. By doing this beforehand, time will be saved and the integrity of the sample will be maintained.

In closing, we stress that sampling can be complex and time consuming and the errors costly. Researchers not experienced in this type of work should seek advice from experts, who would be able to provide the guidance needed for obtaining meaningful results in an efficient manner.

APPENDIX 5-W SELECTING, TRAINING, AND SUPERVISING INTERVIEWERS

In Sec. 5.10.2. in Chapter 5, we introduced the topic of formal surveys and referred to this appendix for additional details about interviewers. The topics of this appendix include the criteria for selecting interviewers, determining the number of interviewers, hiring and training interviewers, and supervising their work.

CRITERIA FOR SELECTING INTERVIEWERS

FSR&D teams need to select their interviewers carefully to help assure that their surveys produce useful data. While the characteristics of good interviewers may vary from country to country, they should generally have a friendly, yet professional, approach to their work.

From worldwide experience, several criteria have emerged to help in choosing good interviewers. These criteria concern personality, language, education, farming knowledge, sex, racial and ethnic background, motivation and honesty, and residence in the area. We briefly expand on these criteria in this section, which draws on the works of Bernsten (1979) and several who are quoted in Kearl (1976).

PERSONALITY

Interviewers need to have a friendly, outgoing personality and a sincere interest in the farmers and the project. Individuals who appear threatening or are negative toward the farmers seldom make good interviewers. On the other hand, overly friendly interviewers may spend too much time visiting with a farmer when they could be interviewing other farmers.

LANGUAGE

The interviewers must speak the language and dialect of the farmers if they are to collect useful data. Not understanding the language or dialect causes errors.

EDUCATION

The interviewers should have enough education to complete the questionnaires without difficulty. Generally, multiple choice questionnaires require less education to administer than do open-ended questionnaires. Several researchers have found that a high school education is adequate for most interviewers. Ogunfowora (1976c) pointed out that highly educated interviewers produce more reliable data, but their turnover rate is high. They use their job as an interviewer to advance their careers.

FARMING KNOWLEDGE

Interviewers need to know the farmers' language and terms for different farming and household practices; otherwise interviewers may obtain erroneous information.

SEX

Sometimes, the sex of the interviewer is crucial. In some cultures women interviewers are needed to interview women, while in other cultures men make better interviewers. Gafsi (1976b) suggested that at least one of the interviewing team be a women.

RACIAL AND ETHNIC BACKGROUND

Usually, the interviewers should be of the same racial and ethnic background as the individuals to be interviewed. Interviewers of different racial and ethnic backgrounds may be perceived as threatening. Also, the differences in backgrounds may result in the collection of erroneous data.

MOTIVATION AND HONESTY

Researchers have found that the best interviewers are highly motivated and honest. Potential interviewers who are only moderately motivated may become discouraged and quit partway through the survey, and some may even falsify the data.

RESIDENCE IN THE AREA

Generally, researchers have found that interviewers should come from the study area. El Hadari (1976c) suggested this because (1) researchers need not worry about their lodging and transportation, (2) such interviewers often know people in the villages, (3) farmers are not as suspicious as when an outsider visits them, and (4) such interviewers know the farmers' language and terms.

NUMBER OF INTERVIEWERS

We cannot provide firm answers about the number of interviewers needed for a survey since the circumstances of each survey are so different. Instead, researchers need to consider several factors in determining the number of interviewers to hire:

- the length of the questionnaire and the time required to complete the interview
- the average number of interviews that can be completed in one working day
- the total number of interviews to be completed
- the number of days or weeks allocated for the job
- travel time between interviews
- the time required for interviewers to travel from their homes or places of living to the area where they will be interviewing
- the interviewers' experience.

To illustrate the calculations, consider that an FSR&D team wants to know how many interviewers are needed to interview 100 farmers in two weeks. The team estimates that each questionnaire takes an hour to administer. From experience, the researchers know that an hour interview takes from two to three hours of the interviewer's time. For this illustration, we will assume that an average interview will require three hours and the interviewers will, therefore, be able to complete three interviews per day. The interviewers are scheduled to work six days per week. Thus, during the two weeks, one interviewer will be able to complete 36 interviews—i.e., 3 interviews/day × 12 days. Therefore, three interviewers would be needed—i.e., 100 interviews ÷ 36 interviews/interviewer = 2.8, which rounds to 3.0. Since interviewers would be hired on a full-time basis and, to allow for contingencies, we would recommend that four interviewers be hired.

We recommend the extra interviewer to allow for

unforeseen problems, such as an illness or when an interviewer quits. On the other hand, experienced researchers caution against hiring too large an interviewer staff. Bernsten (1979) says that by hiring a small number of interviewers, the possibilities of introducing bias into the data are reduced and the quality of the data is improved. Furthermore, hiring a small interviewer staff allows those who are hired to improve their interviewing skills, produce better data, and learn a skill they can use again. Thus, we suggest hiring enough interviewers to do the job, but the FSR&D teams should keep the number of interviewers small enough to produce quality data.

FSR&D teams undertaking surveys for the first time should consult those experienced in this type of work. In this way, the team will learn about problems unique to such surveys; moreover, if those being consulted have conducted surveys previously in the research area, the team will learn about the area as well.

HIRING INTERVIEWERS

Once the candidates have been selected, some researchers test them to assess their skill in handling the interviews and completing the questionnaires. According to Bernsten (1979) these tests assess the candidates' abilities to (1) make simple mathematical calculations, (2) follow instructions, (3) write legibly, (4) speak fluently, (5) understand the farmers' language, and (6) understand farmers' practices.

Alternatives for acquiring interviewers include hiring them permanently or temporarily, or borrowing them from other organizations. The following discussion on this subject is based, in part, on Bernsten (1979) and on statements by Collinson, Norman, and Flinn as contained in Kearl (1976).

Researchers experienced in conducting surveys in rural areas generally favor hiring a permanent staff of interviewers. Several reasons emerge

- Once trained, the interviewers can be used on other surveys, which shortens subsequent training periods and reduces costs.
- Interviewer skills require time to develop and, once trained, the interviewers' skills continue to improve and they produce better results.
- Experienced interviewers require less supervision.
- Researchers might use probationary periods to select the more effective interviewers for permanent positions.
- Permanent interviewers generally have a more professional commitment than temporary interviewers.

Although most researchers favor permanently hired interviewers, some tend to see temporary interviewers as a reasonable approach for FSR&D projects. Bernsten (1979) suggested interviewers might be hired from such sources as agricultural research staff, extension agents, local officials, and teachers. But in selecting them, the researchers should look into a candidate's reputation. Also, with temporary interviewers, the research team spends less time discussing conditions of service—e.g., vacations and insurance—with the candidates (Bernsten, 1979).

Borrowing interviewers from other groups produces mixed results (Flinn, 1976a). By borrowing interviewers from other agencies, the researchers may be creating conflicts between the interviewers' responsibilities to the FSR&D project and their parent organization. The Beirut Seminar Working Group (1976b) pointed out that help can sometimes be obtained from the extension service, since the extension agents are government employees. However, because the extension staff may view research surveys as embodying an evaluation of extension's effectiveness, agents used as interviewers may not be able to maintain their objectivity during the interviews.

TRAINING INTERVIEWERS

Interviewers should complete a training program if they are to adequately conduct the interviews. Overall, the training program is intended to provide interviewers with background on the survey's purpose and to ensure the interviewers know how to administer the questionnaire. The following presentation stems from Bernsten (1979), Byerlee et al. (1980), and Ogunfowora, Flinn, and the Beirut Seminar Working Group in Kearl (1976).

Ideally, all interviewers should ask questions in the same way so the responses would be the same as if all interviews were conducted by the same individual. Thus, the approach to a training program is to produce a team of interviewers who administer the questionnaire uniformly without biasing the results.

Preparation for the training program begins while the questionnaire is still being developed. At this time, researchers need to begin writing an instruction manual for the interviewers. Generally, instruction manuals include

- a short orientation to the team's research and the survey
- how to approach the individuals to be interviewed
- an orientation to the questionnaire
- how to ask questions
- an explanation of each question, including its purpose
- explanations about probing techniques
- how to make any required calculations
- instructions on logistics, pay, and related matters
- instructions on observing and recording observations in a log, whenever these are part of the interviewer's responsibilities.

The detail and complexity of the manual depend upon the level of supervision planned for the interviewers. If interviewers will conduct all interviews in remote villages with little or no supervision, the manual needs to answer most of the questions an interviewer might ask. If interviewers are supervised closely, the interviewers' manual will need less detail.

Next, the researchers need to decide on the length of the training program. Times might range from three days for simple questionnaires of one hour to two weeks when the questionnaires are longer and complex. In either case, the level of training should be geared to the interviewers with the lowest skills. If the interviewers are to conduct the interviews in phases, the training might begin with an intensive session of several days and then be followed with short reviews before each of the subsequent phases.

In addition to training, some researchers use the training sessions to identify those who would be poor interviewers. In fact, some researchers stipulate those attending must successfully complete the training sessions if they are to be hired.

While various approaches are possible, depending on the researchers' objectives and the interviewers' experience, Bernsten (1979) and Flinn(1976b), among others, suggest that training sessions cover the following points:

- an orientation to the research and the survey
- a discussion of the specific objectives of the survey
- the role of the interviewers
- an orientation to survey and FSR&D terminologies
- an orientation to the farming system being studied
- how the individuals to be interviewed were sampled
- how to approach the individuals to be interviewed
- building rapport with the person being interviewed
- a review of each question and its purpose
- techniques for probing and rephrasing questions
- training in special skills such as field measurements and estimating yields
- reviewing and editing the completed questionnaire
- practice interviews in the classroom
- practice interviews in the field
- the timetable for the survey
- administrative details such as pay, terms of hire, and logistics.

In some cases, interviewers may be asked to make observations of farmers, the farm, and its operation. If FSR&D researchers anticipate using interviewers in this role, the training sessions must include some work on how to (1) be effective and efficient observers, (2) keep a field diary, and (3) write up their observations.

SUPERVISING THE INTERVIEWERS

Once the interviewers have been trained, the researchers normally supervise the interviewers throughout the interviewing process, provide the necessary logistical support, keep up the interviewers' morale, and spot-check for falsified interviews. These ideas, and the following discussion on this topic, are taken, in part, from Bernsten (1979) and the inputs of Norman and the Beirut Seminar Working Group to Kearl (1976).

The researcher's supervisory role includes collecting and editing questionnaires; checking the questionnaires for legibility, completeness, consistency, and accuracy; discussing problems with the interviewers; and when necessary, shifting the interviewers' assignments.

Bernsten (1979) commented on the important role researchers play in providing logistical support to interviewers. They make sure interviewers have an adequate supply of questionnaires, pens, and paper; a waterproof container for questionnaires; and provisions such as boots and umbrellas. He further stressed the importance of paying the interviewers on time.

Furthermore, the researchers' presence in the field helps boost interviewer morale. Without this, interviewers may become disheartened and quit. When this happens the researchers must recruit and train replacements, which can be highly disruptive. Thus, researchers should visit interviewers regularly, supervise their work closely, and respond to their personal problems and needs.

The researchers can improve interviewers' morale in other ways such as (1) having interviewers work in teams, (2) choosing interviewers whose families live in the area, (3) providing bonuses for completing the assignments, (4) providing transportation or reimbursements for transportation expenses, and (5) providing rain gear.

Finally, we need to mention a problem that sometimes occurs—the falisification of interviews. Interviewers have been known to complete questionnaires without ever having talked with the intended persons. This creates erroneous data and might nullify the entire survey. Thus, most researchers spot-check the work of their interviewers by visiting with some of the individuals whose questionnaires have been completed. In doing this, the researcher might ask the persons whose names are on the completed list to clarify a particular response. If the researcher finds an individual who says he or she was not interviewed, most researchers will want to varify all of that interviewer's completed questionnaires and interviews. Occasionally, the interviewer will simply have made a mistake in identifying the person; or, the person may have forgotten about the interview. Thus, the researcher should not jump to conclusions until after completing a thorough investigation. Then, if the researcher finds that the interviews were falsified, the interviewer will usually be dismissed.

APPENDIX 5-X FARM RECORD KEEPING

Hildebrand has come out strongly in favor of farm record keeping as an efficient means for gathering data about farmers over time. We discussed some of these aspects in Sec. 5.10.2. in Chapter 5. In Part 1 of this appendix, we include considerable portions of the report of Hildebrand (1979b) on ICTA's farm record project; and in the second part, we include excerpts from Hatch (1980) on record keeping by illiterate farmers.

PART 1: ICTA'S FARM RECORD PROJECT

In this part, we extract from Hildebrand's report (1979b), which includes sections on the history of the project and current procedures. Current procedures include sections on selection of collaborators, managing and analysis of the data, and reports. Also included are sample tables and calculations.

"HISTORY OF THE PROJECT

"From the beginning, the project was conceived as a *crop* record project and was not intended as a *farm* record program. That is, no attempt was made to take full farm inventories, impute depreciation costs of equipment to each crop, enter into household expenses and/or use of farm products, etc.; rather family owned machinery and animal power, family labor and . . . land rent were all charged at the current contract or hired cost for similar items. This characteristic had three important advantages. One is that it held to a minimum the amount of time and bother the farmer had to put into the data gathering process. Second, training of personnel was simplified, and third, the analyses were simplified. This probably was one of the main reasons the project has had the success it has enjoyed. Had it been designed as a full farm record project from the start, it would have been so complicated that it probably would have failed before producing enough data to demonstrate its productivity.

* * *

"The farmers were given simple sheets, [Table 5-X-1] on which they were to write their activities each day for the crop on which they were keeping records. These "Daily Work Sheets" were kept as simple as possible and included only the information that could not be obtained in periodic visits by the technicians. Those details were left for another sheet that the technician filled out during visits with the farmers.

"Following each cycle of work (land preparation, planting, weeding, etc.) the technician collected the daily work sheet after varifying facts with the farmer and took them to the office where he transferred the information to tabulation sheets. Following the end of the harvest and final collection of information, the tab sheets were analyzed to provide the final report for ICTA and to prepare a report for each farmer who participated in the project. Each farmer was given a folder which contained a labor summary, an input summary and a general summary for his own crop showing how much he made or lost that year. The folders were presented in a meeting in which all participated and in which the results were discussed.

* * *

"In order to process the additional amount of data being accumulated by the growth of the project, analysis of the records was modified. First, instead of tabulation sheets that were found to be unwieldy, summary sheets for inputs, [Table 5-X-2]; manual, animal or mechanized labor, [Table 5-X-3]; and a General Summary, [Table 5-X-4], were designed. Although the daily work sheets were different for each of the work areas to account for differences in measures and other characteristics of the farms, the summary sheets were the same for all the zones so they could be analyzed in a standard format. Programs were written for hand-held programmable calculators (HP-67 and TI-59) so that all basic analyses could be made rapidly and efficiently.

* * *

"CURRENT PROCEDURES

"Selection of Collaborators

"At the present time, with many of the field personnel of the Institute involved in the record keeping project, some of the farmers who are collaborators in field trials or Farmers' Tests are also record keepers. This has the advantage of minimizing travel distances and number of farm visits while it also tends to increase communication between the participating farmer and the technician. It has the disadvantage of possibly biasing the sample towards those farmers who are most cooperative. However, any sample of small farmers that includes only those who are willing to keep records will be biased to a certain extent. In practice, bias is kept to a

Table 5-X-1. Daily record sheet, Region IV 19___ (Hildebrand, 1979b).

FARMER'S NAME ____________________ FARM NO. ______ ROAD _____
CROP __________ AREA __________ HAS __________ SECTOR ______________

DATE	WORK DONE	No. OF [UNITS]	MATERIALS USED		No. OF PERSONS		WAGE		WITH FOOD (YES OR NO)
			CLASS	QUANTITY	FAMILY	HIRED	DAY	[UNIT]	

Table 5-X-2. Input summary (Hildebrand, 1979b).

REGION ______________ SUB-REGION____________ NAME ______________________________
TOWN ______________ COUNTY______________ CROP ______________ HECTARE ______________

INPUTS, EACH APPLICATION	HECTARES	QUANTITY		PRICE	VALUE		DATE
		TOTAL	PER HECTARE	PER UNIT	TOTAL	PER HECTARE	

Table 5-X-3. Labor summary (Hildebrand, 1979b).

REGION______________ SUB-REGION____________ NAME ______________________________
TOWN______________ COUNTY______________ CROP ______________ HECTARE ______________
COST OF RENT PER HECTARE______________

WORK	DATE	AREA	FAMILY LABOR		HIRED LABOR		TOTAL		PER HECTARE	
			MAN-DAYS	$	MAN-DAYS	$	MAN-DAYS	$	MAN-DAYS	$

minimum by utilizing farmers who use the same cropping systems and practices the [field] team is modifying, on the assumption that most farmers who use that particular system should be fairly representative of the population. Also, some farmers who are not collaborators in trials and Farmers' Tests are also included in the record keeping project.

"Illiteracy has been somewhat of a problem in selection of collaborators that is worse in some areas than others. However, it has been found that in many families, one of the children has had enough schooling to be able to manage the simple daily work sheet. But it has also been found that even though the farmer is literate, many do not like to fill out the sheet for several reasons. One is that at the end of the day they are tired, and writing represents an especially burdensome task to which they are not accustomed. Others are afraid of getting the form dirty or torn. Some of these make notes on separate paper and when the technician visits, ask him to transfer it to the daily work sheet. The technicians report that in most cases, it is they and not the farmers who ultimately fill out the daily sheet. This requires more frequent visits on the part of the technician than otherwise would be necessary, but also broadens the nature of the sample and minimizes the problem of illiteracy.

"Managing the Data

"In each area office, a folder is made for each crop and each farmer. As the daily sheets are brought in, they are put in the appropriate folder and when the technician from Socioeconomics has the time he transfers the data to the summary sheets.

"For each crop, the technician also fills out an additional sheet which contains such information as planting distances, prices of inputs and product, production, land rent, etc. This sheet becomes a permanent part of the file and part is also transferred to the general summary and the other summary sheets. At the end of the crop year, the technician should have most of the summary sheets filled out and need only add the final data to complete his files.

"Carbon copies of each of the summary forms

Table 5-X-4. General summary (Hildebrand, 1979b).

REGION ______ SUB–REGION __ NAME ______
TOWN ______ COUNTY ______ CROP __ HAS __

ITEM	FARM	HECTARE	CWT
PRODUCTION AND INCOME			
cwt Produced			
Value of Production			
COSTS			
Family labor			
Hired labor			
Animal labor			
Mechanized labor			
Inputs			
TOTAL DIRECT COSTS			
Land rent			
Interest, 5% of direct costs			
Management, 10% of direct costs			
TOTAL COSTS			
PROFIT			
LOSS			

are made so that one copy can be transferred to the [national headquarters] for national analysis purposes while one copy remains in the [regional headquarters] office for use by the local team. The technician also fills out a separate folder on which all the summary forms have been printed and this folder is given to each farmer as a permanent record. By printing the farmers' forms on the inside of the folder, possibility of loss and getting dirty is minimized.

"In order to analyze the data, all the input forms for a particular crop are grouped, all the manual labor forms are grouped, etc. In this manner, the data can be rapidly transferred to the calculators for analysis. The data are also filed permanently in this form for rapid retrieval.

"Analysis of the Data

"One calculator program analyzes all the labor data whether manual, animal or mechanized. The program produces a detailed table of all operations or practices used in the crop, the percent of collaborators who used each practice, the percent of the area on which it was used, the number of man-days (or animal or machine days) used per hectare and the cost per hectare. The modal dates of each operation are determined manually and added to the table, [Table 5-X-5]. This same detailed table also includes

Table 5-X-5. Detail of agronomic practices and inputs used: wheat: Quezaltenango, Region I, 1978 (Hildebrand, 1979b).

No. Collaborators: 53 *Area: 29.23 Has*

	Percent of Collab	*Percent of Area*	*Man-days per ha*	*$/ha*	*Dates*
PRACTICES					
Manual labor					
Incorporating organic matter	2	1	13.06	28.57	1-15/1
•					
•					
Hoeing	43	30	33.57	53.33	1/5-15/6
•					
•					
1st Applic. Herbicide	83	92	4.29	6.79	1-30/7
•					
•					
Animal labor					
Plowing	6	16	3.86	33.54	1-15/6
•					
•					
Mechanized					
Plowing	17	17	1.86	23.87	1-15/3
•					
•					
INPUTS			Quantity		
Seed			*per ha*		*Prices*
Gloria-74	32	40	4.39 cwt	60.63	13.82
Xelajú-66	23	18	4.11 cwt	50.69	12.32
•					
•					
Fertilizers					
1st fertilization 16-20-0	25	17	6.47 cwt	62.11	9.60
•					
•					
2nd fertilization Urea	49	56	2.63 cwt	28.06	10.67
Herbicides					
Hedonal Esther	25	26	2.33 lts	7.00	3.00
•					
•					

appropriate information on the inputs, calculated using another program.

"The labor program also provides information of the proportion of labor that is hired or family, the value of a day's work for both and the average daily wage rate considering all the work done on the crop [Table 5-X-6].

"From the table on the detail of practices and inputs, those most representative of the area are chosen and transferred to a "Typical Technology" table, [Table 5-X-7]. This table demonstrates the technology that is most common or most typical for the zone based first on area, but in cases where one or two non-typical farmers possess a large part of the area, the number of farmers who use the practice or input is also used.

Table 5-X-6. Family and hired labor, per hectare: wheat: Quezaltenango, Region I, 1978 (Hildebrand, 1979b).

No. Collaborators: 53 *Area: 29.23 Has*

Class	*Man-days*		*Total cost*		*Cost/day*
	No.	*%*	*$*	*%*	*$*
		MANUAL LABOR			
Hired	26.87	67	49.79	72	1.85
Family	13.11	33	19.63	28	1.50
Total	39.98	100	69.42	100	1.74
		ANIMAL LABOR			
Hired	0.41	44	2.46	37	6.00
Family	0.52	56	4.19	63	8.06
Total	0.93	100	6.65	100	7.15
		MECHANIZED			
Hired	2.20	99	37.67	100	17.12
Family	0.02	1	0.03	0	1.50
Total	2.22	100	37.70	100	16.98

"Although the typical technology is the most representative of the area, many people prefer to utilize the weighted average cost of production taking the total cost of each practice and dividing it by the total area in the crop. This information is transferred to another table, [Table 5-X-8], in such form that it shows the different major categories of expense. This information is then added to another table that includes yields of the crops, the prices obtained on the average by the farmers using that particular system, and gross income. Net income is calculated deducting both direct and indirect costs, [Table 5-X-9].

"Reports

"To be useful to the [field] teams and the Commodity Programs, the reports of farm records must be made available rapidly. A standard format has been developed to facilitate this task. The report includes a description of the area and of the type of farmer who has been keeping records. This information helps orient readers from outside the area or region. Then, a description of the year as it influenced crop production and prices is included to help judge the results. The information included in this

Table 5-X-7. Typical technology: wheat: Quezaltenango, Region I, 1978 (Hildebrand, 1979b).

No. Collaborators: 53 *Area: 29.23 Has*

ITEM	*Form*	*Percent of*		*Dates*	*$/ha*
		Col	*Area*		
Practices					
Forming beds	Manual	40	34	1-15/6	44.01
Seeding, fert and cover	"	91	84	1-30/6	31.61
Apply herbicide	"	83	92	1-30/7	6.79
2nd fertilization	"	49	55	15-30/7	3.47
Cutting and carrying	"	79	64	15-30/11	27.47
Threshing (hand labor)	"	100	100	1-15/12, 1-15/1	6.80
Threshing (machine)	Mechanized	96	82	same	38.47
Total					158.62
Inputs	*Kind*			*Quantity/ha*	
Seed	Gloria-74	32	40	4.39 cwt	60.63
Fertilizers					
1st Applic.	20-20-0	75	83	6.71 cwt	64.20
2nd Applic.	Urea	49	56	2.63 cwt	28.06
Herbidices					
1st Applic.	2-4-D Esther	25	26	2.33 lts	7.00
					159.89
DIRECT COSTS					318.51
Interest, 5% of Direct Costs					15.93
Management, 10% of Direct Costs					31.85
Land rent					76.77
TOTAL COSTS OF PRODUCTION					443.06

Table 5-X-8. Summary of direct costs for one hectare: wheat: Quezaltenango, Region I, 1978 (Weighted Average) (Hildebrand, 1979b).

No. Collaborators: 53				*Area: 29.23 Has*
ITEM	*Partial*		*Total*	
	$	%	$	%
Manual Labor			99.14	32
Land preparation	34.37	11		
Planting	27.54	9		
Cultural practices	10.59	3		
Harvest	26.64	9		
Animal Power			9.50	3
Land preparation	8.54	3		
Planting	0.86	0		
Cultural practices	0.00	0		
Harvest	0.10	0		
Mechanized			53.86	18
Land preparation	11.20	4		
Planting	1.63	1		
Cultural practices	0.14	0		
Harvest	40.89	13		
Inputs			143.27	47
Seed	52.16	17		
Fertilizer	79.67	26		
Herbicides	11.44	4		
TOTAL	305.77	100	305.77	100

Table 5-X-9. Income and expense, per hectare: wheat: Quezaltenango, Region I, 1978 (Hildebrand, 1979b).

No. Collaborators: 53			*Area: 29.23 Has*
ITEM	*Yield cwt/ha*	*Price $/cwt*	*Value $*
CROP			
Wheat	52.86	11.64	615.29
GROSS INCOME			615.29
COSTS			
Direct costs			305.77
Interest, 5% of direct costs			15.29
Management, 10% of direct costs			30.58
Land rent			76.77
Total costs			428.41
NET INCOME			186.88

description should indicate whether results were better, much better, worse or much worse than average. Usually, the technicians will have this information readily available from their visits with farmers even in the first year of the project in an area.

"Following this general discussion, each crop or crop system is presented utilizing the tables discussed previously. Only short discussions need to be made concerning each table as they are designed to stand on their own.

"From the General Summary sheets, [Table 5-X-4], the frequency distribution of yield, costs and income can be calculated and included in the report in graphic form. The report should also include comparisons with previous years when more than one year's data are available. These comparisons can include yield, prices, income, stability of systems, use of technology, etc. By maintaining a constant record of farms in an area over a period of time, changes in use of technology can be monitored and used for purposes of evaluation even when no benchmark studies of the area were made.

"Finally, it is useful to include an appendix that describes all the practices used on the crops and referred to in the report in their local terminology. This can be written beforehand and does not need to occupy the time of the technician during the critical period following harvest."

PART 2: HATCH'S RECORD KEEPING SYSTEM FOR ILLITERATE FARMERS

The problem of record keeping among illiterate farmers was mentioned earlier. In this part, we present an illustration of some of the work of Hatch (1980) with illiterate farmers in three Bolivian villages. Hatch accomplishes much of what is normally recorded in writing by using gameboards, pictures, and colored chips. Following is a brief illustration of his approach:

> A gameboard for farm enterprise accounting "is the primary instrument of the system. It consists of a piece of thick cardboard—approximately 12 inches square—which is divided into a grid of up to seven columns and up to six rows. Each column corresponds to a single crop enterprise (maximum six crops or crop associations) with the exception of column 7, which may be used to monitor on-farm activities not specifically attributable to a crop, for example fence repairs, collecting firewood, constructing farm structures, weeding and repairing irrigation canals, etc. In contrast, the six rows of the board correspond to the five stages of the crop cycle—land preparation, planting, cultivation tasks, harvest, and marketing—plus one row for recording off-farm employment activities. Since there are seven spaces in this last row, the household could record its off-farm employment by individual days of the week, or it can assign separate spaces to different members of the family. An illustration of the Crop Board is presented in [Fig. 5-X-1].
>
> "For each space of the gameboard in use a nail or hook is inserted. From it may be hung a variety of color-coded counting chips representing different units of production costs, units of product harvested, and units sold. The chips representing production costs are square-shaped; they include

Figure 5-X-1. Game board for farm enterprise accounting (Hatch, 1980).

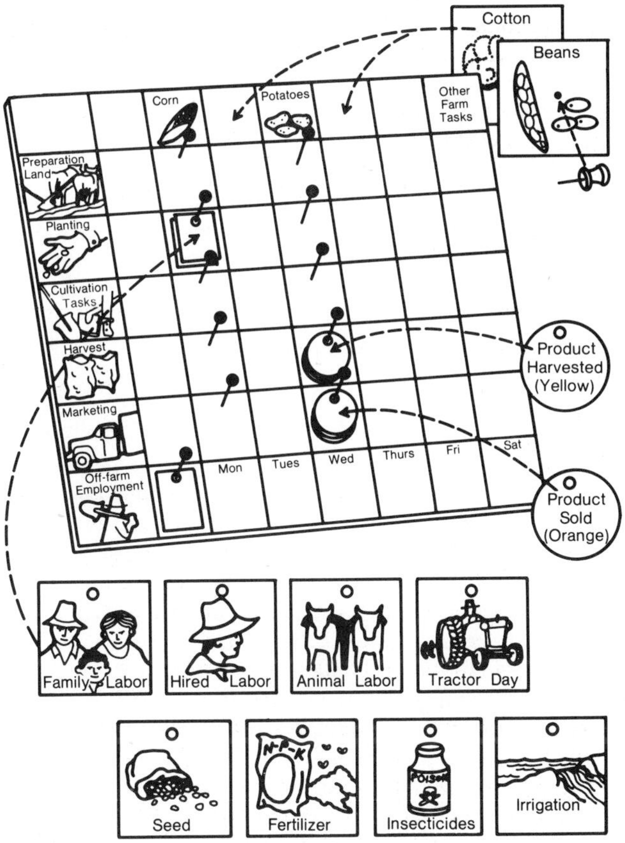

Table 5-X-10. Summary sheet for crop enterprise (Translated from Hatch, 1980).

Name of farmer__________ Cropping year__________

CROP RESULTS

Species and variety__________ Area planted (ha)__________

Activity	Detailed expenditures	Amount	Price	Total value
Land preparation	Family labor			
	Hired labor			
	Machinery			
	Animal labor			
	Other			
	TOTAL			
Planting	Family labor			
	Hired labor			
	Machinery			
	Animal labor			
	Seed			
	Fertilizer			
	Other			
	TOTAL			
Cultivation tasks	Family labor			
	Hired labor			
	Machinery			
	Animal labor			
	Fertilizer			
	Insecticide			
	Other			
	TOTAL			
Harvesting	Family labor			
	Hired labor			
	Machinery			
	Animal labor			
	Other			
	TOTAL			
Other costs	Land rentals			
	Interest payments			
	Other			
	TOTAL			
Total direct costs for the crop				
Crop production				
By-products				
Total value of crop production				
Less direct costs				
Profit (gross margin)				

chips for (1) family labor, (2) hired labor, (3) animal labor, (4) machinery use, (5) seed, (6) fertilizer, (7) insecticides, and (8) irrigation water. The chips for units of harvest are round and yellow in color. So long as consistency is maintained, each chip may represent whatever unit of measurement the participating household may be most familiar with—for example, sack, double-sack load, hundredweight, box, ton, etc. Similar round chips, this time colored orange, are utilized to count units of crop products sold.

"For any given crop the household begins with the land preparation stage and starts to assemble chips in accordance with the day-by-day use of its own and purchased inputs. (EXAMPLE: the family utilized on its corn crop 2 days of oxen for plowing, 4 days of family labor, 2 days of hired labor, and 4 hours of irrigation water during Land Preparation. Thus, a total of 12 chips of four different colors representing different inputs would be hung on the nail corresponding to the column for corn and the row for land preparation.) When the first stage is completed, the collection of chips for the second stage—Planting—begins, and so on. At periodic intervals (every 2-4 weeks) the household is visited by a paratechnician supervisor. He empties the chips from the nail corresponding to any crop stage which has been completed. He sorts and counts the chips. Then he converts the chip-counts into numerical values of quantity, unit price, and total value. These are recorded on a summary sheet—one stage at a time—the format for which is presented in [Table 5-X-10]. At the end of the harvest period the paratechnician summarizes the data from all stages, calculating total income, total costs, and net income."

APPENDIX 5-Y MONITORING AND OBSERVATIONAL ACTIVITIES

In Sec. 5.10.2., Chapter 5, we indicated that ILCA uses monitoring techniques to investigate how livestock production systems change with the introduction of production-oriented projects. In this appendix, we elaborate on ILCA's monitoring of livestock systems and pay particular attention to socioeconomic monitoring.

WHAT IS MONITORING?

ILCA (1980) defined monitoring for livestock systems as a development exercise in detecting and following change in production systems due to development activities and other factors. ILCA's approach is to assign a monitoring team to a project. Then, that team develops the necessary monitoring techniques for the project being undertaken.

For a Kenya livestock project, ILCA (1980) reported on ecological monitoring, herd production monitoring, and socioeconomic monitoring. The ecological monitoring and herd monitoring have similar conceptual foundations as do the techniques discussed in Monitoring, Sec. 5.10.2. For the Kenya livestock program, the socioeconomic monitoring included observation on two levels. On the first level, a skilled observer stays in an area for an extended period making detailed observations of the community and then records those observations. On the second level, the monitoring consists of periodically observing specific attributes of the livestock system and recording those observations (ILCA, 1980). Using these two observational approaches helps identify both the livestock system's characteristics on a wider basis and situations needing further research (ILCA, 1980).

As a result of ILCA's social monitoring of livestock systems in Botswana, ILCA (1980) learned, ". . . the land on which the ranches were sited was not unoccupied, as had been claimed, but in fact had been used by groups of hunters/gatherers who considered that their rights had been usurped." At "full occupancy the ranching block is likely to support up to 1000 people which is considerably more than planned." In addition, these findings dictated further support of such a population would be necessary. Thus, ILCA (1980) needed to consider these and other factors further.

WHY USE OBSERVATION?

FSR&D researchers need to consider observational methodologies for several reasons: (1) with some pastoral societies, researchers may have difficulties conducting traditional experiments; (2) observational methods can be used to check data gathered by other techniques and add another dimension to the data base; (3) observational methodologies, if used properly, can eliminate some of the threats to validity as discussed in Appendix 5-T; and (4) observational methodologies can reveal behavior patterns, organizational forms, and norms not revealed through surveys (ILCA, 1980).

APPENDIX 5-Z DATA MANAGEMENT

We introduced the topic of data management in Sec. 5.12. in Chapter 5 and will now elaborate on additional points that the FSR&D teams may wish to consider. The following sections are on general considerations, tabular sheets, sorting strips, computers, and the use of programmable pocket calculators.

GENERAL CONSIDERATIONS

Before collecting data, the team should decide what types of analyses they will perform on the data. Decisions are needed at this point because analysis procedures have differing data processing requirements and implications. Early knowledge of anticipated analysis techniques reduces the time necessary to process data. For example, if much cross-tabulation is planned, sorting strips might be used. In designing the questionnaire, wherever possible, questions to be cross-tabulated should be placed on the same page. Likewise, pre-numbering and pre-coding of questionnaires may facilitate transferring data to analysis sheets or for preparing data for computer processing.

Several methods are available to tabulate survey information. The methods the teams will use depend on the sample size, the level of knowledge and skill of available assistants, and available time, equipment, and other resources. Questionnaires should be pre-edited for accuracy and completeness, local measurements or categories converted to standard units, and inconsistencies clarified. Questionnaires should be discarded where inconsistencies are too large or non-response is too high.

In Table 5-Z-1, we present Bernsten's (1979) comparison of three common systems for data preparation: namely, tabular sheets, sorting strips, and computers. Within computers, Bernsten also compares the Fortran language with the Statistical Package for the Social Sciences (SPSS).

TABULAR SHEETS

This simple and quick approach requires only a large piece of lined paper with columns and rows. The researcher must first determine how to order respondents—e.g., by village, irrigation type, and farm size. This decision is made based on how the information will be presented in tables or charts. Each question from the questionnaire is listed across the top of the paper—one to each column, leaving the first column for the respondent's name or questionnaire number. The name or number of each respondent is then listed down the far left-hand column. Finally, for each respondent, the value for

Table 5-Z-1. Characteristics of selected data analysis systems (Bernsten, 1979).

System	*Data file preparation*	*Statistical operations*	*Experience required*	*Analytical possibilities*
Tabular sheets	rapid	slow	minimal	minimal
Sorting strips	rapid	slow	minimal	minimal
Computer				
Fortran	moderate	moderate	very specialized	unlimited
SPSS	slow	rapid	specialized	unlimited

each variable is entered in the appropriate row and column. Researchers should verify data transfer, at least by sampling.

As noted in Table 5-Z-1, this simple process quickly provides data for analysis; however, the time it takes to make the statistical calculations can be considerable, if the size of the sample is large. Most researchers consider about 100 respondents the maximum acceptable sample size for this method; also, the analytical possibilities are limited to simple statistics. Analytical possibilities include descriptive frequencies, measures of concentration of the data such as the mean, the mode, or the median; measures of the variability of data such as the range, the mean deviation, or the standard deviation; and measures of the relationship between two or more series of data such as correlation coefficients.

SORTING STRIPS

Sorting Strips add an element of flexibility to tabular sheets. Heavy paper or light cardboard is cut into long and narrow strips and data from one farmer are written across the top of each strip. With data on strips, researchers can easily reorganize farmers according to characteristics of different subgroups. For example, yield per hectare for a certain crop can easily be compared for full-time and part-time farmers by dividing the strips into two piles, one for full-time farmers and one for part-time farmers. Then the average yield for each pile can quickly be calculated. Although this system is also easy to learn and use, it takes longer to prepare the data for sorting strips than when using tabular sheets. The time and other limitations of analytical possibilities are the same for sorting strips as for tabulating sheets. A sample size of 100 respondents or a questionnaire with 140 questions are the practical limits for sorting strips. For larger samples or questionnaires, researchers should consider using a computer.

COMPUTERS

Computers require more time to prepare data for analysis because they must be coded and keypunched onto cards in a predetermined order. Usually, numeric codes must be developed for each question such as 1 = yes, 2 = no; or 1 = corn, 2 = wheat, and 3 = corn and wheat. These codes must be inclusive, yet mutually exclusive.

Coding may be of either two types, pre-coding or post-coding. If computer analysis of the data is decided during planning, an effort should be made to pre-code the answers. Furthermore, the researchers should prepare a code book for their use and the interviewer's use. This book explains the various codes and provides the rationale for each question. Pre-coding allows interviewers to submit coded responses directly, and the questionnaire, once checked, can be key-punched directly. If questionnaires are not pre-coded but post-coded, someone must code the questionnaires separately.

Handling data by computer has great advantages, even though data preparation may take a long time and operating a computer program requires fairly high skills. If the staff members understand computer programs, they can benefit from their rapid statistical operations. Computers reduce the probability of computing errors, as long as data and programs are correct, and they greatly facilitate the testing of sophisticated hypotheses. The computer is able to generate both descriptive and inferential statistics and can provide the researcher with results in a variety of formats.

In spite of computer analysis advantages, researchers should be aware of the tendency to overlook errors or relationships in the data that would be revealed in manual analysis. If the team manually tabulates part of the data before computer analysis, it (1) will have a quick summary of key information, (2) should develop a "feel" for the data, and (3) may note new relationships. Doing such tabulations manually often produces results faster than waiting for computer printouts. Later, the computer can provide the team with (1) checks on its preliminary analyses, (2) more detailed and complex analyses, and (3) predictions and statistical relationships.

USE OF PROGRAMMABLE POCKET CALCULATORS

Since the mid 1970s, programmable pocket calculators have become widely used by researchers in both the field and office. Some of the more popular models are Texas Instruments 59 and Hewlett Packard 65, 67, and 97.

The advantages of using these calculators are

- complex analyses can be completed rapidly
- the unit can be operated without the help of others
- analysis can be carried out without delay wherever the user may be
- a variety of packaged programs are available from the

manufacturers; or the owners can write the programs.

At the present time, a large selection of programs are available from the manufacturers and government and private sources. Programs from Hewlett Packard and Texas Instruments that make the following analyses should interest most FSR&D teams:

- general statistics—factorial, permutation, combination, moments, and skewness
- analysis of variance—one-way and two-way analysis of variance
- distribution function—normal, inverse normal, chi-square, and t and f distributions
- curve fitting—multiple linear regression and polynomial approximation
- test statistics—t statistics, chi-square evaluation, and contingency table
- queuing theory—single and multi-server queues.

Also, organizations can develop programs for their specific needs. For example, ICTA has written the following statistical programs for its use (Hildebrand, 1979c):

"1. Analysis of variance (Anova), split plots, without limits
2. Anova, randomized block, up to 6 replications, no limit on treatments
3. Missing plots, randomized blocks, up to 6 missing plots
4. Anova, without limits
5. Multiple regression, 3 independent variables
6. Multiple regression, 5 independent variables,
7. Quadratic regression, 2 independent variables, with interaction
8. Linear, exponential and quadratic regression for 1 independent variable
9. Duncan's analysis [multiple-range test or mean separation test and experiments with unrelated treatments]
10. Tukey's analysis [randomized blocks, two way classification, and additivity test]
11. Yates method [correction for continuity, adjusted chi-square, and analysis of counts]
12. Farm record analysis for labor
13. Farm record analysis for inputs
14. Several different programs for converting plot data to kg/ha."

APPENDIXES TO CHAPTER 6

APPENDIX 6-A
SELECTION OF VARIABLES FOR CLIMATIC MONITORING OF A RESEARCH AREA

The need for climatic monitoring was given in Sec. 6.3.1., Chapter 6. The FSR&D team should select the climatic variables to be examined and measured carefully. Since climatic variables require frequent tabulating of observations or reading of strip charts, they may soon become so voluminous that their analysis can be unwieldy and interfere with other program operations. For this reason, the data requirements should be reviewed well in advance of the season's activities, and variables discarded or the frequency of observations reduced if they are not clearly needed.

The climatic variables selected for measurement depend upon the general climatic conditions in the research area, how they vary from season to season, and the need for analyzing the results of biological experiments. The physical aspects of the research area are also a consideration. For example, in rain-fed, humid, tropical areas, the team could emphasize the collection of data on the distribution of rainfall, and omit the collection of temperature data; whereas in an arid irrigated area, the team could measure pan evaporation, temperature, relative humidity and wind, and omit collecting rainfall data. Another consideration in selecting climatic variables for monitoring is the importance of the data in explaining how farmers respond to infrequent, but severe, climatic events such as flooding, hurricanes, and droughts.

The ease with which the data can be collected for the needed accuracy must also be considered. For example, the team can easily monitor rainfall, but the intensity is difficult to measure and might be omitted unless it is needed for the analysis of soil erosion or infiltration. Similarly, the monitoring of wind speed usually is omitted except for research areas where it may be needed to calculate evapotranspiration or in areas where strong winds influence crop production or cause other problems on the farm.

Finally, the team should consider the variability of the measured variable over the geographic area and among years. Long-term weather data available from a standard meteorological station *near* the research area are often more reliable than that which could be obtained from a few years' measurement *in* the area.

Following are details on rainfall, irrigation, temperature, wind, evapotranspiration, relative humidity, and solar radiation.

RAINFALL

Rainfall is perhaps the most common and usually most useful climatic variable to monitor. It will often be the only variable measured within a research area. For rain-fed systems, rainfall is a prime determinant of the system, and either an excess or a deficiency can have dramatic effects on both crop and animal production. In irrigated systems rainfall can also be a serious hindrance or a marked benefit. In either case, of the various climatic variables, rainfall generally has the greatest potential for explaining the difference in biological response between years and areas.

Rainfall may, however, be quite variable within a research area thus requiring the placement of gauges near the research plots. This could be especially true if the research area extends inland from a coastline to a mountain range. Also, a substantial variation in rainfall from year to year may cause a change in the timing of farm activities. The farmers' responses to year-to-year variations in climate are important in the analysis of climatic effects on the farming system. This year-to-year variation explains why long-term records are best used for general design and planning and are not necessarily suited for area specific activities.

Measuring rainfall usually is simple and inexpensive. For most projects, simple plastic gauges are adequate. These can be placed throughout the area to monitor anticipated gradations of rainfall, for example, at intervals of 2 to 5 kilometers, or alternatively, near experimental fields. The gauges must be placed in an open area free of interference from buildings or overhanging vegetation and protected against disturbances by people or animals. Readings should be made daily during periods of frequent rains and after each rain during the dry season. If a major storm occurs, the individuals collecting data may need to read the gauges more often to make sure they do not overflow. Data usually are recorded daily and summarized weekly and monthly as needed for the analysis of the experimental results.

IRRIGATION

While irrigation is not a climatic variable, it modifies rainfall and prolongs the "rainy season" or supplies additional water during the growing season. Except for sprinkler systems, which are rare for most small farmers, the amount of irrigation water applied may be difficult and time consuming to measure. Also, fields that can be separated for measurement of water are an exception in the small communities typical of developing countries.

In most small farm irrigation schemes, the system is operated by the government. Usually, the government provides the water free or collects a small charge based on irrigated areas or crop because the cost of charging by volume through metering water to individual small farms is prohibitive. Thus, farmers along the watercourse near the source of supply generally do not, voluntarily, conserve water for others farther down the watercourse.

For FSR&D, small farm irrigation systems should be analyzed for their potential to provide water for crops. They should also be evaluated in terms of time and duration of flow, rotation period, and water quality. More detailed studies may involve the equity of distribution and methods of control and management. Such studies may require extensive effort and require irrigation specialists.

TEMPERATURE

After water, air temperature probably is the most useful climatic factor in the analysis of farming systems. Most temperature monitoring concentrates on understanding the occurrences of critical temperatures that are either too warm or too cool for specific crops. Temperature data can also help explain yearly differences in the maturity of crops in temperate regions when early warm weather hastens maturity, or when prolonged cool weather delays maturity. Critical, as well as optimum, temperatures for plant growth during the season tend to be reasonably similar from year to year. Local temperatures usually can be assessed fairly well from long-term records and can be considered in design. Very important temperatures in crop production are the extremes. The early-season or late-season temperature extremes may need careful evaluation in temperate regions having considerable topographic complexity. In other areas such as the low elevation tropics where air temperatures are relatively uniform, adverse temperatures do not normally occur and weekly means may suffice.

Unless large differences in elevation are encountered, temperatures in a research area tend to be reasonably stable. Thus, where topographical features are relatively uniform, useful data can usually be obtained from a meteorological station located in or near the research area. Maximum and minimum temperatures are usually recorded daily. From these data, the average, high, and low temperatures can be calculated for weekly and monthly intervals.

Soil temperature measurements generally are used less than air temperatures. Soil temperature may be important in temperate regions where germination of seed from some species is greatly retarded in the spring because of low soil temperatures. In the fall, for some crops—e.g., winter wheat—planting after soil temperature falls below a certain temperature reduces root infections. Where soil temperatures are needed for analyzing plant growth effects, reliable soil maximum-minimum thermometers should be used.

WIND

One important consideration of winds in agricultural research is the effect of excessive speeds on crop growth during critical physiological stages. Problems associated with winds include increased sterility at flowering, lodging of grain, and excessive transpiration from strong, hot winds.

Winds that cause production problems usually result from either large infrequent depressional storms such as typhoons or seasonally incessant winds. Depressional storms are erratic. Although, on a global base, the seasons for depressions are well-defined and information on peak periods for specific areas is probably available, the occurrence in a specific location during a critical time for crops may not be frequent enough for farmers' planning. In this case, the team may treat depressional storms as a risk of farming in the area.

Incessant winds, caused by major air mass movements, occur over a more prolonged period and occur seasonally. These winds can be anticipated and included in the design of management practices. These winds are usually stable over a large area, but local intensity may be altered markedly by geologic features such as mountain valleys that may reduce or increase the wind velocities. In the study of incessant winds, the team should identify how the onset and decline of the winds vary from month to month. Usually, such information is not available from long-term records since most stations measure only the general direction of the wind. Therefore, where the team anticipates that winds will have a significant impact on crop or livestock production, the team should monitor the winds. Since wind speed is more difficult and more expensive to measure, simply averaging wind movement for some time period is the more frequent measurement. Although the team can use such data to calculate average wind speed during the recorded period, the peak intensities are the values generally needed in designing new technology.

EVAPOTRANSPIRATION

The evapotranspiration rate may not be an important determinant in some FSR&D activities, but it is a major consideration in scheduling irrigation deliveries and in making water balance computations. Evapotranspiration does, therefore, become an important variable in research programs for irrigated areas in which improving the effectiveness of the irrigation system is one of the prime objectives.

The climatological approach to studying water

balance requires (1) knowledge of the maximum evapotranspiration of a particular crop or potential from a reference crop, (2) rainfall data, and (3) the soil's moisture content. From these data, a researcher can compute the actual evapotranspiration rate, soil moisture deficit, and surplus water by summing inflows and outflows to the system. Researchers customarily establish a reference for evapotranspiration by measuring the rate for an extended surface of a short, actively growing green grass. The grass should be of uniform height, completely shade the ground, and be well supplied with water. Researchers use various methods involving climatic data or other physical measurements to calculate the reference potential evapotranspiration. Alternative approaches include the Thornthwaite, Penman, and Jensen-Haise methods, pan evaporation, aerodynamic methods, energy budget approaches, and direct measurement with a lysimeter. The various methods are described in textbooks, such as Chang (1968) and Rosenberg (1974). Different methods may be used in different situations depending in part upon available climatic data and situations.

The potential evapotranspiration of a particular crop at any stage of development may be calculated from the reference potential evapotranspiration using the crop coefficient for the given crop. Researchers use the crop coefficient to convert the potential evapotranspiration value determined for short grass cover to that for the specific plant under study. The best crop coefficients are obtained experimentally in the research area. Since these data may not be available in an FSR&D research area, crop coefficients can be obtained from published research; for example Doorenbos and Pruitt (1977) have determined crop coefficients for 53 crops under different humidity conditions.

If a team needs water budget analyses for a project and long-range climatic data are available at a nearby meteorological station, the team's researchers can calculate the potential evapotranspiration for a crop from those data if they know the crop coefficient. If local data are not available, pan evaporation can be monitored relatively easily. The approach is to set up a Class "A" evaporation pan according to its specifications and measure the daily evaporation of water with a calibrated gauge.

RELATIVE HUMIDITY

Relative humidity is the ratio of the actual to saturation vapor pressure of water in the air, expressed as a percentage—both at the same temperature. The saturation vapor is the maximum water air can hold in vapor form at any temperature. The saturation deficit is the difference between the actual vapor pressure and the vapor pressure of saturated air at the same temperature. Since one of the factors influencing the rate of transpiration of plants is the vapor pressure deficit of the air, relative humidity is one of several environmental measurements that can be used to estimate evapotranspiration and thereby irrigation water requirements. Blaney-Criddle (Rosenberg, 1974) is one of the more common methods.

Relative humidity is often measured with wet-bulb and dry-bulb thermometers (a psychrometer). The wet-bulb thermometer is kept moist with a wick dipping into distilled water. The actual vapor pressure of water in the air is calculated from the equation:

$$e = e_s - AP(T-Tw)$$

where T and Tw are the respective dry-bulb and wet-bulb temperatures in °C, P is the air pressure, A is a proportionality constant, and e_s is the saturation vapor pressure of Tw (values are obtained from tables). With the value of "e", relative humidity can then be calculated.

SOLAR RADIATION

Although incident solar radiation is critical for photosynthesis and crop development, knowledge about it is not particularly useful for most farming systems programs except as it helps to explain temperature and evaporation. Radiation data can contribute to the more sophisticated analyses of yields, which are used for selection of new crops and modeling of results for transfer to other areas. This is usually beyond the scope of most national farming systems programs, unless the government is part of an international network. Since radiation values tend to be relatively uniform over wide areas, data from the nearest meteorological station are generally satisfactory for farming systems research.

Until recently, accurate radiation measurement required an expensive Eppley-type pyrheliometer, coupled to integrators. Now, some closed-system alcohol distillation, recondensation instruments such as the inexpensive Bellani pyrheliometer are sufficiently accurate for FSR&D.

A simple alternative to the pyrheliometer is an instrument that measures sunshine duration or cloudiness. A nonelectrical sunshine duration meter is available that will record the duration and intensity of sunlight. With sunlight duration the total solar radiation can be estimated. These instruments may be the most practical for FSR&D projects needing measurements of solar radiation.

APPENDIX 6-B
THE LAND EQUIVALENT RATIO (LER)

In our discussion of productivity criteria in Sec. 6.4.2. in Chapter 6, we considered several efficiency criteria. When comparing the efficiencies of intercropping with those of sole cropping, researchers often use the land equivalent ratio (LER) as a measure of biological performance. Intercropping involves growing two or more crops at the same time on the same plot in close association with each other, whereas sole cropping involves growing a single crop by itself on a plot.

In this appendix, we define the ratio, illustrate how to calculate it and comment on its application.

DEFINITION OF LER

Harwood (1979) defined LER as the "area needed under sole cropping to give as much produce as 1 hectare of intercropping or mixed cropping at the same management level, expressed as a ratio. LER is the sum of the ratios or fractions of the yield of the intercrops relative to their sole-crop yields."

EXAMPLE

Hiebsch (1978) provides a simple two-crop example that illustrates the concepts. Let

Y^{I}_{C1} = yield of intercrop C1

Y^{I}_{C2} = yield of intercrop C2

Y^{S}_{C1} = yield of C1 when planted as a sole crop

Y^{S}_{C2} = yield of C2 when planted as a sole crop.

Then, the

$$\text{LER} = \frac{Y^{I}_{C1}}{Y^{S}_{C1}} + \frac{Y^{I}_{C2}}{Y^{S}_{C2}}$$

If the yield of C1 is 2.0 tons/ha when intercropped and 3.0 tons/ha when planted as a sole crop and C2 is 1.0 tons/ha when intercropped and 2.0 tons/ha when planted as a sole crop, then the LER is 1.17. That is,

$$\text{LER} = \frac{2.0}{3.0} + \frac{1.0}{2.0} = 0.67 + 0.50 = 1.17$$

This means that the farmer must use 17 percent more land to obtain the same output when sole cropping as when intercropping. Stated differently, by intercropping the farmer can obtain the same output as for sole cropping and have 15 percent more land available for other use—i.e., 1 − (1 ÷ 1.17) = 0.15.

GENERALIZED FORMULA

The foregoing example can be generalized to handle more than two crops. For this, we let

Y^{I}_{Ci} = yield of Ci when intercropped

Y^{S}_{Ci} = yield of Ci as a sole crop

Ci = Crop i

Then, the

$$\text{LER} = \sum_{i=1}^{n} \frac{Y^{I}_{Ci}}{Y^{S}_{Ci}}$$

where n = number of intercrops.
Since we can write yield as quantity (Q) divided by area (A), we can rewrite this equation as

$$\text{LER} = \sum_{i=1}^{n} \frac{\dfrac{Q^{I}_{Ci}}{A^{I}_{Ci}}}{\dfrac{Q^{S}_{Ci}}{A^{S}_{Ci}}} = \frac{A^{S}}{A^{I}}$$

in which A^S is the area in sole cropping that produces the same output as from the area, A^I, planted as an intercrop, i.e.,

$$\sum_{i=1}^{n} Q^{I}_{Ci} = \sum_{i=1}^{n} Q^{S}_{Ci}$$

When the LER is greater than 1.0, the farmer gains from intercropping and when the LER is less than 1.0 the farmer is better off sole cropping—all else being equal.

APPLICATION OF THE LER

The LER gives satisfactory results when (1) the management levels for intercropping and sole cropping are the same, and (2) one or more of the crops making up the intercropping combination could not have been planted sequentially as a sole crop during the cropping season.

For example, if one of the intercrops could be grown as a sole crop more than once during the season, or could be followed by another crop during the season, then the output from sole cropping would increase and the LER for the intercropping would decrease.

For additional information on this topic and a means for measuring cropping efficiencies when sequential planting is possible, see Hiebsch's (1978) discussion of the area-time equivalency ratio (ATER).

APPENDIX 6-C
A GUIDE FOR LOCATING ON-FARM EXPERIMENTS

As we discussed under Locations of Experiments in Sec. 6.4.3., Chapter 6, one of the FSR&D team's initial tasks in planning on-farm experiments is to establish the basis for selecting experimental locations. In this appendix, we provide a brief guide to some of the relevant factors—the physical, socioeconomic, and cultural conditions—that researchers should consider.

PHYSICAL FACTORS

The team should consider factors such as the type of area—e.g., irrigated or rain-fed, the soil's physical and chemical properties, topography, and microclimate. The emphasis the team places on these and other factors, such as pests, depends on individual situations.

IRRIGATED AND RAIN-FED AREAS

If both irrigation and rain-fed agriculture are practiced within the research area, the team should delineate them. Rain-fed land should be categorized on the basis of monthly and annual rainfall and, if possible, the team should estimate the monthly variability of rainfall. In irrigated areas, the source and quality of water, seasonal availability, and methods of irrigation will determine the types of crops that can be grown. Also, the locations of farms within an irrigated area may depend on whether the irrigation water is perennial or non-perennial and whether the water comes from a reservoir, a river diversion, or the ground.

SOIL'S PHYSICAL PROPERTIES

Farms should be selected to represent major differences in the soil's physical properties. Bases for differentiation include texture, structure, position of the water table, discontinuities in the soil profile, and depth. These properties control the soil's moisture storage capacity, water movement characteristics, rooting depth, and aeration. All these factors are important in determining the growth of crops on both irrigated and rain-fed land. With the aid of a soil sampling auger, a soil surveyor can easily determine most of the soil's physical properties that need to be identified for mapping variations and locating experiments within the research area.

SOIL'S CHEMICAL PROPERTIES

The most significant soil chemical factors for selecting farms for experiments are soil fertility, salt, cation exchange capacity, and pH. Some topographical features that may suggest differences in soil chemical properties are slope, surface drainage characteristics, and water seep spots. Heavily eroded slopes often indicate low soil fertility. Indicator plants such as saltgrass *(Distichlis stricta)* may also assist in identifying certain chemical properties of the soil.

TOPOGRAPHY

Besides using topography as an indicator of soil chemical properties, the FSR&D team can also use topography for contrasting areas according to such factors as flooding and erosion. By studying topographical features, the team can assess the need for flood control measures such as retention ponds and levees, and erosion control measures such as interceptor ditches and terracing.

MICROCLIMATE

The microclimate may be important in research areas with a rolling topography or with poor drainage of the surface air. In climates where temperature sensitive crops are planted as soon as the danger of frost has passed or where early frost may cause damage in the fall, the air movement may indicate where certain crops should be planted. The direction of the slope is another factor in temperate zones. Thus, in northern latitudes, north slopes are more moist than south slopes; and in rain-fed areas in these latitudes, the direction of slope may be a basis for selecting contrasting experimental fields. Or, the direction of the prevailing winds may also cause variations in microclimate that will influence crop growth.

After considering the physical attributes of the research area, the researchers may have enough data to map the principal physical features that influence farmers' management practices. For example, the FSR&D team might map land types that represent the production potential for different types of livestock and crops. Near Cairo, Egypt, major differences in soil texture are found in short distances along the Mansouria Canal. Farmers react to this situation by planting different crops and cropping patterns and by following different management practices. Consequently, on-farm research there is adapted to each situation.

SOCIOECONOMIC AND CULTURAL ENVIRONMENT

With information from the reconnaissance or other problem identification surveys, the team can estimate how receptive the farmers might be to change, and the extent to which local leadership would support project efforts. Where the social behavior varies greatly within a research area, such information may be used to stratify the area into social patterns for selection of individual farms.

Meaningful differences in the economic environment within the research area may be a basis for locating farms. For example, the economic environment influences profit margins and risks of growing cash crops, which, in turn, are influenced by the size and type of the farmers' holdings. The team should consider the cash requirements and risk of alternative production techniques and possible adverse effects if a new technology fails. When evaluating risk, the team should be aware that farmers customarily perceive the risk of a new technology as being greater than it really is. Thus, the FSR&D team and extension need to help farmers obtain a clearer understanding about the technology's variability (Wharton, 1968).

Finally, the cultural environment includes factors such as family customs, the descent system, patron-client relationships, and the extended family. Because these relationships may not be subject to short-term change, the team should be aware of them when locating experiments.

APPENDIX 6-D
FIELD DESIGNS AND STATISTICAL PROCEDURES FOR ON-FARM EXPERIMENTS

In Chapter 6, Sec. 6.4.3. on Experimental Design Characteristics, we introduced the subject of alternative field designs and related topics. This appendix provides considerably more detail on four field design possibilities that are frequently used in FSR&D. We also outline the statistical methods commonly employed in analyzing the results from these experimental designs.

We do not explain these statistical procedures in detail, since adequate descriptions are widely available on experimental designs. Two such textbooks are *Agri-*

cultural Experimentation: Design and Analysis by Little and Hills (1978) and *Experimental Designs* by Cochran and Cox (1957). We include this overview to illustrate design and statistical procedures. If the FSR&D team needs assistance in designing experiments and statistically analyzing the data, the team should seek the help of an agronomist or statistician experienced in these procedures.

Following are sections on (1) problems with on-farm experimentation; (2) four experimental approaches: completely randomized design, randomized complete block design, paired treatments design, and randomized incompleted block design; and (3) data analysis for each of these methods. Included in these sections are discussions of the mean, the standard deviation, the coefficient of variation, analysis of variance, and mean separation.

PROBLEMS WITH ON-FARM EXPERIMENTATION

On-farm experimentation presents several problems for researchers in the design of field trials and tests and the statistical analysis of the results. These problems include

- heterogeneity of many field properties such as drainage, fertility, slope, and previous land use
- the small size of many farms and the need for large plots for farmer-managed tests
- the limited number of plots on a single farm
- the limited capability for randomizing treatments.

On-farm research is frequently characterized by response data having a high degree of within-treatment variability. By within-treatment variability, we mean the variability between the way a dependent variable—e.g., yield or income—responds to a treatment—e.g., cropping pattern—in one situation versus the way it responds to the same treatment in another situation. For example, a treatment that increases the maize yield by 800 kg/ha (from 2,200 kg/ha to 3,000 kg/ha) in one trial might increase the maize yield by only 400 kg/ha (from 1,500 kg/ha to 1,900 kg/ha) in another trial. While the physical quantities are considerably different, the percentage increases are more closely equivalent—i.e., 36 percent and 27 percent, respectively. Because farmers' criteria for acceptability of a new technology are more closely related to percentage increases in yields and income, rather than absolute increases, percentage increase may be the best way to express results.

FIELD DESIGNS

Four basic experimental designs are particularly applicable to on-farm research:

- completely randomized design
- randomized complete block design
- paired treatments design
- randomized incomplete block design.

Table 6-D-1. Summary of field designs used for on-farm research.

Design	*No. of Treatments That Can Be Evaluated*	*No. of Treatments Per Farm*	*Sensitivity to Detecting Treatment Differences*
Completely Randomized	Any number	One	Low
Randomized Complete Block	Any number but usually less than four	Must be the same as the total number being evaluated*	High
Paired Treatments	Two	Two	High
Randomized Incomplete Block	Any number	Fewer than the total number being tested	Medium to high

*We assume each farm is considered as a block.

Table 6-D-1 summarizes these designs and their appropriate uses. The easiest to plan and analyze is the completely randomized design, which we recommend when only one treatment can be tested on each farm. This situation is encountered when farms are too small to accommodate both the farmer's traditional practice and a new cropping practice in the same field. The design may also be applicable when animals are to be introduced as a modification of the existing farming system.

COMPLETELY RANDOMIZED DESIGN

In implementing the completely randomized design, all the treatments to be tested—including the farmers' traditional practice—are randomly assigned to the cooperating farms. Each farm receives one treatment for testing. For example, assume 60 cooperating farmers are in an area and two alternative cropping patterns are to be tested and compared with the farmers' traditional pattern—assumed to be the same on the 60 farms. Each farmer would randomly be assigned one of the three cropping patterns being evaluated—either the traditional pattern or one of the two alternative ones.

The primary disadvantage of using the completely randomized design for on-farm research is the difficulty in detecting differences between the treatments. This is because the variability among farms is often great and this variability may obscure differences among treatments. The more homogeneous the farms in a study area and the greater the response from the alternative treatments, the greater are the chances for identifying treatment differences.

RANDOMIZED COMPLETE BLOCK DESIGN

The problem of detecting treatment differences can be great when considerable variability exists among farms in a region. A good portion of this variability can be identified and removed during data analysis if the experimental units are grouped into blocks. In on-farm research, this can be done in two ways.

The first is useful when only one treatment is being tested on a farm. If the experimental units—e.g., farms—can be broken into subgroups on the basis of some factor that will affect the response variable, more sensitivity in detecting treatment differences will result. In this case, the subgroups are called blocks. A block should consist of farms that are as uniform as possible so that variability between farms within a block is less than variability between farms in different blocks.

As an example, consider the case discussed previously in which three treatments are to be tested on 60 farms. If the 60 farms can be broken into two blocks on the basis of some factor expected to change the results (say, soil texture), this should be done. Farms on predominantly sandy, well-drained soils might constitute one block while those with a heavier soil texture would constitute the second block. Treatments are randomly assigned to the farms in each block. The blocks must contain equal numbers of farms and treatment designations. If not, the randomized complete block design should not be used. Thus, if the 60 farms are separated into two blocks, each block should contain 30 farms with 10 farms randomly assigned to each of the three treatments.

A second, more effective type of blocking can be employed if more than one treatment is being tested on each farm. The farms can be considered as blocks in this case. Because variability within farms is generally much less than variability between farms, this design eliminates considerable variability and makes detecting differences between treatments easier.

Assume two new cropping patterns are to be tested in a region and compared with the traditional pattern. Each of the three patterns is tested on each farm. In this case, each farm is a block. This design's advantage is a more accurate assessment of the treatment differences than any of the other designs previously discussed. However, its disadvantage is that each treatment must be tested on every farm. This may limit the number of treatments that can be tested and compared.

PAIRED TREATMENTS DESIGN

Where only two treatments are under study—for example, the traditional and an alternative practice—a paired treatment design is a useful and easy test. The two practices are placed side-by-side in the same field and the treatments are replicated—i.e., repeated on several farms. Each farm has the same pair of treatments. Pairing the treatments increases the sensitivity of detecting treatment differences. This design is limited, however, to two treatments.

RANDOMIZED INCOMPLETE BLOCK DESIGN

If more than one treatment can be tested on each farm, but the total number of treatments to be tested exceeds the number that can be tested on one farm, the randomized incomplete block design may be useful. Such a design allows some of the variability due to differences between blocks—i.e., farms—to be identified during data analysis. Thus, this is more sensitive in detecting differences between treatments than the completely randomized design.

A potential use of the incomplete block design is illustrated in the following example. Five alternative cropping patterns are to be tested in a research area and compared with the traditional cropping pattern, for a total of six cropping patterns. Each farm will be assigned only two of the six cropping patterns or treatments. Because the traditional pattern is considered as just another treatment, it will be tested on some, but not all, of the farms. This is a potentially complicating point worth emphasizing. Some farmers may be selected to use the traditional pattern and only one alternative for testing—the traditional pattern plus one alternative—while others may be selected to test two alternative patterns. The appropriate experimental design for six treatments with two treatments per block and five replications is given in Table 6-D-2. If the researcher desires to include more farms for greater accuracy, the plan can be repeated any number of times. For 60 farms, the pattern would be repeated four times.

Table 6-D-2. Plan for a randomized incomplete block design.

Replication I		*Replication II*		*Replication III*	
Block	Treatments	Block	Treatments	Block	Treatments
(1)	1,2	(4)	1,3	(7)	1,4
(2)	3,4	(5)	2,5	(8)	2,6
(3)	5,6	(6)	4,6	(9)	3,5

Replication IV		*Replication V*	
Block	Treatments	Block	Treatments
(10)	1,5	(13)	1,6
(11)	2,4	(14)	2,3
(12)	3,6	(15)	4,5

t = 6 k = 2 r = 5 b = 15 $\lambda = 1$ E = 0.6

Key: t = number of treatments
k = number of treatments that occur on each block (farm)
r = number of replications (number of times each treatment occurs in the plan)
b = number of blocks (farms) in the plan
λ = number of blocks in which a given treatment pair (e.g., 1,2) appears

$$\lambda = \frac{r(k-1)}{t-1} = \frac{5(2-1)}{6-1} = 1$$

E = a constant used in the analysis

$$E = t\lambda/rk = (6 \times 1) \div (5 \times 2) = 0.6$$

Whenever possible, minimizing the variability within replications is preferable. Treatments are numbered and the blocks are randomly assigned to the treatment pairs.

Thus, this plan allows a comparison between all the treatments being tested. The randomized incomplete block design is more precise than the completely randomized design because some of the between-farm variability is removed. The design's disadvantages are increased difficulties in (1) setting out the experiments, (2) analysis, and (3) compensating for missing data.

DATA ANALYSIS

The purpose of the data analysis, regardless of which experimental design is used, is to find out whether the new treatments are really better than the farmers' traditional methods. In data analysis, researchers want to find out whether the results of various treatments could have occurred as a result of chance alone, or whether the treatments really do improve production. In the remainder of this appendix, we will illustrate some of the more common statistical procedures for doing this.

COMPLETELY RANDOMIZED DESIGN

In this section, we will show how researchers can conclude about the statistical significance of the results from a completely randomized design. In doing this, we will calculate the mean, the standard deviation, and the coefficient of variation and then discuss sources of variation and methods for testing differences between the means.

For illustrative purposes, let us assume that two new cropping patterns are to be compared with the traditional cropping pattern. Together, they give us three treatments to be evaluated. Since each treatment is to be replicated 10 times and only one treatment is assigned to a farm, 30 farms must be selected for conducting the tests. Results might be summarized as in Table 6-D-3.

Table 6-D-3. Results for a cropping pattern test using a completely randomized design.

	Cropping Pattern *Traditional*	*Alternative 1* Yield (kg/ha)	*Alternative 2*
	1,800	2,020	2,830
	1,760	1,030	3,220
	1,510	1,600	3,670
	890	1,410	2,940
	1,410	2,730	1,980
	1,760	1,860	3,720
	1,300	1,540	3,210
	1,650	1,000	2,140
	900	3,100	1,390
	1,310	940	2,700
Y_i (sum of treatment yields)	14,290	17,230	27,800
$\bar{Y}_i$ (mean of treatment yields)	1,429	1,723	2,780
S_i (standard deviation)	336	729	751
CV_i(coefficient of variation)	24%(CV_1)	42%(CV_2)	27%(CV_3)

Y (sum of all treatment yields) = 59,320
$\bar{Y}$ (mean of all treatment yields) = 1,977

Mean

The mean is the arithmetic average of a group of values and is usually considered the most representative single value for the whole group. $\bar{Y}_i$ is the mean of treatment i and is calculated as follows:

$$\bar{Y}_i = \frac{\sum_{j=1}^{r} Y_{ij}}{r} = \frac{Y_i}{r}$$

Where: $\sum$ = summation symbol

$i = 1, 2, ..., t$

t = number of treatments in the experiment ($t = 3$ in this experiment and we have 3 treatment means: $\bar{Y}_1$, $\bar{Y}_2$, and $\bar{Y}_3$)

$j = 1, 2, ..., r$

r = number of replications in each treatment ($r = 10$ in this experiment)

Y_{ij} = individual yield values

$\sum_{j=1}^{r} Y_{ij} = Y_i$ = the sum of individual yield values from $j = 1$ to $j = r$ in treatment i

As an example, we calculate the mean for the traditional cropping pattern ($i = 1$):

$$\bar{Y}_1 = \frac{\sum_{j=1}^{r} Y_{1j}}{r} = \frac{Y_1}{r}$$

$$= \frac{\sum_{j=1}^{10} Y_{1j}}{10} = \frac{Y_1}{10}$$

$$= \frac{1{,}800 + 1{,}760 + 1{,}510 + 890 + 1{,}410 + 1{,}760 +}{10}$$

$$\frac{1{,}300 + 1{,}650 + 900 + 1{,}310}{10}$$

$$= \frac{14{,}290}{10}$$

$$= 1{,}429 \text{ (kg/ha)}$$

$\bar{Y}$ is the mean of all the yield values of the whole experiment. The following formula is used for calculating $\bar{Y}$:

$$\bar{Y} = \frac{\sum_{i=1}^{t} \bar{Y}_i}{t} = \frac{\sum_{i=1}^{t} Y_i}{rt} = \frac{Y}{rt}$$

Where: Y = sum of all the yield values of the whole experiment. In our example:

$$\bar{Y} = \frac{14{,}290 + 17{,}230 + 27{,}800}{10 \times 3} = \frac{59{,}320}{30}$$

$$= 1{,}977 \text{ (kg/ha)}$$

Standard Deviation (S)

The standard deviation (S) is a measure that gives an estimate of how dispersed the data are about the mean. A formula for computing the standard deviation is

$$S_i = \sqrt{\frac{\sum_{j=1}^{r} Y_{ij}^2 - \frac{\left(\sum_{j=1}^{r} Y_{ij}\right)^2}{r}}{r - 1}}$$

where S_i = standard deviation of treatment i

S_i^2 = variance of treatment i

For the traditional cropping pattern (i = 1) in the above example:

$$S_1 = \sqrt{\frac{(1{,}800^2 + 1{,}760^2 + \ldots + 1{,}310^2) - \frac{(1{,}800 + 1{,}760 + \ldots + 1{,}310)^2}{10}}{9}}$$

$$= 336 \text{ (kg/ha)}$$

S_1 = 336 means to the statistician that about two thirds of the yield values of this treatment can be expected to fall within a range of the mean ± S—i.e., between (1,429 − 336) and (1,429 + 336), which is 1,093 and 1,765. Actually seven yield values of treatment 1 fall within this range and three fall outside of it. This also means that the yields of treatment 1 are likely to conform to the normal distribution.

Coefficient of Variation (CV)

The coefficient of variation allows a comparison of the variability of data from different treatments. It expresses the standard deviation of treatment i (S_i) as a percent of the treatment mean ($\bar{Y}_i$).

$$CV_i = \frac{S_i}{\bar{Y}_i} \times 100 \qquad i = 1, 2, \ldots, t$$

For the traditional pattern (i = 1) in the above example:

$$CV_1 = \frac{336}{1{,}429} \times 100$$

$$= 24\%$$

As illustrated in Table 6-D-3, the variability in yields of the cropping pattern for alternative 2, (CV_3), is not much different from the variability in yields for the traditional pattern, (CV_1). However, the yields from alternative 1, (CV_2), do exhibit considerably more variability when compared with the yields for the traditional pattern, (CV_1).

Analysis of Variance

The completely randomized design has only two sources of variation—variation among experimental units receiving the same treatment—called experimental error—and variation between the treatment means. The researchers should construct an analysis of variance table similar to Table 6-D-4. The data in this table can then be used to find out whether at least one of the treatments is significantly different from the others at the five percent probability level. By five percent probability level, we mean the following: if a test of a particular treatment is repeated hundreds of times, the observed treatment differences can be attributed to chance rather than to the treatment 5 out of every 100 times. In Table 6-D-4, below, we show how to calculate the F value, which will be used in drawing conclusions about the significance of the results.

Degrees of freedom (df) apply to total observations, treatments, and error. The df terms for the first two are one less than the total number of observations and the number of treatments. With 30 observations, we have a total df of 29; and with three treatments, the df for this term is 2. The df for the error term is found by subtracting the df for the treatments from the df for the total observations.

After determining the degrees of freedom, we proceed in the following manner:

1. Calculate the correction term (C):

$$C = \frac{Y^2}{rt}$$

Where Y = the sum of all yield values
r = the number of replications
t = number of treatments

In the above example:

$$C = \frac{59{,}320^2}{10 \times 3} = 117{,}295{,}000$$

2. Calculate sums of squares for treatments (SST):

$$SST = \frac{\sum_{i=1}^{t} Y_i^2}{r} - C$$

Where Y_i = treatment total of treatment i

Table 6-D-4. Analysis of variance data for a completely randomized design.

Source of Variation	Degrees of Freedom (df)	Sum of Squares (SS)	Mean Squares (MS)	Observed F	Required F (5 percent probability)
Total	29	20,961,000			
Treatments	2	10,097,000	5,049,000	12.56	3.35
Error	27	10,864,000	402,000		

In our example:

$$SST = \frac{(14{,}290^2 + 17{,}230^2 + 27{,}800^2)}{10} - 117{,}295{,}000$$

$$= 10{,}097{,}000$$

3. Calculate mean square for treatment (MST):

$$MST = \frac{SST}{df\ (treatment)}$$

$$= \frac{10{,}097{,}000}{2} = 5{,}049{,}000$$

4. Calculate the sum of squares (SS):

$$SS = \sum_{i=1}^{t} \sum_{j=1}^{r} Y_{ij}^2 - C \qquad \begin{array}{l} i = 1, 2, \ldots, t \\ j = 1, 2, \ldots, r \end{array}$$

where

Y_{ij} = individual yield values

C = correction term

$$SS = (1{,}800^2 + 1{,}760^2 + \ldots + 2{,}700^2) - 117{,}295{,}000$$

$$= 138{,}256{,}000 - 117{,}295{,}000$$

$$= 20{,}961{,}000$$

5. Calculate the sum of squares for error (SSE) and the mean square for error (MSE):

$$SSE = SS - SST$$

$$= 20{,}961{,}000 - 10{,}097{,}000$$

$$= 10{,}864{,}000$$

$$MSE = \frac{SSE}{df\ (error)}$$

$$= \frac{10{,}864{,}000}{27}$$

$$= 402{,}000$$

6. Calculate the F value for treatments:

$$F = \frac{MST}{MSE}$$

$$= \frac{5{,}049{,}000}{402{,}000}$$

$$= 12.56$$

This F value of 12.56 is compared with the tabulated F value for 2 df for treatment (numerator) and 27 df for error (denominator). The minimum value required for significance at the 5 percent probability level can be found in texts such as that by Little and Hills (1978). From such a table of F values, we obtain a value of 3.35, which is considerably less than the calculated value of 12.56. Thus, we can say that at *least one of the treatments is significantly different from the others at the 5 percent probability level.* However, we do not yet know which ones are significantly different.

Mean Separation

Several methods are available for testing the *differences between means.* The one described here is the "least significant difference" (LSD), which can be used in two situations:

- LSD can be used to make meaningful comparisons that are planned when the experiment is designed. Thus, researchers will need to choose in advance the particular pairs of treatments they want to compare. They *cannot* decide to compare two treatments after the data have been collected and look promising. The specific comparisons should be based on predetermined characteristics of importance. For example, alternative 1 might be simply a change in the method of cultivation and alternative 2 might be a change in the method of cultivation plus the use of a herbicide.
- LSD can be used, when the F test on treatment means is significant, by comparing adjacent means arranged in order of magnitude. The LSD should not be used to make all possible comparisons among the means.

Table 6-D-5. Comparison for the LSD test between actual and required differences between adjacent means.

Treatment	*Treatment Mean*	*Actual Difference Between Adjacent Means*	*Minimum Required Difference Between Adjacent Means**
		Yield in kg/ha	
Traditional	1,429		
		294	582
Alternative 1	1,723		
		1,057	582
Alternative 2	2,780		

*At 5% probability level.

The LSD is a form of the t test, whose formula at the 5 percent probability level is

$$LSD_{0.05} = t_{0.05}\sqrt{\frac{2\ MSE}{r}}$$

The value of $t_{0.05}$ can be found in a t table by the df for the error and 5 percent probability level (Little and Hills, 1978). For 27 df, $t_{0.05}$ is 2.052. Using the data from Table 6-D-4, the $LSD_{0.05}$ for the treatment means is

$$LSD_{0.05} = 2.052\sqrt{\frac{(2 \times 402{,}000)}{10}}$$

$$= 582\ (kg/ha)$$

In Table 6-D-5, we arrange the treatment means from our experiment in increasing order as required in the LSD test. Only adjacent treatment means that differ by 582 kg/ha or more are significantly different at the 5 percent probability level. As Table 6-D-5 shows, the mean of alternative 1 does not differ significantly from the mean of the traditional practice at this probability level. However, the mean of alternative 2 does differ significantly from the mean of alternative 1 and thus is also better than the mean of the traditional practice because of the order. Based on the results of this analysis, FSR&D researchers should probably discontinue working with alternative 1 and concentrate on alternative 2.

RANDOMIZED COMPLETE BLOCK DESIGN

We will now illustrate some of the characteristics of randomized complete block designs by using another example. Suppose that an FSR&D team plans to test two new crop varieties by comparing them with the farmers' varieties in a series of researcher-managed trials. Conditions allow the team to set up a randomized complete block design by testing the two new varieties and the farmers' variety on each of 18 farms.

After conducting the experiments, the team presents the results of the experiments, as shown in Table 6-D-6. Yields are grouped by treatment and blocked by farm. The team then uses the individual yields and the computed means for the analysis of variance.

Table 6-D-6. Maize yields from farmer-managed tests grouped by treatment and block.

Block No.	*Treatment*			*Block Total* Y_j	*Block Means* $\bar{Y}_j$
	Traditional	*Alternative 1*	*Alternative 2*		
		Yield (kg/ha)			
1	2,200	3,010	2,910	8,120	2,707
2	2,510	3,360	3,030	8,900	2,967
3	2,390	3,600	2,870	8,860	2,953
4	2,820	3,135	3,240	9,195	3,065
5	3,100	4,650	4,340	12,090	4,030
6	1,570	2,250	1,780	5,600	1,867
7	1,990	2,925	2,460	7,375	2,458
8	3,450	4,500	3,640	11,590	3,863
9	3,200	4,200	3,300	10,700	3,567
10	2,000	2,730	4,200	8,930	2,977
11	1,830	2,750	2,980	7,560	2,520
12	2,710	3,320	3,570	9,600	3,200
13	2,310	3,300	2,810	8,420	2,807
14	2,540	3,800	2,670	9,010	3,003
15	2,470	3,600	3,330	9,400	3,133
16	2,280	3,420	2,380	8,080	2,693
17	1,950	2,760	2,640	7,350	2,450
18	1,870	2,900	2,700	7,470	2,490
Y_i	43,190	60,210	54,850	158,250	
$\bar{Y}_i$	2,399	3,345	3,047		$\bar{Y}$ = 2,931

Table 6-D-7. Data for analysis of variance for data in Table 6-D-6.

Source of Variation	*Degrees of Freedom (df)*	*Sum of Squares (SS)*	*Mean Squares (MS)*	*Observed F*	*Required F (5 percent probability)*
Total	53	26,504,000			
Blocks	17	14,085,000	829,000	7.03	1.93
Treatments	2	8,415,000	4,208,000	35.66	3.28
Error	34	4,004,000	118,000		

Analysis of Variance

The summary of the analysis of variance calculations is shown in Table 6-D-7. These values and the calculations are similar to those presented in the foregoing example, except that in the present example, the blocks provide a means for identifying a third source of variation. As before, we will use a 5 percent probability level.

The degrees of freedom for total observations are 53, which is the number of blocks (18) times the number of treatments (3) minus one. The df for the blocks is 17 (18 blocks − 1); the df for the treatments is 2 (3 treatments − 1); and the df for the error term is the difference between the total df and the df attributed to the blocks and treatments.

The calculations follow

1. Calculate the correction term (C):

$$C = \frac{Y^2}{rt}$$

$$= \frac{158{,}250^2}{(18)\ (3)} = 463{,}760{,}000$$

where Y = sum of all yield values in the whole experiment
r = number of replications = 18
t = number of treatments = 3

2. Calculate sums of squares for blocks (SSB):

$$SSB = \frac{\sum_{j=1}^{r} Y_j^2}{t} - C$$

$$= \frac{(8{,}120^2 + 8{,}900^2 + \ldots + 7{,}470^2)}{3} - 463{,}760{,}000$$

$$= \frac{1{,}433{,}534{,}000}{3} - 463{,}760{,}000 = 14{,}085{,}000$$

3. Calculate mean square for blocks (MSB)

$$MSB = \frac{SSB}{\text{df (blocks)}} = \frac{14{,}085{,}000}{17} = 829{,}000$$

4. Calculate sum of squares for treatments (SST)

$$SST = \frac{\sum_{i=1}^{t} Y_i^2}{r} - C$$

$$= \frac{43{,}190^2 + 60{,}210^2 + 54{,}850^2}{18} - 463{,}760{,}000$$

$$= 8{,}415{,}000$$

5. Calculate mean square for treatments (MST):

$$MST = \frac{SST}{\text{df (treatments)}} = \frac{8{,}415{,}000}{2} = 4{,}208{,}000$$

6. Calculate the sum of squares (SS):

$$SS = \sum_{i=1}^{t} \sum_{j=1}^{r} Y_{ij}^2 - C \qquad \begin{array}{l} i = 1, 2, \ldots, t \\ j = 1, 2, \ldots, r \end{array}$$

$$= (2{,}200^2 + 2{,}510^2 + \ldots + 2{,}700^2) - 463{,}760{,}000$$

$$= 26{,}504{,}000$$

7. Calculate the sum of squares for error (SSE) and mean square for error (MSE):

$$SSE = SS - SST - SSB$$

$$= 26{,}504{,}000 - 8{,}415{,}000 - 14{,}085{,}000$$

$$= 4{,}004{,}000$$

$$MSE = \frac{SSE}{\text{df (error)}}$$

$$= \frac{4{,}004{,}000}{34}$$

$$= 118{,}000$$

8. Calculate the F values for blocks and treatments:

$$F(\text{blocks}) = \frac{MSB}{MSE}$$

$$= \frac{829{,}000}{118{,}000} = 7.03$$

$$F(\text{treatments}) = \frac{MST}{MSE}$$

$$= \frac{4{,}208{,}000}{118{,}000} = 35.66$$

For the df(blocks) = 17 and for the df(error) = 34:

$$F_{(blocks)} = 1.93$$

and for the df(treatments) = 2 and for the df(error) = 34:

$$F_{(treatments)} = 3.28$$

Because the calculated F values exceed the required F values, we can say that significant differences in yields at the 5 percent probability level occurred both among the farms and among the treatments.

As before, we can use the LSD procedure to compare the treatment means. The value for $t_{0.05}$, with 34 df for the error term, is 2.032. The LSD for the 5 percent probability level and 18 replications is

$$LSD_{0.05} = t_{0.05}\sqrt{\frac{2\,MSE}{r}}$$

$$= 2.032\sqrt{\frac{2(118{,}000)}{18}}$$

$$= 233 \text{ (kg/ha)}$$

In Table 6-D-8, we arranged the treatment means in increasing order and used the LSD test at the 5 percent probability level to compare adjacent means.

Table 6-D-8 shows that the mean yield of alternative practice 2 exceeds the mean yield from the farmers' traditional practice by more than the difference required by the LSD test at the 5 percent probability level. This is also true for the mean of alternative practice 1 as compared with the mean of alternative practice 2. Thus, according to this statistical analysis of the experimental data, (1) alternative practice 2 is significantly better than the traditional practice, (2) alternative practice 1 is significantly better than alternative practice 2, (3) both alternatives are significantly better than the traditional practice.

Table 6-D-8. Comparison for the LSD test between actual and required differences between adjacent means.

Treatment	*Treatment Mean*	*Actual Difference Between Adjacent Means*	*Minimum Required Difference Between Adjacent Means**
	Yield in kg/ha		
Traditional	2,399		
		648	233
Alternative 2	3,047		
		298	233
Alternative 1	3,345		

*At 5% probability level.

PAIRED TREATMENTS DESIGN

If we assume that alternatives 1 and 2 in Table 6-D-6 were paired treatments on each farm, the data can be analyzed as a paired t test. The data in Table 6-D-6 for alternatives 1 and 2 are reproduced in Table 6-D-9. The differences between paired plots, d, are calculated in the last column. The mean of the differences, $\bar{d}$, is 298.

The variance of the differences, S^2_d, is

$$S^2_d = \frac{\sum_{j=1}^{r} d_j^2 - \frac{\left(\sum_{j=1}^{r} d_j\right)^2}{r}}{r-1}$$

Where r = number of farms (replications)

$$S^2_d = \frac{(100)^2 + (330)^2 + \ldots + (200)^2 - \frac{(100 + 330 + \ldots + 200)^2}{18}}{17}$$

$$= \frac{7{,}748{,}000 - 1{,}596{,}000}{17} = 362{,}000$$

The variance of the mean difference, $S^2_{\bar{d}}$ is calculated as follows:

$$S^2_{\bar{d}} = \frac{S^2_d}{r} = \frac{362{,}000}{18} = 20{,}000$$

Then, the standard error of the mean difference is:

$$S_{\bar{d}} = \sqrt{20{,}000} = 141$$

The t test for significance is calculated as follows:

$$t = \frac{\bar{d}}{S_{\bar{d}}} = \frac{298}{141} = 2.11$$

With 34 df(error) and a 5 percent probability level, the value of $t_{0.05}$ = 2.032. We obtain the df for the error terms as follows: df for total observations is 18 farms × 2 treatments = 36 minus 1 = 35 df (total). From this total, we subtract the df for treatments, which is 2 treatments minus 1 = 1, giving a net of 34.

A t value of 2.11, compared with a minimum value of 2.032, is significant at the 5 percent probability level. Thus, according to this test, the two alternative practices are also shown to be different at the 5 percent probability level. To be able to apply this paired test, the researchers need to design the experiment so that the treatments are side-by-side. By doing so, the paired t test increases the precision; but, the design is limited to paired treatments.

Table 6-D-9. Experimental yields of maize for paired treatments, alternatives one and two.

Farm Number	*Alternative 1*	*Alternative 2*	*Paired Plot Differences* d_j
1	3,010	2,910	100
2	3,360	3,030	330
3	3,600	2,870	730
4	3,135	3,240	-105
5	4,650	4,340	310
6	2,250	1,780	470
7	2,925	2,460	465
8	4,500	3,640	860
9	4,200	3,300	900
10	2,730	4,200	-1,470
11	2,750	2,980	-230
12	3,320	3,570	-250
13	3,300	2,810	490
14	3,800	2,670	1,130
15	3,600	3,330	270
16	3,420	2,380	1,040
17	2,760	2,640	120
18	2,900	2,700	200
Totals	60,210	54,850	5,360
Mean of paired plot differences			$\bar{d}$ = 298

Table 6-D-10. Maize yields arranged for a randomized incomplete block analysis.

Block	*Treatment* 1	2	3	4	5	6	*Block Totals* = B_j
	Maize Yields (tons/ha)						
1	1.8	2.2					4.0
2			1.6	3.0			4.6
3					2.8	2.8	5.6
4	2.3		1.2				3.5
5		1.6			3.1		4.7
6				2.6		3.1	5.7
7	2.4			3.1			5.5
8		1.8				3.1	4.9
9			1.9		3.2		5.1
10	2.6				2.9		5.5
11		2.1		3.1			5.2
12			1.6			3.2	4.8
13	2.3					3.1	5.4
14		2.0	1.8				3.8
15				3.2	3.2		6.4
Y_i	11.4	9.7	8.1	15.0	15.2	15.3	Y = 74.7
$\bar{Y}_i$	2.28	1.94	1.62	3.00	3.04	3.06	$\bar{Y}$ = 2.49

RANDOMIZED INCOMPLETE BLOCK DESIGN

We will illustrate the statistical procedure for a randomized incomplete block design using the format introduced in Table 6-D-2 and the data for an experiment in maize as presented in Table 6-D-10. From Table 6-D-2, we have the number of treatments (t) as 6, the number of treatments per block (k) as 2, the number of replications (r) of each treatment as 5, and the number of blocks (b) in the plan as 15. These values give us $\lambda = 1$ and $E = 0.6$, as computed in Table 6-D-2. Following, are the calculations.

1. Calculate the adjusted treatment means ($\bar{t}_i$):

Treatment effects cannot be estimated directly from the treatment means. Treatment means must first be adjusted for possible block effects. The formula for the adjusted treatment mean ($\bar{t}_i$) is

$$\bar{t}_i = \frac{Q_i}{Er} + \frac{Y}{rt}$$

Where:

$$Q_i = Y_i - \frac{\text{(sum of totals of all block totals } (B_j) \text{ containing treatment i)}}{k}$$

$i = 1,2, ..., t$

$$Q_1 = 11.4 - \frac{4.0 + 3.5 + 5.5 + 5.5 + 5.4}{2} = -0.55$$

Similarly,

$Q_2 = -1.60$
$Q_3 = -2.80$
$Q_4 = 1.30$
$Q_5 = 1.55$
$Q_6 = 2.10$

As a check, the sum of Q_i values should be 0.

$$\sum_{i=1}^{6} Q_i = -0.55 - 1.60 - 2.80 + 1.30 + 1.55 + 2.10 = 0$$

Therefore, the adjusted treatment means for the control ($\bar{t}_1$) and the other five treatments ($\bar{t}_2$ to $\bar{t}_6$) are:

$$\bar{t}_i = \frac{Q_i}{Er} + \frac{Y}{rt}$$

$$\bar{t}_1 = \frac{-0.55}{(0.6)(5)} + \frac{74.7}{(5)(6)} = 2.31$$

Similarly,

$\bar{t}_2 = 1.96$
$\bar{t}_3 = 1.56$
$\bar{t}_4 = 2.92$
$\bar{t}_5 = 3.01$
$\bar{t}_6 = 3.19$

2. Calculate sum of squares for treatments (SST):

$$SST = \frac{\sum_{i=1}^{t} Q_i^2}{Er} \qquad i = 1,2, ...,t$$

$$= \frac{(-0.55)^2 + (-1.60)^2 + (-2.80)^2 + (1.30)^2 + (1.55)^2 + (2.10)^2}{(0.6)\ (5)}$$

$$= \frac{19.21}{3}$$

$$= 6.40$$

3. Calculate sum of squares for blocks (SSB):

$$SSB = \frac{\sum_{j=1}^{b} B_j^2}{k} - \frac{Y^2}{rt} \qquad j = 1,2, ..., b$$

$$= \frac{4.0^2 + 4.6^2 + ... + 6.4^2}{2} - \frac{74.7^2}{(5)(6)}$$

$$= 190.26 - 186.00$$

$$= 4.26$$

4. Calculate the sum of squares (SS):

$$SS = \sum_{i=1}^{t} \sum_{j=1}^{b} Y_{ij}^2 - \frac{Y^2}{rt} \qquad \begin{matrix} i = 1, 2, ..., t \\ j = 1, 2, ..., b \end{matrix}$$

$$= (1.8^2 + 2.2^2 + 1.6^2 + ... + 3.2^2) - \frac{74.7^2}{(5)(6)}$$

$$= 197.27 - 186.00$$

$$= 11.27$$

5. Compute SSE and MSE, as before:

$$SSE = SS - SST - SSB$$

$$= 11.27 - 6.40 - 4.26$$

$$= 0.61$$

$$MSE = \frac{SSE}{df\ (error)}$$

where $df\ (error) = tr - t - b + l$

$$MSE = \frac{0.61}{(6)\ (5) - 6 - 15 + 1} = \frac{0.61}{10}$$

$$= 0.061$$

Then the LSD is calculated from the formula previously presented. The value of $t_{0.05}$ and 10 df of the error term is 2.228 and the resulting LSD for 5 replications is

$$LSD_{0.05} = t_{0.05} \sqrt{\frac{2\ MSE}{r}}$$

$$= 2.228 \sqrt{\frac{(2)\ (0.061)}{5}}$$

$$= 0.35\ (tons/ha)$$

In Table 6-D-11, we arranged the adjusted treatment means in increasing order and compared the adjacent means. The yield increases between adjusted treatment means 2 and 3 and 1 and 2 are statistically significant. Nevertheless, they are probably not worth pursuing since they produced worse results than the control. Treatment 4 is signficantly better than the control. No significant difference was found between treatments 4 and 5 or 5 and 6. However, because of the order, treatments 4, 5, and 6 all must be considered significantly better than the control. Thus, from these data, FSR&D researchers will probably substitute any one of the alternatives 4, 5, and 6 for the control treatment. If they want to differentiate among treatments 4, 5, and 6, they will need to undertake additional experiments.

Table 6-D-11. Comparison for the LSD test between actual and required differences between adjacent means using the LSD 0.05 test.

Treatment	*Adjusted Treatment Mean*	*Actual Difference Between Adjacent Means*	*Minimum Required Difference Between Adjacent Means**
		Yield in tons/ha	
3	1.56		
		0.40	0.35
2	1.96		
		0.35	0.35
1 (control)	2.31		
		0.61	0.35
4	2.92		
		0.09	0.35
5	3.01		
		0.18	0.35
6	3.19		

*At 5% probability level.

APPENDIX 6-E
EXAMPLE OF A PROCEDURE FOR DESIGNING A CROPPING PATTERN EXPERIMENT

This appendix describes an approach followed by the Central Research Institute for Agriculture (CRIA) in designing new cropping patterns in Indonesia (McIntosh, 1980). The procedure outlines the types of information to gather and analyze when planning on-farm experiments (see Sec. 6.4.4. in Chapter 6). Below, we provide some of the details on CRIA's approach, as they relate to partitioning the research area, describing the physical environment, analyzing the socioeconomic environment, evaluating current technology, selecting and managing new cropping patterns, and preparing experimental designs.

PARTITIONING THE RESEARCH AREA

The team first partitions the research area into smaller areas, generally based on the environmental factors for the different farming systems. In addition, the team may partition the research area by cropping systems. The reason for partitioning a research area is to identify farming units with similar management, which in turn facilitates the transfer of technology to farmers operating under similar conditions.

McIntosh (1980) gives several examples in which a research area within a common agroclimatic zone was partitioned for research on the basis of physical features or irrigation features. Following are two of these examples.

SOUTHERN SUMATRA

Part of Indonesia's new transmigration projects involves opening up new lands currently in forests or *Imperata* grass. Much of the land is rolling to hilly and not suitable for food crop production, unless soil conservation practices can be improved. McIntosh (1980) reported the research area was partitioned into three categories based on the following criteria for upland areas:

"Category I. Relatively level land on hill tops
Category II. Sloping land that must be terraced
Category III. Land newly opened from forests (compared with land opened from *Imperata*)."

INDRAMAJU, WEST JAVA

CRIA teams partitioned areas in this region according to the period of irrigation. The lands were lowland, relatively level, alluvial clay soils, three to four months with rainfall greater than 200 mm, and with a long, dry season. Water control problems in the area include flooding during the rainy season and limited irrigation supplies during the dry season. McIntosh (1980) reported the basis for partitioning was

"Category I. Area with 10 months irrigation
Category II. Area with 7 months irrigation
Category III. Area with 5 months irrigation."

DESCRIBING THE PHYSICAL ENVIRONMENT

The physical environment includes land type, soils, topography, climate, and irrigated or rain-fed land. The information, which comes from the analysis of the data base, is used (1) to assist in identifying new crops or shifts in the cropping pattern and (2) to provide options for testing in different agroclimatic zones.

ANALYZING THE SOCIOECONOMIC ENVIRONMENT

The analysis of the socioeconomic environment provides guidelines for determining the possible acceptance of new crops or the redesign of existing cropping patterns. The analysis evaluates current resources, social customs, markets, and infrastructural contraints on the introduction of new cropping patterns.

EVALUATING CURRENT TECHNOLOGY

This includes an analysis of the technology now being used by the farmers—for example, agricultural chemicals, crop varieties, and irrigation practices. The team also analyzes the effectiveness of transfer mechanisms now being used in the research area. The current levels of technology and transfer capabilities are important considerations for selecting the technologies to be tested.

SELECTING NEW CROPPING PATTERNS

In the initial phases of a research effort, improvements to existing cropping patterns that recognize such factors as the farmers' available resources and markets and irrigation deliveries will probably be more successful than new crop introductions. As the team gains knowledge and success with existing patterns, the potential for success with new patterns increases. In altering cropping patterns, the team considers both the agronomic adaptability of the new patterns and the markets for the crops.

Agronomic adaptation of new crops to the local environment such as climate, irrigation water, and disease or insect attacks in the research area is of primary consideration in the design. New designs generally emphasize production of those crops farmers value the highest. Crops of secondary importance are then integrated into the pattern to give additional annual output from the pattern.

The market potential is important when the team considers the introduction of new crops. Suitable introductions are new food crops if surveys show them to be acceptable for local consumption, processing, or export. Before initiating extensive research on a new crop, however, a market analysis should indicate a high probability of its acceptance.

MANAGING NEW CROPPING PATTERNS

Farm surveys are used to identify constraints and opportunities for the farmers in the management of their current cropping patterns. Factors most often affecting farmers' decisions are risk, labor, and cash.

Minimizing risk is important for farmers with limited resources. Farmers usually reduce risk by practices such as low input levels, growing a single crop during a shorter season when weather conditions are least hazardous, growing the most essential food crops, or intercropping.

The labor requirements of a new cropping pattern

may be a constraint to adoption if special seedbed preparation is required, if timing of operations is critical, and for other reasons. If labor is a constraint, the new design should consider minimizing the labor requirements of the pattern or distributing the labor requirements over a longer period.

Most small farmers in developing countries are typified as those with limited resources. These farmers have little money available for the purchase of inputs. Credit may not be available or the farmer may not want to borrow. Thus, patterns should be designed for several input levels. As technologies with lower input requirements are successfully adopted, farmers will be more inclined to adopt technologies requiring higher inputs.

PREPARING EXPERIMENTAL DESIGNS

After an analysis of the physical and socioeconomic environments and the farmers' current practices, the design team outlines a cropping pattern experiment. Typically, a cropping pattern experiment will consist of two to four cropping patterns at one or two levels of management. The farmers' current pattern usually is the control. An experimental approach described by McIntosh (1980) has two cropping patterns (existing and new) at two levels of management (current and new levels of inputs):

1) Farmers' present cropping patterns and current level of inputs—to serve as the control treatment.
2) Farmers' present cropping pattern with additional inputs and improved management—to evaluate the farmers' pattern with reduced input constraints.
3) Newly designed pattern with low inputs—to evaluate the new pattern at input levels approximately the same as the farmers' current input levels.
4) Newly designed pattern without input constraints—to determine the potential of the new pattern.

For additional comments on these alternatives, see Sec. 6.4.1. in Chapter 6.

APPENDIX 6-F FIELD ASSIGNMENTS

In Sec. 6.5.3. in Chapter 6, we talked about assignments for the field teams and said additional information would be provided in the appendix. Thus, this appendix elaborates on the responsibilities of the field team's leaders, researchers, technical assistants, and disciplinary and commodity specialists.

TEAM LEADERS' RESPONSIBILITIES

Field team leaders are responsible for their team's activities, as outlined in the regional planning workshops. When a field team is small, its leader may assume major responsibility in his or her discipline. For larger teams, other researchers may have to assume responsibility in the leader's specialty.

The leader helps the team by allowing each member to participate in writing his or her assignment. In this way, members' responsibilities can be clarified, misunderstandings reduced, and team interdisciplinarity enhanced.

Time permitting, the leader should hold monthly meetings to review the team's progress and to adjust the work plan as needed. Included in the review is an examination of the team's objectives, approaches, findings, and logistical and financial requirements.

With help from the team, the leader assembles reports of each research activity. Besides their current uses, the team can use these summaries to report on its activities at the planning and analysis workshops. The summaries might include

- recommendations for technologies that appear suitable for multi-locational testing
- results and concepts that add to the team's knowledge
- research proposals for the next season.

RESEARCHERS' RESPONSIBILITIES

Each of the field team's researchers is responsible for supervising a set of on-farm experiments, as well as for participating in other research activities. More specifically, each team member should (1) interact continually with the rest of the team, cooperating farmers, and others and should (2) guide and train the technical assistants in experimental procedures. Once the experiments are in progress, the researchers should visit each location at weekly intervals, but not so regularly that the assistants and farmers can anticipate their arrival. The purpose of these visits is to observe progress on the experiments and to help the farmers and the technical assistants. Each researcher is responsible for analyzing, summarizing, and reporting to the team leader as soon as the experiments are completed.

TECHNICAL ASSISTANTS' RESPONSIBILITIES

The technical assistants should be capable and interested in research and should probably live in the research area. Thus, they form an important liaison between the field team and the farmers. They should be loyal both to the field team that employs them and to the community in which they live. These assistants carry out the many routine tasks such as the daily climatic observations, supervision of farm records, monitoring field experiments, and meeting with the farmers and farm families. They also help lay out and harvest the experiments and recruit extra labor when needed. They prepare reports of their activities for regular transmittal—often weekly—to their supervising researcher. Because these assistants are often the best informed on farmers' conditions, they can help in interpreting experimental results.

SPECIALISTS' RESPONSIBILITIES

Disciplinary and commodity specialists are often assigned to regional headquarters and help directly in the field team's work. Ideally, these specialists will have had experience in interdisciplinary research, know on-farm research procedures, and value farmers' inputs.

The specialists generally serve as key resource persons in the region. They assist in the planning and analysis workshops, as described in Sec. 6.5 and Sec. 7.11., respectively. They guide the regular field team members throughout the FSR&D process. Finally, they can be particularly helpful in (1) setting the initial directions of research for the field teams, (2) solving specialized problems, (3) establishing research methodologies, (4) developing instructional materials, and (5) training other members of the regional and field teams.

APPENDIXES TO CHAPTER 7

APPENDIX 7-A
FORMS FOR COLLECTING DATA FOR ON-FARM CROPPING EXPERIMENTS

Adequate forms can simplify and improve data collection and thereby raise the quality of on-farm cropping and livestock experiments. In the section on Cultural Practices and Data Collection in Sec. 7.1.1. on Research on Crops of Chapter 7, we discussed the need for field teams to use the planning workshops to help develop adequate procedures for data collection.

In this appendix, which is a companion to Appendix 7-B on livestock, we provide illustrative forms that cover the background information for cropping experiments, climatic data for the area, monitoring of plant growth and field operations, a checklist of field operations, and a form for the experiment's summary, conclusions, and recommendations. As the field teams gain experience, they will probably find the need to develop additional tables that could, for example, provide details on the more important field operations, including the types of activities, resource requirements, and costs. The specifics of these tables will change depending on whether the experiment is researcher-managed, farmer-managed, or superimposed; however, the general types of information to collect remain about the same.

We drew on Zandstra et al. (1981) for many of the ideas and material contained in this appendix.

Table 7-A-1. Background for on-farm cropping experiments.

Title ______________________________

Identification of experiment

Supervising technician ____________ Date ____________

No. ____________ Farmer ____________

Location ____________ Type ____________

Crops ________ Dates: Begun ________ Ended ________

Cropping pattern ______________________

Objectives ______________________

Justification ______________________

Treatments ______________________

Plot diagram

Summary of management history

Crop	________	________	. . . .
Dates: Planted	________	________	. . . .
Harvested	________	________	. . . .
Yields	________	________	. . . .
Fertilizers (types, amounts, dates)	________	________	. . . .
Insect & disease controls (types, amounts, dates)	________	________	. . . .
Weed control (types, amounts, dates)	________	________	. . . .
Cultural practices	________	________	. . . .
Crop disposal	________	________	. . . .
Other	________	________	. . . .

Climatic characteristics

Annual rainfall ______________________

Monthly rainfall ______________________

Annual temperature ______________________

Monthly temperature ______________________

Other (e.g., major wind directions, cloud cover, floods) ____________

Table 7-A-1 (cont.)

Land and water
Land type__________
Slope __________
Depth of water table __________
Irrigated or rain-fed__________
Irrigation water: source_____ availability_____ quality_____
Drainage __________
Other __________

Soil properties
Soil series __________
Texture (surface) __________
Depth __________
Soil acidity (surface) __________
Soil salinity (surface) __________
Soil structure (profile) __________
Permeability__________
Soil fertility __________
Soil organic matter__________
Other __________

Table 7-A-2. Climatic data for the area (Adapted from Zandstra et al., 1981).

Date	Evaporation (mm)	Temperature (°C)		Solar Radiation (Ly)	Relative Humidity		Maximum Wind Velocity (km/hr)	Rain Gauge Readings (mm)				Comments*
		Max	Min		Max	Min		1	2	3	. . . .	

*Unusual weather events that influence biological production or farmers' management practices.

Table 7-A-3. Monitoring plant growth (Adapted from Zandstra et al., 1981).

Date	Stage of Growth	Plant Characteristics		Pests				Comments
		Foliar	Roots	Weeds	Insects	Diseases	Other	

Table 7-A-4. Monitoring field operations (Adapted from Zandstra et al., 1981).

	Field	Labor (hours)			Agricultural Chemicals			Water		Residue		
Date	Operation*	Hand	Animal	Tractor	Type	Rate (kg/ha)	Application Method	Amt (cm)	Application Method	Type	Amt (kg/ha)	Comments

*See Table 7-A-5 for a checklist of field operations.

Table 7-A-5. Checklist of field operations for FSR&D (Adapted from Zandstra et al., 1981).

Land preparation

01 Clearing residue - removing, piling in heaps, cutting, burning.
02 Field repairs - fixing of bunds, ditches, fences, etc.
03 Plowing - initial or primary tillage operations in the field to break soil surface before secondary tillage. Do not include plowing done to seedbed.
04 Harrowing - process of breaking clods by passing any type of harrow (comb, tooth disk, etc.) over a field. (As in plowing do not include harrowing done to seedbed.)
05 Leveling - final operation before transplanting by passing over plain board on harrowed field to reduce slight soil surface depressions for even water distribution.
06 Furrowing - passing over a finally prepared field with a plow or other tool to prepare furrows at a given (row) distance just before planting.
07 Incorporating - mixing or placing fertilizer, insecticide, pesticide, herbicide into the soil.
08 Intercrop land preparation - any tillage operation to allow planting of secondary crops between main crops.
09 Other land preparation - operations that cannot be classified into any of the above operations. Specify the operation with an explanation at the comments section. Examples: ridging, bedding, etc.

Crop establishment

20 Transplanting - planting of seedlings (often rice) in the pattern plot.
21 Planting - placing crop seeds properly in or on the soil by broadcasting, dibbling, drilling, or other methods for crop establishment.
22 Replanting - planting seedlings or seed in missing hills after first planting.
23 Thinning - removing extra plants to obtain the desired plant density.
24 Ratooning - the crop re-growth and yield obtained after harvesting the plant crop.
25 Soaking (or dipping) - immersing of seeds for pregermination or the treatment of seeds or seedlings with chemicals.
29 Other crop establishment - operations that cannot be classified into any of the above operations (codes 20-25). Provide explanation at the comments portion.

Crop care

30 Fertilizing - application of fertilizer material with particular nutrients that aid in crop growth and development.
31 Pesticide application - spraying of chemicals or broadcasting them in granular form to control destructive insects and diseases.
32 Herbicide application - spraying or broadcasting of herbicides to the plot to control weeds.
33 Nonchemical pest control - operations for control of pests, manual insect control and control of rats, birds, etc.
34 Handweeding - removing weeds from the field manually or by nonmechanical tools such as blades, hoes, etc. (no rotating or oscillating parts).
35 Mechanical weeding - weed control method using hand or engine powered mechanical equipment.
36 Canopy manipulation - bending, clipping, pruning, binding up or in any other way systematically changing the structure of the crop canopy, e.g. bending back of maize.
37 Mulching - placement of straw or similar farm residues on the ground (often to conserve soil moisture or reduce soil temperatures).
38 Hilling-up - plowing between rows of plants with furrow slice thrown toward the base of the plant.
39 Off-barring - plowing between rows of plants with the furrow slice thrown back to back to the centre between plant rows.
49 Other crop care - operations that cannot be classified into any of the above operations (code 30-39). Provide explanation at comments section.

Harvesting

50 Crop cut sampling - sample harvested in a defined area of a plot for yield determination.
51 Manual harvesting - cutting the crop manually using scythe or any other tool.
52 Power harvesting - method of cutting the crop by employing mechanical harvesters.
53 Manual threshing - separating straw from grains without machines, e.g. by foot or by striking a bundle of panicles over slats or by animal trampling.
54 Power threshing - separating grain from straw by using an engine, human or animal powered mechanical thresher.
55 Manual winnowing - separating unfilled grains from developed grains by gravity or natural air current.
56 Power winnowing - separating unfilled grains from developed grains by a mechanical blower.
57 Drying - removal of excess moisture in seeds to meet desired moisture level for storage. This is done by exposure to the sun or in driers or ovens.
58 Hauling - transporting manually and/or mechanically of product from the field to storage or to the market.
59 Shelling - removal of the outer seed cover of a crop like peanut or the maize grain from the husk.
69 Other harvest - operations that cannot be classified into any of the above operations (codes 50-59). Provide an explanation at comments section.
70 Crop failure - If crop failed, enter the date the crop was discontinued and provide explanation at comments section.

Table 7-A-6. Summary, conclusions, and recommendations.

Summary of Results

Conclusions

Recommendations

Sources
Working documents______________________
____

Publications______________________
____

Location of Materials
Personal file______________ Office file______________

APPENDIX 7-B
FORMS FOR COLLECTING DATA FOR ON-FARM LIVESTOCK EXPERIMENTS

Adequate forms can simplify and improve data collection and thereby raise the quality of on-farm cropping and livestock experiments. In Research on Livestock, Sec. 7.1.2. of Chapter 7, we discussed several characteristics of on-farm experiments with livestock.

In this appendix, which is a companion to Appendix 7-A on crops, we provide illustrative forms that cover the background information for livestock experiments, climatic data for the area, inventories and changes in the test herds, description and history of the test animal, monitoring livestock inputs and feeding activities, requirements for family labor, monitoring animal traction, and a form for the experiment's summary, conclusions, and recommendations.

We include these appendix tables as illustrations of the types of data to collect and a means for recording the data. The field teams will need to modify these forms and develop additional ones to meet their specific needs.

Table 7-B-1. Background for on-farm livestock experiments.

Title ______________________
Identification of experiment
Supervising technician ______________ Date ______________
No. ______________ Farmer ______________
Location ______________ Type ______________
Livestock __________ Dates: Begun __________ Ended __________
Livestock patterns ______________________
Objectives ______________________
Justification ______________________
Treatments ______________________
Layout of animal-keeping facilities
Summary of management history

Animals (kind and numbers)	________	________	. . . .
Animal's condition	________	________	. . . .
Output and uses	________	________	. . . .
Breeding and acquisition	________	________	. . . .
Feed and nutrition (types, amounts, dates)	________	________	. . . .
Diseases and parasite controls (types, amounts, dates)	________	________	. . . .
Sanitation	________	________	. . . .
Tending	________	________	. . . .
Disposal of animals and animal products	________	________	. . . .
Other	________	________	. . . .

Climatic characteristics
Annual rainfall ______________________
Monthly rainfall ______________________
Annual temperature ______________________
Monthly temperature ______________________
Other (e.g., water availability and range conditions)______________

Table 7-B-2. Climatic data for the area (Adapted from Zandstra et al., 1981).

Date	Evaporation (mm)	Temperature (°C) Max	Temperature (°C) Min	Solar Radiation (Ly)	Relative Humidity Max	Relative Humidity Min	Maximum Wind Velocity (km/hr)	Rain Gauge Readings (mm) 1	2	3	. . . .	Comments*

*Unusual weather events that influence biological production or farmers' management practices.

Table 7-B-3. Inventories and changes in test herds (number of animals).*

Item	Date	Breeding Herd				Growing Herd				Remarks†
		Mature Animals		Replacements		Nursing Animals		Weaned Animals		
		Females	Males	Females	Males	Females	Males	Females	Males	
Beginning inventory										
Increase (+) or Decrease (−)										
Closing inventory										

*Columns may be subdivided further as needed for age, weight, or other characteristics.
†Indicate reasons for increase or decrease—e.g., animals bought, sold, died, taken for home or community use.

Table 7-B-4. Description and history of the test animal.*

A. *Description of animal at beginning of test*
Breed__________ Sex_____ Age_____ Weight_____
Health ____________________
Nutritional condition ____________________
Utilization ____________________
Other: e.g., management factors such as housing, pregnant in 5th month____________________

B. *History of animal*
Health ____________________
Nutrition ____________________
Production statistics: e.g., quality of product such as the fat content of milk, periodic weight gains, number of days used for traction power per year, number of completed pregnancies, number weaned per litter, quality and productivity of offspring____________________

C. *Other* ____________________

*Prepare form for test herd or separate forms for each test animal as needed.

Table 7-B-5. Monitoring livestock inputs.

Date	Kind of Work*	Labor			Tools & Equipment			Materials			Other			Total Cost	Observations
		Kind	Hours	Cost	Kind	Hours	Cost	Kind	Amount	Cost	Kind	Amount	Cost		

*E.g., herding, watering, feeding, cutting and hauling feed, fencing, milking, cleaning, checking, testing.

Table 7-B-6. Monitoring livestock feeding activities.

Date	Type of Feed*	Quality of Feed	Method of Feeding†	Quantity of Daily Feed	Animal Response	Observations

*E.g., pasture, cured hay, green-cut hay.
†E.g., grazing, roadside, herding.

Table 7-B-7. Requirements for family labor.

Date	Females 14 and under		Males 14 and under		Females 15 and older		Males 15 and older		Total Hours Per Day
	Task*	Hours	Task*	Hours	Task*	Hours	Task*	Hours	

*E.g., herding, watering, gathering or hauling feed, grading, milking, veterinarian skills.

Table 7-B-8. Monitoring animal traction.

Date	Type of Animal	Farm Operation*	Hours of Use	Type of Supervision	Other

*E.g., plowing, furrowing, harvesting, hauling, pumping water.

Table 7-B-9. Summary, conclusions, and recommendations.

Summary of Results

Conclusions

Recommendations

Sources
Working documents________________________
____

Publications________________________
____

Location of Materials:
Personal file______________ Office file______________

APPENDIX 7-C DATA COLLECTION FORM FOR FARMER-MANAGED TESTS

Farmer-managed tests can be more efficiently conducted when farmers and researchers have a mutual understanding of the nature and details of the experiments. This is part of the Farmer-Researcher Relationships mentioned in Sec. 7.3.1. in Chapter 7. One way to help improve this understanding is for the field team to develop data collection forms, in cooperation with the farmers, that describe the experiment and set out the sequence and timing of the more important field operations.

Table 7-C-1 was developed by Colorado State University's Egypt Water Use and Management Project for collecting data on the initial stages of maize production. The table begins with the final cutting of clover, which precedes the maize crop under study, and ends with a scheduled post irrigation soil sample after the first irrigation. We do not include them, but similar tables should be prepared for subsequent activities when they are important to the experiment. Also included in Table 7-C-1 is a column for recording the dates at which the operations took place. By comparing planned with actual activities and dates, the team learns about farmers' conditions and the acceptability of new technologies.

Table 7-C-1. Data collection form for field tests.

Experiment ___Farmer-managed test___ Crop ___Maize___
Farmer ________________ Part Crop ___Annual clover___
Distributory ________________ Season ________________
Treatment ________________

Calendar (Date)*	Crop* Days from Planting	Planned Operation†	Actual Operation
1			
2			
3	-8	Final clover cut	Final clover cut
4	-7	Plowing	
5	-6	Furrowing	Plowing
6	-5		
7	\\\-4\\\	Planting { 25 × 70 cm hills	Furrowing
8	\\\-3\\\	{ 5 seeds per hill	
9 /////	\\\-2\\\		→ (No available labor caused delay)
10 /////	\\\-1\\\		
11 //////	\\\0\\\	Pre irrig. soil sample	→ Planting and soil sample taken
12 /////	\\\1\\\	Planting irrigation	Planting irrigation
13 //////	2		
14	3		
15	4	Post irrig. soil sample	Soil sample taken
16	5		

Table 7-C-1 (cont.)

17	\\\6\\\\		
18	\\\7\\\		Plant emergence observed
19	8		
20	9		
21	10	Thinning for fodder	
22	11		
23	\\\12\\\\		
24 /////	\\\13\\\\		(Delay for better fodder)
25 /////	\\\\14\\\\		
26 /////	\\\\15\\\\	Fert. application 65ka N/ha	Fertilizer not available
27 /////	\\\16\\\\	Pre irrig. soil sample	1st irrig. advanced to next irrig. period
28 //////	17	1st irrig.	
29	18		Thinning for fodder
30	19		Soil sample taken
31	20	Post irrig. soil sample	

*The crosshatched area in the first column indicates when the farmer is expected to receive irrigation water; the crosshatched area in the second column indicates when the farmer actually received the water.

†Except for soil sampling, which is taken by the technical assistants, all operations are by the farmer.

APPENDIX 7-D FIELD DESIGN FOR FARMER-MANAGED CROPPING TEST

In Sec. 7.3.1, Chapter 7, we discussed farmer-managed cropping tests. If the cropping pattern remains the same, the experimental procedures are generally much simpler than if researchers test new cropping patterns. The following design is suitable for a farmer-managed test with a given cropping pattern for three alternative management practices plus the farmers' practice—the control.

The proposed practices for testing could be a fertility experiment for flooded rice consisting of zinc sulfate (1) applied in the seedling bed, (2) broadcast and incorporated into the soil before planting, and (3) applied as foliar spray at vegetative and jointing stages of growth. The researchers would have determined the application rates and methods from prior researcher-managed trials. The farmers' practice does not involve the use of zinc fertilizers.

A field design suitable for the zinc fertilizer experiment described above could be

1) Select the three predominant land types
2) Select a representative cross section of 20 farmer cooperators in each land type (see Farmer Selection in Sec. 7.3.1.).
3) Choose a randomized complete block design, with five replications of each treatment in each land type and one treatment per field. With three land types, this would give an experiment of 60 field plots—i.e., four treatments × three land types × five replications.
4) The plots should be large enough for the researcher to obtain reliable estimates of the farmer's labor and other inputs. Waugh (personal communication) suggests that plots for farmer-managed tests be at least 0.1 ha. Sometimes plots of this size will occupy the farmer's entire field.
5) The researchers would use data collection forms of the type shown in Appendix 7-A to describe the research field and to monitor field operations and plant growth.
6) The researchers can use the statistical methods described in Appendix 6-D for analyzing the results of a randomized complete block design.
7) The same experiment should probably be repeated two or three years in succession on farms within the same land types to evaluate year-to-year variations.

APPENDIX 7-E
FIELD DESIGN OF A FARMER-MANAGED CROPPING PATTERN TEST

We described farmer-managed cropping tests in Sec. 7.3.1., Chapter 7. In this appendix, we provide additional information on how FSR&D teams use such tests to evaluate the suitability of new cropping patterns under farmers' conditions. Farmer-managed tests help the FSR&D team identify management problems with the new technologies that do not surface from the research-managed and superimposed trials.

Farmer-managed cropping pattern tests compare patterns that may differ in crop species, number of crops grown per year, time and method of crop establishment, and other ways. The team needs to analyze the pattern in the farming system and have a fairly complete listing of inputs and outputs. Since these requirements include an analysis of the seasonal labor needs and their relationships with other farming activities, the farmers' practice is generally used for one of the treatments.

Cropping patterns are often evaluated for the major physical subdivisions recognized in a research area—e.g., land and soil types and availability of markets. The team generally uses a randomized complete block design (see Appendix 6-D) to evaluate cropping patterns over several land types. The team chooses representative plots on individual farmers' fields within a given land type for treatment replications—usually four or five times on each land type.

Zandstra et al. (1981) recommend the following design for cropping pattern testing in a research area:

1. Select two or three land types for evaluating the proposed patterns.
2. For example, select three or more cropping pattern designs for each land type. (Some patterns may be tested on more than one land type.) During the first year the research team may want to test five or six newly designed patterns and reduce this number for the subsequent years as it identifies the better patterns.
3. Replicate each cropping pattern—including the control—on at least four fields per land type for the first year. Replications may be increased for subsequent years as the number of patterns is decreased. A three-year test probably is minimum.
4. The farmers' cropping pattern is the "control." The patterns selected as control are usually the predominant one or two patterns for the land type. The control patterns are monitored in the same manner as the newly designed patterns.

We give an example of a cropping pattern design for a three-year test in Table 7-E-1. The table shows a reduction in patterns tested as the better patterns are identified. The "control" pattern is included each year to evaluate the new designs.

We discuss the selection of cooperating farmers and data collection in Sec. 7.3.1. and give the statistical analysis for a completely randomized design in Appendix 6-D.

Table 7-E-1. Design of a three-year cropping pattern test (Adapted from Zandstra et al., 1981).*

Land Type	*Cropping Pattern Number*						*Total Fields (plots)*
	1†	*2*	*3*	*4*	*5*	*6*	
	Number of Replications, Year 1						
1	4		4	5	4		17
2	4	5	4			4	17
3	4		4		4	4	16
Total	12	5	12	5	8	8	50
	Number of Replications, Year 2						
1	4			4	5	5	18
2	4		5	4			13
3	4		5	4		5	18
Total	12		10	12	5	10	49
	Number of Replications, Year 3						
1	4			4		4	12
2	4		6	4			14
3	4		6	4		4	18
Total	12		12	12		8	44

*The table gives the number of replicates or fields of each pattern on a land type for each year of the test.

†Cropping pattern 1 is the farmers' pattern and should be placed on each land type. Where two farmers' patterns predominate, both should be included in the test and placed on each land type.

APPENDIX 7-F
EXAMPLE OF A MIXED CROP-LIVESTOCK RESEARCH PROJECT

As pointed out in Sec. 7.3.2., Chapter 7, livestock and cropping research should be integrated for a mixed farming system. In 1976, researchers of ILCA (International Livestock Centre for Africa) began a mixed crop-livestock project at Debre Zeit, located southeast of Addis Ababa in the highlands of Ethiopia. This appendix, based on ILCA's 1979 and 1980 reports, discusses the major stages in this project as they relate to (1) development of a research base, (2) problem identification, (3) an improvement strategy, (4) experimental results, and (5) problems and opportunities for research.

DEVELOPMENT OF A RESEARCH BASE

ILCA researchers began by studying agricultural activities in Ada District around the town of Debre Zeit. First they reviewed the literature on the area and found that most of the reports documented conditions before the government's land reform program. Because that program altered farming conditions, the researchers began a series of field surveys. These surveys were to collect baseline data on 151 households whose members belonged to 21 farmers' associations.

After ILCA's research and testing program started, the researchers began surveying 42 nonparticipating farmers. At least weekly, ILCA's field staff interviewed different subsamples of these farmers, measured crop conditions, and surveyed market conditions for livestock and agricultural products.

PROBLEM IDENTIFICATION

ILCA's survey of the research area revealed low food crop and livestock production efficiency. On a typical mixed crop and livestock farm of about 2.5 hectares, the household's subsistence required 75 percent of the land. Furthermore, the researchers found family food supplies were low and animals had low genetic potential. Farmers fed crop residues to their cattle, which received only 60 percent of their nutritional requirements. This low feed intake further reduced animal productivity.

AN IMPROVEMENT STRATEGY

In response to these problems, ILCA's researchers introduced and tested an improvement package on 20 small farms near Debre Zeit. The strategy behind the package was to

1) substantially increase the yields of subsistence crops
2) use the land saved from subsistence crops for planting new forage crops
3) use forage to feed genetically improved dairy cattle.

Success of the forage-dairy component of this development depended on positive responses to these questions:

1) Is there a market for dairy products?
2) Is additional family labor available?
3) Can the forage-dairy enterprises give better returns than food crops on a per hectare basis?

To answer the first question, an ILCA team surveyed Addis Ababa's dairy market. The survey showed (1) the urban area's demand for fresh milk, cooking butter, and milk curd could absorb a moderate increase in milk production and (2) moving milk to the urban area presented no problem because Debre Zeit falls within the 120-km radius of the government's milk collection program.

In answer to the second question, the survey teams found family labor was (1) insufficient during harvest of teff *(Eragrostis tef)* and other food crops in November and, to a lesser extent, when the crops must be weeded in September and (2) underutilized during the rest of the year. In response to this situation, ILCA's researchers proposed a dairy improvement program based on stall feeding. Such a program requires minimal labor throughout the year, and women and children can supply that labor.

The answer to the question of whether the dairy-fodder enterprises can compete economically with food crop production had to wait for the results of the on-farm tests.

In describing this program ILCA (1980) reported

> "The main elements were the use of improved seeds and fertilizer for teff and wheat crops with supporting technical advice, for example on the timing of planting, together with the introduction of forage oats and vetch and the purchase of one in-calf Friesian × Borana crossbred heifer from a government breeding ranch."

The farmers participating in the test (1) could buy all required inputs on credit, (2) received advice on how to handle the new enterprises and technologies, and (3) could select only the parts of the improvement package they wanted to try.

EXPERIMENTAL RESULTS

The results of the 20 farmers participating in testing the improvement package are now available for 1977 through 1979. ILCA researchers compared those results with the results of the frequent surveys of the 42 nonparticipating farmers. These farmers have about the same size farms as do the participating farmers. However, the nonparticipating farmers' lands are slightly more favorable in terms of soil fertility and drainage than the lands of the participating farmers.

For 1979, the participating farmers' yields generally declined from 1978 because of lower rainfall and from not using fertilizers (Table 7-F-1). Despite this decline, the

Table 7-F-1. Comparison of average crop yields of farmers testing and not testing the improvement package at Debre Zeit, 1977-1979 (Adapted from ILCA, 1980).

	Participating Farmers (n = 20)			*Nonparticipating Farmers (n = 42)*		
	1977	*1978*	*1979*	*1977*	*1978*	*1979*
	Yields in kg/ha					
Teff	1154	1524	1289	772	984	932
Wheat	627	1209	766	470	850	407
Horse beans	970	874	938	790	878	769
Chickpeas	650	304	182	620	585	420
Forage crops	2900	5900	3500	--	--	--

participating farmers had higher yields than the nonparticipating farmers. This was particularly true for teff, the major crop that was raised on 50 percent or more of each farmer's cropland. Since growing forage crops is part of the improvement package, nonparticipating farmers did not raise these crops.

When compared to nonparticipating farmers, the participating farmers produced about 30 percent more per hectare for food crops. On the other hand, the increases for forage crops were not as high as for teff, but comparable to the increases for other food crops. Furthermore, the potential increase in teff production is limited because of seasonal labor constraints and crop rotation requirements.

ILCA (1980) describes the economic results as follows:

> The "gross margin for dairying [was] US$377.0 in 1978 and US$325.0 in 1979. To this should be added the gross margin for food crop production of US$721.5 per holding in 1978 and US$631.0 in 1979, giving an overall gross margin of US$1098.5 in 1978 and US$956.0 in 1979 for participating farmers. These figures may be converted to gross margins of US$435.5 per ha in 1978 and US$409.5 in 1979, compared with average gross margins of US$317.5 per ha in 1978 and US$259.0 in 1979 for non-participating farmers. This comparison implies an increase in the value of production of 37% in 1978 and [58%] in 1979 attributable to ILCA's innovation package."

The inclusion of milk and fodder production has other favorable effects on the farming system such as

1) Increased fodder production permits farmers to continue keeping cattle, even though grazing areas are declining because of population pressures.
2) Farmers use cattle in the traditional farming system for tractive power, as a source of manure, and for milk and meat. Farmers said that they were glad to have milk for consumption and sale.
3) Livestock ownership gives farmers some security against crop failures.
4) The crops used for fodder production help control weeds and thus improve subsequent crops.

Participating farmers said they liked the dairy program, especially since milk yields and calf performance are greater than the original estimates. As further evidence of the program's acceptance, non-participating farmers would like to join the on-farm tests.

PROBLEMS AND OPPORTUNITIES FOR RESEARCH

Additional problems and opportunities for research emerged from the on-farm tests. Here is a partial list:

1) The farmers' tests indicated that fodder production may not be great enough to meet the household's needs. Thus, alternate ways of increasing fodder must be considered. For example, farmers might use unconventional crops for dry-season feeding, sweet potatoes as a double crop, and fodder-producing bushes on sloping marginal lands.
2) Dairy animals might be used for both milk production and traction.
3) Flooding and waterlogging of farm lowlands are now serious problems. If these hazards could be eliminated, green matter production could be expanded in the dry season, weeds could be better controlled, and fertilizer could be applied more effectively.
4) Farm uplands are now low in fertility. The dairy animals could provide manure to improve soil fertility.

Following the 1980 harvest, sufficient results might be available from these Debre Zeit tests to begin multilocational tests and eventually production programs in the Ethiopian highlands. The ILCA team has identified institutional constraints that need to be overcome if the improvement package is to be introduced more widely. With regard to the dairy enterprise the team found that

1) Milk production from crossbred cows on participating farms dropped in 1979 relative to 1978. The drop was attributed to delays in servicing the cows, which caused calving intervals and dry periods to be extended. Because the artifical insemination service was unreliable, a bull was purchased for the station.
2) Forage seed supplies were lacking; the supply of concentrated feeds was irregular and unreliable; and a shortage of cross-breed dairy heifers existed.
3) Although a government collection system provided dairy farmers within a 120-km radius of Addis Ababa with a marketing outlet, dairying in other parts of Ethiopia cannot be expanded until an adequate marketing and transportation system is developed.

APPENDIX 7-G ESTIMATING NET BENEFITS FROM ALTERNATIVE TREATMENTS

Estimating net benefits from alternative treatments is both important and sometimes difficult to understand. Consequently, we provide this appendix to explain some of the concepts discussed under Net Benefits in Sec. 7.7.3. in Chapter 7. We begin with a section that explains our use of the field price as a basis for estimating the value of output. Next, we discuss assumptions underlying partial budget analysis. The last section explains, in more detail, than in Sec. 7.7.3., the approach for estimating the opportunity costs of the family's consumption of its output. As in Chapter 7, we will illustrate these concepts using the example of alternative fertilizer applications for increasing maize productivity.

VALUES BASED ON FIELD PRICES

When estimating the net benefits of alternative treatments, we must be careful to be consistent in the way we include or exclude the associated benefits and costs. Analysts have choices to make in both the level of detail for the estimates and the point at which ther farmers' output is valued. Perrin et al. (1976) and we, in Part 2 of Chapter 7, have chosen to value the farmers' output in terms of the field price. From Perrin et al. (1976) the field price of the output is the "value to the farmer of an additional unit of production in the field, *prior* to harvest." Assuming the output is sold for money, the money field price is "the market price of the product minus harvest, storage, transportation and marketing costs, and quality discounts." Should farmers turn over their crops for custom–i.e., hired–harvesting, then the field is the logical place to estimate the benefits and costs of producing the crop. The benefit is the crop's value in the field and the costs are those associated with the farmers' efforts in bringing the crop to the point where it is ready for harvest.

However, farmers more frequently harvest their own crops. Then, the more logical approach would be to use the value of the harvested crop, which means taking into account the added costs of harvesting and any harvesting and storage losses until the grain is sold or consumed by the family. The appropriate value of the crop would then be what the farmer could receive if a trader purchased the grain at the farmers' gate–i.e., the farm gate price.

Still another possibility is if the farmer were to take the grain to the local village for sale. The value normally increases, reflecting the costs of storage, transportation, and selling. In this case, the value of the crop is the market price.

To summarize, we have given three alternative points at which the crop could be valued: the farmers' field, the farm gate, and the local market. Since prices should increase as the product is moved from the field to the local market, so do the costs incurred by the farmers. Should the costs to the farmers for their time, effort, and purchase of materials increase in approximately the same amounts as the value of their output–which is often a reasonable assumption–then we should expect the farmers to be largely indifferent whether they sold the field crop to a custom harvester, sold the harvested crop at the farm gate, or moved the crop to the local market for sale. Moreover, for analysis purposes–especially for production oriented research–the field price is a convenient point for analysis.

PARTIAL BUDGET ANALYSIS

In using partial budget analysis, the reader should be aware of three basic principles. First, the analysis assumes that the farmers' current position is acceptable to them; otherwise, they would change. The farmers' situation may not be good by others' standards, but we can assume that the farmers will continue as they are as long as better alternatives are not available.

Second, an alternative that produces a positive net benefit–i.e., additional benefits greater than additional costs–means that the farmers will have an incentive to change. If they do not, then the incentive is not great enough to offset the disadvantages of change.

Third, only the differences among the alternatives count, which means that costs and benefits common to all alternatives can be ignored in choosing among the alternatives. The ability to concentrate on only the differences among the alternatives helps the analysts focus on the most important aspects of the technologies.

These three principles underlie the analysis procedures described in Part 2 of Chapter 7. Thus, we compared alternative treatments with the farmers' current methods of production. If the incremental net benefit was greater than the incremental variable cost, we considered the new treatment as being better than the farmers' method–but not necessarily enough better to interest the farmer in changing. In estimating the marginal rates of return, we used only the values of the increased yields and the costs of fertilizer and labor for its application. We were able to ignore the other costs of production because they were the same. Finally, we eliminated differences in harvesting costs by assuming that the value of the crop standing in the field was a suitable basis for estimating the crop's value to the farmers.

OPPORTUNITY COSTS

In developing countries, many of the small farmers' activities do not involve cash transactions. For this reason, measuring the opportunity cost of the farmers' inputs and outputs becomes important. While many of the farmers' inputs are often not traded and could be used to illustrate how to estimate an input's opportunity cost, we have applied the concept only to the family's labor (Part 2 of Chapter 7). Moreover, the outputs of the farm such as grains, livestock, and household products are often not traded in the market, but they certainly have value. Estimates of the value of farmers' consumption is therefore also important.

Following are two examples that illustrate the basis for estimating the opportunity costs of the farmers' maize consumption. These two illustrations are based on the maize example from Part 2, Chapter 7. Suppose a farm household normally has the following annual maize activities:

Consumption	800 kg
Production	600
Purchases	200 kg

Also, suppose an improved treatment increases the household's annual production by 200 kg. Now, the farmers do not have to go to the market to purchase maize, nor incur the costs–either money or farmers' efforts–to bring the maize home. These two costs, which the farmers no longer have to pay because of their increased production, are benefits from the new technology. Using \$1,200/ton (i.e., \$1.20/kg) as the *market* price for maize and assuming transportation costs of 20¢/kg., the farmer saves $280/yr—i.e., 200 kg/yr × $1.40/kg.[1]

On the other hand, in Chapter 7 we valued the farmers' maize crop as standing in the field. To correctly compare the value of the maize that the farmers purchase in the market with the value of the increased output from the new treatment, we must subtract the cost of harvesting from these estimated savings. If harvesting costs 10¢/kg., the farmers will spend an extra $20/yr (200 kg/yr × 10¢/kg) for the increased production. Consequently, the value of maize to the farmers—which we label as the opportunity cost of the farmers' consumption—is $260/yr ($280/yr savings in not having to buy and transport the maize less $20/yr for having to harvest). This value can then serve in place of the gross field benefit for arriving at the net benefit to the farmer, using terms from Table 7-4 in Chapter 7.

The other situation associated with the farmers' consumption of maize occurs when the farmers are already selling maize and decide to increase their maize consumption because of the increased output from using fertilizers. Because limited income restricts even consumption of basic foodstuffs, an increase in productivity can very likely encourage the family to consume more maize. Below is an illustration of a family's annual maize consumption with and without fertilizers.

	Without fertilizers	*With fertilizers*
Production	900 kg	1,100 kg
Consumption	800	900
Sales	100 kg	200 kg

In this illustration the household is already meeting its consumption requirements and has a surplus for sale. The 200 kg annual increase from using fertilizers results in increased sales of 100 kg and increased consumption of 100 kg.

The extra sales of 100 kg can be handled as normal money transactions. The extra consumption of 100 kg has an opportunity cost, which we can value as the field price of maize. Clearly, if the family did not consume the maize, it would be available for sale. Using a field value of $1,000/ton ($1/kg), the family is foregoing $100/yr (100 kg/yr × $1/kg) by not selling the maize. That is, the opportunity cost of the maize consumed is $100/yr. In this illustration, we have used the field price of maize in making the calculations, thereby assuming that the increase in the value of the harvested grain equals the cost of harvesting.

APPENDIX 7-H ANALYSIS OF CROPPING PATTERN RESEARCH IN INDONESIA

In Sec. 7.8.1. in Chapter 7, we discussed the analysis of farmer-managed tests. In this appendix, we provide an example of how the Central Research Institute for Agriculture (CRIA) in Indonesia carries out analyses for several cropping patterns in partially irrigated and upland areas. The analyses include (1) comparisons of experimental yields and estimated economic returns from both the farmers' traditional patterns and alternative patterns for farmers in partially irrigated areas, (2) calculations of average yields of crops and approximate net returns for check plots and full treatment plots, and (3) estimates of calories and protein produced per hectare from year-round cropping patterns with no fertilizer treatment and with full fertilizer treatment.

These examples are from CRIA's research in Central Lampung, where rice is grown each year in the partially irrigated lowlands near Nambah Dadi village and where rice and cassava are grown throughout the year in upland areas. The rainfall and cropping patterns are show in Fig. 7-H-1. The accompanying text is excerpted from McIntosh (1980):

"PARTIALLY IRRIGATED AREA

"Our first attempts to improve the cropping patterns in this area consisted of introducing earlier maturing rice varieties, reducing turn-around time between planting of rice crops in the wet season and growing secondary crops during the dry season. We found that the supply of irrigation water was less dependable than anticipated and that we could not save enough time to get two lowland rice crops per season. But there appeared to be potential for growing legume crops during the dry season after the rice crops. We had tried gogo rancah (direct seeding of rice on aerobic soil that will be flooded later) in some small plots and found that even though the rats and birds ate the rice because it was out of phase with the surrounding crops, the vigor of the crop of rice (gogo rancah) appeared good. The soil was considerably different from the alluvial clay soils of Indramayu. . . , but other factors made the area suitable for this method of rice culture. First of all, the rainfall gradually increases over time (more than two months) to reach a peak suitable for flooding and rainfed lowland rice culture [see Fig. 7-H-1]. There was sufficient rainfall to permit establishment of an upland crop. However, if the rainfall is high, aerobic conditions to permit continued cultivation of upland crops is possible only if aerobic conditions are provided by extensive drainage facilities. The gogo rancah rice, on the other hand, can be directly seeded on aerobic soil and then flooded. The usual benefits of flooding for control of weeds and improved nutrient availability result. Generally, in well drained soils, such as that existing in Nambah Dadi, excessive drainage hinders gogo rancah rice production. Furthermore, since the soil is not puddled when this cultural method is used, downward per-

[1]Note: The dollar sign represents a monetary unit, not necessarily the U.S. dollar.

Figure 7-H-1. Working calendar of farmers' and pre-production (introduced) cropping patterns. Central Lampung, Lampung. 1977–1978 (Adapted from McIntosh, 1980).

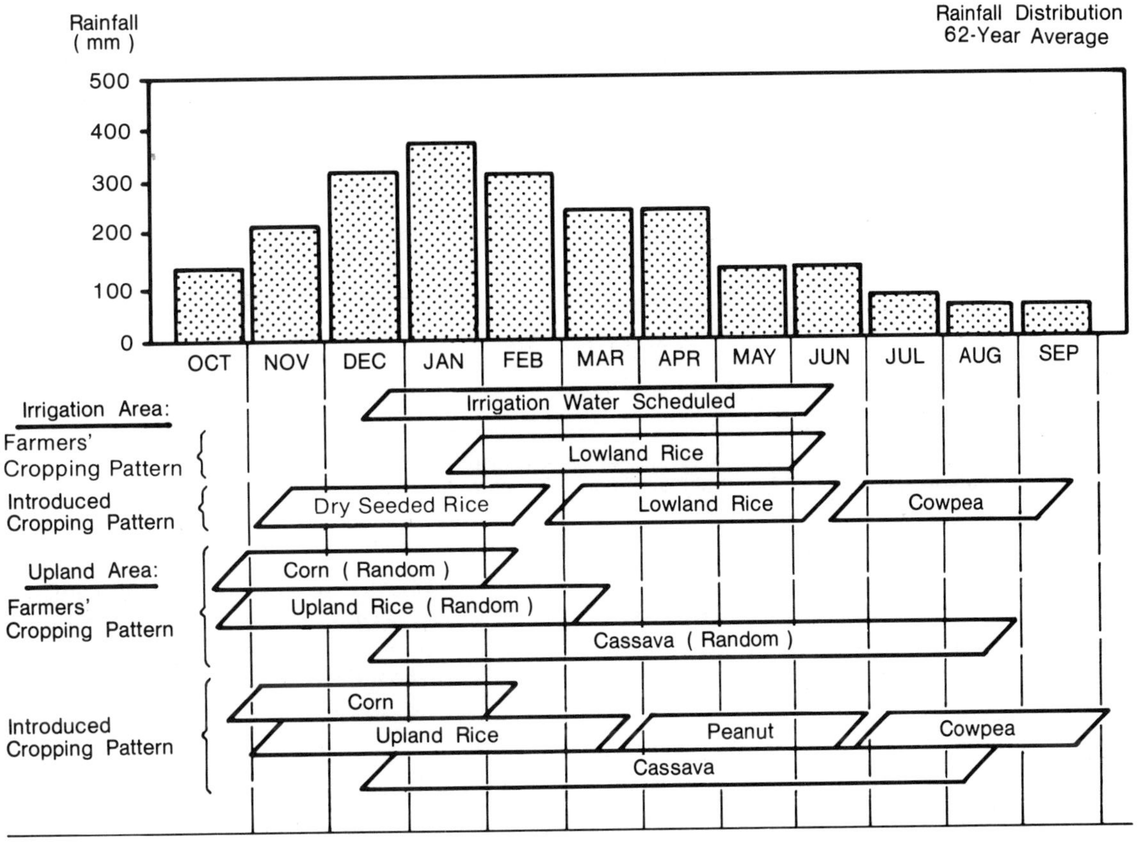

Table 7-H-1. Comparisons of yields and economic returns from farmers' and introduced cropping patterns in pre-production trials in partially irrigated area with 3 hectare plots. CRIA C.S. Proj., Nambah Dadi, Central Lampung. 1977-78 (McIntosh, 1980).

*Cropping Patterns**	*Yields and Cash Surplus†*					
	Yield	*Gross Returns*	*Material Costs*	*Hired Labor Cost*		*Cash Surplus*
	kg/ha			*RP/ha*		
Farmers' C.P.:						
Lowland Rice-	3,743	190,434	24,670	73,537		92,227
Fallow	--	--	--	--		--
				Total	Rp	92,227
Introd. C.P.:						
Gogo Rancah Rice-	5,562	382,387	44,450	128,177		209,760
Walik Jerami Rice-	2,873	146,523	35,300	61,405		49,818
Cowpea	679	74,690	7,300	6,000		61,390
				Total	Rp	320,968

*Farmers' and Introduced Patterns are diagramed in Fig. [7-H-1]. Gogo Rancah Rice is rice directly seeded on aerobic soil at beginning of rainy season. Later it is flooded.

†Yields were measured by sampling from farmers' fields and from within the 3 hectares of contiguous plot area. Cash surplus means Gross Returns minus cash costs of Materials and Labor only.

Table 7-H-2. Average yield of crops and approximate net returns for check and full treatment plots. CRIA C. S. Proj., Bandarjaya, Central Lampung. 1973-74* (McIntosh, 1980).

Fertility Treatment	Corn	Upland rice	Peanut	Rice bean	Cassava Fresh root (ton/ha)	Approximate net return Rp/ha†
	Dry grain-kg/ha					
			Mixed Cropping			
Check	467	690	161	55	12.7	65,000
Full + mulch	1,165	1,358	356	248	28.3	132,000
			Intercropping			
Check	455	769	222	93	14.6	91,000
Full + mulch	1,350	2,724	567	627	23.2	265,000
			Sequential Planting			
Check	606	850	—	153	—	(-6,000)
Full + mulch	2,935	3,536	—	723	—	74,000

*Yields subsequently have varied due to pests and management but have remained basically the same.
†One (1) U.S. dollar equivalent to Rp. 415 until November 1977; thereafter, approximately Rp. 620 per dollar.

colation of water and leaching of plant nutrients are excessive. Fortunately, because of the level topography of the partially-irrigated rice fields and the gradual raising of the water table with irrigation water and rainfall within the system, this problem did not occur. [Table 7-H-1] shows the yields and economic returns from farmers' and introduced cropping patterns in this partially-irrigated area when the gogo rancah technology was used in the improved pattern. The plots were about 3 hectares in size and included fields of several farmers. This preproduction trial was sufficient evidence to many farmers to spontaneously adopt this technology. Thus, the technology was transferred from Java successfully and within the Nambah Dadi area. Considerable research effort and time were saved.

"UPLAND AREAS

"The other two sites were located in upland fields that were being newly opened (Komering Putih) and that had been opened several years before but allowed to revert to *Imperata* (Bandar Agung). [Fig. 7-H-1] shows the predominating cropping pattern used by the farmers. The farmers' patterns appeared to be well adapted to the existing soil, climatic, and market conditions. But the probability of increasing production by use of fertilizer, improved management practices (planting in rows) and introduction of more legumes into the systems appeared to be good [Fig. 7-H-1]. Research had begun in 1973 to evaluate these ideas. [Tables 7-H-2 and 7-H-3] illustrate some of the dramatic results that were obtained. Yield and net return data show the advantage of planting in rows compared with random planting as practiced by the farmers. This was especially important for crops other than cassava when fertilizers were used. The data also show that the practice of farmers to grow combinations of crops was more productive and profitable than growing crops separately in sequence. Most of all these data show the importance of improving soil fertility through use of fertilizer. This is further illustrated in [Table 7-H-3] where the yield data for the intercropping pattern are expressed in terms of calories, protein and paddy rice (gabah equivalent). Thus, if properly managed, the total production per hectare on these nonirrigated and underutilized uplands can exceed that of the irrigated and fertile lowland areas of Java."

Table 7-H-3. Calories and protein produced per hectare from year-round cropping patterns with no and full fertilizer treatments. CRIA C. S. Proj., Bandarjaya, Central Lampung. 1973-74 (McIntosh, 1980).

Cropping Pattern	No Treatment Yield kg/ha	No Treatment Calories K cal/ha	No Treatment Protein kg/ha	Full Treatment Yield kg/ha	Full Treatment Calories K cal/ha	Full Treatment Protein kg/ha
Corn +	455	1,615	42	1,350	4,792	124
Rice /	769	1,840	52	2,724	6,521	185
Cassava /	14,600	17,520	102	23,200	27,840	162
Peanut -	222	1,003	51	567	2,563	145
Rice bean	93	308	23	627	2,075	157
Total		22,286	270		43,791	773
Gabah* equivalent kg/ha/year		9,325	4,060		18,323	11,371

*Gabah × 0.665 = milled rice and average value of 6.8% protein used for conversion from protein to gabah.

APPENDIXES TO CHAPTER 8

APPENDIX 8-A
MEMORANDUM OF AGREEMENT TO ESTABLISH A PILOT PRODUCTION PROGRAM IN THE PHILIPPINES

In Sec. 8.5, Chapter 8, we discussed the use of pilot production programs as a means for learning if new technologies should be extended to a regional or even national scale. The approach requires the participation of many government and private groups, as well as the FSR&D teams and local farmers. Below is a 1976 memorandum of agreement among the relevant groups in the Philippines. This memorandum, taken from Haws and Dilag (1980), illustrates the types of organizations that are parties to the agreement as well as their individual responsibilities.

"MEMORANDUM OF AGREEMENT

"KNOW ALL MEN BY THESE PRESENTS:

"This Memorandum of Agreement entered into by and among the different government agencies and instrumentalities, represented by their respective regional Directors/Provincial Heads/ Managers with full authority to do so:

1. IRRI-PCARR-The International Rice Research Institute - Philippine Council for Agriculture and Resources Research (Rainfed Rice Projects)
2. UPLB-NFAC [University of the Philippines at Los Banos-National Food and Agriculture Council] (National Multiple Cropping Program)
3. National Food and Agriculture Council
4. Bureau of Agricultural Extension [BAEx]
5. Bureau of Plant Industry [BPI]
6. Dept. of Agrarian Reform [DAR]
7. Bureau of Soils
8. National Grains Authority [NGA]
9. Philippine National Bank [PNB]
10. Rural Bank [RB] of Sta. Barbara
11. Agricultural Credit Administration [ACA]
12. Area Marketing Cooperative [AMC] at Iloilo

"WITNESSETH: THAT

"WHEREAS, a pilot extension project on cropping systems on Rainfed Rice with designation 'KABUSUGAN SA KAUMAHAN' - 'KABSAKA-SALUD-ULAN' will be established at Sta. Barbara, Iloilo covering 500 hectares, more or less;

"WHEREAS, the said project will necessarily need the support and assistance of all the above-listed agencies in the performance of activities appropriate to their respective functions;

"NOW, THEREFORE, for and in consideration of the foregoing premises, the above-listed agencies, through their respective representatives, hereby agree to assume and undertake in the said project their respective functions and responsibilities, to wit:

1. IRRI-PCARR (Rainfed Rice Projects) shall
   a. Provide technology on direct seeding
   b. Help oversee implementation of technology
   c. Help train the management and technical group
   d. Provide four (4) motorcycles (IRRI)
2. UPLB-NFAC (National Multiple Cropping Program) shall
   a. Help oversee implementation of technology
   b. Help train the management and technical group
3. NFAC shall
   a. Provide funds for training and operational expenses
   b. Act as liaison between the agencies herein
   c. Monitor progress of the project
4. BAEx shall
   a. Undertake information drive in the project
   b. Assign technical personnel to the project
   c. Supervise the farmers involved in the project regarding technology
   d. Prepare progress and other reports
   e. Provide technical supervision of personnel involved in the project
5. BPI shall
   a. Provide technology or production technology personnel

b. Make available to the project recommended certified seeds
c. Undertake plant pest control as needed

6. DAR shall
   a. Assist in the organization of Samahang Nayon [cooperative] and Compact Farms
   b. Undertake information drive on land consolidation
   c. Train the farmers in the project on their responsibilities as far as Agrarian Reform is concerned
7. NGA shall
   a. Undertake the procurement of rice, corn, sorghum, soybeans and yellow corn
   b. Provide up-to-date market information
8. The RB of Sta. Barbara, PNB and ACA shall
   a. Provide production credit to the project which shall include agricultural inputs and chemicals under the supervised credit system
   b. Undertake collection of loans extended with the assistance of representatives of all the agencies involved
9. AMC shall
   a. Wage information drive on the organization of agricultural cooperatives
   b. Extend production inputs to farmers in the project
   c. Provide free service/delivery of farm inputs to the project area
   d. Arrange custom plowing for the project farmers
   e. Assist in the gathering of harvest of the project for storage and drying purposes
   f. Assist in the collection of loans extended by RB, PNB, and ACA
   g. Provide containers for the produce of the project
10. Bureau of Soils shall
   a. Provide soil analysis to the farmers in the project free of charge
   b. Supervise collection of soil samples in the project area
   c. Provide fertilizer and liming recommendations
   d. Update soil fertility recommendations for project area
   e. Provide information and technical assistance related to soils in the project.

"This MEMORANDUM OF AGREEMENT shall take effect immediately upon approval by the parties hereto and shall remain in force for the duration that the project is in existence and until the project can undertake or handle by itself what the agencies - parties hereto are bound to undertake for the project unless sooner terminated by the parties hereto to take effect thirty days after due notice.

"IN WITNESS WHEREOF, the parties hereto have hereunto signed this Agreement this *23rd* day of November, 1976 at Iloilo City."

APPENDIX 8-B
LETTER OF UNDERSTANDING BETWEEN ICTA AND DIGESA

Our discussion in Sec. 8.6.1., Chapter 8, mentioned the desirability of close association between research and extension. This appendix contains a letter of agreement that proposes to bring these two activities closer together. The agreement between ICTA (research) and DIGESA (extension) in Guatemala was signed on February 8, 1978 (translated by the Consortium for International Development; ICTA and DIGESA, 1978).

"1. REASONS THAT JUSTIFY THE COORDINATION:

"Furthering the national development programs creates the need for those institutions that hold complementary objectives to coordinate their actions and to mutually support and/or mutually strengthen their activities and work programs.

"Both institutions agree that agricultural research, promotion, and training must be integrated into a single effort so that the production alternatives investigated become the technologies used by the farmers.

"ICTA, for its part, conducts research and prepares agricultural alternatives validated under the ecological, social, and economic conditions of the different regions of the country; DIGESA has established a promotional and training program, whose main purpose is to provide services that offer alternative solutions to farmers' problems. The principal role of ICTA is to generate and test technology for transfer to DIGESA and other groups; the role of DIGESA is to transfer technology to the farmers through technical assistance.

"To generate and identify better production alternatives than the ones known and used by farmers, ICTA counts on a group of professionals and technicians at different levels, grouped in support programs and disciplines whose primary objectives are generating, testing, and promoting the use of science and agricultural technology. To reach this goal, ICTA has designed a strategy for carrying out agro-socioeconomic studies to identify the limiting factors in a given production area. Based on this information, ICTA conducts experiments at the Production Centers to generate and evaluate technology that provides solutions for the identified problems and the limiting factors. To make the results consistent, Farm Trials are carried out on strategically selected farms to measure the variability of different environments within a given region. The results so obtained are then subjected to Farmers' Tests to evaluate the generated technology when farmers manage the technologies under their own conditions, which minimizes the risks of implementing new production systems. The technology generated in this way is then transferred to DIGESA for large-scale implementation by the area's farmers.

"To carry out this technology transfer effectively, DIGESA is developing a program of technical assistance staffed by a group of technicians operating within the Regional Units and the Technical Organizations. The 'Production Program for Farmer Education' is divided into six phases, each of which includes a number of projects. The phases are: Motivation (that consists of training children and young people in rural areas), Formation (that refers to the training of adult farmers and their wives), Promotion (that includes technical assistance and supervised credit), Follow-Up (that consists of sustained and prolonged promotion), Support (that includes activities that help production) and Control (that, as its name indicates, refers to the establishment of production standards).

"Thus, coordination will make the activities of both institutions more effective. On one hand, DIGESA will collaborate with ICTA by feeding back possible problems that become apparent during the technology transfer stage. On the other hand, ICTA will collaborate with DIGESA by providing technology that farmers can apply and that has been fully validated under the ecological and economic conditions of the region in quesiton.

"Based on all that has been said, it is fully justified that these institutions, whose interests and actions are oriented to benefit the farmer, should integrate their efforts and contribute resources to carry out these activities together.

"2. OBJECTIVES OF THE COORDINATION:

"2.1 General:

"To integrate efforts and contribute resources, to provide a greater number of production alternatives that have been validated under the proper ecological and social conditions, and to make more effective their transfer to and acceptance by the farmers of the country.

"2.2 Specifics:

"a) To coordinate agricultural research, promotion, and training to raise the technical level of farmers through collaboration by the technicians of both institutions.

"b) To facilitate the generation of technology by providing a greater flow of data on how farmers carry out production.

"c) To increase the number of Farmers' Tests so that more farmers participate and learn the characteristics of the technology, in a way that farmers learn to use the technology and the technology is justified through its long-term use.

"d) To incorporate in the credit assistance already offered, the technological elements generated by working with the farmers' system, the Production Center, the Farm Trials, the Farmers' Tests, promotion, and the farmer.

"e) To coordinate efforts and to contribute resources for the production of technical training materials that relate to the generation and transfer of technology.

"f) To make effective those features of the National Development Plan that are contained in the Technological Development and Human Resources programs.

"3. COMMITMENTS AND RIGHTS OF BOTH INSTITUTIONS

"a) For the better coordination of the activities that this understanding provides, both institutions will name a representative or coordinator, who will check on the growth of activities and will submit periodic reports to the directors of both institutions.

"b) ICTA and DIGESA agree and accept that all their interventions will be made exclusively through the coordinators to be named so that the technical and administrative autonomy of each is maintained. Ideally, the coordinators of ICTA and DIGESA will be familiar with their own and the other's organization.

"c) ICTA and DIGESA, through their coordinators, will select and determine the projects to be carried out jointly. These may be in agricultural research, promotion, and education. Once the development projects are determined and prepared, they will be submitted to the directors for their information and approval.

"d) According to their financial resources

and personnel, both institutions will assign resources and make efforts to carry out effectively the projects that are to be established within this understanding.

"e) According to their abilities and convenience, both institutions agree to put at the disposal of the proposed projects their personnel, housing, equipment, instruments, materials, laboratories, and documents; however, it is understood that all of these will continue to remain the property of the institutions providing them.

"f) ICTA and DIGESA agree to designate and to cover the salaries and travel expenses of the personnel involved in the projects, each for its own personnel.

"g) ICTA and DIGESA will supervise and evaluate the projects when necessary.

"h) The organizations will develop materials for farmers and technicians that facilitate the transfer of technologies adapted and approved under conditions actually encountered in the country.

"This Letter of Understanding will take effect the moment it is signed by the respective authorities and will be valid indefinitely. However, on three months' notice, either institution can terminate or modify the letter with the consent of the other institution."

APPENDIXES TO CHAPTER 10

APPENDIX 10-A
AN AGRICULTURAL RESEARCH PROJECT FOR THE SENEGALESE INSTITUTE FOR AGRICULTURAL RESEARCH (ISRA)

The following material, which relates to the overall organization of FSR&D (Sec. 10.2.2.), contains excerpts from a report (ISRA, 1979) prepared as a collaborative effort by scientists from the Senegalese Institute for Agricultural Research (ISRA) and consultants from the International Agricultural Development Service (IADS). The first part of this appendix, ISRA Organization and Headquarters, is taken from Annex 3 and the second part, Terms of Reference for Farming Systems Staff, is taken from Annex 11, Appendix 1.

"ISRA ORGANIZATION AND HEADQUARTERS

"As a result of earlier reviews of ISRA's organization and programs, [the Secretary of State for Scientific and Technical Research] has decided to reorient ISRA's research. Three decisions have been taken: (1) the research programs will be reorganized into commodity or factor-oriented research projects and farming systems research teams, (2) research will be decentralized and regional research stations (and programs) will be created or strengthened, and (3) ISRA headquarters will be reorganized and relocated from Dakar to St. Louis.

"A. RESEARCH PROGRAMS

"The research programs on the basic food crops, farming systems, and livestock systems are presented in Annexes 4 through 13 [not included]. In these annexes the specific locations for research are presented, suggested organizational patterns for regional research stations are made, and proposals are made for strengthening the physical facilities at some of the regional research stations.

"B. ISRA ORGANIZATION

"Administration

Fig. 10-A-1 shows a "proposed organizational chart for ISRA headquarters. The overall responsibility for ISRA would be in the Director General's hands. He would be assisted by an Assistant Director General. Technical advisors would be available to assist him in making decisions and an Administrative Director would take care of personnel, financial, and other administrative details.

"The Administrative Council would be the chief body with responsibility to oversee ISRA's overall activities. This Council exists now but does not function effectively. A strong effort should be made to revive the Administrative Council. The Committe of Direction and the Scientific and Technical Committee would be advisory bodies constituted to advise the Director General on operating policies and scientific research programs.

"Research Departments

"Under the Director General are six major Departments with line responsibility administering the research programs and supporting services. Each commodity or factor-oriented research program would be organized into a coordinated research program . . . which means that the program coordinator and the program staff would be responsible for formulating plans for research and implementing these plans. The Department Head would function primarily as a coordinator of different research programs and as an administrator. The Departments proposed are Crop Science, Animal Science, Farming Systems, Natural Resources, Economics and Sociology, and Research Support Services. . . .

"Oceanography and Fisheries, and Forestry have been grouped together under one department. This was done to minimize the number of departments and because neither of the two organizations is large enough to justify a separate department at ISRA headquarters level.

"Support Services

"Under the Department of Research Support Services are grouped several necessary services which must be available to a productive research organization.

"*a. Technology transfer and training.* Crop yields at research stations in Senegal are, in many cases, substantially higher than those obtained by farmers. This gap reflects a number of factors, in-

Figure 10-A-1. Proposed organization of ISRA headquarters (ISRA, 1979).

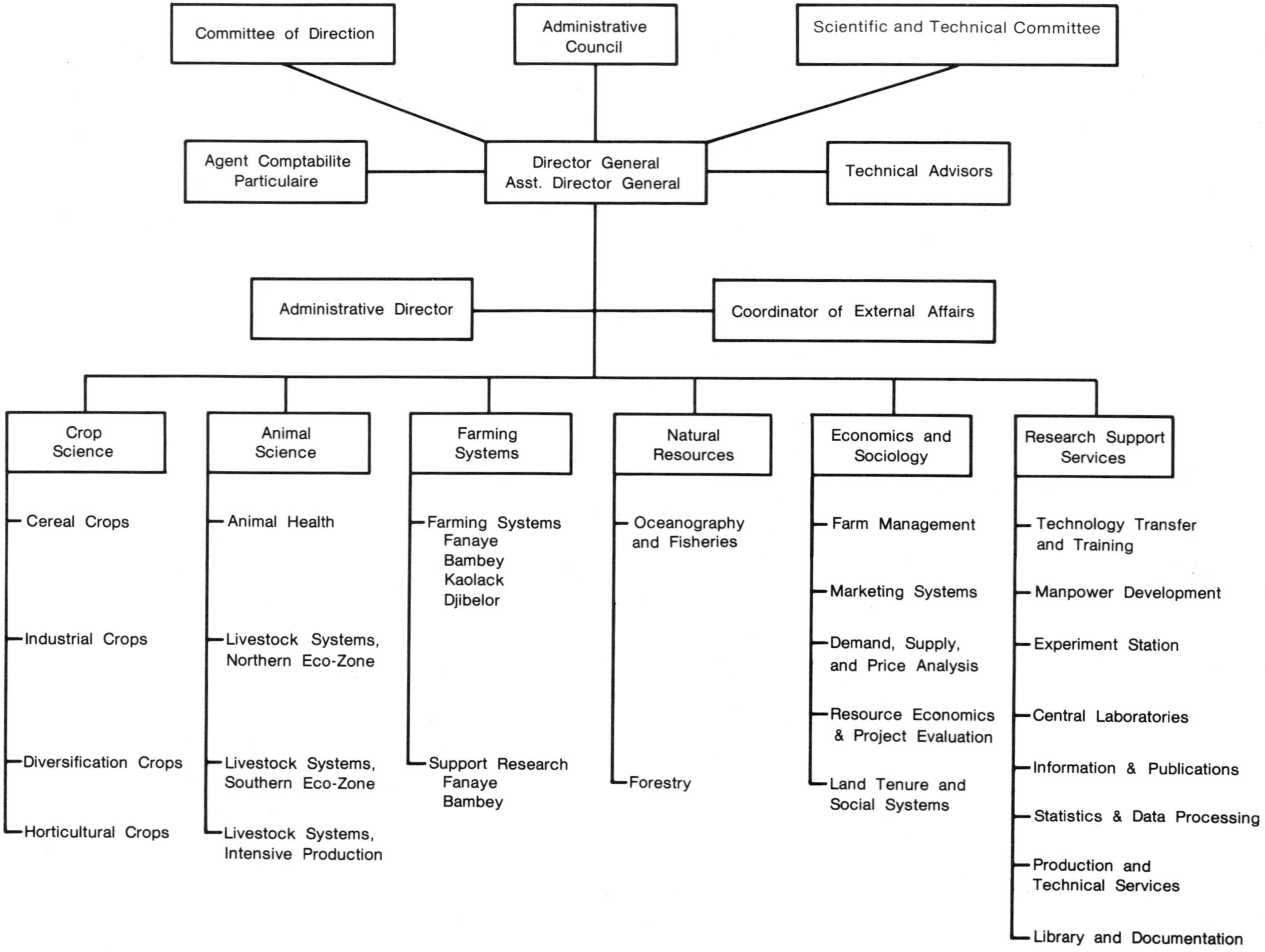

cluding lack of effective use of available technology.

"The transfer of technology from the research station to the extension organizations, farmers' association, and other groups involved in the more widespread application of new materials and practices by farmers is difficult in Senegal because of limitations in transport and communications. The production societies in the different regions must be taken into account in developing procedures for transfer of technology and keeping them functioning effectively.

"In order to ensure attention to the establishment of effective linkages between the research stations and the various production societies in different agricultural regions of Senegal, an office with a small but highly qualified staff would be set up at the ISRA headquarters.

"*b. Manpower development.* To carry out the research programs outlined in this report will require an effort to identify, recruit, train, and post the necessary Senegalese staff. An office in ISRA headquarters would be established to perform these functions and to act as general coordinator for all staff training, both outside and inside Senegal. Currently a relatively low proportion of ISRA's staff members are Senegalese which means that many national scientists must be hired and trained over the next few years. In shifting from a disciplinary to a multidisciplinary approach and in developing national research teams on a problem-oriented basis, it will be necessary to work out carefully the scientific personnel requirements and to match these requirements with training opportunities. Timing or phasing of recruitment and training is critical if these proposed research programs are to be staffed in the shortest possible time.

"A schedule should be set up for the systematic development of agricultural research manpower, including technicians. This schedule should include priorities for those disciplines of research most crucial to the acceleration of development programs.

"Personnel management procedures should be improved, and a system for evaluating and promoting research personnel on the basis of their performance in a job should be established.

"*c. Experiment station development and management.* ISRA has a number of research stations with reasonable experimental field areas, offices, and laboratories. The physical facilities—laboratories and equipment—have not been maintained properly at several of these stations, and they should be rehabilitated to an effective working level as rapidly as possible.

"The expanded scope of ISRA, together with the demands for improved technology for regional development projects, will require the establishment of new research stations and facilities, particularly in Senegal Oriental.

"Most of the research stations of ISRA should be multi-purpose in nature and provide for research on several crops, on non-commodity problems such as soil and water management, and in some cases, for integrated attention to crops and livestock. The stations, therefore, are a basic resource/service facility of ISRA that should be developed, operated, and maintained in a manner that will ensure the quality of experimentation essential for evolving dependable improved technology.

"The intensification of ISRA's major agricultural research activities and the necessary structural changes will facilitate the establishment of a research support unit that will give attention to research station development, operation, and management. . . .

"*d. Central laboratories.* In expanding and decentralizing ISRA research activities it will not be necessary to duplicate expensive laboratory installations in each region, if central research laboratories are utilized. For example, it is proposed that all soil, water, and plant analyses be done by the central laboratories at Bambey. Small, functional laboratories would be necessary at each research location but these should be relatively simple and should be confined to analyzing only the most critical elements. For example, at Fanaye it would be necessary to have available locally equipment for determining irrigation water quality and soil moisture characteristics. More detailed soil physical and chemical characteristics could be obtained by sending soil samples to Bambey. An administrative section in ISRA headquarters would be necessary to administer these services.

"*e. Information and publications.* Information and communication services and public relations must be improved throughout the ISRA system if an expanded research program is to be effective in solving problems of crop and animal production in Senegal. Good public relations is simply doing a thorough job in solving important problems and then letting the public know about it. The job of the research scientist is not completed on any project until the results are made available for the use of farmers and others concerned with food production, processing, marketing, and distribution.

"ISRA should prepare publications, visual aids, and radio and television scripts of its research activities to be used by development societies in their extension work. Moreover, scientific technical reports and the annual ISRA progress reports would be coordinated through this office.

"*f. Statistics and data processing.* In the immediate future, much of ISRA's research will be adaptive, modifying and fitting materials and practices available from national and international sources to specific locations and farming systems in Senegal. Varied trials, or experiments, with a few variables and interactions can be handled routinely by well-trained scientists.

"There will be increasingly complex experiments as the ISRA program is expanded, as more intensive studies of interacting factors involved in decline of yields in present schemes are undertaken, and as integrated crops and livestock programs are activated.

"An office with a small but highly competent staff should be set up at ISRA's headquarters to furnish guidance in the design of experiments and to provide data processing and analysis of the results of more complex experiments. Analyses of more routine experiments should be left to research team leaders and individual scientists.

"In the future it is likely that ISRA would wish to acquire its own computer facilities. If ISRA headquarters is not located in Dakar, it would be unwise to locate computer facilities in ISRA headquarters.

"g. Production and technical services. This section would handle the records and sales associated with ISRA's agricultural production, such as seed production. Also it would make the necessary contracting arrangements for all special services performed by ISRA on behalf of other organizations.

"h. Library and documentation. As with all other research organizations, a good, up-to-date library is essential to ISRA's success. Each research station should have its own working collection of the important journals and texts in that station's area of interest, but a complete library at each station is too expensive. The best solution is to create a central library which can make loans or photocopies available to other stations. Currently the principal library is at Bambey and it should remain there if ISRA's headquarters are shifted to St. Louis. Dakar is, however, the best location for a central library because it is the easiest location for all the other ISRA stations to reach. If ISRA headquarters are to remain in Dakar, the central library should be at headquarters."

"TERMS OF REFERENCE FOR FARMING SYSTEMS STAFF"

This section includes job descriptions for FSR&D team members: namely, the agronomist, entomologist, economist, sociologist, animal scientist, and subject matter specialist.

"AGRONOMIST

"This scientist should hold at least the M.Sc. degree in agronomy or soil science (crop and soil management). He should be knowledgeable about crop production methods which would entail familiarity with the crops being grown, soils, fertilizers, use of agricultural implements, and techniques for weed control. He should also have a good understanding of experimental design and basic statistics.

"The agronomists in the farming systems research program would study the current farming systems utilized by the local farmers to understand exactly what the farmer does and why he does it.

"They would conduct field experiments to determine the optimum techniques for cultivating all the crops in the farming system, both as sole crops and as components of a farming system. They would seek to find ways to modify the farming system (by adding missing inputs, for example) to make it more productive and more economically profitable to the farmer.

"After devising a better farming system, the agronomists (and the team), working closely with the subject matter specialist and other extension personnel, would conduct research in farmers' fields to verify that the experimental system was, indeed, more productive and profitable, and that it was acceptable to the farmers.

"The agronomist at Fanaye should have training in irrigated agriculture and should be able to conduct experiments involving irrigation variables, (determining critical stages for irrigation, irrigation requirements of crops, etc.).

"At least one of the agronomists in each farming systems team should be knowledgeable about herbicide materials and their use.

"ENTOMOLOGIST

"The entomologist should be trained at least to the M.Sc. level in entomology (pest management or applied entomology). He would be responsible for determining which insects were important in the farming systems and for devising methods to limit insect damage in the farmers' fields. As a component of the overall farming system, the pest management practices must be practicable and profitable.

"The entomologist would conduct surveys of insects in his region, collecting the insects, estimating their economic importance, and studying their dynamics. He would conduct research to find techniques to eliminate or limit insect damage to individual crops and the entire farming system sequence of crops. He would study naturally occurring predators, chemicals, and crop manipulations as possible techniques for reducing damage by insects.

"After developing successful experimental control measures, he would verify these techniques by applying them under farmer conditions (working closely with the subject matter specialist, other extension personnel, and farmers).

"ECONOMIST

"The economist in the farming systems team should have at least the M.Sc. degree, probably in the farm management specialty.

"Working closely with the sociologist and the

subject matter specialist, he would study the current conditions in the rural sphere to learn about the farmer and his farming system. He would try to learn what the farmer's practices are, why he uses these practices, and what the constraints are—physical, social, and economic—that limit the farmer's actions.

"The economist would study the effect on the farmer of economic factors such as price and availability of inputs (labor, fertilizers, machinery, seed, etc.), prices paid for farm products, efficiency of markets, and consumer requirements relating to quality of farm products.

"He would work with the farming systems team to assure that practices recommended to farmers were economically sound. Also he would be responsible to follow closely any adoption of new technology to determine what the social and economic effects of adoption are.

"SOCIOLOGIST

"The sociologist should have an M.Sc. degree in rural sociology, or its equivalent.

"His responsibility would be to study the social influences which shape the farmer's decisions relating to his farming system. He (and the economist) should be fully knowledgeable about the problems faced by farmers and should continually pass such information (together with suggestions of how the problem could be solved) on to the appropriate authorities in government.

"The sociologist should investigate the adoption process, focusing particularly on reasons farmers do or do not adopt recommended technological changes in their farming systems. The goal of his research would be to help design technological changes in farming systems which would be practicable, productive, and profitable and which would benefit the most deserving social element.

"ANIMAL SCIENTIST

"This scientist should hold at least an M.Sc. degree in animal science (range management) and he should have a broad understanding of animal management in grazing systems. Moreover, he should be knowledgeable about the use of agricultural by-products as feeds. This scientist should be able to work with small ruminants and mixed species, as well as cattle.

"Broadly, his responsibility would be to determine how animals fit best in a farming system. He should first study the farming system carefully to fully understand what is being done by the farmer and herdsman and the reason for their actions. He should know the constraints to increased productivity.

"Then research should be undertaken to determine better management practices whether they be related to nutrition, herd health, reproduction, or other factors. The aim would be to modify the present system to make it more productive and efficient.

"Once a modification proved to be useful at the experimental level, the animal scientist, and the farming systems team as a whole, would work with farmers and extension personnel to prove that the change was workable and profitable.

"SUBJECT MATTER SPECIALIST

"This specialist should have the equivalent of the M.Sc. degree in a field of agricultural science, preferably general agronomy, or extension education. He would be placed with the farming systems team but his responsibilities would be to act as a bridge between research and extension activities. He could carry on a limited amount of research but this should not be his major effort.

"This man would maintain a complete understanding of the research being conducted by the farming systems team, and other researchers, and he would analyze the implications of the research results for the farmers. He would prepare extension publications explaining research results for use by extension personnel in the development societies. He would organize farmer field days at the experiment station and training courses for extension personnel.

"The subject matter specialist would assist the researchers in locating cooperative farmers for placing experiments in farmers' fields. An important aspect of his job would be to keep the researchers informed of problems occurring in farmers' fields and the research needed to solve field production problems."

APPENDIX 10-B ALTERNATIVE ORGANIZATIONAL DIAGRAMS FOR FARMING SYSTEMS RESEARCH AND DEVELOPMENT

This appendix contains organizational diagrams followed by notes on four distinct approaches to implementing farming systems research and development activities. The first diagram illustrates the approach taken in Guatemala, where the Agricultural Science and Technology Institute (ICTA) was set up as a semi-autonomous research organization. ICTA has broad responsibilities and authority for agricultural research throughout the country. The second diagram illustrates the National Program for Agricultural Research in Honduras. In this case, the research program is within the ministerial structure, with the government giving considerable authority to regional directorates. The third case presents the organization of the Production Investigation Program (PIP) in Ecuador. The PIP is closely linked with experiment station activities and coordinates on-farm research with extension and agricultural credit. The last diagram represents a project approach to FSR&D. In this instance researchers are applying FSR&D methods to an on-farm water management project in Egypt, administered by the Ministry of Irrigation. (See Sec. 10.2. in Chapter 10 for additional discussion on organizational structure.)

NOTES ON THE ORGANIZATIONAL DIAGRAM FOR ICTA (GUATEMALA)

The organizational diagram shown in Fig. 10-B-1 is for the Agricultural Science and Technology Institute (ICTA) in Guatemala—a semiautonomous government corporation responsible for the country's agricultural research program aimed at the small and medium sized farmers. These notes, based on the work of Ortiz (1980), include information on the Board of Directors, the groups reporting to the Technical Production Unit, and teams within the regions.

BOARD OF DIRECTORS

The Board of Directors is the highest authority of ICTA. Its members are the Minister of Agriculture, who serves as president; the Ministers of Finance and Economics; the Secretary General of Economic Planning; the Dean of the Faculty of Agronomy at the University of San Carlos; and a representative from the private sector.

Figure 10-B-1. An organizational diagram for ICTA in Guatemala (Adapted and translated from ICTA, 1976).

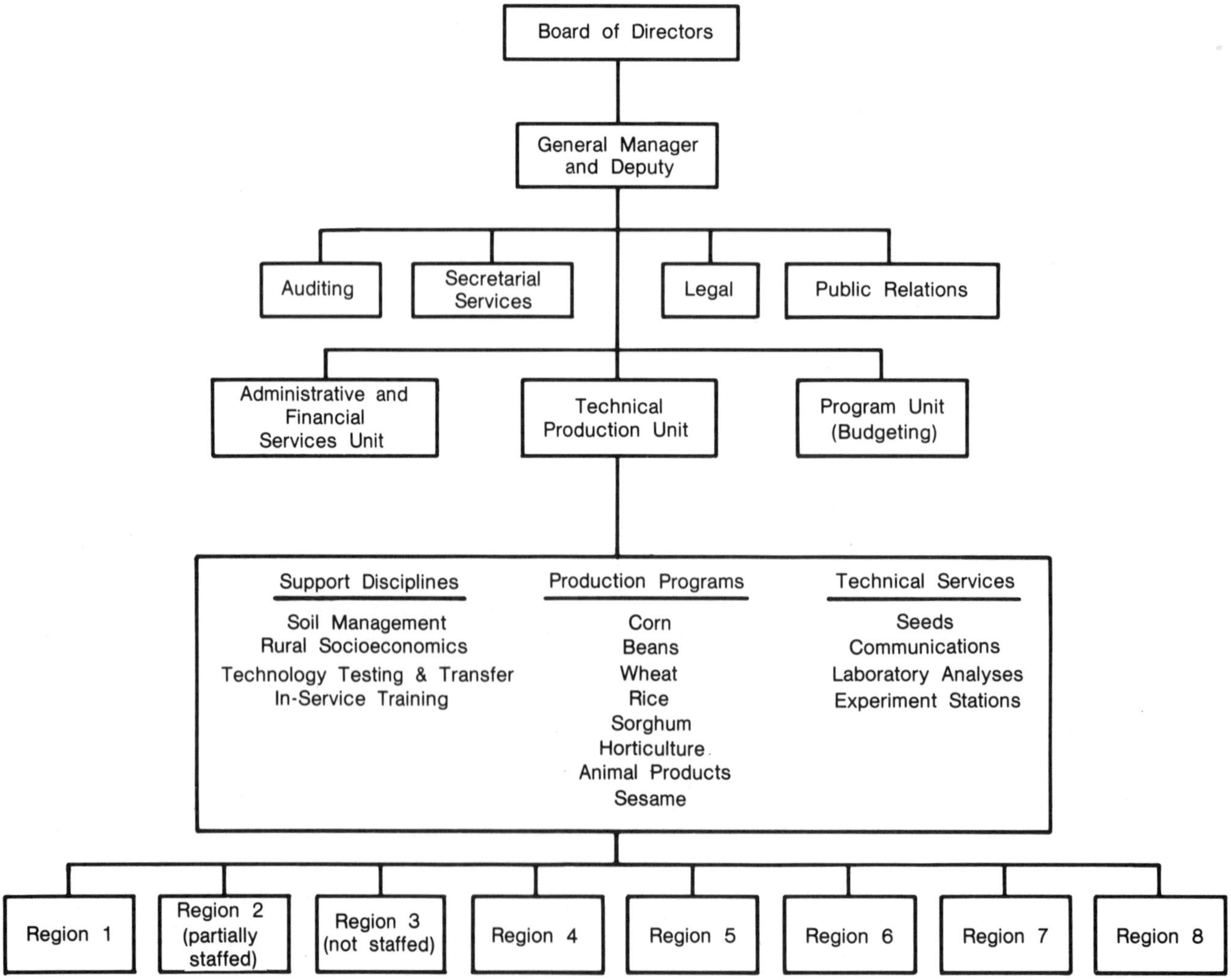

GROUPS REPORTING TO THE TECHNICAL PRODUCTION UNIT

The Production Programs, the Support Disciplines, and Technical Services are represented as fitting within a single box in Fig. 10-B-1. We present these activities in this way to emphasize the interdisciplinary teamwork needed to serve those in the regions. The Production Programs are concerned with identifying, generating, adapting, and testing new technologies for the commodities listed in the diagram. The Support Disciplines and Technical Services assist both the Production Programs and the teams in the regions. Most of the staff in the Support Disciplines are located in the regions.

REGIONAL TEAMS

Fully staffed regional teams operate in all but the second and third regions. A Regional Director, who represents the General Manager and the Technical Director, manages each team and coordinates ICTA's activities in the region. The regional teams function as interdisciplinary units and comprise all those assigned to the region, including staff from the Production Programs, Support Disciplines, and Technical Services, and the in-service trainees. Because of this approach, a technician can simultaneously be a member of a regional team and part of a program or discipline. The experiment station activities within the region also fall within the responsibility of the Regional Director.

NOTES ON THE ORGANIZATIONAL DIAGRAM FOR THE MINISTRY OF NATURAL RESOURCES (HONDURAS)

Generally, agricultural research programs in Honduras are under the General Directorate for Agricultural Operations, while agencies, such as those for coffee and forestry have specific commodity research programs. This discussion is based on a report (PNIA, 1978) by the National Program for Agricultural Research (PNIA).

PNIA is part of the Ministry of Natural Resources (Secretaría de Recursos Naturales), with headquarters in Tegucigalpa—the nation's capital. The directors of research and extension and several disciplinary specialists make the capital their headquarters, but they travel extensively and provide guidance and technical advice to the regional programs.

Seven regional directorates implement the various agricultural programs shown in Fig. 10-B-2. Regionally-based research staff, including commodity specialists, are located at the country's six research stations and form a part of the staff of the regional directorates. Most of these stations are close to the directorates, which facilitates integration of research with extension and other regional activities.

The foregoing arrangement has strong regional emphasis because the regional directorates have the administrative authority and the budgets to implement these agricultural programs. Guidance to the directorates is provided through regional advisory committees and commissions. Because of such strong control over regional programs, the head of the national agricultural research program functions primarily as a technical adviser to these regional programs and has limited power in promoting nationally integrated research.

NOTES ON THE ORGANIZATIONAL DIAGRAM FOR INIAP (ECUADOR)

The organizational diagram shown in Fig. 10-B-3 is for the farming systems activities of the Production Investigation Program (PIP), which is part of the National Institute for Agricultural Research (INIAP) in Ecuador (INIAP, undated). The experiment station is the focal point for the cropping and livestock programs and the supporting disciplines. The PIP screens technological possibilities generated on the experiment station through experiments on farmers' fields. These farmers are selected so that they are representative of specific conditions within the zone of influence of the experiment station. Those technologies found acceptable to farmers are passed on to the three implementing organizations—Projects, Extension, and Agricultural Credit—for wide-spread diffusion. If technologies are found unacceptable, they are returned to the experiment station for further consideration. Coordination between the PIP and Extension is facilitated by having an extension specialist as a member of the PIP teams. In this way, the extension staff is familiar with the technologies being generated in its region.

NOTES ON THE ORGANIZATIONAL DIAGRAM FOR THE ON-FARM WATER MANAGEMENT PROJECT (EGYPT)

The organizational diagram shown in Fig. 10-B-4 illustrates the on-farm water management project in Egypt. The project supports a relatively new activity in Egypt and is receiving substantial technical assistance from expatriate advisers. The solid lines from the Director of the Institute for Water Management represent channels of authority between that office and the Main Office and the three Field Offices. The heads of the Main Office and the Field Offices, in turn, are leaders of the various disciplinary staff listed below them in the diagram. The dotted lines represent the free flow of technical information among the disciplines located in the main office and in the field offices. Together, these teams conduct on-farm experiments and related activities on both improved resource management and improved agronomic practices. Note that because of the importance of irrigation and drainage, two engineers are members of each team. Also agricultural and social sciences, as well as extension, are represented.

Figure 10-B-2. The organizational diagram for agricultural research within the Ministry of Natural Resources in Honduras (Adapted and translated from PNIA, 1978).

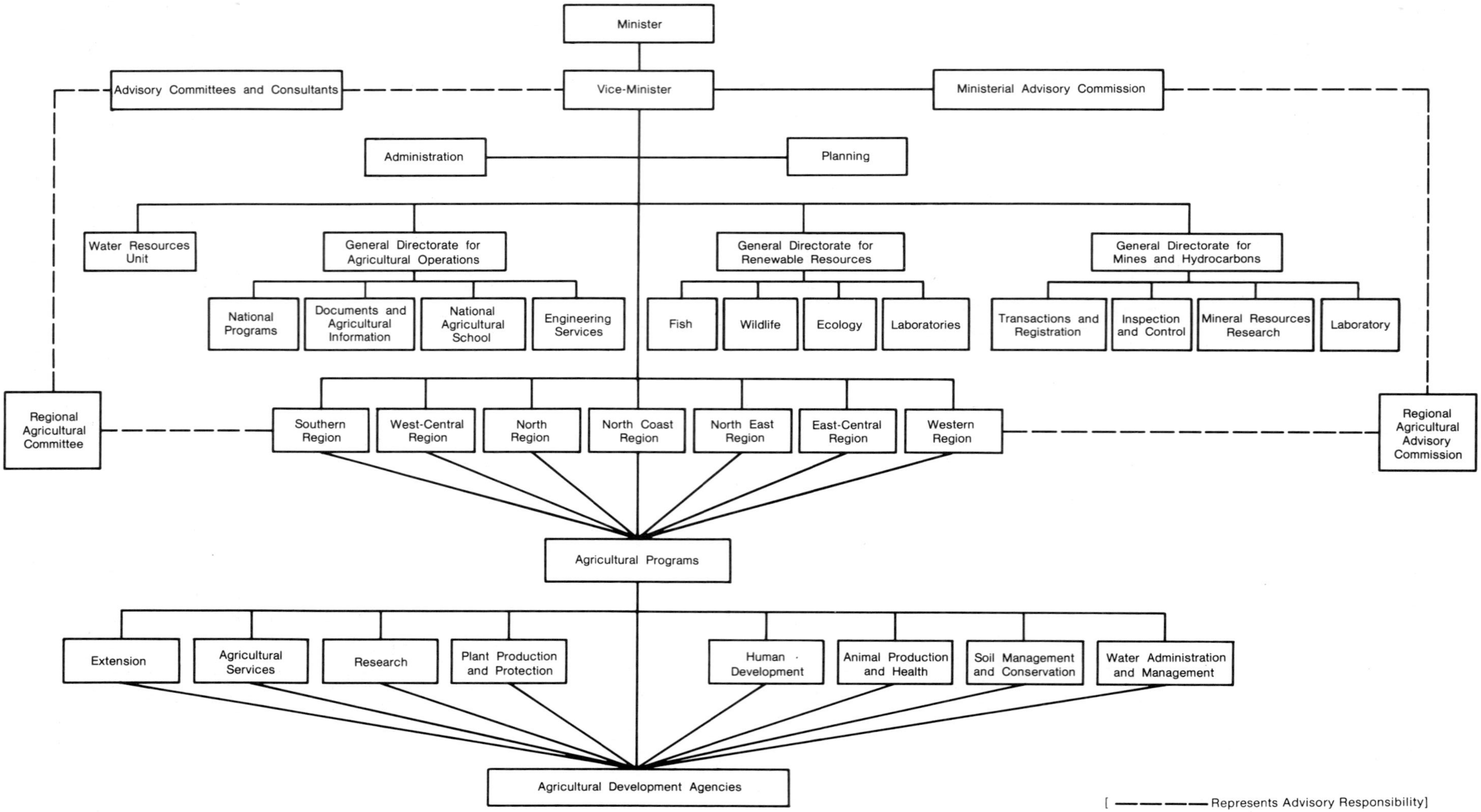

Figure 10-B-3. The structure of the Agricultural Division and functioning of the New Production Investigation Program, PIP (Translated from INIAP, undated).

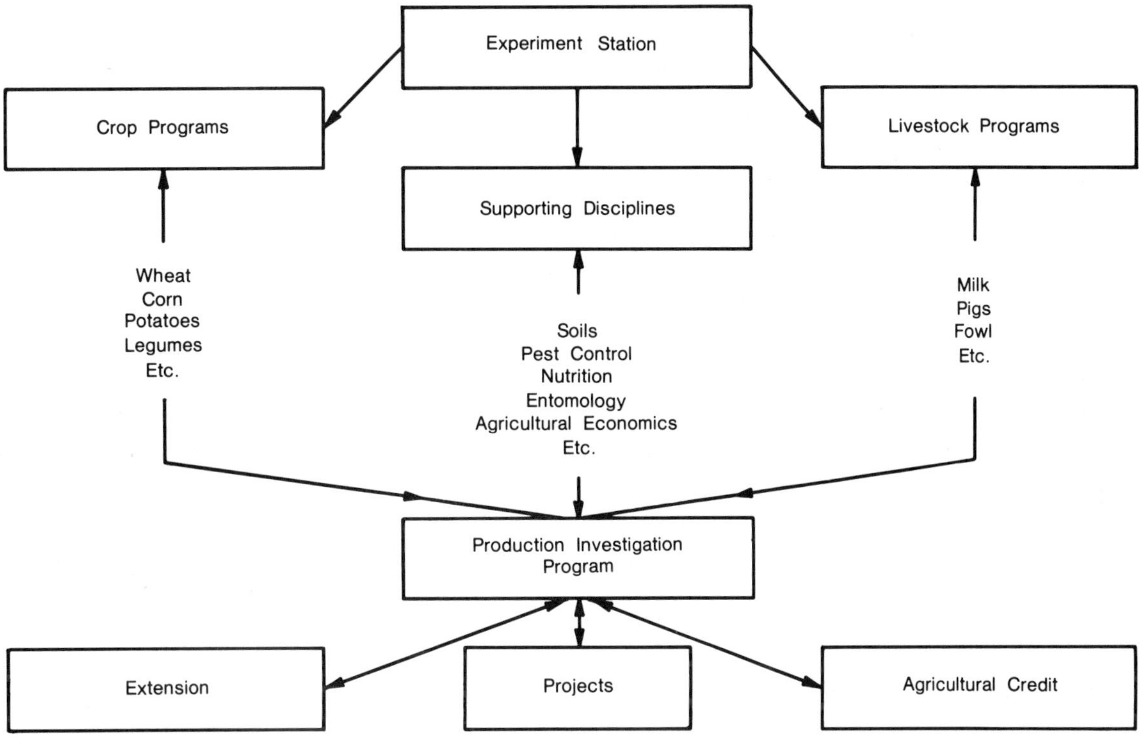

Figure 10-B-4. The organizational diagram for the On-Farm Water Management Project in Egypt (E. Richardson, personal communication, Colorado State University, 1981).

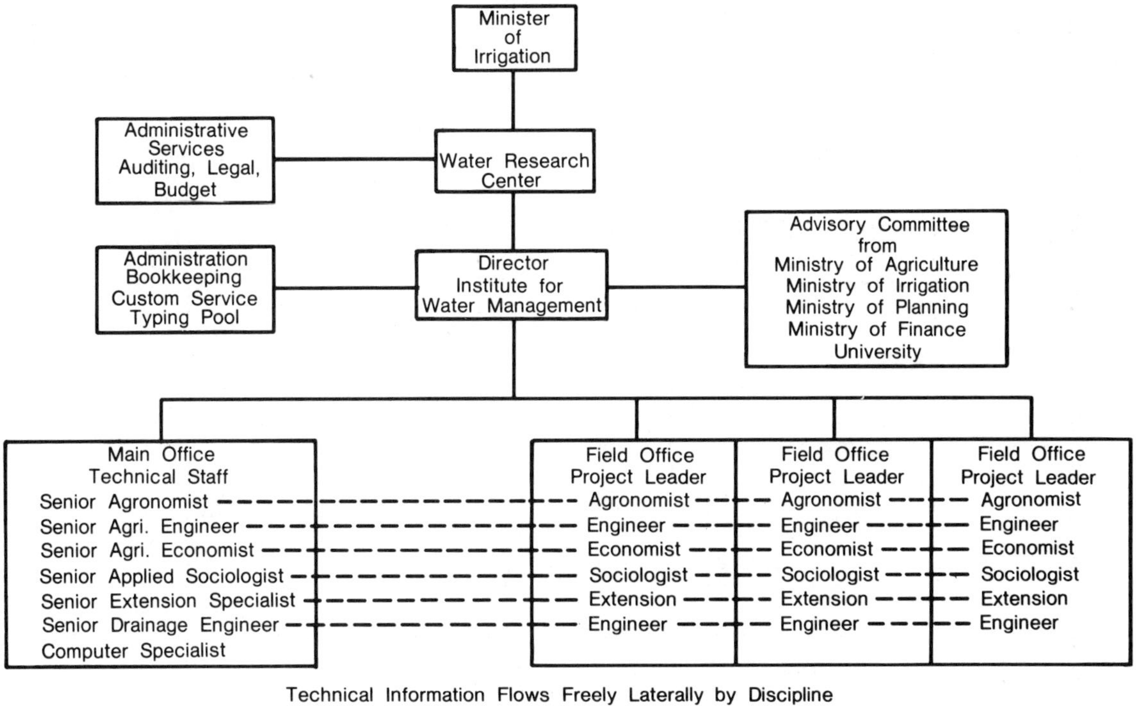

APPENDIX 10-C
TEAM BUILDING

As we discussed in Sec. 10.6.1. in Chapter 10, one of a team leader's responsibilities is to develop effective teamwork. In this regard, a manual on team building by Francis and Young (1979) should interest the reader. This book describes the team-building process, proposes 108 questions that help to probe the nature of a team, and then offers 46 tools for building strengths and clearing blockages to effective teamwork.

The authors define an effective team as "An energetic group of people who are committed to achieving common objectives, who work well together and enjoy doing so, and who produce high quality results."[1] The process of building a team involves (1) the desire to improve, (2) identifying problems and needs, (3) specifying the preferred situation, (4) treating the ailments and building for the future, and (5) reviewing progress.

The questions relate to problems and opportunities concerning (1) effective leadership, (2) suitable team membership, (3) team commitment, (4) team climate, (5) team achievement, (6) relevant corporate role, (7) effective work methods, (8) team organization, (9) critiquing, (10) individual development, (11) creative capacity, and (12) intergroup relations.

Improvements can be effected by having the team engage in some of the 46 exercises presented in the book. These exercises cover the range of topics listed above. Following are two exercises that illustrate the approach. They concern Team Mission and Individual Objectives and Effective Problem-Solving Survey.

"TEAM MISSION AND INDIVIDUAL OBJECTIVES[2]

"PURPOSE

I. To help clarify the team's mission to team members.
II. To test for strengths or problems in the ways that team members relate their individual goals to the team's overall mission.

"TIME

Half an hour for individual preparation. One hour for team discussion. It may require two or more discussion sessions to achieve a satisfactory outcome.

"MATERIALS

I. A copy each of the Team's Mission Sheet and the Individual Objectives Sheet, and a pencil for each participant.
II. A large newsprint pad, felt-tipped markers, and masking tape, or a chalkboard and chalk.

"METHOD

I. The team meets briefly to begin the preparatory work for the main team meeting. The leader distributes copies of the Team's Mission Sheet and the Individual Objectives Sheet to each of the members and asks the members to complete the sheets and bring them to the next meeting.
II. At the second meeting, each team member is asked to read his statement of the team's mission.
III. The individual Team's Mission Sheets are displayed, then the team members review all contributions and work to compile a statement of the team's mission(s) with which the whole team agrees.
IV. Each team member outlines for the team his individual objectives and, in the light of the statement of the team's mission(s), reviews the appropriateness of these objectives.
V. If it is clear that some objectives are no longer appropriate in relation to the team's mission(s), the meeting is adjourned to allow the members to review their objectives. The team schedules a meeting within one month of the original meeting, at which time it will consider objectives again.

"TEAM'S MISSION SHEET

"The major reasons for the existence of this team are to achieve the following:

"INDIVIDUAL OBJECTIVES SHEET

"Over the next (six months, nine months, one year, two years, as appropriate) I wish to achieve the following in my work:

Statement of desired achievement	By what date	How I will know I have succeeded	Part of team's mission it relates to
1.			
2.			

[1]Reprinted from Francis, D., and D. Young, 1979. *Improving Work Groups: A Practical Manual for Team Building.* University Associates, San Diego, Calif. Used with permission.

[2]Ibid.

"EFFECTIVE PROBLEM-SOLVING SURVEY[1]

"PURPOSE

I. To identify strengths and weaknesses in team problem solving.
II. To set agenda for strengthening the weakest characteristics.

"TIME

Forty-five minutes.

"MATERIALS

I. One copy of the Effective Problem-Solving Survey for each participant.
II. A large newsprint pad and felt-tipped markers, or a chalkboard and chalk.
III. Blank paper and a pencil for each participant.

"METHOD

I. The leader distributes one copy of the Effective Problem-Solving Survey, paper, and a pencil to each member. He tells the members to read the instructions and complete the survey. (Five minutes.)
II. After the survey has been completed, the leader asks whether the participants want to report their individual scores orally or to write them down and turn them in anonymously.
III. The leader collects the scores, charts them, and identifies the two items with the lowest scores. (Ten minutes.)
IV. The members discuss the problem items and identify six action steps that could help the group improve in these two areas. The leader charts the suggestions; he makes them available for reference to the next working session. (Thirty minutes.)

"EFFECTIVE PROBLEM-SOLVING SURVEY

"*Instructions.* Please give your candid opinion of your team's most recent problem-solving session by rating its characteristics on the seven-point scales shown below. Circle the appropriate number on each scale to represent your evaluation.

Lacked order and poorly controlled	1 2 3 4 5 6 7	Orderly and well controlled
Confusion about objectives	1 2 3 4 5 6 7	Clear and shared objectives
Organization inappropriate to task	1 2 3 4 5 6 7	Organization was flexible, appropriate to task
Criteria for success not established	1 2 3 4 5 6 7	Clear criteria for success established
Information was poorly evaluated	1 2 3 4 5 6 7	Information was well analyzed
Planning was inadequate	1 2 3 4 5 6 7	Planning was effective, thorough
Action was ineffective	1 2 3 4 5 6 7	Action was effective, adequate
No attempt to learn from the experience	1 2 3 4 5 6 7	Thorough review to help team learn from experience
Time was wasted	1 2 3 4 5 6 7	Time was well used
People withdrew or became negative	1 2 3 4 5 6 7	Everyone participated positively"

APPENDIX 10-D SUMMARIZED CHECKLIST FOR SUCCESSFUL INTERDISCIPLINARITY

This appendix provides two sets of questions to be used as a checklist for promoting successful interdisciplinary teamwork for FSR&D activities (see Sec. 10.6. in Chapter 10). The first set concerns topics related to the initiation of program (or project) activities and the second set concerns topics related to program (or project) operations.

CONCERNING INITIATION OF PROGRAM ACTIVITIES

- Has the appropriate mix of required disciplines been identified?
- Have the most competent persons been chosen for the core team?
- Are those chosen fully dedicated to team goals?
- Is the team leader satisfied with the team?

[1]Reprinted from Francis, D., and D. Young, 1979. *Improving Work Groups: A Practical Manual for Team Building.* University Associates, San Diego, Calif. Used with permission.

- Will the leader be able to balance task-oriented and people-oriented styles in providing creative interaction among team members?
- Is the leader sympathetic to the different paradigms of team members?
- Is the leader competent in one of the key disciplines for the team?
- Is the leader fully dedicated to team and program goals?
- Has adequate time been provided at the outset of activities for team building and refinement of program objectives?
- Has a communications plan for the team been prepared?
- Has physical space been provided for effective team interaction?
- Have any problems of spatial separation been adequately considered?
- Does the team's institution fully support the team and the program?
- Do the team members understand both the basis for the evaluation of team and individual accomplishments and the accompanying reward system?
- Have provisions been made for team liaison with others who are needed to support the team technically?
- Have adequate support services been provided for the team?

CONCERNING PROGRAM OPERATIONS

- Are all members contributing as part of the team?
- Is the leader able to adapt to changes in the team's needs?
- Are new team members effectively integrated into the team?
- Is provision made for additional team and program needs as they arise?
- Is the team able to accomplish agreed upon goals by accomplishing stated objectives and tasks?
- Do team members feel adequately informed?
- Are team members receiving adequate feedback on the performances of the team and themselves and are the rewards in keeping with performance?
- Are communications among team members open?
- Do team members have sufficient freedom to be creative?

APPENDIX 10-E
USAID'S LOGICAL FRAMEWORK

The United States Agency for International Development (USAID) uses the Logical Framework as one of the bases for preparing and justifying project loans and grants. This framework (1) relates project goals, purposes, inputs, and outputs with means for verification and (2) requires statements of important assumptions. Consequently, this approach can also serve as a basis for evaluating project results, as described in Sec. 10.8.3. of Chapter 10.

The following excerpts summarize some of the more important features of USAID's approach (Turner, 1979):

"THE LOGICAL FRAMEWORK

"1. A key element in project planning and evaluation is the establishment of a logical framework for the project design which:

a. Defines project inputs, outputs, purpose, and higher goal in measurable or objectively verifiable terms.

b. Hypothesizes the causal (means-end) linkage between inputs, outputs, purpose, and goal.

c. Articulates the assumptions (external influences and factors) which will affect the causal linkages.

d. Defines the indicators which will permit subsequent measurement or verification of achievement of the defined outputs, purpose, and goal."

A matrix format of the logical framework appears as Table 10-E-1.

"2. The logical framework methodology embodies the concept of causality; i.e., the causal linkage or hierarchy in which resource inputs are intended to produce outputs, outputs are expected to result in the achievement of project purpose, and project purpose is expected to contribute substantially to the higher goal. The concept of causality, in turn, rests on the basic premise that each level in the hierarchy can be shown to be not only necessary but also sufficient to cause the next higher level to be achieved. Since each causal linkage is subject to external factors beyond the control of project management, each linkage must be tested to assure that a given target level (e.g., outputs), in concert with the assumptions at that level are necessary and sufficient to achieve the next level (purpose).

* * *

"CONTENT OF LOGICAL FRAMEWORK"

"1. GOAL – NARRATIVE SUMMARY

a. Goal is a general term characterizing the programming level beyond the project purpose; i.e., the next higher objective to which the project is intended to contribute. It provides the reason for dealing with the problem which the project is intended to solve. Goal denotes a desired result to which an entire program of development may be directed. Goals are established at top program management levels. Project managers need to understand these programming goals even though their contribution in formulating them may be limited.

b. Generally, a goal is not achieved by one project alone, but is established with the intent that success in a variety of project and nonproject activities will be necessary for its achievement. In this

Table 10-E-1. Project design summary: Logical framework (Turner, 1979).

Life of Project: From FY ____________ To FY ____________
Total U.S. Funding: ____________ Date Prepared: ____________
Project Title & Number: ____________

AID 1020-28(1-72)

NARRATIVE SUMMARY	OBJECTIVELY VERIFIABLE INDICATORS	MEANS OF VERIFICATION	IMPORTANT ASSUMPTIONS
Project or Sector Goal: The broader objective to which this project contributes:	Measures of Goal Achievement:		Assumptions for achieving goal targets:
Project Purpose:	Conditions that will indicate purpose has been achieved: End of project status:		Assumptions for achieving purpose:
Outputs:	Magnitude of Outputs:		Assumptions for achieving outputs:
Inputs:	Implementation Target (Type and Quantity)		Assumptions for providing inputs:

respect, the relationship between goal (the end) and project purpose (the means) is causal and partial. Causal relationships become more direct and complete when descending to the *output* and *input* levels. The establishment of a goal is thus only one final stage in a logically progressing series of hypotheses:

(1) *If* this goal is desirable, *then* what project purpose will be necessary to achieve it?

(2) *If* this project purpose will assist goal achievement, *then* what outputs will be necessary to achieve the project purpose?

(3) *If* these outputs are to be provided, then what inputs will be required?

"2. GOAL – OBJECTIVELY VERIFIABLE INDICATORS

The indicators of goal achievement may be quantitative, qualitative, or behavioral, or a mixture of these criteria. Satisfactory measures of achievement are those which indicate a realistic *causative* relationship between project purpose and goal and confirm that the project purpose contributes to the achievement of the goal. Measurement indicators such as the number of local citizens taking part in an election, increased per capita income over a prior period, increased value of exports, and the number of job vacancies at a particular level in government and the private sector, provide a realistic picture of a situation at any given time. The scope of a single project will not usually be comprehensive enough to be the total cause of achievement of the goal. Other projects and nonproject factors may also have a signifant influence on goal achievement.

"3. GOAL – MEANS OF VERIFICATION

State the kinds and sources of data needed to support the indicators which have been cited as measures of goal achievement.

"4. GOAL – IMPORTANT ASSUMPTIONS

Achievement of the goal (and indeed the project purpose and outputs as well) is based on the expectation that certain events or actions outside the scope of the project will occur. These external factors need to be stated clearly as important assumptions regarding goal achievement and evaluated periodically to assure their continued validity. 'Increasing agricultural productivity,' for example, may be a realistic goal. However, achievement of that goal may be dependent on motivating the farm labor force; establishing marketing regulations, distribution centers, and national price structure; and acts of God, such as weather, etc., factors clearly outside the design of the project. The degree of confidence that is placed on the assumptions about these factors depends on familiarity with the cooperating country, knowledge of the sector of concentration, cooperating country performance, etc. A project design is only as sound as the strength of its weakest important assumption. As the project is implemented and the hypothesized causal linkages are tested, the confidence level in the causality between purpose and goal should increase. If this does not occur, the evaluation process should then focus attention on the explicit assumptions.

* * *

"6. PROJECT PURPOSE – NARRATIVE SUMMARY

The project purpose is the specific desired result of the project, not merely the sum total of outputs. A well conceived project has an explicitly defined purpose that contributes causally to the goal in a logical and direct manner. In turn, the combined effect of project outputs contributes in a logical and direct manner to achievement of the project purpose. This purpose represents the solution to a specific development problem and may be derived by inverting the statement of the problem into a statement of the appropriate solution.

"7. PROJECT PURPOSE – OBJECTIVELY VERIFIABLE INDICATORS

a. The statement of the End-of-Project Status conditions (EOPS) is a description of the set of terminal conditions that will exist when the project purpose is successfully achieved. This description takes the form of objectively verifiable indicators, either quantitative, qualitative, or behavioral in character which reflect the end of the project status conditions. In projects which have an institutional purpose, the end-of-project status conditions would include the actual performance of the institution, rather than its readiness (the latter would be output indicators). Indicators of institutional performance would include self-sufficiency, effectiveness in producing goods and/or services, efficiency, creativity, and initiative.

b. In projects that emphasize immediate accomplishments, the end-of-project status conditions expected often are direct results of project goods and/or services. Did the birth rate fall? Did exports rise? Did enough private enterprises (or cooperatives) *survive* to form a critical mass that will continue to grow without AID support? Do fewer children drop out of school as a result of the new instructional methods and textbooks? Did per hectare crop yield increase?

* * *

"9. PURPOSE – MEANS OF VERIFICATION

State here the sources and the specific types of evidence which will be used to verify conditions marking End-of-Project Status.

"10. PURPOSE – IMPORTANT ASSUMPTIONS

As noted in [4], an assumption describes a situation or a condition which must be assumed to exist, if and when a project is to succeed, but over which the project management team may have little or no control. An example is: Increased crop yield (project purpose) will contribute to expanded export of agricultural crops (sector goal) only if price and market conditions are favorable (assumption).

"11. OUTPUTS – NARRATIVE SUMMARY

Project outputs are the planned results produced by the management of specific inputs. In analyzing project outputs, be aware of the distinction between the *kind* and the *magnitude* of the specific results that competent project management is reasonably able to produce. Producing trained cooperating country staff for certain key posts is an output. *The output indicators would state the number of trained staff placed in specified key posts within a particular time frame.*

"12. OUTPUTS – OBJECTIVELY VERIFIABLE INDICATORS

The magnitude of outputs, targeted and expressed in a manner allowing verification, reflects evidence of successful completion of the managerial actions (input-output linkage) that were necessary to produce the output in the first instance. In the case just given, participant training would be the link. Examples of outputs and appropriate targeted output indicators include:

Outputs	*Output Indicators*
a. Trained indigenous personnel for key posts in Radio Correspondence (R/C) course;	a. Cooperating country personnel trained for, and assigned to 15 previously identified key posts by 1978;
b. Courses prepared and taped;	b. 18 courses prepared and taped by end of 1978;

* * *

"13. OUTPUTS – MEANS OF VERIFICATION

State the data source and kind of data for verifying each output indicator.

APPENDIXES TO CHAPTER 11

"14. OUTPUTS – IMPORTANT ASSUMPTIONS

* * *

b. Given outputs such as trained manpower (either through participant training or on-the-job training), a critical assumption may be that the government will formally establish appropriate positions and will budget funds to payroll them.

"15. INPUTS – NARRATIVE SUMMARY

Inputs are the goods and services provided by the Mission, the Bureau, the Office, other donors, and/or the cooperating country with the *expectation of producing certain definable outputs*. The inputs to a project may consist of personnel, equipment, commodities, training, funding, contract services, etc., in almost any combination. These inputs may be provided by the United States (directly or through contractors, participating agencies, or voluntary agencies), the cooperating country, or other donors. With respect to personnel the important factor is the services which each person is to perform rather than simply the assignment of an individual to the project; i.e., the fact that an adviser is at post is not a statement of the input expected from that adviser.

"16. INPUTS – OBJECTIVELY VERIFIABLE INDICATORS

For each element of the above input, list budget categories such as commodities (perhaps broken out into subgroups), participant training, advisory services (direct-hire or contract), and their quantities and approximate expenditure level.

"17. INPUTS – MEANS OF VERIFICATION

This cell of the matrix may not have to be completed if inputs consist of AID Mission-furnished items for which AID records provide accounting. However, other inputs such as those by the cooperating country, voluntary agencies, and third countries, should have confirming data sources shown.

"18. INPUTS – IMPORTANT ASSUMPTIONS

Assumptions at the input level are usually limited to questions of whether the inputs will be available on time. Project designers may use this cell of the matrix to record 'Beginning of Project Status conditions'; the project specific baseline conditions which are the obverse of the terminal or 'End of Project Status condition.'"

APPENDIX 11-A
HONDURAN TRAINING PROGRAM IN FSR&D

In 1980 the developers of the Honduran training program in FSR&D published their experiences from the previous two years (PNIA, 1980). This publication served as a guide for the Honduran training program in FSR&D for 1980 and possibly subsequent years. The training program, primarily for researchers, is of the type we described in Sec. 11.2.1. Below is our translation of this document:

"MINISTRY OF NATURAL RESOURCES NATIONAL PROGRAM FOR AGRICULTURAL RESEARCH: CENTRAL UNIT

"ACTIVITIES MANUAL FOR IN-SERVICE TRAINING

"Agricultural Office for the West-Central Region

1980

Comayagua *Honduras*

"PREFACE

"The In-Service Training Manual combines the experiences of the past two years (1978-1979) of technicians of the Central Unit of the National Program for Agricultural Research. This compilation of national experiences served as a guide for in-service training in the National Agricultural Research Program in 1980.

"The compilation of the information, format, and editing of this report was completed by Engineers Mario Nuñez and Alvaro Díaz, with the help of the following professionals:

Dr. Mario Contreras
Dr. Frank Peairs
Dr. Franklin Rosales
Eng. Juan Carlos Torchelli
Dr. Robert K. Waugh

"We thank, in particular, the personnel of the 1978-1979 In-Service Training Program and ICTA (Guatemala) for their valuable assistance in the development of this manual.

FRANKLIN E. ROSALES
Central Unit Chief

"I. INTRODUCTION

"During the past three years significant changes have occurred in the Agricultural Research Program.

"In 1977, a work team was formed made up of national and foreign technicians. This group was soon joined by an IADS [International Agricultural Development Service] mission that carried out a general diagnosis and proposed a reorganization, including a general methodology, a new organizational scheme, and developmental strategies.

"The new approach culminated with the editing and publication, in January 1978, of the document 'Agricultural Research in Honduras.' This proposal received its official approval from Memorandum M. No. 5-11-78, sent by the Minister of Natural Resources, Rafael Leonardo Callejas, to the Regional Directors on February 17, 1978.

"The following concepts were outlined in the document:

'In recognition of the need for a Research Program that is accountable at the national level and effective in the development of the assigned responsibilities, we propose a reorganization of the program to maximize its potential. This involves experimentation and development of techniques by working with farmers under farmers' conditions. Attention will be focused on a multidisciplinary approach that encompasses the entire production system (biological, socioeconomic, and cultural), and an effective coordination of the program with other activities of the sector (Extension, Human Resources, Planning, etc.) at the administrative and field level.' In addition: 'The Program will be structured as follows:

A) A headquarters will be located in Tegucigalpa with the following responsibilities: represent the Ministry, develop the Plan of Operations, national coordination, and technical administrative support for the Program.

B) A Central Research Unit, located in Comayagua and dependent on the Program's Headquarters, will be formed by a multidisciplinary team with responsibilities for research, training, development of operational plans, and support and supervision of research projects.

C) Regional Units will be composed of a Program Coordinator, Production Systems Teams, and Technical Support Teams. The Production Systems Teams' responsibilities will be the identification and diagnosis of problems at the field level, experimentation, and orientation of research at the experiment stations. The Technical Support Teams will be responsible for carrying out traditional research; however, the thrust of these teams will be determined by alternatives introduced by the Production Systems Teams.

D) Research Projects, based on priorities and availability of resources, will be directed at solving farmers' problems. The present projects will be continued, but with changes introduced gradually according to the needs of each case.'

"During 1978 the ideas were taken to the field by the Central Unit in Comayagua, which set up farm trials in three areas of the Region: El Rosario, San Jerónimo and La Paz.

"These trials were part of a pilot experiment to refine the fieldwork methodology and to make adjustments for possible application to other regions. The trials were designed after reconnaissance and other survey work in the region, and were carried out from January to April, 1978. The objective of the trials was to identify the most common productive systems in the selected zones and to determine the most important problems to which agricultural research could contribute.

"The work in Comayagua and the sub-region of La Esperanza contributed much to enrich the national team's experience, as did the frequent contact with technicians from other institutions of the region, such as CATIE and ICTA.

"During 1978, in-service training started informally in Comayagua, based on this new approach. The experience was very positive, because the three regions were staffed with technicians who had participated in the training.

"This new activity was formalized in February 1979. The course combined theoretical aspects (short courses, seminars, and workshops), with practical activities. Attention was centered on area descriptions and the design, installation, direction, harvest, and analysis of farm trials both on the farm and at the experiment station.

"The practical phase took about 75 percent of the nine-month training program. The theoretical phase was designed to strengthen deficient aspects of previous university education (statistical analysis, experimental design, economic analysis, technical communication); to relate theory to practical experiences in production systems (disease and insect control, weed control, soil-water-plant relationships); and to involve the participants through presentation of papers and through discussions in seminars and workshops.

"II. OBJECTIVES OF THE IN-SERVICE TRAINING

"A) General Objectives

"Provide a general orientation and an opportunity for practical experience to the new technicians who enter the Program and who may carry out graduate studies in the future.

"In the past, newly hired technicians were located at an Experiment Station along with a researcher in charge of a crop. In this way, the majority of the trainees gained experience with a particular commodity.

"The present direction of the Research Program requires other preparation. In effect, technicians are needed with a broader perspective formed by close contact with the problems of the Honduran farmer. It is necessary to direct the technicians toward the study of integrated systems, and to train them to identify and analyze problems, and propose more effective alternatives. The technicians need to understand the complex interaction of agronomic and socioeconomic factors that affect farm production in a region. These technicians will form the base for extending these work procedures to all regions of the country as rapidly as possible, and in the process establish Regional Research Teams that operate at the farm level and focus on production systems.

"B) Specific Objectives

"The in-service training program developed specific skills that include the ability to

1. Identify and analyze agricultural production problems through a multidisciplinary approach that allows for understanding production systems as complex units involving biological and socioeconomical interactions.
2. Establish research priorities in agreement with the national and regional needs and policies that do not exceed available resources.
3. Install, conduct, and analyze field trials to arrive at appropriate conclusions and recommendations.
4. Communicate effectively with the farmers who are the ones who will profit most from the Agricultural Research effort, and cooperate with the extension workers and other regional technicians.
5. Analyze technical information and transmit it correctly in oral and written form.
6. Apply existing technical knowledge, e.g., for the production of basic grains.
7. Develop a positive attitude for effective teamwork.

"III. METHODOLOGY AND ORGANIZATION

"During training an attempt was made to integrate practical with theoretical aspects by giving priority to 'doing rather than talking.' Trainees were given the opportunity to do farm work and take the initiative in solving farmers' problems. At the same time, the trainees developed a common understanding and a practical capacity to work under farmers' conditions. Theoretical activities emphasized the conceptual aspects of the approach; and field application emphasized reasoning rather than memorization.

"During training, an attempt was made to stimulate active and responsible participation by all trainees. The activities located in Comayagua covered a nine-month period from February 1st to October 31st and included two phases: a practical phase that took up 75 percent of the time and a theoretical phase that took the other 25 percent.

"The practical phase was carried out in three zones of the Comayagua region; El Rosario, San Jerónimo, and La Paz. A group of trainees was responsible for the work in each zone. Each zone was assigned a technical supervisor from the permanent technical team of the Research Program in Comayagua. On-farm research carried out by the trainees constituted part of the program's regular Plan of Operations. But in this case, the trials, in addition to the results, served the essential objective of training the Program's new personnel.

"The theoretical phase was the same for all students and was concentrated during slack periods rather than during the time of maximum activity on the farms (planting and harvesting). Besides the trainees, others such as researchers, extension workers, and other technicians in the region took advantage of the theoretical instructions. The in-service Training Coordinator, with support from technicians of the Program's Central Unit, was directly responsible for the in-service training activity.

"IV. PRACTICAL ACTIVITIES

"The practical activities included

A) Characterization and diagnosis
B) Farm trials
C) Farmers' tests
D) Farm records
E) Rainfall records
F) Field days

"Activities A, C, and F were developed in cooperation with the extension worker in the area.

"A) Characterization and Diagnosis

"This phase of the work, carried out at the beginning of training during February and March, consisted of the following steps:

1. Exploring the region. The group of trainees, guided by their technical supervisors, visited the different zones in the region, observing major characteristics and talking with extension workers, individual farmers, and agrarian reform groups. The step served as the first contact with the realities of the region for the training group.
2. Technical information management. During this phase trainees were provided climatic data and other technical information for the region. They analyzed and summarized it with the help of supervisors. The work was done in small groups and culminated with the preparation of a brief report.
3. Questionnaire. This diagnostic instrument helped to obtain technical and social information and provided the means for the team to become acquainted with the farmers. With the help of their supervisors, the trainees developed a questionnaire, used it to gather data in the field, and then tabulated the results. This activity culminated in the development of a technical report on the results and conclusions of the survey.

"B) Farm Trials

"This was the central focus of all the practical activities. It was carried out in groups of three trainees with each group assigned to one of the following zones: La Paz, El Rosario, San Jerónimo. A principal aim was to encourage an interchange of experiences among the teams working in the zones. At times the three groups worked together, especially during periods of planting and harvesting. Farm trials were divided into the following phases:

1. Trial design. By 1980 we had two years of research experience conducted in the three zones. This provided the basis on which to plan new trials. With the participation of the trainees, trials were designed by the technicians of the Central Unit of the Program at technical meetings. The process started with a proposal from the technicians responsible for each area. The trials were in response to the main problems identified in each area. The simplest designs compatible with the objective of the experiments were used. Each trainee group managed a maximum of 15 trials on a maximum of five farms.
2. Selection of collaborators. The trainees were in charge of this activity with the aid of the zonal supervisor and the extension worker. The selected farms had the following characteristics:
   a) Farms were accessible year-round by car or were, at most, 10 minutes by foot from the main road.
   b) Farms were representative of the zone under study in soil, climate, and production system characteristics.
   c) The farmers were receptive and had a good

understanding of the objectives of the project.

This phase was very important for assuring good relations with the collaborating farmers. The team needs to spend sufficient time talking with these farmers because the farmers need to be thoroughly convinced of the trials' usefulness. The farmers also need to know in detail each trial's objectives, its management, and what their responsibilities as collaborating farmers will be.

In the farm trials, the farmers generally supplied the land, did the land preparation, and helped in planting and harvesting. The product of the trials belonged to the collaborator. Experimental inputs were supplied by the Program, except such things as: the farmers' seed for certain control plots, or some treatments the collaborators do on their own crops that are considered necessary for the trials.

3. Installation, management, observation, and harvest of trials. The annex [not included] contains a Methodology Guide for Farm Trials that can be used as orientation during this phase of the work. At the beginning of the work the group for each zone received the required implements and equipment for which they were responsible: a vehicle, a portable sprayer, two pitchforks, two machetes, a measuring tape, two field books, a roll of rope, boots, glasses, gloves, and a protective mask. Other supplies, such as fertilizers, insecticides, fungicides, herbicides and stakes were in the warehouse, and were obtained through the Training Coordinator.

   Each group carried a field book in which it prepared a schematic map of each farm, the trial design, the details of the treatments, and up-to-date observations. The group kept two copies of the field book: one copy was for the field; the other copy stayed in the office and was updated continually.

   The Course Coordinator and the Zonal Supervisors checked the field books at unscheduled times and verified their contents. The trial plots were conveniently marked and the corresponding treatment noted by a ticket located at the left of each plot. In addition to making the necessary observations, each group was provided with a camera to take pictures of the trials at the most critical times. These pictures were kept as part of the Program's records for later reference.

   Each group in the zone met weekly with the corresponding technical supervisor to inform him about the development of the work, to solve operational problems, and to plan for the coming week.

   Each group in the zone submitted a brief monthly report to the Training Coordinator. The report contained information on the jobs completed during that period and problems and comments that were relevant to the development of field work.

4. Analysis of the trials. The trainees received help for the analysis of trials through a basic statistics course, an economics analysis course, and a counseling workshop on statistics. Each group presented a written report containing the results and conclusions of the trials.

   At the end of the program each trainee made an oral presentation on the results and conclusions concerning a group of trials.

"C) Farmers' Tests

"In some zones where field work was undertaken, certain technologies that were studied as part of the farm trials need to be verified by farmers' tests before being recommended for general use. For this, plots of greater size were used. The farmer supplied the inputs and managed the plots with the technical support of the extension worker and the researcher.

"The trainees' activities in this case consisted of collaboration in the organization of the farmers' tests, in advising on their installation and management, in their use as demonstrations, and in their final evaluation.

"D) Farm Records

"Farm records constituted a means for recording in a simple form all of the tasks carried out on a crop or cropping sytem. Items included were descriptions of the area, hand labor used, type and quantity of inputs, and their corresponding costs.

"For the Research and Extension Programs these records were kept to

1. Learn about the common technology practiced by private farmers and those in the agrarian reform sector.
2. Orient the research by furnishing information on the principal technical and socioeconomic problems limiting production.
3. Make it possible to evaluate how the new technology will be accepted and what its results will be, by studying the information obtained during various years.

"At the same time, the farm records had the following advantages for the producers themselves:

1. They made it possible for the producer to get a good idea of how much the individual farm activities will pay off.
2. For each crop, they furnished information about the annual use of family and hired labor, number of days worked, and how much was spent.
3. They improved the knowledge about investment for use in planning future requests for credit.

"In 1980, each trainee was to be responsible for at least one farm record for which the trainee will receive adequate help from the Research Program's economist.

"E) Rainfall Records

"Rain gauges were installed in three zones and managed by the farmers. Each group of trainees visited these farmers periodically to verify that the records were being kept and to record data in the field book.

"F) Field Days

"The group in each zone organized at least one field day for technicians of the Research Program and other Programs to show them the farm trials and to discuss preliminary results or specific problems. The group received support from the supervisor and, whenever possible, help from the extension worker.

"The field day was planned well in advance and supported with adequate materials for demonstration.

"V. THEORETICAL ACTIVITIES

"Theoretical activities included the following aspects:

A) Short courses
B) Conferences
C) Workshops
D) Seminars
E) Consultations and lectures

"A) Short Courses

"Short courses corrected the educational deficiencies of the trainees in specific subjects, and reoriented and updated the trainees in important agronomic disciplines.

"National and international technicians from CATIE, CIAT, CIMMYT, etc. were in charge of these courses, which were programmed for three days. The objective of the courses was to combine theoretical and practical aspects of the material. Trainees were expected to participate actively.

"For 1980 the following short courses were planned

1. Training introduction: institutional overview
2. Basic statistics
3. Economic analysis for agricultural research
4. Soil-plant-water relationships
5. Agricultural systems
6. Weed control
7. Climatology in Honduras (with emphasis on the Comayagua Valley)
8. Soil conservation
9. Insect and disease control
10. Technical communication
11. Research administration

"B) Conferences

"These consisted of brief presentations on specific subjects with invited national and foreign technicians in charge.

"Presentations were approximately one hour long, with an additional hour allowed for questions and discussion.

For 1980 the following conferences were programmed

1. How to obtain clean bean seed
2. Methodology for regional diagnoses
3. Sorghum and maize oidium (powdery mildews)
4. Small-scale agriculture
5. International research programs
6. Agricultural research and national development
7. Minimum tillage
8. Appropriate rural technology
9. Diseases of basic grains in Honduras
10. Farm records in ICTA, Guatemala.

"C) Workshops

"These were working sessions between the trainees, the supervisors, and other technicians from within and outside the Program.

"In general, the workshops were directly related to the planning of activities and to discussions of the field work. The workshops therefore had both a practical and a theoretical character, with active participation expected of the trainees.

"For 1980 the following workshops were programmed

1. Reconnaissance of the region; first impressions
2. Development of surveys
3. Management of regional data
4. Tabulation of survey data
5. Selection and initial design of farm trials and selection of collaborators
6. Selection and final design of farm trials and selection of collaborators
7. Statistical counseling
8. Discussions of farm trial results, 1980.

"D) Seminars

"These were brief expositions where the trainees were in charge. The seminars were about subjects proposed by the training program directors or the trainees themselves. The purpose of the seminars was to improve the trainee's ability to understand and synthesize a subject and present it in a clear and concise form. Before each seminar a written summary was presented about the subject. The trainees were evaluated for their clarity of concepts, ability to synthesize, quality of presentation, and supportive material.

"The following subjects were some of those proposed for the 1980 Seminars:

1. Management of agrochemicals
2. Maize/bean systems
3. Maize/millet systems
4. Characterization and potential of an agricultural zone
5. Appropriate technology
6. Implements for intermediate tillage

7. Energy in agriculture
8. Post-harvest problems in basic grains
9. Soil conservation
10. Irrigation in the Comayagua Valley.

The duration of the presentations was one hour, with 30 minutes for questions.

"E) Consultations and Lectures

"During the practical and theoretical activities, the trainees consulted with the Training Coordinator, the technical supervisors, and the technicians of the Research Program's Central Unit.

"A small library was organized in Comayagua and was available to the trainees. It contains selected bibliographies about subjects that were taught during the training program.

"VI. EVALUATION

"During 1980, an evaluation of the main theoretical and practical activities of the trainees was carried out.

"This evaluation had the following objectives:

1. To carry out an adequate follow-up of the activities and to estimate if program goals were achieved.
2. To stimulate adequate participation on the part of each of the trainees, and to identify their problems as soon as possible.
3. To take corrective measures using objective criteria.

"This evalution included

"A) Theoretical Knowledge

"At the end of each short course a brief written examination was taken. Questions were designed to evaluate the trainee's understanding of the fundamental concepts presented on the subject.

"B) Communication

"1. Oral communication. In evaluating the seminars, the expositions during the field days, and the final presentation of the results for the year's trials, the following factors were considered: general approach, clarity of concepts, ability to synthesize, presentation, participation in discussions, and reference materials.

"2. Written communication. The reports for each activity were evaluated for general approach, clarity of concepts, ability to synthesize, editing, and drawings.

"C) Field Work

"This included management of trials, adequate interaction with the farmer and the extension worker, organizational ability, personnel management, ability to work as a team, and keeping field books.

"D) Responsibility

"The following were evaluated: attendance, punctuality, fulfillment of responsibilities within the time required, and dedication to the training program."

APPENDIX 11-B OUTLINE OF AN IN-SERVICE TRAINING PROGRAM IN FSR&D PREPARED BY ICTA FOR DIGESA

As we noted in Sec. 11.2.2. of Chapter 11, FSR&D activities and the extension service benefit from effective training in FSR&D. In this appendix, we present an outline of an extension training program in Guatemala for the General Directorate for Agricultural Services (DIGESA) prepared by the Agricultural Science and Technology Institute (ICTA). The material in this appendix reproduces the summary of the original project (ICTA-DIGESA, 1979) without change:

"AGRICULTURAL PUBLIC SECTOR INSTITUTIONAL AND TECHNOLOGICAL RELATIONSHIPS PROJECT. ICTA-DIGESA
Jutiapa, Guatemala February, 1979 Summary

"COURSE OBJECTIVES

"The general objective of the course is to improve the technical capacity of [DIGESA].

"In the same way as in similar courses, training will be implemented in such a way that the trainee will have skills for:

1. Applying known technical knowledge in the decisions to be taken during the production process.
2. Identifying and analyzing agronomic problems and proposing alternative solutions.
3. Programming, registering and analyzing the results of farm enterprise and agricultural research.
4. Evaluating the validity and applicability of the technical innovations.
5. Knowing the type of jobs and the methodology of research in the specific programs.
6. Having the ability to communicate with the farmers, agricultural technicians or other developmental agency representatives.

"Another specific objective is that training serves to select technical personnel for [farming systems] teams. During the course qualities and aptitudes of the trainees will be identified.

"GOALS

"The following goals are contemplated for 1979:

Goals For Technical Training During 1979

	Number
Agricultural meetings	14
Field days	4
Seminars	10

Monographs	10
Conference hours	370
Written reports	85
Training plots	10
Farm trials	43
Test parcels	30
Farm registers	30
Precipitation registers	10

"TRAINING SITE

"Training will be carried out in the Jutiapa department. This subregion has typical characteristics of the eastern part of the country including poor rain distribution. Crops grown include: rice, sorghum, tomatoes, onions, watermelon, hot peppers and tobacco.

"Trainees will operate in areas with climatic differences and where traditional farming systems are very diverse. These sites are the municipalities of 1) Jutiapa and El Progreso, 2) Quesada, and 3) Asunción-Mita.

"COURSE DURATION

"The course will last 10 months. It will start March 1, 1979 and be completed in December. It is essential that the course last for 10 months. The reasons are

a. During March and April the trainees will become familiar with ICTA and identify/characterize their work area.
b. The harvesting of first (May) and second (August-September) crops will proceed under the same work plan followed in 1978. Some second crops will be planted when the first crops are still in the field. Others will be planted in May and will be harvested in November or December, for example: creole, sorghum and peas.
c. The trainees need more technical support during the first cropping cycle. During the second cropping cycle they can work more independently because of the experience gained during the initial months.
d. The technicians and the trainees have a chance to obtain and analyze the data and present the results.
e. They have more time to relate theory to practice. Sometimes the theory requires that it be given at an opportune time, for instance: to reinforce statistical theory, with the experiment that will be designed, and then the experiment is carried out and data analyzed.
f. Trainees need time to familiarize themselves with the farmers, farm families and the communities in the general area where work will be carried out.

"TRAINING AREA

"The course will have two areas of action:

1. *Training in Agricultural Production*
   All the trainees must participate in this training to learn about the technologies of the basic crops used in farm trials, test plots, and training plots.
2. *Training in Specific Areas*
   Training will include specific emphasis on the phases of farming systems including concepts, philosophy, and principles from technology generation to validation stages. Additional training will be conducted in statistics, technical communications, farm management principles, agricultural technology, group dynamics, audio visual aids, and technology transfer techniques.
   Field work will include farm records, interviews, *sondeos*, farm visits, work plots, field days, farm tests, field plots, report writing, and monographs."

APPENDIX 11-C TRAINING IN FSR&D AT SELECTED INTERNATIONAL CENTERS FOR AGRICULTURAL RESEARCH

As we noted in Sec. 11.4. of Chapter 11, the IARCs and at least one regional center are a good resource for training FSR&D staff. This appendix contains brief accounts of some of the training programs relevant to FSR&D that are offered at four of the international centers and one regional center. The international centers are the International Center for Tropical Agriculture (Colombia), International Maize and Wheat Improvement Center (Mexico), International Livestock Centre for Africa (Ethiopia), and International Rice Research Institute (Philippines); the regional center is the Tropical Agricultural Research and Training Center (Costa Rica). The material is representative of the types of training programs that we believe will be of interest to those responsible for training at the national level. Below are sections on program objectives, approach, training categories, production training, payment of costs, and how to apply. Because of data limitations, we are unable to include information on all of these sections for each organization.

The interested reader may want to see the following references for further description of these centers' research and training programs: CATIE (1979), CIAT (1980), CIAT (undated), CIMMYT (1978a), CIMMYT (1978b), CIMMYT (1978c), ILCA (undated), IRRI (undated), IRRI (1978), Fonseca (1979), and Perdon (1977). The addresses of these and a few other organizations concerned with crop and livestock research that have training programs in production and farming systems are provided at the end of this appendix.

INTERNATIONAL CENTER FOR TROPICAL AGRICULTURE (CIAT)

OBJECTIVE

CIAT's general training objectives are to (1) transfer new technologies to the national and local level so these technologies can be validated, adapted, and passed on to the farmer as rapidly as possible and (2) prepare young researchers for careers with national institutions.

APPROACH

CIAT's training programs center around three interrelated concepts: (1) critical mass, (2) learning by doing, and (3) training in specific skills.

A large enough group of individuals from priority areas must participate in the training to guarantee the application of the methodology, work patterns, and perspectives acquired at CIAT.

Learning by doing is an effective way to train young researchers in new technology. The trainees not only acquire understanding of theoretical concepts, but they also learn by fully participating in the application of new technologies in the field and the laboratory.

Training in specific skills is to ensure that the trainees will be able to successfully apply CIAT-generated technologies in their home institutions.

TRAINING CATEGORIES

CIAT divides training into seven categories: (1) postdoctoral fellows, (2) visiting research associates, (3) research scholars, (4) postgraduate research interns, (5) postgraduate production interns, (6) special trainees, and (7) short course participants. All training is commodity based and run by individual research programs or support units.

These commodity programs and support units include beans, cassava, rice, tropical pastures, data services, seed production, documentation services, station operations, and communications.

PRODUCTION TRAINING

CIAT's production training is administered within the context of formally organized short courses that cover either 4 to 6 weeks or 4 to 6 months. About 200 trainees participated in such programs in 1979: 66 in beans, 46 in cassava, 34 in rice, 37 in seed production, and 18 in documentation services.

These programs generally cover a wide range of topics such as (1) physiology, entomology, agricultural economics, and agronomy as related to crops, such as beans, cassava, and rice; (2) production processing, testing, multiplication, storage, and distribution as related to seed production; and (3) documentation services.

PAYMENT OF COSTS

Costs of training fall into three categories: those related to international travel, and to direct and to indirect costs incurred at CIAT. The sponsor is expected to provide the trainee with an economy-class air ticket to and from Cali, Colombia, a baggage allowance, and a reasonable amount for expenses enroute. CIAT will provide trainees while at the Center with a "monthly stipend sufficient to cover food, housing, and reasonable personal expenses. In addition, the Center provides trainees with an allowance for training resources as well as for local travel expenses that are directly related to their training program. Furthermore, all trainees are provided with medical and hospital insurance covering the period of their stay at CIAT" (CIAT, undated). Indirect costs of administration and training facilities are also incurred by CIAT.

The trainee's sponsor is normally asked to pay CIAT for its direct and indirect costs "prior to initiation of training whenever the period of training does not exceed six months. For longer periods CIAT normally bills the sponsor of the trainee at periodic intervals of his stay at the Center" (CIAT, undated).

HOW TO APPLY

Initial contacts regarding the training of an individual at CIAT are made by the candidate, the candidate's organization, or a sponsoring organization. The candidate for short-course programs must have the ability to speak and write Spanish or English. Furthermore, the candidate must be a staff member of a national institution or development organization working on one or more of the commodities that concern CIAT.

Before final acceptance, the applicant must complete a medical examination and the examining physician must return CIAT's medical form directly to the center.

INTERNATIONAL MAIZE AND WHEAT IMPROVEMENT CENTER (CIMMYT)

OBJECTIVE

At CIMMYT, the training objective is to eliminate the barrier between the researcher and the farmer so that the researcher can develop improved agricultural technologies to meet the farmers' needs and transfer results effectively to the farmers' fields.

APPROACH

CIMMYT's training programs strongly stress the approach of learning by doing and the discipline of working long hours in the fields.

The production training programs require the trainees to (1) lay out on-farm trials on farmers' fields, (2) perform all the work throughout the cropping season, (3) analyze the on-farm trial results, and (4) prepare recommendations for farmers.

TRAINING CATEGORIES

CIMMYT's research is based on two major commodities—maize and wheat. Both commodity programs have separate, but parallel training. These programs are generally divided into five categories: (1) in-service training, (2) master's degree program, (3) predoctoral fellows, (4) postdoctoral fellows, and (5) visiting scientists. Following are descriptions of two of the in-service training programs.

Maize and Wheat Production

In-service training in CIMMYT is largely focused on the maize and wheat commodities. Twice a year these programs are offered to 25 to 50 selected young professionals from Asia, Africa, and Latin America. The duration of the training is one full crop cycle, with six to nine months residence in Mexico.

The in-service trainee learns a variety of subjects such as economics, farm machinery, plant physiology, genetics, pests and diseases, communication, production management, and interdisciplinary teamwork.

The major objective of this program is to prepare young professionals to use the techniques of agricultural research in an interdisciplinary manner to solve the problems in farmers' fields.

The general requirements for the in-service candidates are

(1) experienced in national crop programs
(2) nominated by a national institution
(3) able to spend 6 to 9 months in Mexico
(4) have the ability to speak and write Spanish or English.

Special Course for Economists

CIMMYT has recently initiated a course for economists that is located in Mexico and Kenya. The course is designed to give economists understanding and experience in applying economic concepts to farming systems situations. Trainees learn about the farming systems approach, biological aspects of crop production, data collection, experimental methods, technologies of improved maize and wheat production, economic policy issues, and analytical techniques in economics.

The objectives of this training are to (1) increase the ability of economists to work and communicate with technical scientists as part of interdisciplinary teamwork in FSR&D and (2) use appropriate economic methods for evaluating alternative farming systems patterns and practices.

The major difference between the programs in Mexico and Kenya is their duration. The Mexican program lasts three months, while the Kenyan program may take up to two years. This latter program is part of CIMMYT's Eastern African Economics Programme that places trainees with collaborating scientists at selected research stations.

INTERNATIONAL LIVESTOCK CENTRE FOR AFRICA (ILCA)

OBJECTIVE

ILCA (undated) reported that its general training "objective is to improve the capability of practicing research scientists in the methods and value of systems research for livestock development. More specifically it is hoped that at the end of the course the participants will (a) be familiar with problem identification and techniques of data collection in: *environmental assessment*, including simple methods of assessing conditions and trends in natural vegetation, *animal production*, including survey methods of determining animal productivity and distribution, through aerial survey, stock inventory and continuing study, *household economics*, including methods of recording labour, output and consumption data; (b) be familiar with current analytical and modelling techniques."

SHORT COURSE

Once a year, ILCA offers a 4-month short training course to 12 African scientists. This course is divided into two parts: (1) 4 weeks of lectures on livestock systems in general and (2) 12 weeks of field work in specific topics such as sampling methods, animal production, stock inventory, and mathematical modeling.

REQUIREMENTS

All applicants should have the following qualifications:

(1) a B.S. degree
(2) a minimum of 3 years' practical experience in livestock research or in planning livestock development projects
(3) the ability to speak and write English or French.

INTERNATIONAL RICE RESEARCH INSTITUTE (IRRI)

OBJECTIVE

Training plays an important role in IRRI's activities. Perdon (1977) reported that the general objective is to improve the technical proficiency of national research and extension personnel and encourage developing countries to participate in the training of these individuals.

TRAINING CATEGORIES

Two types of training programs are offered at IRRI: (1) a research-oriented training program and (2) a production-oriented training program. The research training program includes degree (M.S. and Ph.D.), non-degree, and postdoctoral fellows. Production training courses include the following: rice production for 6 months, cropping systems for 6 months, genetic evaluation and utilization for 4 months, and agricultural engineering for 2 weeks.

The 6-month rice production training program is designed to improve the trainee's knowledge and specific skill in rice production and the trainee's ability to communicate this knowledge and skill to others. This program covers a wide range of topics such as plant physiology, plant breeding, agronomy, engineering and meteorology, pests and control, soil and fertilizers, diseases and control, field experimentation and statistics, technical writing, communication and extension teaching, economics, and grain processing.

The 6-month cropping systems training program is designed to teach (1) the newest technology for growing crops and (2) the related crop sciences. The trainees also learn the specific skills necessary to apply this technology in crop production. This course is divided into 12 topics: (1) weather and climate, (2) water management, (3) soils, (4) land preparation, (5) plant physiology, (6) weed management, (7) field experimentation and statistics, (8) calculations, (9) cultural requirements, (10) cropping systems and crop intensification, (11) economics, and (12) communication and extension.

The 4-month genetic evaluation and utilization training program is focused on four types of rice cultures: (1) irrigated, (2) rain-fed, medium deep-water, (3) deepwater, and (4) upland. The objective of this program is to improve the capability of trainees in the methods of developing new, improved rice varieties under a wide range of environments.

The 2-week agricultural engineering training course is designed to (1) demonstrate the application and (2) evaluate the design of IRRI's farm machinery.

APPROACH

Trainees spend approximately half of the time in classroom activities, and the other half in field training. The classroom activities are covered by staff members from IRRI and the University of the Philippines at Los Banos (UPLB), and guest speakers from government and private organizations. The field training requires the trainees to plan and manage research plots at the IRRI farm. During the crop season, the trainees conduct tests, sample yields, analyze their data, and make recommendations based on results from their experimental plots.

The concept of training by teaching is also incorporated in IRRI's training approach. All trainees must set up a 2-week training course for local extension agents, scientists, and farmers.

TROPICAL AGRICULTURAL RESEARCH AND TRAINING CENTER (CATIE)

OBJECTIVES

CATIE's training program is in response to member countries' interest in having their technicians trained in agricultural research, technology transfer, education, and farming systems. The latter embodies "a new outlook toward perfecting research in systems methodology under local conditions to generate new technological alternatives, which, at the same time would serve as instruments of change, helping in the massive transfer of technology already validated" (Fonseca M., 1979).

TRAINING CATEGORIES

A major effort of CATIE's training is the master's program in cooperation with the University of Costa Rica. This program, which emphasizes an interdisciplinary approach, offers four major research areas: (1) problem identification, (2) generation of alternative production systems, (3) on-farm testing, and (4) technology transfer and evaluation.

Since 1980, CATIE has developed other training programs such as short courses, workshops, seminars, and inservice training. These programs are in response to member countries' interests in having CATIE also provide rapid training possibilities.

POSSIBLE CONTACTS FOR TRAINING IN FSR&D

Below are some organizations that are engaged in various facets of FSR&D that may have production and farming systems training programs of interest to national agricultural research organizations.

International Center for Agricultural Research in the Dry Areas (ICARDA)
P.O. Box 114/5055
Beirut, Lebanon

International Center for Tropical Agriculture (CIAT)
Apartado Aereo 6713
Cali, Colombia

International Crops Research Institute for the Semi-Arid Tropics (ICRISAT)
1-11-256 Begumpet
Hyderabad 500016, A.P., India

International Institute for Tropical Agriculture (IITA)
P.M.B. 5320
Ibadan, Nigeria

International Livestock Centre for Africa (ILCA)
P.O. Box 5689
Addis Ababa, Ethiopia

International Maize and Wheat Improvement Center (CIMMYT)
Apartado Postal 6-641
Mexico 6, D.F., Mexico

International Potato Center (CIP)
Apartado 5969
Lima, Peru

International Rice Research Institute (IRRI)
P.O. Box 933
Manila, Philippines

Tropical Agricultural Research and Training Center (CATIE)
Turrialba, Costa Rica

West Africa Rice Development Association (WARDA)
P.O. Box 1019
Monrovia, Liberia

APPENDIX 11-D A CROPPING SYSTEMS TRAINING PROGRAM AT IRRI

In our discussion in Sec. 11.4. of Chapter 11, we spoke of some of the production and farming systems training courses at the various IARCs. In this appendix, we have reproduced without change the contents and objectives of a six-month cropping systems training program at IRRI (undated). This illustrates the type of training programs being offered. We now quote IRRI's report:

"THE INTERNATIONAL RICE RESEARCH INSTITUTE
Office of Rice Production Training and Research
Los Banos, Laguna, Philippines

"INSTRUCTIONAL TOPICS of
CROPPING SYSTEMS TRAINING PROGRAM
(Sept. 14, 1977 to March 17, 1978)

"To achieve the five general objectives, the participants will learn the latest technology of growing crops and develop the required skills to apply this technology in profitable crop production. The participants will learn the related crop sciences to provide an understanding of the new technology, and will learn the art of communication to insure the application of the acquired knowledge.

"The following pages describe the instructional topics and the specific behavioral objectives required for the trainees to achieve during the six-month IRRI Cropping Systems Training Program.

"Weather and Climate

"Weather elements. The trainees after attending the lecture on weather elements are expected to be able to:

1. Identify and explain important cc weather elements such as precipita radiation, temperature, day length, weather fronts, pressure, relative humidity, etc.;
2. Describe the relationships among weather elements;
3. Discuss the influence of weather elements on crop production;
4. Relate and explain how to make use of meteorological data reports in the current crop season.

"Water Management

"The trainee will be able to:

1. Describe the water cycle;
2. List the ways by which water is lost;
3. Describe the ways of minimizing water losses;
4. Discuss water requirements of major crops;
5. Describe irrigation practices and relate them to efficiency of water use and crop growth;
6. Describe water regimes in soils;
7. Compute the amount of water needed to irrigate an area to specified moisture content of the soil;
8. Describe different types of water pumps;
9. Describe how to operate and maintain water pumps;
10. Compute for the desired HP of water pump for an area given the needed data;
11. Describe different types of windmills;
12. Describe how to operate and maintain windmills.

"Soils

"The Soil Body. The trainees must be able to:

1. Enumerate the factors involved in soil formation;
2. Explain how these factors affect soil formation;
3. Describe the process of soil formation;
4. Identify the basic components of soil;
5. Describe the criteria of classification for soil series, soil types and clay types.

"Physical Properties of the Soil. The trainees must be able to:

1. Define topography, relief, soil horizons, color, texture, consistency, structure, porosity, bulk density, particle density, soil tilth, light soils and heavy soils;
2. Relate these properties to workability of the soil, to chemical properties of the soil and soil productivity and soil fertility;
3. Relate physical properties of the soil to cropping systems.

"Chemical Properties of the Soil. The trainees must be able to:

1. Define soil pH, cation exchange capacity, base saturation, active acidity, reserved acidity, anion exchange capacity, redox potential, ox-

idized and reduced zones of submerged soil;
2. Relate the items in No. 1 to soil fertility and productivity;
3. Explain how soil pH affects nutrient availability, nutrient deficiency and microbial activity;
4. Explain the relationship among soil pH, cation exchange capacity and base saturation;
5. Enumerate the chemical and physiochemical changes of soils due to submergence and relate them to soil fertility and productivity;
6. Discuss chemical and physicochemical changes brought about by submergence of lowland soil and relate these changes to cultural practices for higher crop production.

"*Biological Properties of Soil.* The trainees must be able to:
1. Cite the role of soil microbes under various soil conditions;
2. Enumerate the various microbial populations in soils and their relative number in the soil;
3. Explain how the C:N ratio affects microbial population and nitrogen mineralization;
4. Discuss how soil microbes affect soil fertility and productivity;
5. Discuss nitrogen mineralization and factors affecting it;
6. Discuss denitrification in submerged soil;
7. Cite practices to minimize denitrification.

"*Soil Fertility.* The trainees must be able to:
1. Define and differentiate soil fertility and productivity;
2. Describe and cite advantages and limitations of the various methods of assessing soil fertility;
3. Enumerate the essential nutrient elements and the forms by which they are absorbed by plants;
4. Enumerate the criteria of essentiality of nutrient elements;
5. Cite the sources and ways by which these essential nutrient elements are lost or removed from the soil;
6. Discuss soil conditions which favor availability of essential nutrient elements.

"*Soil Sampling.* The trainees must be able to:
1. Enumerate the advantages and limitations of soil tests as a method of assessing soil fertility;
2. Enumerate the steps involved in proper soil sampling;
3. Sample the soil of an area properly;
4. Cite the precautions in soil sampling;
5. Know the other needed information about the area sampled which should go with the soil sample;
6. Know how to interpret and put to use the results of soil tests.

"*Land Preparation*

"*Objectives, Methods and Implements.* The trainees must be able to:
1. Define primary tillage and secondary tillage;
2. Explain the reasons for land preparation;
3. List and describe the steps in land preparation;
4. List and describe various methods of land preparation;
5. Evaluate what method is suited for a set of soil conditions in terms of efficiency, cost and kind of implements to use;
6. Cite and evaluate the advantages and disadvantages of using animal power and tractor power;
7. Cite the conditions when to use a plow, or a tiller, or a rotovator;
8. Identify the various farm implements and parts and cite the functions of each;
9. Describe a well-prepared land;
10. Know the reasons why puddling is necessary in lowland rice paddy;
11. Differentiate between puddled and upland soil in terms of soil structure, soil aeration, water holding capacity and soil tilth after harvest;
12. Perform actual land preparation using the various methods.

"*Plant Physiology*

"*Mineral Nutrition.* The trainees must be able to:
1. List the essential elements and describe their major functions in the plant;
2. Cite the preferences of crops to some forms of nutrient elements;
3. Indicate the amount of nutrients removed by a crop;
4. Describe the nutrient uptake at various growth stages of crops;
5. Describe the various mechanisms by which nutrient elements are taken in by plants.

"*Nutritional Disorders.* The trainees must be able to:
1. Describe the nutrient element deficiency symptoms of the major crops;
2. Cite conditions which cause deficiency and/or toxicity of certain nutrient elements;
3. Describe some consequences or manifestations in the plant when certain nutrient elements are deficient;
4. Describe measures to prevent deficiency of certain nutrient elements;
5. Prescribe measures to correct deficiency.

"*Weed Management*

"*Weeds and Their Control.* The trainees must be able to:
1. Define weeds;
2. Identify and classify weeds common in lowland and upland cultivated land;
3. Describe the effects of weed competition on crop yield;
4. Explain the interaction of crops on weed competition;
5. Identify weaknesses and tolerance of weed species in a crop-weed competition;

6. Enumerate the principles of weed control;
7. Enumerate and describe methods of weed control in mono-crop and multi-crop systems;
8. Differentiate pre-emergence from post-emergence herbicide;
9. Differentiate systemic from contact herbicides, selective from non-selective herbicides;
10. List conditions needed for effective use of herbicides;
11. Design cropping systems for effective ecological control of various weed species;
12. Evaluate herbicides according to effectiveness, physical forms and relative cost.

"Field Experimentation and Statistics

"Principles of Experimentation. The trainees must be able to:

1. Enumerate methods of acquiring new knowledge;
2. Describe each method;
3. Describe the steps of scientific experimentation;
4. Differentiate inductive from deductive reasoning;
5. Formulate objectives for specified field experimentation;
6. Apply properly the concepts required for scientific experimentation.

"Planning of Field Experiments/Applied Research. The trainees must be able to:

1. Formulate experimental objectives;
2. Perform library research to gather current findings on the variables to be tested;
3. Formulate treatments and other variables;
4. Determine plot size, shape of plots, block orientation, number of replications;
5. Choose a suitable design and construct field layout diagrams with proper notations;
6. List materials to use, compute and weigh or measure the desired amounts with proper labels;
7. Illustrate in a diagram the sampling techniques to be followed;
8. Prepare all necessary forms and record books for recording data and other observations on the experiment.

"Statistical Designs. The trainees must be able to:

1. Enumerate the statistical designs commonly used in field experimentation;
2. Characterize each design and cite advantages and disadvantages of each design;
3. Cite the conditions for each design;
4. Follow the conditions needed and note the restrictions when using a particular design.

"Analysis of Variance. The trainees must be able to:

1. Compute and convert the collected data into units commonly used in crop production;
2. Record the data on standard forms or tabulations for statistical analysis;
3. Compute for missing observations;
4. Enumerate and compute for statistical parameters of a given design and set of data;
5. Perform analysis of variance for a given set of data;
6. Compare treatment means statistically;
7. Perform simple correlation and regression analysis;
8. Derive inferences from the data after analysis of variance;
9. Draw conclusions and make recommendations based on the analyzed data;
10. Write a report following the format of scientific writing.

"Establishing Field Experiments. The trainees must be able to:

1. Select a good field and adequate area for the experiment;
2. Lay out the experiment on the field, construct levees when needed, apply the treatments, furrow the field and seed the test crops;
3. Maintain the experiment and perform all the field operations indicated in the treatments at the specified time.

"Sampling Techniques and Collection of Data. The trainees must be able to:

1. Define according to experimental objectives, the kind of data to collect;
2. Determine the sample size for the various crops;
3. Determine when to collect data for various crops in a cropping system;
4. Record the data on the final form at the time of data collection;
5. Cite and follow correctly the procedures for sampling and collecting data;
6. Describe how to collect data objectively.

"Calculations

"Computation Exercises on Chemical Inputs. The trainees must be able to:

1. Familiarize themselves with the formulations of the fertilizers, insecticides, fungicides and herbicides commonly used in cropping systems;
2. Compute for the amounts of these materials given the recommended rates and area of the field;
3. Compute for the amounts of active ingredient given the amounts of these materials and percent content active ingredient per unit area and per hectare basis;
4. Calibrate sprayers to desired amount of spray solution for a specified area;
5. Compute for the amounts of materials given the percent active ingredient, the percent toxicant of final solution and the volume of spray solution;
6. Compute for relative costs of the various materials based on recommended rates.

"Cultural Requirements (for all crops under study)

"Recommended and Promising Varieties. The trainees must be able to:

1. Describe the ideal plant type of the various crops;
2. Describe the current recommended varieties;
3. Describe the crops/variety adaptations to environmental conditions;
4. Cite varieties/selections/accessions in trials at other places.

"Stand Establishment. The trainees must be able to:

1. Describe and operate equipment such as ridger or furrower and seeders;
2. Describe spatial arrangement in a mono-crop and multi-crop planting;
3. Know the recommended plant populations per unit area for the various crops in a mono-crop and multi-crop planting;
4. Compute for the actual amount of seeds needed for a unit area given the percent germination, rate of seeding and spacing.

"Growth Stages. The trainees must be able to:

1. Enumerate and cite the duration of the different growth stages of the various crops;
2. Cite the implications of growth stages to field operations such as cultivation, fertilizer application and control of weeds, insect pests and diseases;
3. Cite the importance of the growth stages to crop yield;
4. Enumerate and explain the effects of environmental factors on the duration of growth stages and on crop development.

"Fertilizer Management. The trainees must be able to:

1. Identify the various fertilizer materials available for commercial use;
2. Describe the fertilizer materials according to the nutrient elements supplied, percent nutrient content, physical and chemical properties and behavior in soils;
3. Describe the response of various crops to application of fertilizer;
4. Cite total nutrient removal per crop of the various crops;
5. Cite recommended fertilizer management practices employed for the various crops;
6. Cite recommended rates of fertilizer for the various crops.

"Pests and Their Control. The trainees must be able to:

1. Identify the different insect and other pests of the various crops;
2. Describe the life cycle of the various pests, especially the stage inflicting damage to crops;
3. Cite environmental factors which affect fluctuations of pest populations;
4. Cite host-plant/variety resistance to pests;
5. Cite the principles involved in pest control;
6. Describe the different methods used in insect control and evaluate their effectivity;
7. Cite the effects of crops and cropping systems on pest-predator-parasite balance and insect pest control;
8. Identify and cite the recommended chemicals for insect control.

"Diseases and Their Control. The trainees must be able to:

1. Define the terms disease, symptoms, signs, host, vector, etc.;
2. Differentiate the terms used in naming a disease or terms used in describing disease symptoms;
3. Identify and describe the typical symptoms of the major diseases of various crops;
4. Classify broadly the causal organisms of the various diseases;
5. Cite environmental factors which affect the development of the various diseases;
6. Cite host/variety resistance to diseases;
7. Identify physiological diseases and allelopathy resulting from intensive cropping;
8. Cite the principles and methods used in the control of crop diseases;
9. Cite the recommended control measures of the major diseases of the various crops.

"Harvesting, Harvest Processing and Storing. The trainees must be able to:

1. Describe the different methods of harvesting, threshing, drying and storing;
2. Cite the advantages and disadvantages of the different methods of harvesting, threshing, drying and storing;
3. Identify and be able to use properly the available equipment for harvesting, threshing, drying and storing the produce;
4. Cite the moisture contents of cereals at harvesting and at storing;
5. Describe the steps in moisture determination using the available moisture testers;
6. Compute the crop yield per unit area based on specified moisture content.

"Cropping Systems and Crop Intensification

"The trainees must be able to:

1. Define the terms used in cropping systems;
2. Cite factors affecting specific cropping systems;
3. Describe cropping systems based on rice;
4. Cite current findings on cropping systems in terms of interplant relations, crop yields, insect pests, disease and weed control;
5. Cite common practices in cropping systems;
6. Cite prospective research areas in cropping systems.

"Economics

"The trainees must be able to:

1. Cite and apply the principles of economics in cropping systems;

2. Discuss efficient use of farm resources;
3. Evaluate the effects of inputs on intensive cropping systems;
4. Discuss the interaction between fixed and variable inputs;
5. Evaluate the economic interaction between crops in a cropping pattern;
6. Compute the returns using techniques in farm planning and budgeting given a series of crop production data;
7. Determine profitability of cropping systems;
8. Cite prospective research areas on the economics of cropping systems.

"*Communication and Extension*

"The trainees must be able to:

1. Enumerate the elements of communication;
2. Describe how elements of communication affect the communication process;
3. Discuss the communication process using a model;
4. Formulate instructional/behavioral objectives;
5. Cite and discuss the different channels of communication;
6. Cite and discuss barriers to the learning process and communication;
7. Cite, discuss and evaluate the effectivity of the various extension methods;
8. Write a simple communication to instruct a specified audience to do a certain practice in cropping systems;
9. Enumerate and apply the basic rules in writing;
10. Describe the format of scientific reporting."

APPENDIX 11-E SIX PRINCIPLES FOR TECHNICAL CHANGE

In Sec. 11.6., Chapter 11, we suggested several references for preparing training material for FSR&D teams. One of these references referred to Margaret Mead's work on introducing change. In this appendix, we have extracted six of seven principles that Mead (1955) presented on this subject.[1] The principles were derived from the findings of psychiatrists and psychologists and apply to those who work on a daily basis in villages, schools, agricultural demonstration stations, and the like:

"1. The agents of change, the teacher, the agricultural extension worker, the nurse, must realize that their own behaviour, beliefs, and attitudes are not universal and axiomatic. They must realize that their ways of counting, reckoning time, judging conduct, expressing enthusiasm or disgust, are—like the behaviour, beliefs and attitudes of those whom they are helping to change—learned and traditional.

"2. The beliefs and attitudes of the people among whom they are working must be seen as having functional utility. For each individual, they give continuity to his personality—permit him to feel that he is a named, identified person. . . . So, he will hold on to his beliefs and practices because they help him to direct his daily behaviour and solve his daily problems of relating himself to other people. If the teacher or extension agent recognizes such clinging to old beliefs and practices as having real usefulness for an individual, rather than interpreting it as evidence of stubborness, unco-operativeness, ignorance, inability to learn, etc., he will be better able to introduce changes.

"3. Any change must be examined from the point of view of the individuals who are exposed to the change. Where a change may seem to the expert to be merely a better way of feeding cattle, or of disposing of waste, to the people it may seem to be a rejection of the commands of the gods, or a way of giving their welfare and safety into the hands of sorcerers. . . . It is, therefore, useful always to ask: How does this change look to those whom it will directly and indirectly affect?

"If this question is asked, it will assist the experts in forecasting some of the difficulties which may occur, and in devising ways of compensating for them.

"4. The experts must be on their guard against the apparently logical solution that, because all the aspects of the life of a people are interrelated, the way to deal with a change in any one aspect of living is to make a complete blueprint for changes in the whole. . . . There is no available body of knowledge which makes it possible to predict in advance the way in which individuals will respond even to one far-reaching change, so that it is necessary to avoid master plans, . . .

"5. Any significant change in the life of an individual tends to introduce some degree of instability or disharmony in the way his life activities, his beliefs and attitudes, are organized. Such instability can be described psychologically, as emotional tension.

"A significant change results in tension either because old behaviour is found to be inadequate or by creating new situtations for which new behaviour must be acquired. . . . Even if the individual is willing to give up his old responses for new ones, he will be in a

[1]Extracts from *Cultural Patterns and Technical Change.* © UNESCO 1955. Reproduced by permission of UNESCO.

state of tension while he unlearns the old responses. . . .

"Although we may expect that the existence of such tensions will be accompanied by changes in the individual's behaviour which tend to reduce the original tension, the dissipation of such tensions may take a considerable time or may not be successful. . . .

". . . even when changes in an individual's life are very painful for him to achieve, if he has a strong wish to change, then the successful resolution of his tensions is quicker and more likely, and there are fewest returns to old and undesired responses. Also there may be less tension and reluctance to change if the new procedures deal with the recognized difficulties and sufferings of the people and will directly help to resolve those difficulties and reduce those sufferings which have long been endured as inevitable.

"It follows that there is less reason to fear mental health disturbances among those populations in which the individuals affected by the change have themselves desired the change. . . .

"6. * * *

The possible consequences of frustration are very numerous and they are not by any means all bad. The nature of the consequences depends partly upon the severity of frustration. . . .

"Many or most of the frustrations experienced in daily life are not harmful. . . . They alter the way in which the individual's energy is distributed among his daily tasks, help him reorient himself, learn and grow, by (a) leading to intensification of his efforts; (b) forcing him to re-examine the situtation in which he is blocked; (c) leading to his finding some way around the difficulty, often a new way.

"It is conceivable, then, that it is not harmful . . . to stimulate needs, desires, demands among peoples who now feel no such needs or desires, or who at least are not conscious of them or articulate about them. Such stimulation does, of course, create instability, disharmony and tension. . . . But if means are made available, then the created frustrations can become the basis for new, desired, and self-perpetuating behaviour. . . . The danger of cultivating 'felt needs', which are unrealized under existing conditions, can be diminished by keeping close to local conditions, to that which is immediately feasible, so that training teachers precedes building local schools, importing a minimum supply of a new seed precedes the demonstration of its superiority."

REFERENCES CITED IN THE APPENDIXES (By Chapter)

2. INTRODUCTION

Adrien, P.M., and M.F. Bumgardner. 1977. Landsat computers and development projects. Science 197:466–470.

Anderson, F.M., and J.C.M. Trail. 1978. Initial application of modeling techniques in livestock production systems under semi-arid conditions in Africa. ILCA Working Doc. No. 8. ILCA, Kenya.

Angus, J.F., and H.G. Zandstra. 1979. Climatic factors and the modeling of rice growth and yield. IRRI, Los Banos, Philippines.

Barlow, C., S. Jayasuriya, V. Cordova, N. Roxas, L. Yambao, C. Bantilan, and C. Maranan. 1979. Measuring the economic benefits of new technologies to small rice farmers. IRRI Res. Paper Ser. IRRI, Los Banos, Philippines.

Barton, R.F. 1970. A primer on simulation and gaming. Prentice-Hall, Inc., Englewood Cliffs, New Jersey.

Beneke, R.R., and R.D. Winterboer. 1973. Linear programming applications to agriculture. Iowa State University Press, Ames, Iowa.

Flinn, J.C., and S. Jayasuriya. 1979. Incorporating multiple objectives into plans for low-resource farmers. Paper No. 79-01. IRRI, Los Banos, Philippines.

Hart, R.D. 1980a. One farm system in Honduras: a case study. *In* Activities at Turrialba 8:1:3–8. CATIE, Turrialba, Costa Rica.

______. 1980b. Region, farm and agroecosystem characterization: the preliminary phase in a farm system research strategy. A paper presented at the 72nd Ann. Meeting of the A. Soc. of Agron. Detroit, Mich. 30 Nov. - 5 Dec. 1980.

______. 1979. Agroecosistemas: conceptos básicos. CATIE, Turrialba, Costa Rica.

Heady, E.O., and W. Candler, 1958. Linear programming methods. Iowa State University Press, Ames, Iowa.

Hill, R.W., D.R. Johnson, and K.H. Ryan. 1978. A model for predicting soybean yields from climatic data. Proc. Workshop on Crop Simulation. Clemson Univ., Clemson, S.C.

Hillier, F.S., and G.J. Lieberman. 1974. Operations research. Holden-Day, Inc., San Francisco, Calif.

Jayasuriya, S. 1979. New cropping patterns for Iloilo and Pangasinan farmers: a whole farm analysis. IRRI Sat. Sem. 21 July 1979. IRRI, Los Banos, Philippines.

Kenlen, H.V. 1976. Simulation of influence of climatic factors on rice production. p. 345–358. *In* K. Takahashi and M.M. Yoshino (eds.). Climatic change and food production. Univ. of Tokyo Press, Tokyo.

Low, A.R.C. 1975. Small farm improvement strategies: the implications of a computer simulation study of indigenous farming in Southeast Ghana. Oxford Agrar. Studies 14:3–19.

MacDonald, R.B., and F.G. Hall. 1977. LACIE: a proof-of-concept experiment in global crop monitoring, LACIE NASA/JSC-NR 7-00031. NASA, Houston, Texas.

McCarl, B.A. 1978. A farm linear programming analysis of dry-land and wet-land food crop production in Indonesia. IBRD, Washington, D.C.

Nelson, R.S.M. 1974. Linear programming for smallholder agriculture. Farm Mgmt. Branch, Land and Farm Mgmt. Div., Min. of Agric., Kenya.

Odum, H.T. 1971. Environment, power and society. John Wiley & Sons, New York.

Ritchie, J.T. 1972. Model for predicting evaporation from a new crop with incomplete cover. Water Resources Res. 8:1204–1213.

Roumasset, J.A. 1976. Rice and risk: decision-making among low-income farmers. North-Holland Publishing Co., Amsterdam.

Sanders, J.O., and T.C. Cartwright. 1979a. A general cattle production systems model I: structure of the model. Agric. Sys. 4:27.

______. 1979b. A general cattle production systems model II: procedures used for simulating animal performance. Agric. Sys. 4:289.

Tamisin, M.M., D.T. Franco, E.B. Manalo, and H.G. Zandstra. 1979. Modeling of potential evapotranspiration and solar radiation for different regions of the Philippines. IRRI Sat. Sem. 13 Oct. 1979. IRRI, Los Banos, Philippines.

Thodey, A.R., and R. Sektheera. 1974. Optimal multiple cropping systems for the Chiang Mai Valley. Agric. Econ. Rep. 1. Chiang Mai Univ., Chiang Mai, Thailand.

Valdes, A., G.M. Scobie and J.L. Dillon (eds.). 1979. Economics and the design of small-farmer technology. Iowa State University Press, Ames, Iowa.

Wardhani, M.A. 1976. Rational farm plans for land settlement in Indonesia: a study using programming techniques. Dev. Studies Center, Australia National Univ., Canberra, Australia.

3. CONCEPTUAL FRAMEWORK

Byerlee, D., M.P. Collinson, R.K. Perrin, D.L. Winkelmann, S. Biggs, E.R. Moscardi, J.C. Martinez, L. Harrington, and A. Benjamin. 1980. Planning technologies appropriate to farmers: concepts and procedures. CIMMYT, El Batan, Mexico.

CATIE. 1979. CATIE, tropical agricultural research and training center. CATIE, Turrialba, Costa Rica.

CGIAR. 1976. Consultative group on international agricultural research. CGIAR, New York.

CIAT. 1980. CIAT report, 1980. CIAT, Cali, Colombia.

CIMMYT. 1978. CIMMYT review, 1978. CIMMYT, El Batan, Mexico.

CIP. 1977. Annual report: CIP. CIP, Lima, Peru.

CRIA Cropping Systems Working Group. 1979. [Draft] Network methodology and cropping systems research in Indonesia. CRIA, Bogor, Indonesia.

EMBRAPA. 1978. EMBRAPA ano 6. EMBRAPA, Brasilia, DF, Brazil.

Gilbert, E.H., D.W. Norman, and F.E. Winch. 1980. Farming systems research: a critical appraisal. MSU Rural Dev. Paper No. 6. Dep. of Agric. Econ., Michigan State Univ., East Lansing, Mich.

His Majesty's Government Department of Agriculture. 1979. Work plan, winter 1978–79. Cropping Sys. Prog., Agron. Div. HMG Dep. of Agric., Khumaltar, Lalitpur, Nepal.

ICARDA. 1978. Farming systems research programme, 1978–79. ICARDA, Beirut, Lebanon.

ICRISAT. 1979. ICRISAT research highlights. ICRISAT, Hyderabad, India.

ICTA. 1976. Objetivos, organización, funcionamiento. Pub. Mis., Folleto No. 3. ICTA, Guatemala.

IITA. 1978. Annual report for 1978. IITA, Ibadan, Nigeria.

ILCA. 1980. ILCA the first years. ILCA, Addis Ababa, Ethiopia.

IRRI. 1977a. Research highlights for 1977. IRRI, Los Banos, Philippines.

______. 1977b. Symposium on cropping systems research and development for the Asian rice farmer. IRRI, Los Banos, Philippines.

ISRA/GERDAT. 1977. Recherche et développement agricole: les unités expérimentales du Sénégal. CNRA Sem. 16–21 May 1977. Bambey, Senegal.

Perrin, R.K., D.L. Winkelmann, E.R. Moscardi, and J.R. Anderson. 1976. From agronomic data to farmer recommendations: an economics training manual. Inf. Bull. 27. CIMMYT, El Batan, Mexico.

PNIA. 1981. Funcionamiento del programa nacional de investigación agropecuaria y su integración en un sistema tecnológico. Secretaría de Recursos Naturales. PNIA, Tegucigalpa, D.C., Honduras.

Technical Advisory Committee (TAC). Review Team of the Consultative Group on International Agricultural Research. 1978. Farming systems research at the international agricultural research centers. The World Bank, Washington, D.C.

WARDA. 1978. A quinquennial report of WARDA research activities, 1973–78. WARDA, Monrovia, Liberia.

Zandstra, H.G., E.C. Price, J.A. Litsinger, and R.A. Morris. 1981. A methodology for on-farm cropping systems research. IRRI, Los Banos, Philippines.

4. TARGET AND RESEARCH AREA SELECTION

Byerlee, D., M.P. Collinson, R.K. Perrin, D.L. Winkelmann, S. Biggs, E.R. Moscardi, J.C. Martinez, L. Harrington, and A. Benjamin. 1980. Planning technologies appropriate to farmers: concepts and procedures. CIMMYT, El Batan, Mexico.

Collinson, M.P. 1979. Understanding small farmers. A paper presented at a conf. on Rapid Rural Appraisal. 4–7 Dec. 1979. IDS, Univ. of Sussex, Brighton, UK.

CRIA Cropping Systems Working Group. 1979. [Draft] Network methodology and cropping systems research in Indonesia. CRIA, Bogor, Indonesia.

IRRI. 1974. An agro-climatic classification for evaluating cropping systems potentials in Southeast Asian rice growing regions. IRRI, Los Banos, Philippines.

Jodha, N.S., M. Asokan, and J.G. Ryan. 1977. Village study methodology and research endowments of the selected villages in ICRISAT's village level studies. Occ. Paper 16. Econ. Prog., ICRISAT, Hyderabad, India.

5. PROBLEM IDENTIFICATION AND DEVELOPMENT OF A RESEARCH BASE

Babbie, E.R. 1973. Survey research methods. Wadsworth Publishing Co., Belmont, Calif.

Beal, G.M. and D.N. Sibley. 1967. Adoption of agricultural technology by the Indians of Guatemala. Rur. Soc. Rep. No. 62. Dep. of Soc. and Anthro. Iowa State Univ., Ames, Iowa.

Beirut Seminar Working Group. 1976a. Choice of sampling methods. p. 29–30. *In* B. Kearl (ed.). Field data collection in the social sciences: experiences in Africa and the Middle East. Agricultural Development Council, New York.

______. 1976b. Extension agents. p. 117. *In* B. Kearl (ed.). Field data collection in the social sciences: experiences in Africa and the Middle East. Agricultural Development Council, New York.

Berelson, B. 1971. Content analysis in communications research. Hafner, New York.

Bernsten, R. 1979. [Draft] Design and management of survey research: a guide for agricultural researchers. CRIA/IRRI Cooperative Program, Bogor, Indonesia.

Biggs, S.C. 1978. Planning rural technologies in the context of social structures and reward systems: Asian report #3. CIMMYT/ICRISAT, New Delhi, India.

Byerlee, D., M.P. Collinson, R.K. Perrin, D.L. Winkelmann, S. Biggs, E.R. Moscardi, J.C. Martinez, L. Harrington, and A. Benjamin. 1980. Planning technologies appropriate to farmers: concepts and procedures. CIMMYT, El Batan, Mexico.

Campbell, D.T. 1975. Reforms as experiments. p. 71–101. *In* E. L. Struening and M. Guttentag (eds.). Handbook of evaluation research. Sage Publications, Beverly Hills, Calif.

______, and J.C. Stanley. 1966. Experimental and quasi-experimental designs for research. Rand McNally College Publishing Company, Chicago.

Collinson, M.P. 1979. Understanding small farmers. A paper presented at a conf. on Rapid Rural Appraisal. 4–7 Dec. 1979. IDS, Univ. of Sussex, Brighton, UK.

______. 1976. The cost of greater accuracy. p. 32–33. *In* B. Kearl (ed.). Field data collection in the social sciences: experiences in Africa and the Middle East. Agricultural Development Council, New York.

Cook, T.D., F.L. Cook and M.M. Mark. 1977. Randomized and quasi-experimental designs in evaluation research: an introduction. p. 103–139. *In* L. Rutman (ed.). Evaluation research methods: a basic guide. Sage Publications, Beverly Hills, Calif.

______, and D.T. Campbell. 1976. The design and conduct of quasi-experiments and true experiments in field settings. p. 223–326. *In* M.C. Dunnette (ed.). Handbook of industrial and organizational psychology. Rand McNally College Publishing Company, Chicago.

Cyert, R., and J. March. 1963. A behavioral theory of the firm. Prentice-Hall, Englewood Cliffs, New Jersey.

DeBoer, A.J., and A. Weisblat. 1978. Livestock component of small-farm systems in South and Southeast Asia. Presented at the Bellagio Conference. 18–23 Oct. 1978. The Rockefeller Foundation, New York.

Duff, B. 1978. The potential for mechanization in small farm production systems. Presented at the Bellagio Conf. 18–23 Oct. 1978. The Rockefeller Foundation, New York.

El Hadari, A.M. 1976a. Association with village leaders promotes confidence. p. 53–54. *In* B. Kearl (ed.). Field data collection in the social sciences: experiences in Africa and the Middle East. Agricultural Development Council, New York.

______. 1976b. Offending a key local person can be costly. p. 173. *In* B. Kearl (ed.). Field data collection in the social sciences: experiences in Africa and the Middle East. Agricultural Development Council, New York.

______. 1976c. Advantages to hiring interviewers from the region. p. 124–125. *In* B. Kearl (ed.). Field data collection in the social sciences: experiences in Africa and the Middle East. Agricultural Development Council, New York.

FAO. 1976. A framework for land evaluation. Soils Bull. 32. FAO, Rome.

Flinn, J.C. 1976a. "Borrowed" staff bring problems. p. 130. *In* B. Kearl (ed.). Field data collection in the social sciences: experiences in Africa and the Middle East. Agricultural Development Council, New York.

______. 1976b. Content of the training program. p. 133. *In* B. Kearl (ed.). Field data collection in the social sciences: experiences in Africa and the Middle East. Agricultural Development Council, New York.

Gafsi, S. 1976a. Listening to farmers' problems is productive. p. 168–169. *In* B. Kearl (ed.). Field data collection in the social sciences: experiences in Africa and the Middle East. Agricultural Development Council, New York.

______. 1976b. Personal qualities of enumerators. p. 124. *In* B. Kearl (ed.). Field data collection in the social sciences: experiences in Africa and the Middle East. Agricultural Development Council, New York.

Gucelioglu, O. 1976. Record book stimulates recall. p. 81. *In* B. Kearl (ed.) Field data collection in the social sciences: experiences in Africa and the Middle East. Agricultural Development Council, New York.

Gladwin, C.H. 1979. Cognitive strategies and adoption decisions: a case study of nonadoption of an agronomic recommendation. *In* Economic Development and Cultural Change 28:1:155–173.

Harris, N. 1966. The cultural ecology of India's sacred cattle. Current Anthro. 7:51–66.

Hart, R.D. 1980a. Region, farm and agroecosystem characterization: the preliminary phase in a farm system research strategy. A paper presented at the 72nd Ann. Meeting of the A. Soc. of Agron. Detroit, Mich. 30 Nov. - 5 Dec. 1980.

______. 1980b. One farm system in Honduras: a case study. *In* Activities at Turrialba 8:1:3–8. CATIE, Turrialba, Costa Rica.

Harwood, R.R. 1979. Small farm development: understanding and improving farming systems in the humid tropics. Westview Press, Boulder, Colo.

Haskins, J.B. 1968. How to evaluate mass communications: the controlled field experiment. The Advertising Research Foundation, New York.

Hatch, J.K. 1980. A record keeping system for rural households. MSU Rural Dev. Paper No. 9. Dep. of Agric. Econ., Michigan State Univ., East Lansing, Mich.

Hildebrand, P.E. 1979a. Summary of the *sondeo* methodology used by ICTA. Proc., conf. on Rapid Rural Appraisal. 4–7 Dec. 1979. IDS, Univ. of Sussex, Brighton, UK.

______. 1979b. The ICTA farm record project with small farmers: four years experience. ICTA, Guatemala.

______. 1979c. Incorporating the social sciences into agricultural research: the formation of a national farm systems research institute. ICTA, Guatemala, and The Rockefeller Foundation, New York.

Hill, P. 1972. Rural Hausa: a village and a setting. Cambridge University Press, Cambridge, UK.

House, E.R. 1980. Evaluating with validity. Sage Publications, Beverly Hills, Calif.

ILCA. 1980. ILCA the first years. ILCA, Addis Ababa, Ethiopia.

Kearl, B. (ed.). 1976. Field data collection in the social sciences: experiences in Africa and the Middle East. Agricultural Development Council, New York.

Kerlinger, F.N. 1973. Foundations of behavioral research. Holt, Rinehart and Winston, Inc., New York.

Kinnear, T.C., and J.R. Taylor. 1979. Marketing research: an applied approach. McGraw-Hill Book Co., Inc., New York.

Lele, U.J. 1975. The design of rural development. International Bank for Reconstruction and Development. Washington, D.C.

Lin, N. 1976. Foundations of social research. McGraw-Hill, New York.

Lipton, M., and M. Moore. 1972. The methodology of village studies in less developed countries. IDS disc. paper No. 10. Univ. of Sussex, Brighton, UK.

Martius-von Harder, G. 1979. How and what rural women know: experiences in Bangladesh. p. 406. *In* S. Zeidenstein (ed.). Studies in family planning: learning about rural women. Vol. 10. The Population Council, New York.

McCown, R.L., G. Haaland, and C. DeHaan. 1979. The interaction between cultivation and livestock production in semi-arid Africa. *In* A.E. Hall, G.H. Cannell, and H.W. Lawton (eds.). Agriculture in semi-arid environments. Ecolog. Studies, Vol. 34. Springer-Verlag, Berlin, Heidelberg, New York.

McDowell, R.E., and P.E. Hildebrand. 1980. Integrating crop and animal production: making the most of resources available to small farms in developing countries. Presented at Bellagio Conf. 18–23 Oct. 1978. The Rockefeller Foundation, New York.

McSweeney, B.G. 1979. Collection and analysis of data on rural women's time use. p. 379. *In* S. Zeidenstein (ed.). Studies in family planning: learning about rural women. Vol. 10. The Population Council, New York.

Mellor, J.W. 1966. The economics of agricultural development. Cornell University Press, Ithaca, New York.

Mencher, J.P., K. Saradamoni, and J. Paniker. 1979. Women in rice cultivation: some research tools. p. 408. *In* Zeidenstein (ed.). Studies in family planning: learning about rural women. Vol. 10. The Population Council, New York.

Morgan, W.B., and J.C. Pugh. 1969. West Africa. Methuen, London.

Mosher, A.T. 1966. Getting agriculture moving. The Agricultural Development Council, New York.

Nabila, J.S. 1976a. Incentives only if required by tradition. p. 164. *In* B. Kearl (ed.). Field data collection in the social sciences: experiences in Africa and the Middle East. Agricultural Development Council, New York.

______. 1976b. Accepting local custom puts respondents at ease. p. 170–171. *In* B. Kearl (ed.). Field data collection in the social sciences: experiences in Africa and the Middle East Agricultural Development Council, New York.

______. 1976c. Diplomacy takes time, is worth it? p. 172. *In* B. Kearl (ed.). Field data collection in the social sciences: experiences in Africa and the Middle East. Agricultural Development Council, New York.

Netting, R. 1965. Household organization and intensive agriculture: the Kofyar case. Africa 35:422–429.
Norman, D.H. 1976a. Incentives had questionable value. p. 164–165. *In* B. Kearl (ed.). Field data collection in the social sciences: experiences in Africa and the Middle East. Agricultural Development Council, New York.
______. 1976b. Factors that influence recall. p. 67–68. *In* B. Kearl (ed.). Field data collection in the social sciences: experiences in Africa and the Middle East. Agricultural Development Council, New York.
Ogunfowora, O. 1976a. Winning cooperation takes time and patience. p. 52–53. *In* B. Kearl (ed.). Field data collection in the social sciences: experiences in Africa and the Middle East. Agricultural Development Council, New York.
______. 1976b. Some sources of reluctance to cooperate. p. 167. *In* B. Kearl (ed.). Field data collection in the social sciences: experiences in Africa and the Middle East. Agricultural Development Council, New York.
______. 1976c. High educational standards increase turnover. p. 120–121. *In* B. Kearl (ed.). Field data collection in the social sciences: experiences in Africa and the Middle East. Agricultural Development Council, New York.
Quinn, N. 1976. A natural system used in Mfantse litigation settlement. A. Ethnolog. 3:331–351.
Rahim, S.A. 1976. Communication and policy for development: an annotated bibliography. East-West Communication Institute, Honolulu, Hawaii.
Rappaport, R.A. 1968. Pigs for the ancestors. Yale University Press, New Haven, Conn.
Rogers, E.M. 1976. Communication and development: critical perspectives. Communication Research 3:2:99–240.
______. and F.F. Shoemaker. 1971. Communication of innovations. The Free Press, New York.
Rosengren, K.E. 1980. Advances in content analysis. Sage Publications, Beverly Hills, Calif.
Ruthenberg, H. 1971. Farming systems in the tropics. Clarendon Press, Oxford, UK.
Safai, M. 1979. Circumventing problems of accessibility to rural Muslim women. p. 405. *In* S. Zeidenstein (ed.). Studies in family planning: learning about rural women. Vol. 10. The Population Council, New York.
Sahlins, M.D. 1957. Land use and the extended family in Moala, Fiji. A. Anthropologist 59:3:449–462.
Simon, H. 1969. A behavioral model of rational choice. Q.J. Econ. 69:99–118.
Slonim, M.J. 1960. Sampling. Simon and Schuster, New York.
Smith, E.D. 1977. Assessment of the capacity of national institutions to introduce and service new technology. *In* Symp. on Cropping Sys. Res. and Dev. for the Asian Rice Farmer. IRRI, Los Banos, Philippines.
Smock, A.C. 1979. Measuring rural women's economic roles and contributions in Kenya. p. 385. *In* S. Zeidenstein (ed.). Studies in family planning: learning about rural women. Vol. 10. The Population Council, New York.
Tollens, E. 1976. Why stratification permits economies. p. 33–34. *In* B. Kearl (ed.). Field data collection in the social sciences: experiences in Africa and the Middle East. Agricultural Development Council, New York.
Tversky, A. 1972. Elimination by aspects: a theory of choice. Psyc. Rev. 79:281.
van Raaij, J.G.T. 1974. Rural planning in a savanna region. Univ. Pers., Rotterdam.
Zandstra, H.G., E.C. Price, J.A. Litsinger, and R.A. Morris. 1981. A methodology for on-farm cropping systems research. IRRI, Los Banos, Philippines.
Zeidenstein, S. (ed.). 1979. Studies in family planning: learning about rural women. Vol. 10. The Population Council, New York.

6. PLANNING ON-FARM RESEARCH

Chang, J. 1968. Climate and agriculture. Aldine Publishers, Hawthorne, New York.
Cochran, W.G. and G.M. Cox. 1957. Experimental designs. John Wiley & Sons, New York.
Doorenbos, J., and W.O. Pruitt. 1977. Crop water requirements. FAO Irrigation and Drainage Paper 24. FAO, Rome.
Harwood, R.R. 1979. Small farm development: understanding and improving farming systems in the humid tropics. Westview Press, Boulder, Colo.
Hiebsch, C.K. 1978. Comparing intercrops with monocultures. *In* Tropical soils annual report for 1976–77. Soil Sci. Dep., North Carolina State Univ., Raleigh, N.C.
Little, T.M., and F.J. Hills. 1978. Agricultural experimentation. John Wiley and Sons, Inc., New York.
McIntosh, J.L. 1980. Cropping systems and soil classification for agrotechnology development and transfer. *In* Proc. Agrotech. Transfer Workshop 7–12 July 1980. Soils Res. Inst., AARD, Bogor, Indonesia and Univ. of Hawaii, Honolulu.
Rosenberg, N.J. 1974. Microclimate: the biological environment. John Wiley & Sons, New York.
Wharton, C.R., Jr. 1968. Risk, uncertainty, and the subsistence farmer: technological innovation and resistance to change in the context of survival. Presented at the Joint Meetings of the A. Econ. Assoc. and the Assoc. for Comp. Econ. 28 Dec. 1968. Chicago.

7. ON-FARM RESEARCH AND ANALYSIS

ILCA. 1980. ILCA the first years. ILCA, Addis Ababa, Ethiopia.
______. 1979. Highlands. ILCA, Addis Ababa, Ethiopia, (mimeo).
McIntosh, J.L. 1980. Cropping systems and soil classification for agrotechnology development and transfer. *In* Proc. Agrotech. Transfer Workshop. 7–12 July 1980. Soils Res. Inst., AARD, Bogor, Indonesia and Univ. of Hawaii, Honolulu.
Perrin, R.K., D.L. Winkelmann, E.R. Moscardi, and J.R. Anderson. 1976. From agronomic data to farmer recommendations: an economics training manual. Inf. Bull. 27. CIMMYT, El Batan, Mexico.
Zandstra, H.G., E.C. Price, J.A. Litsinger, and R.A. Morris. 1981. A methodology for on-farm cropping systems research. IRRI, Los Banos, Philippines.

8. EXTENSION OF RESULTS

Haws, L.D., and R.T. Dilag, Jr. 1980. Appendix A. *In* Development and implementation of pilot production programs. Presented at the Cropping Sys. Conf. 3–7 March 1980. IRRI, Los Banos, Philippines.
ICTA and DIGESA. 1978. Carta de entendimiento entre El Instituto de Ciencia y Tecnología Agrícolas (ICTA) y La Dirección General de Servicios Agrícolas (DIGESA). ICTA, Guatemala.

10. IMPLEMENTATION

Francis, D., and D. Young. 1979. Improving work groups: a practical manual for team building. University Associates, San Diego, Calif.

ICTA. 1976. Objetivos, organización, funcionamiento. Pub. Mis., Folleto No. 3. ICTA, Guatemala.

INIAP. [undated] Programa de investigación en producción. Quito, Ecuador.

ISRA. 1979. Senegal agricultural research project. Government of Senegal, Dakar, Senegal.

Ortiz D., R. 1980. [Draft]. Generation and promotion of technology for production systems on small farms: ICTA's approach and strategy in Guatemala. University of Florida Press, Gainesville, Fla.

PNIA. 1978. Agricultural research in Honduras. Secretaría de Recursos Naturales. PNIA, Comayagua, Honduras.

Turner, H.D. 1979. [Draft]. Aid handbook on evaluation. USAID, Washington, D.C.

11. TRAINING

CATIE. 1979. Objectives-organization-functions. CATIE, Turrialba, Costa Rica.

CIAT. 1980. CIAT Report 1980. CIAT, Cali, Colombia.

______. [undated]. Postgraduate training opportunities at CIAT. CIAT, Cali, Colombia.

CIMMYT. 1978a. This is CIMMYT 78. CIMMYT, El Batan, Mexico.

______. 1978b. CIMMYT review 1978. CIMMYT, El Batan, Mexico.

______. 1978c. CIMMYT Today: CIMMYT Training, No. 9. CIMMYT, El Batan, Mexico.

Fonseca M., S. 1979. Director of CATIE addresses ministers of agriculture from Mexico, Central America, Panama and Dominican Republic. *In* Activities at Turrialba 7:4:2–5. CATIE, Turrialba, Costa Rica.

ICTA – DIGESA. 1979. Agricultural public sector: institutional and technological relationships project. ICTA-DIGESA, Jutiapa, Guatemala.

ILCA. [undated]. Announcement of livestock systems research training course: 1 July - 3 Oct. 1981. ILCA, Addis Ababa, Ethiopia.

IRRI. 1978. Research highlights for 1977. IRRI, Los Banos, Philippines.

______. [undated]. Instructional topics of cropping systems training program. 14 Sept. 1977 - 17 March 1978. IRRI, Los Banos, Philippines.

Mead, M. (ed.). 1955. Cultural patterns and technical change. Mentor Publishing Company, New York.

Perdon, E.R. 1977. The IRRI 6-month training in rice production. IRRI, Los Banos, Philippines.

PNIA. 1980. Manual de actividades de capacitación en servicio. Secretaría de Recursos Naturales. PNIA, Comayagua, Honduras.

INDEX

Acceptability index, 141
Acceptance, farmer, 90, 95, 96, 158, 279. *See also* Technology, farmer adaptation
acceptability index, 141
analysis of, 139-141
measurements of, 125, 135
predictive approach to, 124
sociocultural analysis, 141, 142
Activities in FSR&D process
development of research base, 5, 28, 72-83, 150
experiment station collaboration, 28, 30, 31, 89, 96, 97, 104, 150
extension of results, 6, 28-30, 150, 152-156
extension's collaboration, 28, 30, 31, 35, 149-152
planning on-farm research, 5, 6, 28, 87-106, 150
problem identification, 5, 28, 61-72, 150
research, (on-farm) and analysis, 6, 28, 29, 111-114, 150
target and research area selection, 5, 27, 28, 41-56, 150
timing of, 31-33, 35
Adrien, P.M. and M.F. Bumgardner, 232
Advisory committee, 180, 181
Agency for International Development, U.S. *See* USAID
Agricultural extension service. *See* Extension organizations
Agricultural Science and Technology Institute (Guatemala). *See* ICTA
Agroclimatic environments
definition of, 213
as zones, 98, 103, 336
Alvin, P.T., 99
Analysis of data. *See* Research results analysis; Statistical analysis
Analysis, research. *See* Research results analysis; Statistical analysis
ACSN (Asian Cropping Systems Network—Philippines), 49, 154, 202, 237
Anderson, F.M. and J.C.M. Trail, 232
Angus, J.F. and H.G. Zandstra, 232
Applied research. *See also* Basic research
"downstream" research, 37
FSR&D emphasis on, 168
Asian Cropping Systems Network (Philippines). *See* ACSN

Babbie, E.R., 303, 305
Barlow, C., et al., 232
Barton, R.F., 233
Baseline data, definition of, 213
Basic research. *See also* Applied research
"upstream" research, 37
Beal, G.M. 262
Beal, G.M. and D.N. Sibley, 65, 142, 262, 263
Beirut Seminar Working Group, 303, 305, 308, 309
Benefits. *See* Net benefits
Beneke, R.R. and R.D. Winterboer, 232
Benor, D. and J.Q. Harrison, 157
Berelson, B., 284
Bernsten, R., 278-280, 299, 300, 303-309, 316, 317
Biggs, S.C., 265
Biological factors
animals, 64, 91, 102
definition of, 213
diseases and nutritional disorders, 64, 296, 394, 396
insects, 64, 96, 396
other pests, 64, 91, 296, 396
plants, 64, 91, 102, 394
weeds, 64, 394, 395
Biological feasibility
considerations in setting priorities, 71
definition of, 213
Biological system, definition of, 213
Brazilian Agricultural Research Corporation (Brazil). *See* EMBRAPA
Brown, M.L., 125, 215
Budget analysis. *See* Partial budget analysis
Built-in evaluation, 191
Byerlee, D., et al., 23, 27, 44, 76, 215, 238, 243, 244, 259, 278, 279, 299, 302, 308

Campbell, D.T., 298
Campbell, D.T. and J.C. Stanley, 299
Capital input. *See* Economic environmental factors
Case studies. *See also* Data, primary
advantages and disadvantages of, 81
form of frequent interview survey, 81
CATIE (Tropical Agricultural Research and Training Center—Costa Rica), 22, 114, 190, 237, 389
FSR&D program, 239
mathematical modeling, 231, 232
training program, 191, 197, 202, 203, 384, 387, 392, 393
CENTA (National Agricultural Technology Center—El Salvador), 78
Central Research Institute for Agriculture (Indonesia). *See* CRIA
Centro Agronómico Tropical de Investigación y Enseñanza (Costa Rica). *See* CATIE
Centro Internacional de Agricultura Tropical (Colombia). *See* CIAT
Centro Internacional de Mejoramiento de Maíz y Trigo (Mexico). *See* CIMMYT
Centro Internacional de la Papa (Peru). *See* CIP
Centro Nacional de Tecnología Agropecuaria (El Salvador). *See* CENTA
CGIAR (Consultative Group on International Agricultural Research—U.S.), 190, 202, 237-239
Chang, J., 323
CIAT (International Center for Tropical Agriculture—Colombia), 18, 115, 387
FSR&D program, 237-239
superimposed livestock trial, 115, 116
training program, 197, 198, 201, 203, 389, 390, 392
CIMMYT (International Maize and Wheat Improvement Center—Mexico), 18, 23, 62
analytical concepts manual, *From Agronomic Data to Farmer Recommendations*, 123
FSR&D program, 36, 237-239
recommendation domain, 44
training program, 197, 199, 202-204, 387, 389-391, 393
workshop on methodological issues, 169, 170
CIP (International Potato Center—Peru), 237, 393
FSR&D program, 238, 239
Climatic analysis, definition of, 213
Climatic monitoring and records
cropping-systems research, 341, 342
evapotranspiration, 222, 323, 346
example of data collection form, 255
livestock-systems research, 345, 346
microclimate, 325
on-farm research, 94, 126, 128, 321-323
rainfall, 255, 321, 322, 341, 345, 346 (*See also* Water)
relative humidity, 323, 346
solar radiation, 323, 346
temperature, 322, 346
wind, 255, 322, 346
Cloud, K., 69, 75
Cochran, W.G. and G.M. Cox, 326
Coefficient of variation (C.V.), 102, 113, 114, 119, 329
definition of, 213
Collaborating farmers, definition of, 213
Collecting data. *See* Data collection
Collinson, M.P., 37, 44, 45, 243, 245, 293, 306, 308
Colombian Agricultural Institute (Colombia). *See* ICA
Commodity-oriented research, definition of, 17, 18, 213
Commodity specialists, 338
definition of, 213
Component technology. *See also* Technology, development
definition of, 92, 213
Conducting on-farm crop research. *See also* Research results analysis
analysis and reporting of results, 113, 120, 121, 328-335
cultural practices and data collection, 112, 113, 115, 119
field design of experiments, 101, 112, 114, 115, 118, 119, 325-328
field selection, 100, 111, 112, 114, 118, 324, 325
farmer-researcher relationship, 117, 118
farmer selection, 116
incentives and agreements, farmer, 116, 117
measuring crop yields, 113, 115, 120
monitoring progress, 113, 115, 119, 120
Conducting on-farm livestock research, 113-116, 121, 122
Confidence interval, definition of, 213
Contreras, M.R., et al., 204
Cook, T.D. and D.T. Campbell, 299
Cook, T.D., et al., 297, 298
Cooperative agreements, examples of, 359-362
Costs, fixed and variable, 129, 131, 132
CRIA (Central Research Institute for Agriculture—Indonesia)
cropping pattern research analysis, 237, 354-356
example of design for cropping pattern experiment, 335-337
research design conditions, 97
selection of FSR&D areas, 54, 55, 246-248
Crop yields, measuring, 113, 115, 120
Cropping patterns
definition of, 213
example of analysis of, 354-356
example of research design for, 154, 155, 335-336
Cropping systems
definition of, 4, 16, 17, 213
format for describing, 67, 267-270
time required, 31, 32
Cropping systems research. *See also* Conducting on-farm crop research
checklist of field operations, 344
definition of, 17, 213
example forms for collecting data, 341-345
farmer-managed tests, 101, 116-121
researcher-managed trials, 101, 111-113
superimposed trials, 101, 114, 115
Cultural practices, 294, 295, 396
definition of, 213
in on-farm research, 112, 113, 115, 119
Cyert, R. and J. March, 266

Data collection
combining methods of, 81, 82
guidelines for preliminary survey, 293-296
limitations of informal methods, 73, 74
primary data, formal and informal methods, 54, 55, 73-81
rural settings, 72, 278-281
sampling procedures, 79, 304-306
secondary data, 54, 72, 73
Data collection forms, examples of
climate, 255, 342, 346
for cropping experiments, 286, 287, 294, 295, 315, 341-345

economic resources of farmers, 269, 270, 294, 295, 310–312
for farmer-managed tests, 348, 349
for farming systems, 245, 285–289, 293–296, 310, 311
for livestock experiments, 288, 289, 345–348
monitoring progress, 342, 347
physical resources, 255, 295
Data management, 82, 316–318
computers, 317
programmable pocket calculators, 317, 318
sorting strips, 317
tabular sheets, 316, 317
Data, primary. *See also* Data, secondary; Primary information; Secondary information
case studies, 81
farm record keeping, 79, 80, 309–315
frequent interview surveys, 78
monitoring, 80, 81
participant observation, 76, 77
questionnaire, 78, 79
reconnaissance surveys, 74–76, 284–289
single interview surveys, 77
sondeo, 76
sources of, 49, 54, 55, 73
Data, secondary. *See also* Data, primary; Primary information; Secondary information.
accuracy and reliability of, 73, 283, 284
adequacy of, 73
clarity of terminology for, 73
recency of, 73
relevancy and specificity of, 73
sources of, 49, 54, 55, 72, 73
DeBoer, A.J. and A. Weisblat, 275
Decision making, by farm household
decision trees, 265–267
example of sociological research methods in Guatemala, 262–264
influenced by, 65
socio-cultural-economic values, 69, 70, 103, 263–265
Decision trees, 142, 265–267
DeDatta, S.K., et al., 98
de Haan, C., 32, 49, 125
Delgado, C.L., 33, 121
DIGESA (General Directorate for Agricultural Services—Guatemala)
cooperative agreement with ICTA, 157, 166, 360–362
extension training program in Guatemala, 200, 388
Dillon, J.L., 15
Dillon, J.L. and J.B. Hardaker, 204
Dirección General de Servicios Agrícolas (Guatemala). *See* DIGESA
Disciplinary research, definition of, 213
Disciplinary specialists, 338
definition of, 213
Dominance analysis, 131, 132
Doorenbos, J. and W.O. Pruitt, 323
Double cropping
definition of, 213
example of in Indonesia, 164, 165
Duff, B., 275

Economic analyses of research results, 125, 126, 128–139
Economic environmental factors. *See also* Socioeconomic factors
capital, 64, 66, 269, 294
costs, crop or livestock, 286–289, 294
credit, 64, 91, 261, 270
definition of, 213, 214
labor, 64–66, 91, 261, 269, 294–296
land, costs and ownership, 65, 295
market factors, 64, 65, 259–261
storage and processing facilities, 64, 260, 396
tools, equipment, and supplies, 64, 91, 261, 294
transportation, 64, 260
Economic Development Institute (World Bank), 143
Economic feasibility
considerations in setting priorities, 71
definition of, 214
factors for determining, 131–139
Education. *See* Training
El Hadari, A.M., 279, 307
EMBRAPA (Brazilian Agricultural Research Corporation—Brazil)
FSR&D program in Brazil, 201, 237
Empresa Brasileira de Pesquisa Agropecuária (Brazil). *See* EMBRAPA
Enterprises, definition of, 16, 214
Environmental factors. *See also* Biological factors; Economic environmental factors; Physical factors; Sociocultural factors; Socioeconomic factors
definition of, 63, 214
Errors. *See* Measurement error; Sampling, error
ESFS. *See* Extension specialist in farming systems
Evaluation of FSR&D projects
built-in, 191
caveat for, 192, 193
developing procedures for, 192
groups involved in, 35
impact, 191, 192
Logical Framework of USAID, 192, 376–379
time required for, 35
types of, 191, 192
Experiment station
collaboration in FSR&D, 5, 28–31, 89, 96, 97, 104, 150, 178, 179
extension's ties with research, 149, 150, 156, 157, 163, 166, 360–362
support—a research method, 93, 96, 97
Experimental variables, definition of, 214
Extension agents, 150–152
Extension assistants, 150–152
Extension director and staff, 151
Extension officer, regional and staff, 150, 151
Extension organizations
budget problems of, 158
collaboration in FSR&D, 28–31, 100, 104, 149–151, 156
definition of, 149
organization problems of, 158
research ties of, 150, 156, 157, 163, 166, 360–362
single-commodity orientation of, 157, 158
staff and proposed organization of, 149–151
training for, 155, 157, 388, 389
Extension of results
groups involved in, 35, 152
multi-locational testing, 149, 152–155
overview of, 6, 28, 29, 397
pilot production programs, 149, 152, 155, 156, 359, 360
Extension specialist in farming systms (ESFS), 149–152, 154, 155, 157, 182, 184, 201
definition of, 214
primary link between research and extension, 151
responsibilities of, 149, 150
FAO (Food and Agriculture Organization of the United Nations), 99, 190, 204, 238
agroclimatic zones, 98
land evaluation principles, 258
Farm family. *See* Household
Farm organizations and local leaders
involvement, 34, 35, 53, 151, 260, 278, 279
Farm populations. *See* Homogeneous farmer groups
Farm records. *See also* Record keeping
definition of, 214
Farmer-managed cropping tests, 116–121
analysis and reporting of results, 120, 328–335
cultural practices and data collection, 119
farmer agreements and incentives, 116, 117
farmer and field selection, 102, 116, 118
farmer-researcher relationship, 117, 118
field design of experiments, 101, 118, 119, 325–328, 349, 350
measuring crop yields, 120
monitoring progress, 119, 120
plot size, 119
replication, 118, 119, 349, 350
Farmer-managed livestock tests, 121, 122
data collection and analysis, 121, 122
methodology development, 121
research possibilities, 121
Farmer-managed tests, 116–122
analyzing, 140, 141
definition of, 214
form for collecting data, 348, 349 (*See also* Farming systems, forms for recording data)
incomplete tests, 102, 120
overview, 95, 96
purpose of, 28, 29, 97, 116
research design considerations, 100, 101
Farmer-managed tests, analysis of
cause of farmer's changes in experimental design, 140
climatic effects, 140
extent of farmer's changes, 140
farmer's allocation of time, 140, 141
probability of farmer's changes in future tests, 140
Farmer-researcher reslationship, 117, 118, 178, 182, 279
Farmer's environment. *See also* Biological factors; Economic environmental factors; Physical factors; Socioeconomic factors; Sociocultural factors
biological setting, 64, 91, 102, 103
definition of, 214
economic setting, 64, 91, 103
physical setting, 64, 91, 102, 103
sociocultural setting, 64, 91, 103
Farmer's technical expertise and experience, 65, 66, 71, 263, 264, 270. *See also* Management practices of farmers
Farming, definition of, 16, 214
Farming enterprises, definition of, 66, 67
Farming household. *See* Household
Farming, small-scale, definition of, 16
Farming subsystems. *See also* Farming systems; FSR&D
cropping, definition of, 16, 17
definition of, 16
livestock, definition of, 17
mixed, definition of, 17
Farming systems. *See also* FSR&D
conceptual models for, 67, 68
definition of, 16, 214
description of, 64–67, 285–289, 293–296
forms for recording data, 67, 245, 255, 269, 270, 285–289, 293–296, 310–312, 315, 341–343, 345–349
hypotheses for improving system, 68, 69
Farming systems research and development. *See* FSR&D
Feedback, 28, 31, 87, 88, 156, 163
Fertilizer trials, example of economic analysis, 126–139
Field design of experiments, 100–102, 112, 114, 115, 118, 119, 325–328, 349, 350. *See also* Research design; Statistical design
Field (experimental) selection, 100, 111, 112, 114, 118, 324, 325
Field operations records, 28, 310, 311, 343, 344. *See also* Farming systems, forms for recording data
Field plot. *See* Field (experimental) selection
Field team
authority and accountability of, 33
building teamwork of, 185, 190, 374, 376
composition, size, and skills of, 33, 178, 183, 184, 337, 338, 368, 369
definition of, 214
and extension, 179, 182, 184
and farmer, 178, 182, 183
interdisciplinarity model, 185–188
interdisciplinary aspect of, 182, 184–190, 375, 376
involvement in FSR&D process, 35
management of, 122, 123, 182, 184
and research scientist, 33, 178, 179, 182–184, 337
responsibilities of members, 178, 179, 337
Field team leader, responsibilities, 122, 179, 180, 185, 186, 337
Financial feasibility, 71, 124, 139. *See also* Economic feasibility
definition of, 214
Flinn, J.C., 308, 309
Flinn, J.C. and S. Jayasuriya, 232
Fonseca M., S., 389, 392
Food and Agriculture Organization of the United Nations. *See* FAO
Francis, D. and D. Young, 186, 374
Frequent interview survey, 78. *See also* Surveys
definition of, 214
Friedrich, K.H., 78
FSR&D (farming systems research and development)
activities, 4–6, 27–31, 41–56, 61–83, 87–106, 111–144, 150, 156
characteristics of, 3, 4, 18–20
conceptual framework of, 27–38
and "conventional" research, 14, 15, 166
cost-effectiveness of, 8, 169, 170
definition of, 3, 4, 13–16, 214
extension's role in, 149–152, 166
interdisciplinary approach of, 19, 27, 166, 185
national and international programs, 237–239
organizational structure of, 174–178, 365–373
purpose and scope of, 13
strategies of, 35–38
FSR&D Guidelines
objective of, 20, 21
processes and procedures of, 5, 6
scope of, 21–23
users of, 13, 21
Funding, outside sources of, 190, 191

Gafsi, S., 280, 307
General Directorate for Agricultural Services (Guatemala). *See* DIGESA
GERDAT (Group for Studies and Research in the Development of Tropical Agronomy—France), 190, 204, 237
Gilbert, E.H., et al., 23, 37, 169, 170, 237
Gines, H.C. and H.G. Zandstra, 112, 113

Gittinger, J.P., 142, 143
Gladwin, C.H., 65, 142, 265, 266
Goals
farmer, 65, 71
national and FSR&D, 163-165
research, 375-378
Government. *See* Policy
Grant, E.L., et al., 142
Group for Studies and Research in the Development of Tropical Agronomy (France). *See* GERDAT
Gucelioglu, O., 82, 296, 297
Guidelines. *See* FSR&D Guidelines

Harrington, L., 36, 37, 169, 170
Harris, N., 261
Hart, R.D., 81, 204, 231, 232, 272, 274
Harwood, R.R., 13, 16, 99, 164, 204, 216, 272, 324
Haskins, J.B., 299
Hatch, J.K., 80, 283, 309, 314, 315
Haws, L.D. and R.T. Dilag, Jr., 155, 359
Heady, E.O. and W. Candler, 232
Hernandez X., E., 204
Hiebsch, C.K., 324
Hildebrand, P.E., 27, 32, 70, 141, 169, 192, 289, 309-314, 318
Hildebrand, P.E. and S. Ruano, 141
Hill, P., 277
Hill, R.W., et al., 232
Hillier, F.S. and G.J. Lieberman, 233
His Majesty's Government Department of Agriculture, 237
Homogeneous farmer groups, 27
alternative to subareas, 44, 243
definition of, 214
method used in Zambia, 243-245
recomendation domain, 44, 243
House, E.R., 299
Household
definition of, 16, 214
as integrating unit, 64

IADS (International Agricultural Development Service–U.S.), 365, 383
IARCs (International Agricultural Research Centers), 164, 238, 389, 393
training activities, 8, 191, 202, 203, 205
"basic research" organizations, 37
ICA (Colombian Agricultural Institute–Colombia)
superimposed livestock trial, 115, 116
ICARDA (International Center for Agricultural Research in the Dry Areas–Lebanon), 75, 237, 392
FSR&D program, 238, 239
ICRISAT (International Crops Research Institute for the Semi-Arid Tropics–India), 55, 82, 237, 392
FSR&D program, 238
research area selection, 250, 251
ICTA (Agricultural Science and Technology Institute–Guatemala), 44, 318
acceptability index, 141, 169
cooperative agreement with DIGESA, 157, 166, 360-362
extension training program developed for DIGESA in Guatemala, 200, 388, 389
farm record project, 80, 309-314, 387
impact evaluations of FSR&D project, 192
organization of FSR&D in Guatemala, 176, 237, 369, 371
sondeo approach, 76, 289-293
time requirements for problem identification and planning, 32
training programs, 197, 198, 201, 383, 384

IITA (International Institute for Tropical Agriculture–Nigeria), 201, 392
FSR&D program in Nigeria, 237, 238
ILCA (International Livestock Centre for Africa–Ethiopia), 38, 114, 121, 389, 392
conceptual model for herbivore, 270, 272
development path strategy, 71, 72
example of mixed systems research in Ethiopia, 165, 350-352
FSR&D program, 237-239
livestock systems, 32, 33, 80, 81, 316
mathematical modeling, 22, 126, 231
training program, 201, 203, 391
whole farm analysis, 125
Impact evaluation, 191
Implementing FSR&D activities
cost-effectiveness, 8, 169, 170
emphasizing interdisciplinary aspects and team, 174, 179, 182, 184-190, 375, 376
evaluating success of project/program, 191, 193, 376-379
first step: whether or not to adopt FSR&D approach? 163-170
funding, outside sources of, 190, 191
governmental support, 8, 174
off-site management considerations, 180-182
organizing, 6, 7, 174-178, 365-369
project vs. program approach, 173, 174
staffing and determining roles, 7, 169, 178-180, 182-184
timing, 6, 31-33
training, 7, 8, 197-205
Incentives
for farmers, 116, 117, 279
for staff, 182, 188
Informal surveys, definition of, 214. *See also* Surveys
Information, primary. *See also* Data, primary
definition of, 46, 215
sources of, 49
Information, secondary. *See also* Data, secondary
definition of, 28, 46, 216
sources of, 49
Infrastructure, definition of, 214
INIAP (National Institute for Agricultural Research–Ecuador)
organization of FSR&D in Ecuador, 371, 373
Insect control studies, 96
Institut Sénégalais de Recherches Agricoles (Senegal). *See* ISRA
Instituto de Ciencia y Tecnología Agrícolas (Guatemala). *See* ICTA
Instituto Colombiano Agropecuario (Colombia). *See* ICA
Instituto Nacional de Investigaciones Agropecuarias (Ecuador). *See* INIAP
Intercropping, 165, 286
definition of, 214
efficiency compared with sole cropping by land equivalent ratio, 99, 323, 324
Interdisciplinary
aspect of FSR&D, 19, 27, 166, 185
definition of, 214
model, 185-188
teamwork, 184-190, 375, 376
International Agricultural Research Centers. *See* IARCs
International Center for Tropical Agriculture (Colombia). *See* CIAT
International Crops Research Institute for the Semi-Arid Tropics (India). *See* ICRISAT
International Institute for Tropical Agriculture (Nigeria). *See* IITA

International Livestock Centre for Africa (Ethiopia). *See* ILCA
International Maize and Wheat Improvement Center (Mexico). *See* CIMMYT
International Potato Center (Peru). *See* CIP
International Rice Research Institute (Philippines). *See* IRRI
Interviewing process. *See also* Surveys
analyzing content of informal interviews, 284
how to conduct interviews, 279, 280
incentives for interviewees, 279
interviewers–selecting, hiring, training, and supervising of, 306-309
number of interviewers needed, 307, 308
problems in interviewing, 280, 281
IRRI (International Rice Research Institute–Philippines), 18, 23, 62, 140, 155, 237, 389
climatic zone grouping in Southeast Asia, 47, 48, 248, 249
FSR&D program, 193, 238
mathematical modeling, 22, 231
physiographic classifications in Southeast Asia, 47, 49, 50, 249
soils description, 49
time requirements for FSR&D programs, 31, 32
training program, 197, 199, 201-203, 391-397
Irrigation. *See* Climatic monitoring and records, rainfall; Physical factors, water
ISRA (Institut Sénégalais de Recherches Agricoles–Senegal)
organization of FSR&D in Senegal, 176, 237, 365-369
training materials, 204
Iterative process
definition of, 214
in FSR&D, 61, 87

Jayasuriya, S., 126, 232
Jodha, N.S., et al., 55, 250

Kearl, B., 278, 296, 303, 305, 307-309
Kenlen, H.V., 232
Kerlinger, F.N., 264
Kinnear, T.C. and J.R. Taylor, 78, 299, 301
Kuhn, T.S., 186

Labor. *See also* Economic environmental factors, labor
family, 65, 131, 265, 269
form for recording data, 269, 347
and management practices, 131
Land equivalent ratio (LER), 99, 323, 324
Land, evaluation of, 258, 259, 295
Land type, 64, 102, 154, 155, 255-257, 342, 349, 350. *See also* Physical factors, topography
definition of, 215
example of data form, 255
Lele, U.J., 259
LER. *See* Land equivalent ratio
Lin, N., 264
Linear programming, 126
definition of, 215
Lipton, M. and M. Moore, 296
Little, I.M.D. and J.A. Mirrlees, 143
Little, T.M. and F.J. Hills, 111, 213, 216, 217, 326, 330
Livestock patterns, definition of, 215
Livestock systems, definition of, 16, 17, 215
Livestock systems research. *See also* Conducting on-farm livestock research
definition of, 17, 215

example forms for collecting data, 345-348
farmer-managed tests, 121, 122
researcher-managed trials, 113, 114
superimposed trials, 115, 116
time required, 32, 33
Logical Framework, USAID evaluation procedure, 192, 376-379
Low, A.R.C., 233

MacDonald, R.B.and F.G. Hall, 232
Management factors
definition of, 215
as household resource, 66
Management, off-site
other institutions, 180, 181
support for field teams, 181
technical review, 181
Management practices of farmers, 93-95, 98, 264, 270
cropping systems, 245, 270, 286, 287, 343, 344, 348, 349
livestock systems, 288, 289, 347
Marginal rate of return, 132-134
Martius-von Harder, G., 283
McCarl, B.A. 232
McCown, R.L., et al., 275
McDowell, R.E. and P.E. Hildebrand, 67, 68, 270-273, 275
McIntosh, J.L., 23, 68, 69, 97, 335, 336, 354-356
McSweeney, B.G., 281
Mead, M., 204, 397
Measurement error, 77, 82, 297. *See also* Sampling, error
Mellor, J.W., 259
Mencher, J.P., et al., 283
Merry, U. and M.E. Allerhand, 186
Milthrope, F.L. and J. Moorby, 99
Minimum acceptable return, 90, 133, 135
Mitchell, R., 99
Mixed intercropping, definition of, 215
Mixed systems
definition of, 17, 215
example of model, showing linkages, 275-278
Mixed systems research
definition of, 17, 215
example of, Ethiopia, 121, 122, 165, 350-352
integration of tests in, 121
Model
conceptual, for analysis of farming system, 67, 68, 270-278
interdisciplinarity, 185-188
mathematical, of IRRI, ILCA, and CATIE, 22, 231, 232
references on FSR&D modeling, 232, 233
simulation, of IRRI, 231
Monitoring, 95, 316. *See also* Data, primary
climatic data, 80. (*See also* Climatic monitoring and records)
livestock systems, 80, 81, 316, 347
on-farm experiments, 80, 342, 343, 347
research progress, 113, 115, 119, 120, 342, 343, 347
Monoculture planting, definition of, 215
Morgan, W.B. and J.C. Pugh, 277
Mosher, A.T., 259
Multidisciplinary, definition of, 215. *See also* Interdisciplinary
vs. interdisciplinary, 185
Multi-locational testing, 29, 30
definition of, 153, 215
example of, 154, 155
and extension, 149, 150, 152-154
groups involved in, 150, 153, 154
Multiple cropping system, definition of, 215

Nabila, J.S., 279, 280
Nair, K. 92
National Agricultural Technology Center (El Salvador). *See* CENTA
National headquarters team
  composition of, 34
  involvement in FSR&D process, 35, 176–178
National Institute for Agricultural Research (Ecuador). *See* INIAP
National Program for Agricultural Research (Honduras). *See* PNIA
Natural resources management, 95. *See also* Biological factors; Physical factors
Nelson, L., et al., 112
Nelson, R.S.M., 232
Net benefits, 128–132, 135–138, 352–354
Neter, J.W., et al., 213, 216
Netting, R., 103, 265
Nonexperimental variables, definition of, 215
Norman, D.H., 15, 82, 95, 119, 150, 168, 279, 296, 308, 309
Nygaard, D., 75

Odum, H.T., 231
Ogunfowora, O., 279, 280, 307, 308
Opportunity(ies)
  definition of, 61
  early identification of, 56, 57, 98
  "targets of opportunity," 115
Opportunity cost, concept of, 128, 129, 131, 353
Ortiz D., R., 176, 370

Pantastico, E.B., et al., 153
Paradigm, 185, 186
  definition of, 185, 215
  Partial budget analysis, 125, 128–131, 353
  definition of, 215
Participant observation, 76, 77. *See also* Surveys
Pelz, D.C. and F.M. Andrews, 189
Perdon, E.R., 389, 391
Performance criteria, 69, 99
Perrin, R.K., et al., 23, 123, 127–135, 137, 138, 144, 238, 353
Personnel. *See* Staff(ing)
Physical factors
  climate, 46, 47, 64, 91, 102, 248, 249, 255, 321–323, 325
  definition of, 215
  land type, 64, 102, 154, 155, 255–257, 342, 349, 350
  soil, 49, 50, 65, 91, 102, 249, 255, 325, 342, 393
  topography, 47, 49, 50, 65, 91, 249, 325 (*See also* Land type)
  water, 47–49, 64, 65, 91, 248, 249, 255, 296, 321, 322, 324, 325, 341, 342, 345, 346, 354, 355
Pilot production program, 29, 30
  definition of, 215
  example of, 155, 156, 359, 360
  and extension, 150, 152, 155
  groups involved in, 150, 155
  size of, 155
Plan, planning. *See* Research planning; *See also* Research design
Plot. *See* Field (experimental) selection
Plot size, 112, 114, 119
PNIA (National Program for Agricultural Research—Honduras)
  organization of FSR&D in Honduras, 237, 371, 372
  training program in Honduras, 383–388
Policy
  advisory committees, set by, 180, 181
  cost-effectiveness of FSR&D, 8, 169, 170
  FSR&D and national development, 44–46, 164, 165
  governmental support, 8, 44–47, 174
  how much change? 166, 397, 398
  issues, 163
  and personnel, 7, 169, 177, 178, 180–182
  project or program approach? 173, 174
  relevance of FSR&D to small farmers' needs, 163, 165, 166
  research priorities and national goals, 163
  society vs. farmers, 71, 143
Predictive approach, 124
Primary information. *See also* Data, primary; Data, secondary; Secondary information
  definition of, 46, 215
  example of, Zambia, 243–245
Priorities
  research vs. national goals, 163
  setting, 70–72, 88, 89
  and technical review committee, 181
Problem analysis, 69, 70. *See also* Problem identification
  approaches to, 69, 70, 267
  performance criteria, 69
Problem identification
  analysis of farming systems, 67–79
  analysis of problems and opportunities, 69, 70
  categorizing and ranking problems, 62
  description of farmers' environment, 64
  description of farming systems, 64–67, 267–269
  extension's role in, 150–152
  groups involved in process of, 35, 150
  iterative process, 61
  includes identification of opportunities, 61
  overview of process, 5, 28, 61, 62
  primary data collection, 62
  reconnaissance survey, 62, 74, 76
  secondary data collection, 61
  sequence of activities, 61, 62
  setting priorities, 70–72
  time required, 32, 35
Problem solving, potential for
  biological potential, 71
  economic and financial feasibility, 71
  resource availability, 71
  sociocultural acceptability, 71
Problems. *See also* Opportunity(ies)
  definition of, 61
Productivity criteria, 99, 323, 324
Program approach. *See also* Project approach
  definition of, 173, 215
  example of, Honduras, 193
  interdisciplinary-commodity organizational modes, 176
  management-centered organizational modes, 175, 176
  organizational structure in, 174–178
  vs. project approach, 173, 174
Programa Nacional de Investigacíon Agropecuaria (Honduras). *See* PNIA
Project approach. *See also* Program approach
  definition of, 173, 215
  example of, in Philippines, 193
  organizational structure in, 174, 175
  vs. program approach, 173, 174

Questionnaire
  design of, 78, 79, 299–303
  pretesting of, 79, 302, 303
Quinn, N., 266

Rahim, S.A., 264
Rainfall. *See* Climatic monitoring and records, rainfall; Physical factors, water
Rain-fed farming, 324, 325. *See also* Climatic monitoring and records, rainfall; Physical factors, water
  definition of, 215
Random sample. *See also* Sampling, random
  definition of, 215
Rappaport, R.A., 261
Ratoon cropping, definition of, 215
Recommendation domain, 44, 243
  definition of, 44, 215
Reconnaissance surveys, 74–76. *See also* Surveys
  basic aspects of, 54, 55, 74
  definition of, 216
  examples of forms for collecting data, 284–289
  objectives of, 74
  sondeo, 76
  step in problem identification, 62, 74, 76
  types of information collected in, 76, 284–289
Record keeping, 309–315. *See also* Data, primary
  advantages and uses of, 80, 297
  example of farmer record form, 310
  by farmer, 80, 296, 297, 309–311
  for illiterate farmers, 311, 314, 315
  types needed in cropping systems research, 341–345
  types needed in livestock systems research, 345–348
Reddin, W.J., 186
Regional headquarters team
  composition of, 33, 34
  involvement in FSR&D process, 35, 176–178
Relay intercropping, definition of, 216
Replication, 112, 114, 118, 119, 349, 350
  definition of, 216
Research activities, 92–94. *See also* Climatic monitoring and records; Management practices of farmers; Technology, development; Technology, farmer adaptation
Research area
  abandoning area or research, 56, 166, 167
  criteria and methods for selection of, 51–53
  definition of, 51, 216
  extension's role in selection of, 150–152
  groups involved in selecting, 35, 53, 150
  Indonesia, data used to select, 54, 55, 246–248
  information collection for selecting, 54, 55
  information needed for selecting, 53, 54
  information sources, 54, 55
  selection of, 5, 27, 51–56, 246–248
  selection done in stages, 55, 246, 249–251
  staffing for selection of, 53
  time needed for selection of, 32, 35, 56
Research base. *See also* Data collection; Data, primary; Data, secondary
  data management, 82
  development of, 28, 72–82
  extension's role in developing, 35, 150, 151
  groups involved in developing, 35, 150
  methodologies for collecting data, 72–82
  time needed for development of, 32, 35
Research, conduct of. *See* Conducting on-farm crop research; Conducting on-farm livestock research
Research design. *See also* Field design of experiments; Research planning
  choosing farmers, 99, 100
  complexity of, 100
  deciding on conditions of, 97, 98
  example, cropping pattern experiment, 335–337
  experimental design characteristics, 100, 101, 325–335
  incomplete experiments, 102, 120
  location of experiments, 100, 324, 325
  methods of analyzing research results, 102
  number of experiments, 100
  searching for improvements, 98, 99
  statistical procedures for field design, 325–335
Research methods, 73, 77, 93, 95, 97, 262, 263
  experiment station support, 93, 96, 97
  farmer-managed tests, 93, 95, 96, 116–122
  informal, 73–77
  monitoring, 80, 93
  record keeping, 80, 93
  researcher-managed trials, 93, 95, 111–114
  superimposed trials, 93, 96, 114–116
  surveys, 77, 78, 93. (*See also* Surveys)
Research planning
  appraising resources and technology, 88
  categorizing, reviewing, and setting priorities, 87–89
  conducting regional workshops, 103–106
  considering alternatives and consequences, 89, 92
  design, 91, 97–102
  developing assumptions and hypotheses, 88, 89
  establishing research collaboration, 89
  estimating values, 91, 92
  gathering additional data, 102, 103
  groups involved in, 35, 150–152
  identifying appropriate research activities, 92–95
  overview of process, 5, 6, 28, 87
  research methodologies, 77, 78, 80, 81, 93, 95–97
  time required for, 32, 35
  understanding farmers' conditions and perspectives, 89–91
Research results analysis
  acceptability index, 141
  acceptability to farmers, 124, 125, 133, 135, 141
  acceptability to society, 143
  of biological performance, 126–128
  dominance analysis, 131, 132
  economic analysis, 123, 126, 128–139
  of farmer-managed tests, 139–141
  financial feasibility, 124, 139
  fixed and variable costs, 129, 131, 132
  groups involved in, 35, 150, 152
  of long-term investments, 142, 143
  marginal rate of return, 132–134
  minimum acceptable return, 133, 135
  net benefits, 128–132, 135–138, 352–354
  opportunity cost analysis, 128, 129, 131, 353
  overview, 6, 29, 35
  partial budget analysis, 125, 128–131, 353
  predictive approach, 124
  risk, 133, 135, 136
  sensitivity analysis, 129, 136
  sociocultural feasibility, 141, 142
  whole farm analysis, 125
Research site. *See* Field (experimental) selection; Research area
Research team. *See* Field team

Research termination, 56, 166, 167
Researcher-managed cropping trials, 111–114
analysis and reporting of results, 113, 328–335
cultural practices and data collection, 112, 113
field design of experiments, 101, 112, 325–328
field selection, 111, 112
monitoring progress, 113
plot size, 112
replications, 112
Researcher-managed livestock trials, 113, 114
Researcher-managed trials, 111–114
definition, 216
overview, 95
purpose of, 28, 29, 97, 111
research design considerations, 100, 101
Resource feasibility, definition of, 216
Resources. *See* Biological factors; Economic environmental factors; Physical factors; Socioeconomic factors
Richardson, E., 373
Risk
analysis of, 133, 135, 136
and farmer, 88, 117, 265, 296
importance in decision making, 265
Ritchie, J.T., 232
Robertson, C., 16
Rogers, E.M., 264
Rogers, E.M. and F.F. Shoemaker, 264
Rosenberg, N.J., 323
Rosengren, K.E., 284
Roumasset, J.A., 233
Row intercropping, definition of, 216
Ruthenberg, H., 16, 214, 277

Safai, M., 283
Sahlins, M.D., 265
Sampling
error, 77, 82, 297
frame, 303, 304
methods, 304–306
nonrandom, 304, 305
procedures and problems, 79, 303–306
purpose of, 79, 303
random, 305, 306
size of sample, 306
two levels minimize error, 82, 297
Sanders, J.O. and T.C. Cartwright, 232
Sassone, P.G. and W.A. Schaffer, 143
Secondary information. *See also* Data, primary; Data, secondary; Primary information
definition of, 28, 216
Selection criteria for experimental field, 100, 102, 103, 111, 112, 114, 115, 118, 324, 325
Selection criteria for farmers, 99, 100, 116, 310
Selection criteria for research area
accessibility, 52
biological environment, 54
farmer-contact agencies, 53
physical environment, 54
production systems and land use, 54
proximity to experiment station, 52
representativeness, 52, 246, 248
size, 52
socioeconomic environment, 54
support of leaders, 53
Selection criteria for target area
broad application, 45
environmental conditions, 45, 46
physical characteristics, 45
policy compatibility and options, 44–46
program cost, 45
rapid payoff, 45
Semi-arid Food Grain Research and Development project (Upper Volta), 238
Senegalese Institute for Agricultural Research (Senegal). *See* ISRA
Sensitivity analysis, 129, 136
Sequential cropping, definition of, 216
Shifting cultivation, definition of, 216
Significance level, definition of, 216
Simmons, E., 77
Simon, H., 266
Single crop system, definition of, 216
Single interview survey. *See also* Surveys
advantages and disadvantages of, 77
definition of, 216
Slash and burn, definition of, 216
Slomin, M.J., 303, 305
Small-scale farming, definition of, 16, 216
Smith, E.D., 259
Smock, A.C. 282
Social scientists, definition of, 216
Sociocultural acceptability, 70, 141, 142
considerations in setting priorities, 71
definition of, 216
factors in planning research, 91, 325
influenced by, 260, 261, 325
Sociocultural factors, 64, 91, 103, 261, 262. *See also* Sociocultural acceptability
customs, 64, 91, 261, 262, 325
definition of, 216
religion, 261
sex and age group rights and obligations, 64
social institutions, 262
Socioeconomic factors, 64, 103, 325. *See also* Sociocultural factors
information channels, 64, 260
definition of, 216
division of labor, 64–66
government regulations, 64, 260
land use customs, 64
support services, 260
Soil. *See* Physical factors, soil
Sole cropping
definition of, 216
efficiency compared with intercropping by land equivalent ratio, 99, 323, 324
Sondeo. See also Surveys
definition of, 216
used by ICTA in Guatemala, 289–293
Staff(ing), 7, 176
constraints, 169, 180
extension, 149–151, 179, 180
field teams, 33, 176, 178, 183, 184
management of, 181, 182
national headquarters team, 34, 176–178
others, 34
regional headquarters team, 33, 34, 176, 178
research area selection teams, 53
target area selection teams, 49–51
Standard deviation, 329
definition of, 216
Statistical analysis
analysis of variance, 113, 329, 330, 332
coefficient of variation (C.V.), 102, 113, 114, 119, 329
confidence interval, 113
least significant difference (LSD), 330, 331, 335
mean, 328, 329
mean separation, 330, 331
standard deviation, 329
validity, 297–299
Statistical design
completely randomized design, 326
paired treatments design, 326, 327
randomized complete block design, 326, 327, 349, 350
randomized incomplete block design, 326–328
Staudt, K., 19, 77
Strategies. *See* FSR&D, strategies of
Stratification, 27
definition of, 216
Stonaker, H.H., et al., 115
Study. *See* Surveys
Subareas, definition of, 216. *See also* Target area and subareas
Subject matter specialists, 150, 151, 369
Subsistence farmers, 90, 91, 164
definition of, 216
Superimposed cropping trials
conducting experiments, 115
field design of experiments, 101, 114, 115, 325–328
field selection, 114
monitoring progress, 115
plot size, 114
statistical analysis methods, 328–335
Superimposed livestock trials, 115, 116
CIAT example, 115, 116
Superimposed trials, 114–116
definition of, 216
overview, 96
purpose of, 28, 29, 97, 114
research design considerations, 100, 101
Support services, 166, 181, 365–368, 370, 371
definition of, 216
Surveys. *See also* Case studies
advantages and disadvantages of formal methods, 77, 78, 80, 81
advantages and disadvantages of informal methods, 73, 74
formal methods, 77–81
frequent interview, 78
guidelines for types of data to collect, 293–296
implementation of, 79, 306–309
informal methods, 73–77, 284
participant observation, 76, 77
reconnaissance, 54, 74–76, 284–289, 293–296
single interview, 77
sondeo, 76, 289–293
validity of information, 297–299
variations of formal surveys, 79–81
System, definition of, 216
Systems approach, 165
definition of, 216

Tamisin, M.M., et al., 232
Target area and subareas, 5, 27, 43–45. *See also* Homogeneous farmer groups
criteria and methods for selection of, 44–46
definition of, 43, 44, 217
extension's role in selection of, 150–152
groups involved in selecting, 35, 44, 57, 150
information needed for selecting, 46–49
information sources, 49
selection of, 5, 27, 43–51
selection of, in Indonesia, 246–248
time needed for selection of, 32, 35, 51
Target populations, definition of, 217. *See also* Homogeneous farmer groups
Technical Advisory Committee (TAC), 23, 37, 92, 213–217, 237, 238
Technical assistants, 33, 178, 183, 184, 337
Technology
definition of, 217
development, 92, 97
diffusion (transfer) of, 149, 151–156, 158, 163
farmer adaptation, 92, 94, 103, 124, 125, 133, 135, 262, 263. (*See also* Acceptance, farmer)
modification and testing of, 153–155, 158
Test plot. *See* Field (experimental) selection
Thodey, A.R. and R. Sektheera, 232
Thom, D.J., 49, 51, 100
Time requirements
for cropping systems, 31, 32
dependent on several factors, 6, 31, 33
example, Philippines cropping systems program, 31, 32
for livestock systems, 32, 33
for research area selection, 32, 35, 56
for target area selection, 32, 35, 51
timetable example, 32, 35, 97
Tollens, E., 305
Townsend-Moller, D., 178
Trail, T., 204
Training
for extension staff, 157, 199, 200, 388, 389
for FSR&D staff, 7, 8, 197–202
graduate degree, 201, 203, 204
ICTA's program for DIGESA, 198, 200, 388, 389
international and regional centers, 202, 203
IRRI programs, 202, 203, 391–397
materials, 204
meetings, 202
non-degree study, 201
PNIA's program in Honduras, 199, 383–388
programs of CATIE, CIAT, CIMMYT, and ILCA, 389–392
for researchers, 197–199
seminars, 202
short courses, 201, 202
for technicians, 200, 201
university programs in the United States, 203, 204
workshops, 202
Treatment, definition of, 217
Tropical Agricultural Research and Training Center (Costa Rica). *See* CATIE
Turner, H.D., 376–379
Tversky, A., 266

United Nations Industrial Development Organization (UNIDO), 143
United States Agency for International Development (U.S.). *See* USAID
USAID (United States Agency for International Development–U.S.), 190–192, 376–379
USDA Soil Survey Staff, 49

Valdes, A., et al., 232
Validity of inferences, 297–299
van Raaji, J.G.T., 277
Variable
definition of, 217
experimental, 214
nonexperimental, 215
Variance, definition of, 217
Venema, L.B., 16

WARDA (West Africa Rice Development Association–Liberia), 237, 393
FSR&D program, 239
Wardhani, M.A., 233
Water. *See* Physical factors, water
Waugh, R.K., 33, 141, 151, 197, 199, 200, 349
Weather. *See* Climatic monitoring and records

Weiner, M.L., 191
West Africa Rice Development Association (Liberia). *See* WARDA
Wharton, C.R., 325
Whole farm analysis, 125
  definition of, 217
Whole farm approach, 13–16, 157, 158
  definition of, 217
Winkelmann, D.L., 70, 166
Wolf, E.R., 65
Women, special considerations of, 75, 178, 281–283
Workshops
  research analysis, 143, 144
  research planning, 103–106
  training, 202

Yield measurements. *See* Crop yields, measuring

Zandstra, H.G., 32
Zandstra, H.G., et al., 23, 32, 67, 92, 98, 119, 153–155, 190, 238, 256, 257, 267–270, 341–346, 350
Zeidenstein, S., 281
Zuberti, C.A., et al., 135